Teacher's Wraparound Edition

Technology

Design and Applications

by

R. Thomas Wright

Professor Emeritus, Industry and Technology

Ball State University

Muncie, Indiana

and

Ryan A. Brown

Teacher, Technology Education

Indianapolis, Indiana

Publisher

The Goodheart-Willcox Company, Inc.

Tinley Park, Illinois

www.g-w.com

Teacher's Wraparound Edition

Contents

Technology: Design and Applications T4

Features of the Teacher's Wraparound Edition T8

Features of the Teacher's Manual T12

Features of the Teacher's Resource Binder
and Teacher's Resource CD T14

Features of the Student Activity Manual T16

Introduction T17

       Technology—A Societal Universe T17

       Technology Education: A Background T18

       Technology and Industry T18

       Systems T18

       Organizing the Content T21

       Scheduling the Program T22

       Working with the Instructional System T24

       Introduction to Section 1 . . . . T29

       Introduction to Section 2 . . . . T35

       Introduction to Section 3 . . . . T45

       Introduction to Section 4 . . . . T55

       Introduction to Section 5 . . . . T70

Standards for Technological Literacy Correlation Chart T74

Scope and Sequence Chart T86

Basic Skills Chart T92

Technology: Design and Applications

This completely new comprehensive technology education text will help students learn how technology affects people and the world in which we live. Review these pages to see what this learning package includes. The following are some of the key features of the *Student Edition*:

➤ Fully correlated with ITEA's new national Standards for Technological Literacy.

➤ Covers the scope of technology, resources and technology, creating technology, technological contexts, and technology and society.

➤ An activity-based program.

➤ Full-color illustrations, referenced within the text, help students associate visual images with written material.

➤ Logical organization, colorful headings, and readable typeface facilitate reading comprehension.

79

Technology Headline

Floating Cities

There is growing concern around the world that within the next several decades, the world's population will be so big, there will not be enough living space for all the people. Some scientists and technologists have suggested building colonies in space. This may be one solution to this problem, but now there is another solution, which may prove to be more realistic for the near future. Plans are already underway for a very large ocean vessel called *Freedom Ship*, which will serve as a floating city!

This nearly mile-long ship will be home to more than 60,000 people, and visitors will be welcome to vacation there. Its 25 floors will contain everything a city needs, including a library, a hospital, schools, an airport, a golf driving range, a marina, bicycle paths, one of the world's largest shopping malls, offices, warehouses, light manufacturing and assembly enterprises, and 200 open acres for recreation. The city will travel, circling the globe every two years, and it will make stopovers at various islands and shores.

Residents of this city will be able to participate in nearly any activity available in traditional cities. They will be able to go to restaurants, nightclubs, casinos, and movie theaters. Tennis courts, bowling alleys, swimming pools, gyms, ice skating rinks, and fishing areas will be available for all residents. In addition, each home

will have satellite television and Internet access.

Besides being comfortable and exciting, this self-contained floating city will also be environmentally friendly. All waste created on the ship will be reused in some way. There will be no need for a sewage treatment plant or any sewage to spill into the ocean, as all sewage will be burned and put in the flower beds. Instead of being dumped into the ocean, waste oil will be used to generate electricity. Finally, all used paper, metal, and glass will be recycled and sold on the ship. It is estimated that residents of this city will produce 80 percent less waste than people living on land.

Although *Freedom Ship* might sound like a dream come true, it will most likely be a home only to the world's wealthiest people, at least at first. Suites on this ship currently range from $121,000 to $11 billion! The people who live here, however, will not have to pay any local taxes, including income tax, real estate tax, sales tax, business tax, and import duties.

This enormous ship is expected to begin sailing the world's oceans within the next few years. If this floating city proves to be successful, you can anticipate more in the coming decades. This is only the beginning, and it might not be very long before you are living on the ocean!

Technology Headlines features in each section highlight an emerging technology.

Chapter 2
Technology As a System

Did You Know? features in every chapter offer bits of trivia about the content covered in the chapter.

Did You Know?

➤ The solar system is a very large natural system. It consists of the Sun, nine planets, and sixty-eight satellites of the planets.

➤ A computer's operating system is the first software seen when a computer is turned on. It is the last software seen when the computer is turned off. The operating system directs all the programs used on a computer. It manages the hardware and software elements of the computer system. This system provides a way for applications to deal with the hardware.

➤ Not all computers have operating systems. Computers controlling simple objects, like microwave ovens, do not need operating systems. The computer in a microwave oven runs a single program all the time.

35

Objectives

The information given in the chapter will help you do the following:

➤ Define and explain a *system*.
➤ Explain the differences between a natural system and a technological system.
➤ Summarize how a system works.
➤ List and describe the seven major inputs to technological systems.
➤ Recognize and give examples of the major types of processes used in technological systems.
➤ Identify and describe the major types of outputs of a technological system.
➤ Name the two types of feedback systems.
➤ Describe some major goals for technological systems.

Key Words

These words are used in this chapter. Do you know what they mean?

closed-loop system
desired output
energy
feedback
feedback system
goal
input
intended output
knowledge
machine
material
natural system
open-loop system
output
pollution
process
system
technological system
time
undesirable output

Objectives identify the topics covered and goals to be achieved by students.

Key Words list new vocabulary covered in the chapter, enhancing student recognition of important concepts.

---

346 Section 3 Creating Technology

Figure 15-22. This model was built from various wood sheet materials. (Custom Modeling & Graphics Studio, Inc.)

models. These materials can also be used as flat panels on industrial models. All of the paper sheet materials can be cut with hobby or utility knives. See Figure 15-23. They can be attached to each other with tape or almost any type of glue. There are several different

New Terms appear in bold, red italics where they are defined.

Figure 15-23. Paper sheet materials can easily be cut with a utility knife.

Safety

When using a utility knife, always adhere to the following precautions:

➤ Use sharp blades.
➤ Wear eye protection.
➤ Keeps your hands out of the path of the blade.
➤ Cut in a direction away from your body.
➤ Cap the blade of the knife when it is not in use.

Safety notes identify activities that can result in personal injury if proper procedures or safety measures are not followed.

types of paper sheet materials. Figure 15-24. They are different in size, thickness, color, and flexibility. The paper materials used in modeling are thicker than normal notebook paper. Two of the thinnest paper products used are bristol board and chipboard.

Chapter 15 Modeling Solutions 347

Figure 15-24. Paper sheet materials come in various sizes and thicknesses. Each one has different applications.

Bristol board is similar to common poster board. It is usually found in ___" thick. This product can easily be curved, but is not very rigid. It is smooth and shiny on both sides. It board can be found in different ___, but is most common in bright white. *Chipboard* is a little different than bristol board. It has a gray color and is produced from two sheets of gray paper pulp. It can be found in thicknesses ranging from 1/32" to 1/8". Chipboard is more rigid than bristol board, but is not as attractive. Chipboard is very common; it is often used for backers in calendars, notepads, and packages of paper.

The next two types of paperboard materials are illustration board and matboard. Both of these materials are around 1/16". They are both sturdy and come in various colors. *Illustration board* is colored the same throughout the board. It comes in handy ___ designers do not need to color ___ they cut. This material can ___ rawn on to add features to ___. *Matboard*, or museum

board, comes not only in many colors, but also in various textures. The color on matboard is only placed on the top of the board. The inside of matboard is always white.

The two thickest paper sheets are corrugated cardboard and foam core board. Each of these materials has two outer sheets attached to an inner core of material. *Corrugated cardboard* has outer sheets of heavy paper. The inner core is made of heavy paper bent into ridges. *Foam core board* has an inner layer of foam covered on each side with paper. Both materials are very sturdy and come in sizes from 1/8" to 1/2". They can be bent if they are scored with a knife first. Corrugated board and chipboard are less expensive than other sheet materials and are used for models in early stages. See Figure 15-25. Foam core board is used for final mock-ups and can be used for some prototypes.

Figure 15-25. Early development models are created with chipboard and corrugated board. (Keith Nelson)

Technology Explained

Technology Explained features briefly explain how technological devices and systems work.

To use this type of system, two wells are drilled. Warm water is brought to the heat pump. The unit extracts heat from the water in the winter. It can add heat to the water (cool the home) in the summer. The altered water is then pumped back into the ground. This type of unit will work anywhere. It uses the difference in temperature between the water and outside air.

Another use of geothermal energy is in power plants, Figure C. These pumps work much differently than residential heat pumps do. First, the earth's magma heats water to at least 350°F. A well and pipes bring the hot water to the power plant. Some of the water is allowed to enter a set of steam separators. Here, a part of the water is allowed to flash

Figure C. Here is a diagram showing the parts and operation of a geothermal power plant.

into steam. The remaining hot water is flashed into steam in a second separator.

The pressurized steam is moved to a turbine through pipes. Here, it turns the turbine connected to an electric generator. The output of the generator

is fed into the power grid through wires.

A condenser pulls the cooled steam from the turbines and allows it to change back into water. A gas emissions treatment unit removes sulfide gas. The hot water is allowed to cool in a cooling tower. Then, the water is returned to the geothermal reservoir through an injection well.

Like most technological systems, there are some drawbacks to geothermal power. The water contains chemicals that can deposit on the walls of the pipes. This will restrict the flow of the hot and cold water. Also, the plants are more expensive to build than conventional power plants are.

Technology Explained

geothermal energy: Energy derived from the natural heat present in the earth.

People are actively seeking new sources of energy. One of these sources is actually very old. It is the natural heat present in the earth. It is the natural heat present in the earth. It is the earth's magma, heating the earth's crust and water below, develops this heat. The effect of this heating can be seen in geothermal areas, such as Yellowstone National Park, Figure A. Eruptions of volcanoes, such as

Mount Saint Helen's, are another example of the earth's geothermal energy.

One of the ways to use this energy is with a geothermal heat pump in a home, Figure B.

Figure A. Here is a geyser at Yellowstone National Park.

Geothermal heat pump

Figure B. This is a diagram of a residential geothermal heating unit.

material in them. Curtains, insulating tiles, and curved walls improve the sound quality of a room.

Summary

Humans have used materials throughout history. Materials are key to our existence. They are used for almost everything we need and want. Materials are found in our clothing and shelters. They become the products and fuels we use. Materials are the food and fibers sustaining our lives.

All materials are either renewable or exhaustible. Renewable resources can be grown and harvested. Proper management will give us a constant supply of these resources. Exhaustible resources have a limited supply. Once used, they cannot be replaced. They require careful use if future generations are to have them too.

materials must be grown or they then must be harvested. Finally, they are processed into industrial or standard materials.

Engineering materials are an important group of materials. They have structures. These materials are the materials used in many products.

Metals, ceramics, polymers, and composites are engineering materials. Each material has its own set of properties. These properties include the physical, mechanical, thermal, chemical, electrical and magnetic, optical, and acoustical qualities of the material. The combination of properties in a material dictates its range of use. Materials are a basic input to technological actions as systems. They are used in agricultural, energy conversion, information processing and communication, construction, manufacturing, medical, and transportation technologies.

Summary provides a review of major concepts covered in the chapter for the students.

Curricular Connections in every chapter offer suggestions of activities and assignments that connect the content to other subject areas.

Curricular Connections

Social Studies

Make a display or poster presenting the major technologies shaping each age. Include the Stone Age, Bronze Age, Iron Age, Industrial Age, and Information Age.

Science

Select a major type of technology, such as agricultural, medical, or transportation. Identify and explain a major scientific discovery aiding in the development of the technology you selected.

Mathematics

Select a type of technology and explain how people using that technology use mathematics. For example, in construction technology, surveyors use trigonometry to measure the distances across rivers.

Social Studies

Trace the development of a major type of technology over recorded history. You may select a technology such as housing, transportation, or food preservation.

Activities

Activities at the end of each chapter encourage students to apply concepts to real-life situations and develop skills related to chapter content.

1. Prepare a chart with the five major ages (Stone, Bronze, Iron, Industrial, and Information) across the top. List the major types of technologies (agricultural, communication and information, construction, energy and power, medical, manufacturing, and transportation) along the side. List the major technological advancements that have happened for each age in each technology. You may want to work in groups to complete this activity.

2. Build a model of a technological advancement contributing to the growth of one of the technology areas (agricultural, communication and information, construction, energy and power, medical,

Test Your Knowledge

Do not write in this book. Place your answers to this test on a separate sheet of paper.

1. Name the three major contexts in which technology can be viewed.
2. How did the Stone Age get its name?
3. Copper and tin were major metals used in the _____ Age.
4. Select the statement best describing the Iron Age:
   A. The period of history when the focus moved from farming to manufacturing.
   B. The period of history when the focus moved from manufacturing to information processing.
   C. The period of history when villages and towns started to appear and metal tools replaced stone ones.
   D. The period of history when building techniques improved, road systems were developed, and new cooking ware and agricultural tools were created.
5. The steam engine was a major development in the _____ Age.
6. The computer is the driving technology in the _____ Age.
7. We live in a (local, national, global) society.
8. Summarize some ways in which technology has changed our expectations of things, such as communication and transportation.
9. Technology, in itself, is neither good nor bad. True or False.
10. Define each of the following:
    A. Agricultural technology.
    B. Communication and information technology.
    C. Construction technology.
    D. Energy and power technology.
    E. Manufacturing technology.
    F. Medical technology.
    G. Transportation technology.

Test Your Knowledge questions help students review the topics and the material covered in the chapter.

Activity 4C

Energy Conversion Technology

635

Introduction

Energy is defined as the ability to do work. Much of this work involves converting energy from one form to another. Water is stored behind a dam (potential energy). When the water is allowed to fall through pipes to a turbine (kinetic energy), the energy of falling water turns the turbine (mechanical energy). The turbine converts the motion into electrical energy. The electrical energy is transmitted over power lines, which have resistance. The resistance produces heat (thermal energy). At the destination, the electrical energy is converted to light, heat, or mechanical energy. In this activity, you will explore one type of energy conversion—light energy into heat energy.

Equipment and Supplies

> Heat lamp and ~~stand~~

~~these~~ cans or #10 cans (from the school cafeteria) ~~painted~~ various colors (black, red, ~~blue~~...). These cans ~~should~~... A rubber gr~~ommet~~... ~~sh~~ould be in e~~ach~~... ~~readin~~g up to 200...

Section Activities correlate to the Student Activity Manual and provide suggested laboratory activities requiring creativity and critical thinking skills.

TSA Modular Activity

This activity develops the skills used in TSA's Manufacturing Challenge event.

Manufacturing Challenge

Activity Overview

In this activity, you will obtain discarded materials from local business or industry, and design products that can be developed with those materials. You will design a manufacturing process to produce the product, perform market research for the product, and prepare a comprehensive report.

Materials

> Discarded materials from local business or industry (will vary)
> Paper
> Pencil
> Three-ring binder
> Computer with CAD and word processing software

Background Information

Obtaining materials. Contact local businesses to obtain scrap material. You can write letters, make telephone calls, or visit businesses in person. Your local telephone book will provide many leads.

Design. After obtaining materials, use brainstorming techniques to develop a list of potential products. Create rough design sketches for some of the products. Consider the available tools and materials when selecting the best idea.

Drawings. Create working drawings using a CAD system. Use multiview drawings with as many views as needed to fully describe the product. If your product is small enough, create the drawings at full scale.

Prototype. Produce a prototype of the product. As you develop the prototype, consider the most effective manufacturing sequence. Are there design changes that could improve the manufacturing process? If design changes are made as you develop the prototype, be sure to update the drawings to reflect those design changes.

Production plan flowchart. Develop a production plan flowchart that illustrates the manufacturing sequence for the product.

TSA Modular Activities features are additional activities intended to develop skills used in TSA's competitive events.

Features of the Teacher's Wraparound Edition

A variety of instructional elements appear in the margins of each page of this Teacher's Wraparound Edition. These elements are intended to guide you in reviewing and reinforcing the chapter content.

All answers to Test Your Knowledge questions are also found in the margin of the corresponding page. This allows you to read through the questions and have the answers at your fingertips.

211

Technology Headline

Interactive Television

Have you ever wished you could check your e-mail while watching your favorite TV show? You may soon be able to do this and many more exciting things with interactive television. This new technology combines TV with the Internet and allows you to interact with the show you're watching.

Interactive TV will give viewers access to Internet sites and instant messaging on their TV screens. You will be able to use the Internet and watch TV at the same time by activating a picture-in-picture screen. A program guide will be available, which will organize channels into categories, making viewing easier. You will be able to choose a channel by clicking on words and graphics.

With interactive TV, you will be able to play along with TV game shows, see the local weather, and access real-time statistics during sports events. Just by using your remote control, you will be able to print recipes from the cooking show you are watching, shop on-line, and order concert tickets. You will even be able to listen to your favorite radio station and view movie trailers.

In addition to these great features, interactive TV will allow you to watch your favorite shows anytime you want. It will include digital video recording, which will allow you to search for and record shows, using an electronic program guide. Instant replay, fast-forward, and skip ahead features will also be available with this technology.

Many TV and cable networks are already developing applications to be used with interactive TV. This new technology will become more common in the coming years. It may not be very long before interactive TV is in your household!

Discussion Starter

Refer the students to the section title "Creating Technology." Have the students complete the following sentence: "Creating technology means…." Ask volunteers to read their responses to the class. Discuss the responses and record accurate ones on the board. Have the students use the responses to make a concept web of ways to create technology.

Group Activity

Write the word *invention* on the board. Ask the students to work in small groups to brainstorm their own definitions for the term.

Writing Assignment

Ask the students to use the text and any available outside resources to research and write descriptive definitions of the eight steps of the design process. Ask volunteers to read their definitions.

Discussion Starters. Questions and activities that relate the topics of the section to the students and encourage discussion.

Group Activities. Activities that require small groups of students to think collaboratively.

Writing Assignments. Ideas for homework or in-class assignments that encourage critical or creative thinking.

review the statements from the clients and create a problem statement.

Problem Statements

A *problem statement* is a clear definition of the problem. For example, a problem may be to "design a device that will evenly spread paint on a wall." This is a clear definition. It is an open-ended statement stating a problem, not a solution. A statement like "design a paint roller" states the solution. This statement is too narrow. It would be hard to develop a new

design with a statement to design a paint roller. The first statement allows the freedom for the designer to go in many directions. See **Figure 11-9.**

Criteria

After the problem statement is written, the designer will list the criteria for the design. *Criteria* are elements of the problem needing to be solved. They are some of the requirements, or parameters, placed on the development of the product or system. Criteria are listed before the designer starts to develop ideas. They are identified

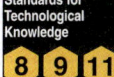

Standards for Technological Knowledge

8 9 11

Resource
Transparency 11-3: Define the Problem

Brainstorm
Have the students select several products and write the possible problem statements used to design the product.

Tech Terms
Requirement— Something demanded or needed.

Standards Correlations. Correlate the text material to the current Standards for Technological Literacy.

Resource. Suggests related material available in the Student Activity Manual or Teacher's Resources.

Brainstorm. Activities that encourage collaborative thinking and discussion.

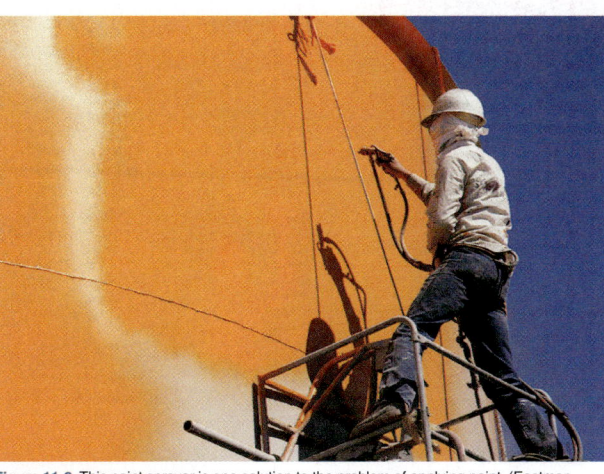

Figure 11-9. This paint sprayer is one solution to the problem of applying paint. (Eastman Chemical Company)

TechnoFact: Paint is made up of a dye, a binder to hold it together, and appropriate thinners to make it easy to use. As technology has advanced, so have the dyes, binders, and thinners in paint. Primitive paints were made from animal fat, indigenous dyes, clay, and charcoal. For much of the nineteenth and part of the twentieth centuries, lead was used in the making of paint. Nowadays, there are a number of substances (latex, for instance) added to paint to allow it to be used in a variety of situations.

...ctives

...mation given in the chapter ...ou do the following:

...y the major contexts in ...echnology can be viewed.

➤ Recognize the major ages in human history.

➤ Give examples of the positive and negative impacts technology has on society.

➤ List the seven common functional areas of technology.

Key Words

55

These words are used in this chapter. Do you know what they mean?

agricultural technology
Bronze Age
communication and information technology
construction technology
energy and power technology
First World country
Industrial Age
Information Age
Iron Age
manufacturing technology
medical technology
Stone Age
Third World country
transportation technology

Chapter Outline. Provides you with a quick overview of the main points covered in the chapter.

Chapter Outline
Did You Know?
Objectives
Key Words
Historical Context
The Stone Age—Up to about 3500 B.C.
The Bronze Age—About 3500 to 1000 B.C.
The Iron Age—1000 to 500 B.C.
The Industrial Age—1750 to the Late 1900s
The Information Age—Late 1900s to Present Day
Societal Context
Scope of Society
Technology and the Individual
Technology and Society
Functional Context
Summary
Curricular Connections
Activities
Test Your Knowledge

Demonstrate. Ideas for in-class demonstrations of technological procedures.

Research. Ideas for assignments that include researching a topic related to the chapter material.

Discussion. Technology-related notes, topics, and questions that can be covered and discussed during class to reteach or reinforce learning.

Figure Discussion. Information to expand on key chapter figures.

Job Classification	Typical Jobs or Occupations	Job Classification	Typical Jobs or Occupations
Executive, Administrative, and Managerial	Accountant Public relations manager Construction and building inspector Construction manager Farmer Farm manager Financial manager Health services manager Hotel manager Industrial production manager Restaurant and food service manager	Administrative Support, Including Clerical	Bank teller Communications equipment operator Computer operator Hotel desk clerk Receptionist Mail clerk Shipping and receiving clerk Mail carrier Bookkeeping clerk Brokerage clerk Statement clerk File clerk Library assistant
Professional and Technical	Aircraft pilot Flight engineer Engineer Engineering technician Drafter Computer programmer Mathematician Food scientist Geologist Lawyer Judicial worker Urban and regional planner Social worker Schoolteacher Librarian Dentist Veterinarian Dental hygienist Broadcast and sound technican Photographer Actor	Service	Janitor Groundskeeper Pest controller Chef Dental assistant Medical assistant Barber Flight attendant Correctional officer Firefighter
		Installations and Repairs, Including Mechanical	Computer repairer Aircraft mechanic Automotive mechanic Musical instrument tuner Bricklayer Stonemason Carpenter Roofer
Marketing and Sales	Cashier Insurance sales agent Manufacturer's sales representative Real estate agent Real estate broker Retail salesperson		Assembler Production supervisor Fishing vessel supervisor Butcher Inspector

Demonstrate
Show the students how to use the Internet to access the Department of Labor's Occupational Outlook Handbook.

Research
Have the students use the Internet and the Occupational Outlook Handbook to find the educational, training, and skill requirements for a job that interests them.

Brainstorm
Ask the students to list the job categories in Figure 8-16 appealing to them and why these categories do.

Tech Terms
Job—An activity, such as a trade or profession, for which a person is paid.

---

134 Section 2 Resources and Technology

Figure 6-6. Can you explain energy, power, and work in terms of these gondola operators?

The Nature of Energy

Energy is everywhere. People and machines use it. Energy is used to move things, light areas, lift weight, and warm spaces.

One form of energy can be changed into another. See **Figure 6-7.** Stored chemical energy in a flashlight battery can be converted into light energy. Your body uses the stored chemical energy in food to do work. A television set changes electrical energy into light and sound energy. An automobile engine changes the chemical energy in gasoline into heat and mechanical energy to power the car.

Energy cannot be created, however, nor can it be destroyed. As we change its form, it can be wasted. For example, the lightbulb is an energy converter. Its purpose is to change electrical energy into light energy. In doing this task, not all the electrical energy ends up as light energy. Some

of it becomes heat energy. The electrical energy that becomes heat energy is not destroyed. It could be considered wasted, however. This energy is not in a form that is useful in the situation, nor is it wanted.

Forms of Energy

Not all energy is alike. See **Figure 6-8.** It appears in several different forms:

▶ Light or *radiant energy.* This is the energy in the atomic motion present in sunlight and fire.

▶ *Heat energy.* This is the energy in the increased activity of molecules

Figure 6-7. These wind generators change mechanical power into electrical power.

Resource
Transparency 6-3: Forms of Energy

Discussion
Discuss how an automobile engine changes the form of energy found in gasoline.

Figure Discussion
Discuss how the windmill in Figure 6-7 uses inexhaustible energy.

Figure Discussion
Ask students to answer the question posed in Figure 6-6.

Tech Terms
Conservation of Energy—The scientific principle suggesting the total energy of an isolated system remains constant regardless of changes within the system. Energy can be made nor destroyed—only changed in form.

TechnoFact: Heat energy is related to variations in temperature between a system and its environment or between elements of the same system.

Chapter 6 Energy and Technology 143

Figure 6-17. The solar cells on the top of this sign convert solar energy into electricity.

Solar Energy

The Sun has been an energy source since the beginning of time. Plants use the Sun's light to make food in a process called *photosynthesis*. Early in history, people learned to use *solar energy*. It was used to dry food and warm buildings. When you hang a wet swimsuit outside to dry, you are using solar energy to do work. Today, the Sun is being used to heat water for homes and businesses. Campers can use solar energy to cook food. The two major commercial uses for solar energy technology are to heat water and produce electricity. See **Figure 6-17.**

Alternate Energy Sources

People are becoming increasingly concerned about the use of nonrenewable energy resources. They are turning to alternate energy sources to reduce our primary dependence on fossil fuels. Individuals, government agencies, and private companies are studying these sources. Large-scale, efficient methods to harness and use alternate energy sources are being explored. Like waterpower, these sources will always be available. We will never run out of them. The more important alternate energy sources are solar, wind, tidal or wave, geothermal, and biomass energy.

Wind Energy

Wind is an indirect use of solar energy. It can be used to do work. A sailboat uses **wind energy** to push it through the water. Ranchers use wind energy to pump water for their cattle. In Holland, windmills have been used for centuries to pump water from lowlying areas. Wind, like the Sun, is free and increasingly harnessed for electricity production. See **Figure 6-18.**

Tidal and Wave Energy

Energy is present in the ocean. The ocean can produce two types of energy: *thermal energy* from the Sun's heat, and *mechanical energy* from the tides and waves.

Oceans cover more than 70 percent of Earth's surface. This makes the oceans huge solar collectors. The Sun's heat primarily warms the surface. This water is warmer than the

TechnoFact: The typical Dutch windmill is called a *tower windmill*. It has a large tower of stone, brick, or wood. At the top of the structure are four to six arms attached to a central hub that can revolve. The 20′ to 40′ long arms have sails constructed of lightweight wood. A fan keeps the wheel facing the wind.

Resource
Reproducible Master 6-3: Power Systems in Your Community and Their Energy Sources

Extend
Discuss how larger windmills, such as the modern propeller-like wind turbines, are joined together as a significant source of electric energy.

Reinforce
Discuss alternate energy resources as inexhaustible resources.

Extend. Activities that challenge students to go beyond the information presented in the chapter.

Reinforce. Directs students to previously covered material relevant to the current topics being discussed.

---

560 Section 4 Technological Contexts

Standards for Technological Knowledge

19

Resource
Reproducible Master 23-1: Secondary Manufacturing Processes

Resource
Transparency 23-1: Producing Finished Products

Example
Give the students some examples of materials produced by mechanical, thermal, and chemical processes.

Tech Terms
Mechanical Processing—Manufacturing processes that cut, grind, or crush a material to create a new form.
Thermal Processing—Manufacturing processes that use heat to change the form or composition of materials. These processes include smelting metallic ores (making copper and steel) and fusing silica sand into glass.
Chemical (and Electrochemical) Processing—

Example. Use to illustrate an important point in chapter material.

Tech Terms. Emphasize important terms or present technology-related terminology.

TechnoFact. Bits of trivia related to the material covered in the chapter that will increase student interest in the topic.

Primary Processes

Primary processing includes three major groups of processes. These were discussed in Chapter 5. The following briefly reviews what you learned there:

➤ Mechanical processing. This includes cutting, grinding, or crushing the material to produce a new form. These processes include cutting lumber and veneer, making cement, crushing rock into gravel, and grinding wheat to make flour.

➤ Thermal processing. This is using heat to change the form or composition of materials. These processes include smelting metallic ores (for example, making copper or steel) and fusing silica sand into glass.

➤ Chemical (and electrochemical) processing. This is using chemical actions to change resources into new materials. These processes include refining aluminum from bauxite and producing most plastics from fossil fuels.

The output of primary processing is called *standard stock*. The materials are available in standard sizes. Most plywood is produced in 4′ x 8′ sheets. Sheet metal is often sold in 24″ x 96″ sheets. Plastics are produced in standard pellets. Sugar is produced in various granular sizes (such as standard, extra fine, and confectionery). Lumber is sold in a number of standard sizes. Many of us have heard people talk about 2 x 4s. This is a standard lumber size for the construction industry.

Standard stock must be further processed before it is useful. A sheet of

Figure 23-4. Lumber and plywood take on added value when they are used to build homes and furniture. (Sauder)

plywood is of little value to a person. It becomes useful when it is made into a desk or doghouse. See **Figure 23-4.** Likewise, few of us have use for plastic pellets, but these pellets can be converted into automobile trim, bowls, and fabric. After this happens, most people place a higher value on the plastic.

Secondary Processes

Secondary processing activities can be grouped under six headings. These are casting and molding, forming, separating, conditioning, assembling, and finishing. See **Figure 23-5.**

Casting and Molding

The first three secondary processing activities (casting and molding, forming, and separating) give materials specific sizes and shapes. In one type of process, the material is first made a liquid. The liquid is poured or forced into a *mold*. The mold has a cavity in the shape of the part or product. Inside the mold, the material hardens. This

TechnoFact: The machines developed in the early part of the Industrial Revolution were too bulky to fit into the home. Therefore, a new form of manufacturing was needed. Special buildings were built to house the machines and the workers who operated the machines. These buildings were called *factories*.

Features of the Teacher's Manual, Teacher's Resource Binder, and Teacher's Resource CD

The Teacher's Manual, Teacher's Resource Binder, and Teacher's Resource CD provide all the resources needed to instruct your technology education class. All materials are grouped together by section for easy reference.

Standards for Technological Literacy Correlation Chart allows you to design your class around standards identified by the International Technology Education Association by indicating where relevant topics are covered in the text.

Basic Skills Chart identifies activities in the text and supplements that encourage the development of reading, writing, verbal, math, science, and analytical skills.

Scope-and-Sequence Chart provides assistance in outlining your courses.

Teaching Materials are listed by section and by chapter to assist in planning daily lessons.

Introductions are presented to highlight the main concepts of each section.

Objectives of Each Section identify the topics covered and goals to be achieved by students.

T74   Technology: Design and Applications

Standards for Technological Literacy Correlation Chart

The International Technology Education Association (ITEA) and its Technology for All Americans Project developed *Standards for Technological Literacy: Content for the Study of Technology* to identify the essential core of technological knowledge and skills for students in grades K–12. This work defined twenty separate standards, divided into five broad categories. Within each standard, benchmark topics are defined for four different grade levels:

- Grades K–2
- Grades 3–5
- Grades 6–8
- Grades 9–12

The following chart lists the standards and the benchmark topics for grades 6–8. Adjacent to each standard and benchmark topic are the chapter and page references identifying material in *Technology: Design and Applications* relating to the item.

Standards for Technological Literacy

Standard	Chapter/Pages
1. Students will develop an understanding of the characteristics and scope of technology.	Chapter 1: pages 24–26, 29–30 Chapter 3: pages 56–63 Chapter 9: pages 218–221, 224–225 Chapter 23: pages 576–577
New products and systems can be developed to solve problems or to help do things that could not be done without the help of technology.	Chapter 1: pages 24–26 Chapter 9: pages 218–219
The development of technology is a human activity and is the result of individual or collective needs and the ability to be creative.	Chapter 1: pages 25–26 Chapter 3: pages 56–63 Chapter 9: pages 220–221
Technology is closely linked to creativity, which has resulted in innovation.	
Corporations can often create demand for a product by bringing it onto the market and advertising it.	Chapter 3: pages 56–63 Chapter 9: pages 224–225
2. Students will develop an understanding of the core concepts of technology.	Chapter 1: pages 29–30 Chapter 23: pages 576–577 Chapter 1: pages 24–31 Chapter 2: pages 36–39, 43–45, 48–49 Chapter 10: pages 234–247 Chapter 11: pages 262–265 Chapter 14: pages 312–314 Chapter 16: pages 375–376 Chapter 18: pages 413–416

Name_____ Date _____
Period _____ Score _____

Alternative Activity

Communicating without Words

The Challenge
Develop a system allowing you to communicate without words.

Background
People need to communicate. They are often, however, far apart. Many times, they cannot hear one another. For this reason, communication technology was developed. Many of these systems do not rely on "hearing" the message. They use signs and symbols to communicate. Below are some examples of ways people communicate without hearing words. The system you develop must be able to deal with all words in the English language, and it must use technology. It cannot simply be body signals. There must be technology (devices people design and make) involved.

Smoke signals

Morse code

Symbols

Semaphore

Design Brief
Develop your communication system using any or all of the following materials:
➤ 1/4″ dowels
➤ Colored paper
➤ Colored pencils
➤ Felt tip markers
➤ Paper clips
➤ Poster board
➤ String

Suggested Schedules and Lengths of Each Section provide recommendations for organizing the content to assist in planning daily lessons.

Preparing to Teach Each Section lists recommended tasks to be completed before presenting the section to your students.

Presenting Each Section suggestions are listed to assist you in teaching each section.

Adjustments for Special Needs and Gifted Students include suggestions for accommodating students of all ability levels and encouraging everyone's contribution to group activities.

Alternative Activities allow for greater flexibility in designing your classroom program.

Implementing the Laboratory Activities recommendations are provided for each section to assist in presenting the laboratory activities in the classroom.

Evaluation Procedures list suggestions for measuring student understanding of concepts in each section.

Procedures for Directing Each Suggested Activity provide detailed suggestions for implementing each laboratory activity.

Answer Keys are provided for Test Your Knowledge questions in the text and Study Questions in the Student Activity Manual.

Reproducible Masters are available for easy conversion to overhead transparencies or classroom handouts.

185

Global Warming

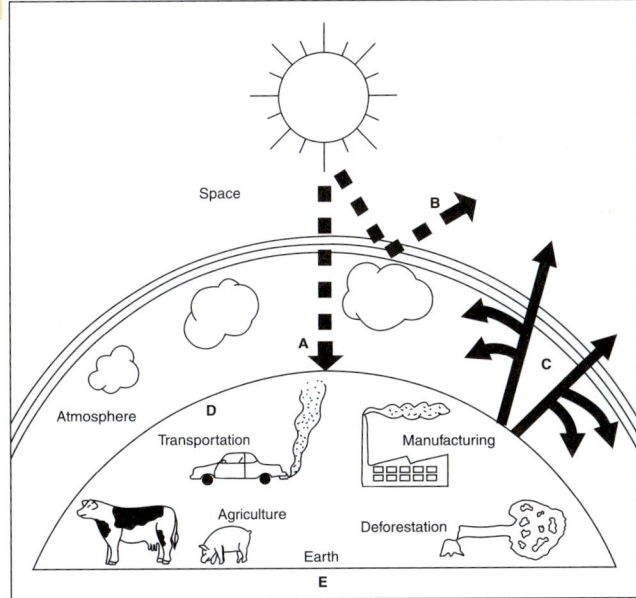

A. About 130 watts per square foot of solar energy reach the Earth's atmosphere at any one moment.
B. Around 30 percent of this energy reflects back into space.
C. An ozone layer high above Earth shields Earth's inhabitants from harmful ultraviolet solar radiation.
D. The use of fossil fuels for energy and the destruction of Earth's forests are major reasons for increased carbon dioxide in the atmosphere. Other pollutants are methane from swamps, landfills, and cud-chewing animals. Chlorofluorocarbons (synthetic gases used in refrigerants, spray cans, and insulation) are reducing the protective ozone layer.
E. All these pollutants together form a "cover" in the atmosphere, which traps heat energy, thus warming Earth.

Reproducible Master 26-1

The Teacher's Resource Binder includes all the resources found in the Teacher's Manual, plus Color Transparencies. The Teacher's Resource CD contains all the resources available in the Teacher's Resource Binder, plus the G-W Test Creation Software and Internet Resources.

Color Transparencies coinciding with the textbook reinforce important technologial concepts. (Available in the Teacher's Resource Binder and on the Teacher's Resource CD only)

Types of Knowledge

Scientific

Studies and describes the natural world

Humanities

Studies and describes human beliefs, values, and actions

Technological

Studies and the ade

Descriptive

Securing Natural Resources

Genetic Resources

Planting/Growing

Harvesting

Exhaustible Resources

Locating

Extracting

Technology: Design and Applications
Copyright Goodheart-Willcox Co., Inc.

Transparency 5-1

- **Need or Opportunity**
- **Define the Problem**
- **Gather Information**
- **Develop Solutions**
- **Model the Solution**
- **Select the Best Solution**
- **Document the Solution**
- **Evaluate the Solution**
- **Improve the Solution**

The G-W Test Creation Software computerized test preparation allows quick and easy creation of testing materials. It contains approximately 1200 questions correlated to the chapters of the textbook. (Available on the Teacher's Resource CD only)

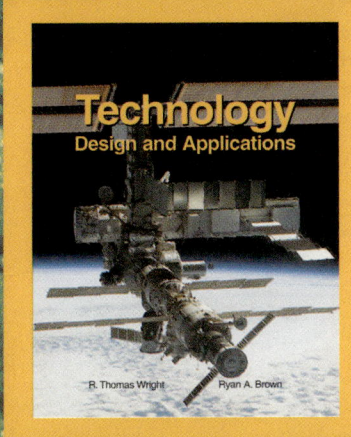

Technology
Design and Applications

Teacher's Resource CD

R. Thomas Wright   Ryan A. Brown

Internet Resources list websites related to the content covered in the textbook. (Available on the Teacher's Resource CD only)

Features of the Student Activity Manual

The Student Activity Manual goes hand-in-hand with the *Technology: Design and Applications* textbook. It includes activities, worksheets, and study questions that correlate to the information discussed in the text, providing the hands-on learning experiences that successful technology education requires. The Student Activity Manual encourages students to review and apply important technological concepts.

Name _____
Period _____

Date _____
Score _____

Activity 4-2
Model Bridge Design Contest

Design Brief

Your task is to design a 12" long 3 2" high model bridge. You will have ten pieces of 1/4" 3 1/4" 3 12" material. Your model will be compared with others built in the class, but you may use any design. The bridge holding the most weight is the best design. Each bridge will be tested to its breaking strength on a test stand, similar to the one at the bottom of this page.

Procedure

1. Research bridge designs in the school library.
2. Design and draw a full-scale bridge framework.
3. Cut the parts needed to make the two sides of the bridge.
4. Glue the bridge sides together.
5. Assemble the two sides with spacers holding the sides 1 1/2" apart.
6. Test the strength of the bridge, recording the maximum weight it supported.
7. Compare the results of your bridge with those of other class members.
8. Determine the "best design" (the bridge holding the heaviest load before breaking).

Test Stand

Test weight

Model bridge

Sand tray

Test stand

Technology: Design and Applications Student Activity

Activities allow students to apply technological concepts in fun and practical ways. Many of the activities correspond to the Section Activities in the textbook, but some are additional activities that can be completed independently from the text.

Study Questions reinforce the main concepts of the material in the textbook. These can be used as chapter reviews, chapter quizzes, or homework assignments. They will help you discover where your students may need additional help.

Name _____
Period _____

Date _____
Score _____

Chapter 12
Study Questions

Researching Problems

Textbook pages: 268–283

1. List the three major types of research. _____

2. A group of people using certain products are called _____.

3. The research process is similar to the _____.

4. Raw facts and figures are known as _____.

5. Aesthetic features relate to the:
   A. Function of the product.
   B. Cost of the product.
   C. Look of the product.
   D. Safety of the product.

6. Analyzed data is called _____.

Matching Test

Match the image with the type of graph.

7.

8.

9.

A. Line graph.
B. Pie graph.
C. Bar graph.

Technology: Design and Applications Student Activity Manual 69

Introduction

As the world begins the new millennium, people in many countries are at a similar turning point. They see society around them plunging into the Information Age brought about by new communication and transportation systems. The world has become a global village, with each nation economically interconnected and affected by events everywhere.

Preparing young people to participate in this global, technology-driven society will require new types of educational programs. To meet this challenge, technology education is becoming an important part of the general education of all youth. Ideally, development of an understanding of technology and its impacts should start in the lower elementary grades. Actual practice, however, finds most students are being introduced to these fundamental concepts beginning in sixth or seventh grade. This instructional system—the textbook, Student Activity Manual, and Teacher's Resources—has been designed to present a broad, introductory, hands-on presentation of technology to young learners in this age group. The system is designed to help students develop technological literacy, which can be described as an ability to locate, interpret, and use information about tools, materials, and processes.

Technology—A Societal Universe

Technological literacy and its associated competencies were the basis for the fundamental changes that occurred during the Industrial Age and are also the basis for the changes now happening in the Information Age. Yet, few people can define and describe this pervasive phenomenon we call *technology*.

Technology is the product of the human mind and its engineering spirit. It is, therefore, the result of human volition and control. All technology has been developed by people to satisfy human needs and wants and is subject to human monitoring. Technology is not waiting to be discovered, but rather to be created and made available for people to use.

This phenomenon is the application of knowledge, tools, and skills to solve problems and extend human capabilities. This could be easily described as the work of engineers, inventors, and technologists in applying knowledge to create and commercialize devices and systems to meet human goals.

Technology is (1) human knowledge and practices (2) that use tools, materials, and systems (3) to produce artifacts (human-made things) and other outputs (such as pollution and scrap) (4) in order to modify or control the natural and human-made environments, **Figure IN-1.**

Technology human knowledge and practices that use tools, materials, and systems to produce human-made things and other outputs in order to modify or control the natural and human-made environments.

Figure IN-1

Technology Education: A Background

Education about the practical world is as old as humankind. Early experiences were almost always trade-oriented. In the last 100 years, educators have recognized that broad, general (not job-specific) education about tools and materials is important. Out of this belief, a new program took form—first as manual training, and later as industrial arts. These subjects, however, drew their content almost totally from the skilled trades and focused on learning "basic" skills and constructing take-home projects. In recent years, a new discipline, technology education, has emerged to replace the skill-oriented, technical instruction of industrial arts.

This new type of education helps students develop technological literacy through the study of the creation, use, and behavior of technical systems (tools, machines, materials, and techniques) and the interrelationship of these elements and systems with humans, societies, and the environment. A technology education program with this focus helps students of all ages to be able to do the following:

➤ Know and appreciate the importance of technology to people and society.

➤ Use tools, materials, and processes effectively and safely.

➤ Apply designing and problem solving processes.

➤ Understand the various contexts in which technology is applied.

➤ Deal with the forces shaping the future.

Technology and Industry

This instructional program deals with two major universals found in all societies—technology and industry. As already described, technology includes all materials, products, and systems humans have developed to extend their potential for modifying and controlling the natural and human-made environments. It is knowledge of efficient and appropriate action, knowledge of "doing." Technology includes all human-made systems that allow us to grow, harvest, and process foods; communicate information and ideas; convert and apply energy; manufacture goods; construct buildings and civil structures; prevent and cure illnesses and injuries; and transport goods and people.

Industry, on the other hand, is part of a fundamental societal institution. Humans, over time, have developed five basic institutions to meet their needs. These are the family, religion, education, government, and the economy. Industry is part of the economic institution, which provides the goods and services making our lives more comfortable.

Systems

Technology and industry are systems that integrate parts into structures that behave in predictable ways. These systems have five major parts, as seen in **Figure IN-2:**

1. Inputs (resources).
   A. People.
   B. Materials.
   C. Capital (tools).
   D. Information.
   E. Energy.
   F. Finances.
   G. Time.

2. Processes.
   A. Production.
   B. Management.
3. Outputs.
   A. Desired.
   B. Ancillary.
4. Feedback.
5. Impacts.
   A. Desirable or undesirable.
   B. Immediate or delayed.
   C. Planned or unplanned.

Figure IN–2

System Inputs

People are fundamental to all technological systems. Only through a desire to meet human needs and wants have these systems developed. People provide the need for the system and also bring human talent to the system. They direct and operate these systems by their knowledge, skills, and attitudes.

Materials are the resources found in nature. We use them to make products, construct structures, and power our many activities. The supplies of many resources are limited and, therefore, must be used wisely.

Capital is the physical means of technology. It includes all buildings, machines, tools, and equipment needed to house and operate technological systems.

Information is the human-produced resource for all technological systems. People generate, retrieve, and apply it. Information can be cataloged under three types of knowledge. Scientific knowledge describes the natural world and provides an understanding of the natural phenomena impacting human life and activities. Technological knowledge describes the human-made world and is used to develop systems extending human potential. Humanities knowledge allows people to assess the value and need for human-made systems and interactions.

Energy is a unique natural resource that allows work to be accomplished. Like materials, many energy sources are finite and must be carefully used.

Finances are part of a system of exchange for human time, effort, and knowledge; material and energy resources; and capital goods. Money is used in modern societies to facilitate this exchange.

Time is the measurable period during which an action, such as harvesting crops, making products, constructing buildings, using energy, or moving cargo, takes place.

System Processes

Each technological system is a dynamic interaction of practices or actions. Production processes are basic to these systems. These processes are used to change inputs into outputs. Agricultural processes transform seeds, plants, and animals into foods and other products. Communication processes transform information into media messages. Construction processes convert materials and manufactured goods into buildings and other structures. Energy conversion processes change energy resources into heat, light, sound, and motion. Medical processes use medicines and devices to treat illnesses and promote health. Manufacturing processes change materials into goods. Transportation processes use vehicles and equipment to move people and goods.

Management processes are also important to technological endeavors. They involve setting goals and courses of action (planning), structuring the resources (organizing), directing and supervising transformation activities or other work (actuating), and monitoring results and making corrective actions (controlling).

System Outputs

All human-made systems have a reason to exist. People established them to meet goals. These goals may be products, structures, media messages, services, or changed conditions. We may call these visible elements the *desired outputs*. Every system, however, also has other outputs. Technological systems generate scrap (usable materials), waste (unusable materials), and pollution (environmentally destructive materials and conditions). These ancillary outputs must be reduced or eliminated, if possible.

Feedback

Feedback is the information used to monitor and adjust technological systems so they are operating at their highest efficiencies. Feedback monitors desired and ancillary outputs to determine the efficiencies of the production and management processes. This information indicates what changes are needed to ensure the system is adequately using system inputs.

Impacts

All technology was developed with a purpose in mind. Therefore, each product of technology was designed to have an impact. Impacts are the effects of technology on people, society, and the environment. Most technology is developed to make people's lives better. People may live in more comfort, have more information available, or move across the globe with more ease by using technologies. These are examples of positive, immediate impacts. Not all impacts, however, are positive. The use of technology may pollute the environment, displace workers, or create health risks. These are negative, undesirable impacts.

Not all impacts are immediate. Some take years to become apparent. For example, many people blame television for contributing to several societal problems, including increased violence, a declining ability to communicate verbally, and reduced creative ability in our youth. If these can rightfully be attributed to television, they would be delayed, undesirable, unplanned impacts. It is easy to see that impacts of technology can be assessed from three major perspectives: desirable or undesirable, planned or unplanned, and immediate or delayed.

Organizing the Content

Many courses in schools are built on a conceptual view of a specific body of knowledge, a set of desired outcomes, and a perception of the role of the subject matter in a student's education. Teachers, curriculum writers, and instructional material developers have individual views of their content areas. Hopefully, these views converge in a workable structure that will help students learn about the fields of study. Whatever view is used, it should meet some common criteria.

First, the content should have a logical organization that is apparent to the learner. This will allow students to understand the relationship between segments of the course and help them transfer the learning to other subjects and life situations.

Second, the overall material for the course should be organized into discrete, stand-alone sections. This allows students who enter the course after it starts to easily join the class.

Third, each section should present selected content and have meaningful hands-on, minds-on activities. This will allow students to apply information they learn to authentic activities.

Last, instruction should allow students to develop knowledge and problem solving skills. This will provide the foundation for lifelong learning.

With these principles in mind, *Technology: Design and Applications* is organized into five basic sections, as shown in **Figure IN-3:**

1. The first chapters introduce students to the scope of technology as a human-developed system that extends people's potential. We then discuss the system components (inputs, processes, outputs, and feedback).

2. The next section introduces the resources of technology, with chapters on tools; materials; energy; information; and people, time, and money.

3. The third section presents the design process as a way of creating technology and explores each step designers use in developing products and technological systems.

4. The fourth section explores common contexts in which technology is found. These include agricultural, construction, energy conversion, information and communication, manufacturing, medical, and transportation technologies.

5. The last section explores the role of technology in people's lives and the future of technology.

Each chapter presents a set of objectives, a well illustrated and easily read presentation of basic information, a set of key terms, some activity ideas, and study questions. Throughout the text are fully developed activities challenging students to apply their knowledge. Additional study questions and activities are included in the Student Activity Manual.

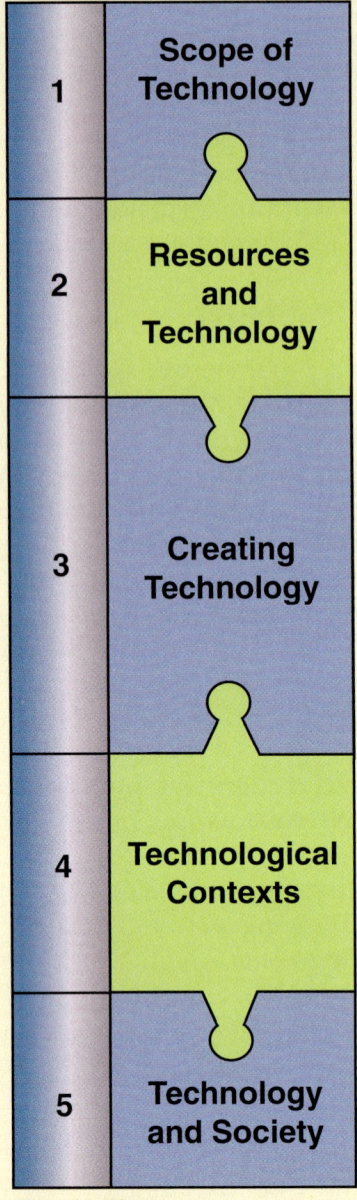

1	**Scope of Technology**	What Is Technology? Technology As a System Contexts of Technology
2	**Resources and Technology**	Tools and Technology Materials and Technology Energy and Technology Information and Technology People, Time, Money, and Technology
3	**Creating Technology**	Invention and Innovation The Design Process Identifying Problems Researching Problems Creating Solutions Selecting and Refining Solutions Modeling Solutions Testing Solutions Communicating Solutions Improving Solutions
4	**Technological Contexts**	Agricultural and Related Technology Construction Technology Energy Conversion Technology Information and Communication Technology Manufacturing Technology Medical Technology Transportation Technology
5	**Technology and Society**	Technological Impacts Technology and the Future

Figure IN–3

Scheduling the Program

One of the most difficult tasks every teacher faces is scheduling content presentations, laboratory activities, and evaluation sessions meeting the students' needs and time constraints on the course. Curriculum writers and instructional material developers can provide basic guidelines. By their very nature, however, these can only be guidelines. Specific courses and groups of students vary greatly. Teachers must modify schedules provided in instructional materials by considering such factors as the length of the course, the length of class period, the students' abilities and backgrounds, their own teaching styles, and the laboratory activities to be used. It is important, however, to present a complete view of the content area. If you do not, students will receive a narrow concept of *technology*. For example, emphasizing the design section, while leaving out the context sections exploring various technological systems, will lead students to believe design is the most important aspect of technology. Likewise, failing to address the impacts of technology will leave students poorly prepared to assess the value of technological advancements.

Some proposed schedules for 9-week, 12-week, 18-week, and 36-week *Introduction to Technology* programs are presented in **Figure IN-4**.

Some Proposed Course Schedules

Chapter Title and Number	Length of the Course in Weeks			
	9	**12**	**18**	**36**
Section 1 – Scope of Technology				
1. What Is Technology? 2. Technology As a System 3. Contexts of Technology	1 week	1 week	2 weeks	2 weeks
Section 2 – Resources and Technology				
4. Tools and Technology 5. Materials and Technology 6. Energy and Technology 7. Information and Technology 8. People, Time, Money, and Technology	1 week 1 week	1 week 1 week	2 weeks 1 week	2 weeks 2 weeks
Section 3 – Creating Technology				
9. Invention and Innovation 10. The Design Process 11. Identifying Problems 12. Researching Problems 13. Creating Solutions 14. Selecting and Refining Solutions 15. Modeling Solutions 16. Testing Solutions 17. Communicating Solutions 18. Improving Solutions	2 weeks	4 weeks	6 weeks	10 weeks
Section 4 – Technological Contexts				
19. Agricultural and Related Technology 20. Construction Technology 21. Energy Conversion Technology 22. Information and Communication Technology 23. Manufacturing Technology 24. Medical Technology 25. Transportation Technology	1 week 1 week 1 week	1 week 2 weeks 1 week	2 weeks 2 weeks 2 weeks	2 weeks 3 weeks 2 weeks 3 weeks 3 weeks 2 weeks 3 weeks
Section 5 – Technology and Society				
26. Technological Impacts 27. Technology and the Future	1 week	1 week	1 week	2 weeks

Figure IN–4

Working with the Instructional System

This *Introduction to Technology* program is an instructional system with several components. These components are the teacher, students, textbook, Student Activity Manual, and Teacher's Resources. Each component plays a unique role in helping students develop a level of technological literacy and competence. Later, students may use their knowledge in adult citizenship, career, family, and consumer roles.

Role of the Teacher

Teachers and curriculum developers generally have several instructional systems available to meet their needs. Some systems try to provide *all* the answers. They are designed to be "teacher-proof." Textbook writers carefully plan and orchestrate each class period. The teacher becomes an "actor," who is expected to deliver predetermined lines from the script and conduct prescribed activities. This type of material has certain attractions. It is predictable and easy to follow. This approach, however, diminishes the professional role of the teacher and is difficult to modify when it does not meet the needs of specific groups of students.

By contrast, the system upon which this program was developed places the teacher in the *key* role in the educational process. This system assigns the teacher at least three distinct roles:

➤ *Curriculum planner.* The textbook, Student Activity Manual, and Teacher's Resources provide suggested objectives, content, course outlines, study questions, and student activities. These elements are recommendations and suggestions. They are general guides that may be modified to fit the local situation. All modifications, however, should maintain the program's overall view of technology.

➤ *Technical expert.* The program presents students with a conceptual view of technology and its impact on their ways of life. Only through hands-on investigation, practice, and exploration can they develop a truly meaningful understanding of this vital force in their life. The textbook and Student Activity Manual provide many appropriate activities for the class. The presentation of the safe and appropriate techniques to complete these activities is the teacher's responsibility.

➤ *Manager of education.* Teaching technology education has moved a long way from the "tell them, show them, test them" mode of the past. It now challenges students to become actively involved in designing, problem solving, interpreting, assessing, and other vital, higher-level activities. This movement makes the teacher the manager of education. In this role, teachers plan, organize, actuate, and control. They should do all of the following:

1. Plan the program. Establish the major goals of the course and the time schedule for the class.
2. Organize the learning. Adopt structures and procedures for carrying out the various learning activities.
3. Actuate the planned activities. Guide and motivate students as they plan and participate in specific learning events.
4. Control the learning process. Evaluate student achievement, your effectiveness, and program adequacy in light of the program goals.

Role of the Students

Society builds schools and hires teachers so students can learn important information and develop critical skills. It wants to equip its young members to meet the challenges of citizenship. Therefore, students must be active players in the learning process. Technology education must challenge students to do the following:

➤ Participate in the class activities.

➤ Work up to their potentials.

➤ Develop cooperative attitudes.

➤ Explore their creative, mental, and physical abilities.

Opportunities to engage in meaningful design and problem solving activities and to make significant choices in the class are an essential part of the process. Laboratory activities are one way to make this happen. The text and Student Activity Manual contain suggested activities that will make the content more meaningful. These activities, however, can be replaced as student interest and local conditions dictate. *Caution:* Check carefully that each activity makes the content of the course more meaningful, while helping to develop student technical abilities. Only use activities supporting the goals of the course and specific units of study. There is no reason to select activities simply because the students like them, the school needs the product, plans and procedure sheets are available, or they have been used before. Also, select activities providing a balance between individual and group endeavors. Students need to develop both individual skills (goal setting, good work habits, and creative thinking) and group competence (leadership, followership, cooperation, and responsibility to others).

Accommodating Individual Differences

Each student has his or her unique set of abilities and a particular learning style. Education has often tried to teach students as if a single set of strategies and a specific attainment goal could accommodate the needs of all students. This approach has proved to be less than fruitful. The modern educational challenge is to meet overall educational goals, while addressing the vast range of individual differences present in most classrooms. Technology education has the power to meet this challenge, if it is approached through a design and problem solving approach where students work in groups to meet stated objectives. This approach requires the teacher to accept several philosophical tenets:

➤ Understanding the process of developing, operating, and using technology is more important than learning specific facts about technology.

➤ The study of technology is focused on the "why" of technology, rather than on how to use specific tools or machines.

➤ Technology education contributes to a society that requires group participation a great deal more than individual, competitive action.

➤ Society needs and can accommodate people with diverse interests and abilities and, therefore, does not require people with uniform abilities, goals, or knowledge.

➤ Society also needs people who understand technology and can help design our technological future through citizen, consumer, and career roles.

➤ Members of society should develop habits enabling them to contribute their best efforts within the constraints of time and resources.

Developing programs having this vision requires a unique view of students and subject matter content. This vision requires instruction to be conceptual rather than fact-focused, group-oriented rather than individual-oriented, and flexible rather than rigid in its approach. Instruction should be structured so each student who applies himself or herself can taste success. This means there must be opportunities for students to develop and use their specific talents. The chart in **Figure IN-5** lists some of the ways this instructional system will create these opportunities as students explore technology.

Helping Students Gain Basic Knowledge of Technology

Textbooks	Films
Filmstrips	Video recordings
Slide series	Field trips
Guest speakers	Research

Helping Students Develop Problem Solving Abilities

Product and structure design	System development
Process planning	Material testing

Helping Students Develop Creativity

Product and structure design	Brainstorming
Communication message design	Advertising design
Package design	Vehicle design

Helping Students Apply Technology

Product and structure design	Constructing structures
Manufacturing products	Designing messages
Energy system design	Material testing
Vehicle design	

Helping Students Assess Technology

Product design review	Material testing
Energy system analysis	Research
Environmental impact analysis	Future projections

Helping Students Develop Communications Skills

Class discussions	Group interaction
Drawing and sketching	Product design presentation
Laboratory summaries	Written and oral reports

Helping Students Develop Leadership Skills

Decision making	Goal setting
Goal assessment	Managing activities
Cooperative group activities	

Figure IN–5

Role of the Textbook

An ever changing world challenges schools to remain relevant. Each citizen is expected to grow intellectually throughout her or his life. This requires a strong foundation in the basics: sciences, technologies, and humanities. Moreover, individuals must develop the skills needed to seek out new information. This development rests on the ability to read. The textbook plays a vital role in these challenges. It presents students with the rudiments of technology and encourages them to develop the ability to comprehend what they read. Each student should, then, have a copy of the course text. She or he should be expected to read it and encouraged to take it home. Class time is not for reading the textbook. Rather, it is the time for technology to come alive in laboratory activities, presentations and discussions, films and videotapes, and other interactive processes. *Technology: Design and Applications* opens the world of technology to all students. It is colorfully illustrated and interestingly written. The textbook is the source of information for class discussions and activity development.

Role of the Student Activity Manual

The Student Activity Manual gives students guidance and support as they engage in their study of technology. The manual is a consumable including the following features:

➤ Laboratory activity sheets for selected activities included in the textbook.

➤ Alternative, problem-centered activities for most chapters in the textbook.

➤ Study questions to support the textbook.

The teacher can conduct the course without the Student Activity Manual. To do so, however, requires more work and preparation time. This manual is, in every respect, a guide for completing the mental and physical activities envisioned for the class.

Role of the Teacher's Resource

The Teacher's Resource goes well beyond an answer key for the textbook and Student Activity Manual study questions. It also does the following:

➤ Offers a rationale for the program that can be used to inform students, parents, guidance personnel, and administrators of the value for a study of technology.

➤ Suggests approaches and procedures for developing and presenting each section, including the following elements:
   1. An introduction.
   2. Objectives.
   3. Suggested length and schedule.
   4. Teacher preparatory activities.
   5. Suggestions for managing student activities.

➤ Presents ways of selecting and presenting laboratory activities.

The Teacher's Resource is not a script. It is a course development guide and, therefore, should be read cover-to-cover before the course is started. This resource does not attempt to provide a cookbook approach to each activity, with the ingredients and procedures carefully prescribed. Nor have the authors developed the program for an "ideal" or "average" class, with the expectation that all other classes will, somehow, fit this mythical mold. You, as the primary educational agent in the classroom, should read and study the guide to do the following:

➤ Develop an overall view of the goals, content, and activities for the class.

➤ Prepare long-range plans, including the need for content presentation development, media selection and development, activity preparation and modification, equipment and supply purchases, and professional growth.

➤ Schedule daily learning activities.

➤ Prepare daily lesson and activity plans.

➤ Design an evaluation system.

➤ Plan a classroom management schedule.

Sections As Content Organizers

In many school subjects, content is presented through a series of discrete topics that are generally the themes of specific chapters. This approach causes students to perceive the course as a group of isolated concepts and series of facts. A more useful approach is to organize content into larger units. This is the approach suggested for this instructional system. Each section or unit of study has an overall uniting theme integrating a number of chapters and activities. Using this approach, students will more likely see technology as a unified and purposeful human activity. Each component of the total instructional system must be placed in its proper context, if learning is to be maximized. No single element can do the whole job. All the information cannot be presented. The student cannot solve all the problems. The textbook cannot provide all the content, nor make it concrete. The laboratory activities cannot be the reason for the class. If each component is used to do what it can do well, however, the class will be challenging, fun, and informative. That's your challenge.

Section 1

Scope of Technology

Focus: Introduction to technology, technological systems, and contexts in which technology is applied.

➤ Chapter 1: What Is Technology?

➤ Chapter 2: Technology As a System

➤ Chapter 3: Contexts of Technology

Teaching Materials

Text

Chapter 1, pages 22–33
  Chapter Activities, page 32
  Test Your Knowledge, page 33
Chapter 2, pages 34–53
  Chapter Activities, page 52
  Test Your Knowledge, page 53
Chapter 3, pages 54–73
  Chapter Activities, page 72
  Test Your Knowledge, page 73
Section Activities
  Activity 1A—Using Technology to Make a Product, pages 74–76
  Activity 1B—What Is Technology?, page 77

Student Activity Manual

Activity 1-1—Using Technology to Make a Product, pages 7–10
Activity 1-2—Marble Counter, page 11
Activity 1-3—Technological Systems, pages 13–14
Chapter 1 Study Questions, page 15
Chapter 2 Study Questions, page 17
Chapter 3 Study Questions, page 19

Teacher's Resources

Alternative Activity—Diamond Kite, pages 47–52
Reproducible Masters
  2-1—The Universal Systems Model, page 53
  2-2—Types of Results for Technological Systems, page 54
  2-3—Systems, page 55
Transparencies (Binder/CD only)
  1-1—Technology Is Knowledge
  1-2—Human History
  1-3—Technology Includes Tools
  1-4—Types of Knowledge
  2-1—Technological Systems
  2-2—Inputs to Technological Systems
  2-3—Processes of Technological Systems
  2-4—Outputs of Technological Systems
  2-5—Feedback in Technological Systems
  3-1—Functional Contexts

Introduction

This section sets the stage for the entire study of technology. It first introduces the class to the term *technology*; then it explains the difference between technology and science. This section presents technology as a system to extend human potential. It describes designing, producing, using, and assessing as the major actions of technology. This section discusses technological systems in terms of inputs, processes, outputs, feedback, and societal influences. Finally, it introduces agriculture, construction, energy conversion, information and communication, manufacturing, medicine, and transportation as major contexts in which technology is applied.

Objectives of the Section

At the end of Section 1, each student should be able to do the following:

➤ Define *technology*.
➤ Describe the difference between technology and science.
➤ Explain technology as a system.
➤ List and summarize the parts of a technological system.
➤ Give examples of production and management processes.
➤ Discuss the goals of technology.
➤ Identify and describe the major contexts in which technology is applied.

Suggested Schedule and Length of the Section (Five to Ten Days)

Section 1 should introduce students to technology. You should have them complete the following assignments:

1. Read Chapters 1 through 3 in their textbooks and answer associated study questions.
2. Participate in presentations and discussions on technology, technological systems, and technological contexts.
3. Complete one or two laboratory activities and follow-up sheets (in the Student Activity Manual).

Schedule for Section 1—9-Week and 12-Week Classes	
Day 1	School start-up activities Textbook and Student Activity Manual distribution Discussion: "What will we study in this class?" Assign Chapter 1: "What Is Technology?" to be read outside class.
Day 2	Discussion: "What is technology?" Assign Chapter 2: "Technology As a System" to be read outside class. Introduce activity.
Day 3	Work on activity. Assign Chapter 3: "Contexts of Technology" to be read outside class.
Day 4	Discussion: "Technological systems and contexts." Work on activity.
Day 5	Complete and summarize activity and section.

Schedule for Section 1—18-Week and 36-Week Classes

Day 1	School start-up activities Textbook and Student Activity Manual distribution Discussion: "What will we study in this class?"
Day 2	Discussion: "What is technology?" Assign Chapter 1: "What Is Technology?" to be read outside class. Introduce first activity.
Day 3	Work on activity.
Day 4	Work on activity. Assign Chapter 2: "Technology As a System" to be read outside class.
Day 5	Discussion: "What is a system?" Summarize and discuss activity.
Day 6	Introduce second activity. Assign Chapter 3: "Contexts of Technology" to be read outside class.
Day 7	Work on activity.
Day 8	Discussion: "What are technological contexts?" Work on activity.
Day 9	Complete and summarize activity and section.

Preparing to Teach the Section

You have several important tasks to complete before this learning unit can be presented to students.

1. Carefully read the Teacher's Resource to develop an understanding of the scope and sequence of the course.

2. Read the textbook cover-to-cover to develop an understanding of the content.

3. Prepare presentation and discussion plans and make notes for four 10 to 15 minute sessions on the following topics:
   A. What is technology? (Textbook, Chapter 1)
   B. Technology as a system. (Textbook, Chapter 2)
   C. What are technological contexts? (Textbook, Chapter 3)
   D. Technology—A summary.

4. Select one or two appropriate laboratory activities from the following options:
   A. Textbook, pages 74–77.
   B. Student Activity Manual, pages 7–14.
   C. Teacher's Resource, pages 47–52.
   D. Other sources of your choosing.

5. Make necessary material and supply preparations for presenting the laboratory activity or activities.

6. Work through the laboratory activity or activities and student study questions presented in the textbook, Student Activity Manual, and Teacher's Resource.

Presenting the Section

All sections in this program follow a basic format. They include the following elements:

1. Discussions and presentations to introduce and reinforce basic information. Plan to limit these sessions to 10- or 15-minute periods distributed across several days.

2. Student study activities included in the Key Words and Test Your Knowledge sections for each textbook chapter and study questions in the Student Activity Manual.

3. Laboratory activities, which shift the basic information from the abstract to the concrete. Refer to the suggested activities contained in the textbook, activities in the Student Activity Manual, and alternative activities in the Teacher's Resource. Remember to develop and demonstrate safe procedures for completing each activity.

4. Laboratory summary sheets to ensure that the students can relate the activities to the course content. These sheets are contained in the Student Activity Manual.

Adjustments for Special Needs and Gifted Students

The five activities suggested for this section lend themselves to group work. One way to adjust instruction for differing abilities and varying interests is to organize the groups with a mix of students in each one. Each group can include gifted, average, and special needs students. Consider grouping those who think logically and sequentially with random, intuitive thinkers. Some students may have highly developed artistic abilities. Think about pairing them with others who have little or no artistic talent.

Give special attention and individual help to those students who require more structure. If the activity is highly structured for all students, however, the gifted students may lack the challenge they need to do their best work.

Evaluation Procedures

Technology education stresses the process and creative answers. Quality (craftsmanship) of the final product is of secondary importance. Therefore, students should be evaluated on the following criteria:

1. Contributions to group activity. This evaluation is possible with daily logs (diaries) the students maintain, teacher observations, and student feedback.

2. Laboratory sheets, sketches, and designs completed during the activity.

3. Participation in discussions conducted during the laboratory activity.

4. Scores on quizzes, chapter questions, or other activities.

Implementing the Laboratory Activities

Five different activities have been developed for this section. These are simply suggested activities, and any activity introducing technology can be substituted. The time constraints for this section restrict the program to one or two relatively short activities.

Procedures for Directing Each Suggested Activity

Using Technology to Make a Product, Textbook, pages 74–76, and Student Activity Manual, pages 7–10

This activity will allow students to explore technology by, first, producing a product without tools and, then, making the same product using technological aides. The activity is designed to be prescriptive in nature, with the procedure provided for the students.

During *phase one* of the activity, each student forms and decorates a 1/4″ × 4″ × 4″ clay tile using his or her hands. Have students complete the following steps:

1. Read the directions for Activity 1A-1 on pages 74–75 of the textbook.
2. Form as many tiles as he or she can in a set time period, such as ten minutes.
3. Place his or her tiles with the other ones the class members produce.
4. Complete the first section of the Activity 1-1 follow-up sheet, found on page 7 of the Student Activity Manual.
5. Discuss the problems of producing a product without technology.
6. Recommend ways to develop technology (tools) to increase the speed of producing the tiles.

During *phase two*, divide the class into groups of five students. Demonstrate the process the students will use to make tiles using technology. Have each student record the steps on the activity procedure sheet found on page 8 of the Student Activity Manual.

Each group should complete the following steps:

1. Read the directions for Activity 1A-2 on pages 74–76 of the textbook.
2. Collect the supplies:
   - Two 1/4″ × 3/4″ × 12″ wood strips
   - Rolling pin
   - Tile cutter
   - Tile decorating press
   - Potter's clay
   - Sponge
   - Can of water
3. Create groups of students by counting off numbers from one to five.
4. Produce 12 tiles using these procedures.
5. Analyze the tiles.
6. Complete the activity follow-up sheet, found on page 9 of the Student Activity Manual.
7. Discuss the problems of producing the products using technology.
8. Complete the activity summary, found on page 10 of the Student Activity Manual.
9. Recommend ways to improve the technology (tools) used to produce the product.

What Is Technology?, Textbook, page 77

This activity is designed to allow the students to use their creativity to design an energy conversion system. They will use the mechanical energy from wind to lift a weight. There is no right answer. To use this activity, divide the class into groups of two to four students. Have each group complete the following steps:

1. Define the problem they are attempting to solve.
2. List any restrictions they face, such as materials that can be used (a list is included with the activity description in the textbook) and the strength of the wind.
3. Develop sketches of two to five ways to solve the problem.
4. Select and refine the best solution.
5. Build a prototype of the solution using only the materials listed in the textbook. (Limiting the materials introduces a common restriction all product designers face.)
6. Demonstrate the solution.
7. List ways to improve the design.

Marble Counter, Student Activity Manual, page 11

This activity is designed to be an open-ended problem solving activity. The students should be provided the design brief and encouraged to develop a unique solution to the problem. The activity can be done individually; however, it will work better with groups of two to four students. Have the students complete the following steps:

1. Define the problem they are attempting to solve.
2. List any restrictions they face, such as marble size and cost of the solution.
3. Develop sketches of two to five ways to solve the problem.
4. Select and refine the best solution.
5. If time permits, build a prototype of the solution.

Technological Systems, Student Activity Manual, pages 13–14

This activity is designed to allow the students to explore the components of technological systems. It can be done individually or in small groups. Have the students complete the following steps:

1. Read the directions for Activity 1-3 on page 13 of the Student Activity Manual.
2. Complete Activity 1-3 worksheet.
3. Discuss their answers with the rest of the class.

Diamond Kite, Teacher's Resource, pages 47–52

This activity is included in the Teacher's Resource as a substitute for the clay tile activity in the textbook or a second activity for 18- and 36-week classes. If this activity is used, the students will work in small groups, using only pencils, rulers, a backsaw, and scissors, to produce a kite with a "low-tech" production system. Then, they will use templates and jigs to make another kite, using a "high-tech" system. You will need to make the following jigs and fixtures for each group to use:

➤ A template to mark out the cover.
➤ An alignment jig that will position the spine and spar so they can be assembled without measuring.
➤ A jig to hold the spine and spar so the notches can be cut quickly. Also, the laboratory sheets for the clay tile will need to be modified to ask questions about the kite and its quality.

Section 2

Resources and Technology

Focus: Exploration of the inputs to technological systems.

➤ Chapter 4: Tools and Technology
➤ Chapter 5: Materials and Technology
➤ Chapter 6: Energy and Technology
➤ Chapter 7: Information and Technology
➤ Chapter 8: People, Time, Money, and Technology

Teaching Materials

Text

Chapter 4, pages 80–105
  Chapter Activities, page 104
  Test Your Knowledge, page 105
Chapter 5, pages 106–127
  Chapter Activities, page 126
  Test Your Knowledge, page 127
Chapter 6, pages 128–148
  Chapter Activities, page 147
  Test Your Knowledge, page 148
Chapter 7, pages 150–164
  Chapter Activities, page 161
  Test Your Knowledge, page 162
Chapter 8, pages 166–185
  Chapter Activities, page 184
  Test Your Knowledge, page 185
Section Activities
  Activity 2A—Tools and Materials As Resources, pages 186–192
  Activity 2B—Tools As a Resource, page 193
  Activity 2C—Materials As a Resource, pages 194–197
  Activity 2D—Energy As a Resource, pages 198–203
  Activity 2E—Energy and Machines As Resources, page 204
  Activity 2F—Information As a Resource, pages 205–209

Student Activity Manual

Activity 2-1—Tools and Materials As Resources, pages 21–23
Activity 2-2—Tools and Energy As Resources, page 25
Activity 2-3—Procedure and Tool Use, page 27
Activity 2-4—Materials As a Resource, pages 29–30
Activity 2-5—Product Analysis, page 31
Activity 2-6—Energy As a Resource, page 33
Activity 2-7—Energy As a Resource, pages 35–36
Activity 2-8—Information As a Resource, pages 37–38
Activity 2-9—Developing an Information Chart, page 39
Chapter 4 Study Questions, pages 41–42
Chapter 5 Study Questions, pages 43–44
Chapter 6 Study Questions, pages 45–46
Chapter 7 Study Questions, pages 47–48
Chapter 8 Study Questions, page 49

Teacher's Resources
Alternative Activity—Materials As a Resource, pages 69–70
Alternative Activity—Information As a Resource, pages 71–72
Reproducible Masters
4-1—Measurement, page 73
4-2—Basic Mechanisms, page 74
4-3—From Tool to Mechanism to Machine, page 75
4-4—Tool Relationships, page 76
4-5—Tool Usage Chart, page 77
4-6—Tool Safety, page 78
5-1—Properties of Materials, page 79
6-1—Nature's Energy Storehouse, page 80
6-2—How a Nuclear Power Station Operates, page 81
6-3—Power Systems in Your Community and Their Energy Sources, page 82
7-1—Data, Information, and Knowledge, page 83
7-2—Technological Knowledge, page 84
8-1—Management, page 85
8-2—Functions of Management, page 86
8-3—Levels of Authority and Responsibility, page 87
8-4—Preparing a Company Organizational Chart, page 88
8-5—Preparing an Organizational Chart for Your School, page 89
Transparencies (Binder/CD only)
5-1—Securing Natural Resources
5-2—Producing Industrial Materials
6-1—Energy Key Terms
6-2—Types of Energy
6-3—Forms of Energy

Introduction

All technological systems use resources to extend human potential. Technological systems use tools, mechanisms, and machines to grow and harvest crops, manufacture products, convert and use energy, construct structures, heal illnesses, communicate information, and transport people and goods. Using human-made artifacts separates technological systems from the natural systems studied in science classes.

Objectives of the Section

At the end of Section 2, each student should be able to do the following:

➤ Discuss humans as toolmakers and users.

➤ Describe the difference between a tool, mechanism, and machine.

➤ List and give examples of the major types of tools humans use to extend their potentials.

➤ Name and describe the major types of machines.

➤ Explain natural materials.

➤ Recognize and explain the major types of engineering materials.

➤ Summarize the difference between renewable and exhaustible materials.

➤ Define *energy* and summarize several of its sources.

➤ Give examples of the major forms of energy.

➤ Cite and discuss the definition of *energy conversion*.

➤ Explain the application and conservation of energy.

➤ Apply the use of symbols and signs in communication information.

➤ Identify the types of jobs available in technological systems.

➤ Summarize the role of money in operating technological systems.

Suggested Schedule and Length of the Section (Ten to Twenty Days)

Section 2 introduces students to the resources (inputs) of technological systems. You should have them complete the following assignments:

1. Read Chapters 4–8 in their textbooks and answer associated study questions.
2. Participate in several discussions and presentations.
3. Complete two or more laboratory activities and follow-up sheets that will reinforce the study of the following topics:
   A. The use of tools and machines.
   B. The types and properties of materials.
   C. The role of energy in technological systems.
   D. Information in technological systems.
   E. The opportunities for people to work in and use technological systems.

Schedule for Section 2—9-Week and 12-Week Classes	
Day 1	Discussion: "Resources and technology." Assign Chapter 4: "Tools and Technology" to be read outside class. Introduce first activity (tools or materials activity).
Day 2	Discussion: "Tools and technology." Work on activity.
Day 3	Work on activity. Assign Chapter 5: "Materials and Technology" to be read outside class.
Day 4	Discussion: "Materials and technology." Work on activity.
Day 5	Complete and summarize activity. Assign Chapter 6: "Energy and Technology" to be read outside class.
Day 6	Discussion:"Energy and technology." Introduce second activity (energy, information, or people activity).
Day 7	Work on activity. Assign Chapter 7: "Information and Technology" to be read outside class.
Day 8	Discussion: "Information and technology." Assign Chapter 8: "People, Time, Money, and Technology" to be read outside class. Work on activity.
Day 9	Discussion: "People, time, money, and technology." Work on activity.
Day 10	Complete and summarize activity. Review and summarize section.

Schedule for Section 2—18-Week and 36-Week* Classes

Day 1	Discussion: "Resources and technology." Assign Chapter 4: "Tools and Technology" to be read outside class. Introduce first activity (tools or materials activity).
Day 2	Discussion: "Tools and technology." Work on activity.
Day 3–4	Work on activity. Assign Chapter 5: "Materials and Technology" to be read outside class. Discussion: "Materials and technology."
Day 5	Complete and summarize activity. Assign Chapter 6: "Energy and Technology" to be read outside class.
Day 6	Discussion: "Energy and technology." Introduce second activity (energy activity).
Day 7–9	Work on activity. Assign Chapter 7: "Information and Technology" to be read outside class.
Day 10	Discussion: "Information and technology." Complete and summarize activity. Assign Chapter 8: "People, Time, Money, and Technology" to be read outside class.
Day 11	Discussion: "People, time, money, and technology." Introduce third activity (information or people and technology).
Day 12–14	Work on activity.
Day 15	Complete and summarize activity. Review and summarize section.

* For a 36-week class, add another activity and spread the discussions over the four weeks scheduled for this section.

Preparing to Teach the Section

Before introducing this section to the students, you should complete the following tasks:

1. Study Section 2 of the Teacher's Resource.
2. Read Chapters 4–8 in the textbook to review the content.
3. Prepare 10–15 minute presentations and discussions on the following topics:
   A. Types of tools. (Textbook, pages 82–94)
   B. Mechanisms and machines. (Textbook, pages 94–102)
   C. Types and properties of materials; obtaining materials. (Textbook, Chapter 5)
   D. Nature, sources, and supplies of energy. (Textbook, Chapter 6)
   E. Information and technology. (Textbook, Chapter 7)
   F. People, time, and money in technology. (Textbook, Chapter 8)
   G. Summary of resources and technology.

4. Select appropriate laboratory activities from the following options:
    A. Tools and technology.
        (1) Textbook, pages 186–193.
        (2) Student Activity Manual, pages 21–27.
        (3) Other sources of your choosing.
    B. Materials and technology.
        (1) Textbook, pages 186–192, 194–197.
        (2) Student Activity Manual, pages 21–23, 29–31.
        (3) Teacher's Resource, pages 69–70.
        (4) Other sources of your choosing.
    C. Energy and technology.
        (1) Textbook, pages 198–204.
        (2) Student Activity Manual, pages 25, 33–36.
        (3) Other sources of your choosing.
    D. Information and technology.
        (1) Textbook, pages 205–209.
        (2) Student Activity Manual, pages 37–39.
        (3) Teacher's Resource, pages 71–72.
        (4) Other sources of your choosing.
    E. People, time, money, and technology.
        (1) Sources of your choosing.
5. Make necessary material and supply preparations for presenting the laboratory activities.
6. Work through the laboratory activities and student study questions presented in the textbook, Student Activity Manual, and Teacher's Resource.

Presenting the Section

This section includes time for short, informative discussions and presentations supported by meaningful laboratory activities. The discussions should be limited to 15 minutes and reinforce the information in the textbook. Summarize all discussion and laboratory work through the use of the Test Your Knowledge questions accompanying the textbook chapters and study questions in the Student Activity Manual. On the last day of the section, summarize the concepts that have been presented. Direct the students' attention to what they have learned over the past two to four weeks.

Adjustments for Special Needs and Gifted Students

The thirteen activities suggested for this section are designed to introduce the primary resource inputs to technological systems. They lend themselves to either individual or group work. One way to adjust instruction for differing abilities and varying interests is to assign the elected activities to groups with a mix of students. This can group gifted, average, and special needs students, as well as those who think logically and sequentially with those who are random thinkers. Special attention and individual help should be given to those students who require more structure. If the activity is highly structured for all students, however, the gifted students may lack the level of challenge needed for them to do their best work.

Evaluation Procedures

The emphasis of this section is on understanding how resources are used within a technological system. The focus should be on the importance of tools, materials, energy, information, people, time, and money in implementing technology. Specific tool nomenclature, craftsmanship of the final products, memorization of material properties, and other similar criteria are of secondary importance. The goal is to understand that materials have a range of properties that must be evaluated when developing products. Therefore, the students should be evaluated on the following criteria:

1. Contributions to group activity. This evaluation is possible with daily logs (diaries) the students maintain, teacher observations, and student feedback.
2. Laboratory sheets, sketches, and designs completed during the activities.
3. Participation in discussions conducted during the laboratory activities.
4. Scores on quizzes and chapter questions.

Implementing the Laboratory Activities

A number of different activities have been developed for this section. These can be found at the end of Section 2 (Chapters 4–8) of the textbook, in Section 2 of the Student Activity Manual, and in the Teacher's Resource. These are suggested activities, and any activity introducing the concept that tools, materials, information, energy, people, time, and money are essential for technology can be substituted. The time constraints for this section restrict the program to a few relatively short activities.

Procedures for Directing Each Suggested Activity

Tools and Materials As Resources, Textbook, pages 186–192, and Student Activity Manual, pages 21–23

This is a prescriptive activity introducing tools as an essential input for technology. The activity presents a specific set of procedures to the students. The system used to make the product should be demonstrated. Then, divide the class into three groups:

➤ Group 1: Makes the game board.
➤ Group 2: Makes and stains the pegs.
➤ Group 3: Makes the storage hole cover.

Have each group complete the following steps:

1. Carefully read the directions for making the product.
2. Assign a step or several steps to each member of the group.
3. Obtain the supplies needed to make the product (listed in the textbook).
4. Produce 24 parts, using the procedures listed in the textbook.
5. Analyze the output (quality control).

When all parts are produced, give each group eight sets of parts. Have each group assemble the products by completing the following steps:

1. Carefully read the directions for the assembly procedure.
2. Assign a step or several steps to each member of the group.
3. Assemble eight products, using these assembly procedures.

Tools As a Resource, Textbook, page 193

This activity is designed to be an open-ended, problem solving activity. The students should be provided the design brief and encouraged to develop a unique solution to the problem. The activity can be done individually. It will work better, however, with groups of two to four students. Have the students complete the following steps:

1. Define the problem they are attempting to solve.
2. List any restrictions they face, such as marble size and cost of the solution.
3. Develop sketches of two to five ways to solve the problem.
4. Select and refine the best solution.
5. If time permits, build a prototype of the solution.

Tools and Energy As Resources—Rubber Band Car, Student Activity Manual, page 25

This activity is designed to be less structured than the game activity provided in the textbook. The students should be provided the drawings for the rubber band car and challenged to complete the following steps:

1. Develop a bill of materials for the product.
2. Prepare a procedure sheet for building the product, using the form on page 27 of the Student Activity Manual.
3. List any restrictions they face.
4. Build a prototype of the racer.
5. Discuss the procedures they used, ways to build the product more efficiently, and other similar topics.

Materials As a Resource—Material Properties and Applications, Textbook, pages 194–197, and Student Activity Manual, pages 29–30

This is a prescriptive activity to introduce students to material properties. A set of material samples should be provided, containing a variety of materials, such as sheet aluminum, copper, steel, acrylic plastic, plywood, and lumber. Divide the class into groups of two to four students and have each group complete the following steps:

1. Perform all the appropriate tests the procedures in the textbook outline.
2. Record test results on the form supplied on page 29 of the Student Activity Manual.
3. Present their test findings to the class.
4. Discuss the implications of the data for material selection and application.

Product Analysis, Student Activity Manual, page 31

This activity is designed to allow the students to analyze products in terms of the materials they contain. To use this activity, divide the class into groups of two to four students. Have each group follow the directions at the top of the activity sheet for Activity 2-5. The groups should complete the following steps:

1. Select a product in the technology education laboratory.
2. Identify four materials the product contains.
3. List the materials and their uses in the chart on page 31 of the Student Activity Manual.

4. Indicate possible reasons for selecting each material. These can include reasons such as appearance, cost, ease of manufacture, and strength.
5. Present their findings and encourage the class to react to their reports.

Materials As a Resource—Testing Material Hardness, Teacher's Resource, pages 69–70

This is a prescriptive activity in which the students will determine the relative hardness of three selected woods. They will follow a set procedure. Divide the class into groups of two to three students. Provide each group with a hardness tester you have built. (The actual construction of the tester is open to modification.) The key parts are the following:

➤ A 3/8″ diameter by 36″ steel rod that has one end rounded.
➤ A PVC pipe with a 5/32″ diameter hole drilled 2″ from one end. The pipe should have a diameter allowing the steel rod to slip through it easily.
➤ A 1/8″ retainer pin that keeps the rod from dropping down the PVC pipe before the test is conducted.
➤ A 24″ space between the retainer pin and specimen.
➤ Parallel strips on the tester base to ensure the specimen is located directly below the steel rod.

Have each group conduct the tests by completing the following steps:
1. Carefully read the directions for the procedure.
2. Complete a face grain and end grain test for each of the three specimens.
3. Record the data.
4. Draw conclusions about the following:
   ➤ Relative hardness of end grain and face grain of the same species of wood.
   ➤ Relative hardness of the three species of wood.

Energy As a Resource—Motor Schematic, Textbook, pages 198–203, and Student Activity Manual, page 33

This is a prescriptive activity allowing the students to make a common energy converter—an electric motor. The activity presents a specific set of procedures to the students. It may be done individually or by groups of students. If using groups, assign three students to each group. Have the students complete the following activities:
➤ Student 1—Lay out and produce the base.
➤ Student 2—Make the armature.
➤ Student 3—Make two motor bearings.
➤ All—Assemble and test the motor.

Energy and Machines As Resources, Textbook, page 204

This activity is designed to allow the students to use their creativity to design a system using a simple machine and human strength. There is no correct answer. To use this activity, divide the class into groups of two to four students. Have each group complete the following steps:
1. Define the problem they are attempting to solve.
2. List any restrictions they face, such as materials that can be used and the strength of the person using the machine.
3. Develop a sketch of two to five ways to solve the problem.
4. Select and refine the best solution.

5. Build a prototype of the solution, using only the materials listed in the textbook. (Limiting the materials introduces a common restriction all product designers face.)
6. Demonstrate the solution.
7. List the ways they can improve their designs.

Energy As a Resource—Wind-Powered Generator, Student Activity Manual, pages 35–36

This activity is an open-ended, problem solving activity challenging the students to develop an energy converter that changes wind energy into electrical energy. To carry out this activity, do the following tasks:

1. Divide the class into groups of two to three students.
2. Give each group the following basic materials:
    A. Low voltage DC motor (available at hobby and electrical supply stores).
    B. 24" of magnet wire.
    C. Low voltage voltmeter.
3. Allow each group access to assorted materials, such as aluminum foil, dowels, thin plywood, strips of pine or balsa wood, poster board, and copper or galvanized wire.
4. Have each group complete the following steps:
    A. Sketch at least three ways they can build the converter.
    B. Select and refine the best idea.
    C. Build a prototype of the idea.
    D. Test the prototype.
    E. Modify the prototype based on test results.
    F. Discuss ways to improve the device to make it more efficient.

Information As a Resource—Note Taking, Textbook, pages 205–209, and Student Activity Manual, pages 37–38

This activity is designed to allow students to read, review, and synthesize information. Generally, this activity is done individually, but the students can work in groups of two. To use the activity, have the students complete the four steps outlined on page 205 of the textbook. As an alternative, this activity can be assigned as homework.

Developing an Information Chart, Student Activity Manual, page 39

This is a prescriptive activity, designed to allow students to analyze sources and types of information. They will locate written information (procedures), graphic information (such as sketches and drawings), and charts and tables (bill of materials). They should follow the directions contained at the top of page 39 in the Student Activity Manual.

An alternate activity is to give the students a simple product and have them develop the three types of information. The product may be a package, a baked good, such as a cookie, or some other simple item.

Another alternate activity is to have the students develop pictorial directions for baking a cake. Each step should be shown using only pictures. Quantities can be shown on a measuring cup or spoon, but all other information must be shown in drawings.

The students should use the form provided on page 39 of their Student Activity Manuals.

Information As a Resource—Evolution of Information Processing, Teacher's Resource, pages 71–72

This is an open-ended activity in which students practice and compare five methods of recording information that have been developed over time:

1. Clay tablets.
2. Pen, ink, and paper.
3. Movable type.
4. Typewriting.
5. Information processing.

The students will first record a message, then be asked to change it slightly. They will compare and contrast each method in terms of the following criteria:

1. Ease of producing the message.
2. Ease of revising the message.

Section 3

Creating Technology

Focus: Presentation of the creation of technology through invention and design. Examination of each of the eight steps of the design process.

➤ Chapter 9: Invention and Innovation
➤ Chapter 10: The Design Process
➤ Chapter 11: Identifying Problems
➤ Chapter 12: Researching Problems
➤ Chapter 13: Creating Solutions
➤ Chapter 14: Selecting and Refining Solutions
➤ Chapter 15: Modeling Solutions
➤ Chapter 16: Testing Solutions
➤ Chapter 17: Communicating Solutions
➤ Chapter 18: Improving Solutions

Teaching Materials

Text

Chapter 9, pages 212–231
   Chapter Activities, page 228
   Test Your Knowledge, page 229
Chapter 10, pages 232–252
   Chapter Activities, page 249
   Test Your Knowledge, page 250
Chapter 11, pages 254–267
   Chapter Activities, page 266
   Test Your Knowledge, page 267
Chapter 12, pages 268–283
   Chapter Activities, page 282
   Test Your Knowledge, page 283
Chapter 13, pages 284–309
   Chapter Activities, page 308
   Test Your Knowledge, page 309
Chapter 14, pages 310–329
   Chapter Activities, page 328
   Test Your Knowledge, page 329
Chapter 15, pages 330–359
   Chapter Activities, page 358
   Test Your Knowledge, page 359
Chapter 16, pages 360–380
   Chapter Activities, page 377
   Test Your Knowledge, page 378
Chapter 17, pages 382–409
   Chapter Activities, page 405
   Test Your Knowledge, page 406

Chapter 18, pages 410–424
  Chapter Activities, page 423
  Test Your Knowledge, page 424
Section Activities
  Activity 3A—Invention and Innovation, page 426
  Activity 3B—Communication Design, pages 427–429
  Activity 3C—Design for Space, pages 430–431

Student Activity Manual

Activity 3-1—Shading and Rendering, page 51
Activity 3-2—Orthographic Sketching, page 53
Activity 3-3—Communication Design, pages 55–58
Activity 3-4—Locker Design, page 59
Activity 3-5—Bio-Related Feeder, page 61
Chapter 9 Study Questions, pages 63–64
Chapter 10 Study Questions, pages 65–66
Chapter 11 Study Questions, page 67
Chapter 12 Study Questions, page 69
Chapter 13 Study Questions, page 71
Chapter 14 Study Questions, page 73
Chapter 15 Study Questions, page 75
Chapter 16 Study Questions, page 77
Chapter 17 Study Questions, page 79
Chapter 18 Study Questions, page 81

Teacher's Resources

Alternative Activity—House Design, page 105
Activity Sheet—Identify the Problem, pages 107–108
Activity Sheet—Research the Problem, page 109
Activity Sheet—Create Solutions, page 110
Activity Sheet—Select and Refine the Solutions, page 111
Activity Sheet—Model and Test the Solution, page 112
Activity Sheet—Communicate the Solution, page 113
Activity Sheet—Improve the Solution, page 114
Reproducible Masters
  9-1—Important Inventions in the Twentieth Century, page 115
  11-1—The Design Process: Defining the Problem, page 116
  11-2—Technological Design Constraints, page 117
  12-1—The Design Process: Gathering Information, page 118
  13-1—The Design Process: Developing Possible Solutions, page 119
  14-1—The Design Process: Selecting Promising Solutions, page 120
  14-2—The Design Process: Refining Selected Ideas or Solutions, page 121
  14-3—The Design Process: Selecting the Best Solution, page 122
  16-1—The Design Process: Testing Design Solutions, page 123
  17-1—The Design Process: Interpreting the Solution, page 124
  17-2—Making an Oblique Drawing, page 125
  17-3—The Design Process: Presenting Solutions for Approval, page 126
Transparencies (Binder/CD only)
  10-1—Technological Design
  10-2—Designing
  10-3—Types of Design
  10-4—Steps of the Design Process
  11-1—Problems

11-2—Questions to Help Define the Problem
11-3—Define the Problem
12-1—Gather Information
13-1—Sketching
13-2—Develop Solutions
13-3—Rough Sketches
13-4—Thumbnails
13-5—Pictorial Drawings
14-1—Select the Best Solution
14-2—Rendering
14-3—Detailed Sketches
15-1—Models
15-2—Model the Solution
15-3—Functions of Models
15-4—Types of Models
15-5—Graphic Models
15-6—Mathematical Models
15-7—Physical Models
16-1—Evaluate the Solution
17-1—Document the Solution
17-2—Isometric Drawing
17-3—Oblique Drawing
17-4—One-Point Perspective
17-5—Two-Point Perspective

Introduction

This section examines the process used to design technological devices and systems. It is essential that the study of technology includes the study of design. The first stage in creating any example of technology involves the design process. All fields of technology have different processes they use to design products, systems, and structures. When they are examined, however, many common elements exist. This book provides a design process that is a combination and revision of many different processes. This process is a general set of steps that can be used to design any type of product or system. The steps in this process include the following:

1. Identifying problems.
2. Researching problems.
3. Creating solutions.
4. Selecting and refining solutions.
5. Modeling solutions.
6. Testing solutions.
7. Communicating solutions.
8. Improving solutions.

Objectives of the Section

At the end of Section 3, each student should be able to do the following:

➤ Define *design*.
➤ List and explain the steps in the design process.
➤ Identify and state a design problem.

➤ Create a design brief.

➤ Gather information to solve a problem.

➤ Produce and prepare rough and refined sketches to develop design solutions.

➤ Refine design solutions through detailed sketches.

➤ Design and build models of design solutions.

➤ Use engineering drawings to communicate design solutions.

➤ Describe and apply appropriate techniques to test design solutions.

➤ Present design solutions for approval.

Suggested Schedule and Length of the Section (Two to Ten Weeks)

Section 3 introduces the students to the design process. You should have them complete the following assignments:

1. Read Chapters 9–18 in their textbooks and answer associated study questions.
2. Participate in several discussions and presentations about invention and innovation, design, and the eight steps of the design process (identifying, researching, creating, selecting and refining, modeling, testing, communicating, and improving).
3. Complete the scheduled activities and follow-up sheets supporting the concepts of the design process.

Schedule for Section 3—9-Week and 12-Week* Classes	
Day 1	Discussion: "The design process." Assign Chapter 10: "The Design Process" to be read outside class. Introduce first activity, "Communication Design."
Day 2	Discussion: "Identifying problems." Work on activity. Assign Chapter 13: "Creating Solutions" to be read outside class.
Day 3	Discussion: "Creating solutions." Work on activity.
Day 4	Work on activity. Assign Chapter 14: "Selecting and Refining Solutions" to be read outside class.
Day 5	Discussion: "Selecting and refining solutions." Work on activity.
Day 6	Work on activity.
Day 7	Complete and summarize activity. Assign Chapter 15: "Modeling Solutions" to be read outside class.
Day 8	Discussion: "Modeling solutions." Introduce second activity, "Bio-Related Feeder."
Day 9	Work on activity.
Day 10	Complete and summarize activity. Review and summarize section.
* For a 12-week class, use an additional activity and cover three more of the ten chapters. Spread the discussions over the time scheduled for this section.	

Schedule for Section 3—18-Week and 36-Week* Classes

Day 1	Discussion: "Invention and innovation." Assign Chapter 9: "Invention and Innovation" to be read outside class. Introduce first activity, "Invention and Innovation."
Day 2	Work on activity.
Day 3	Complete and summarize activity. Assign Chapter 10: "The Design Process" to be read outside class.
Day 4	Discussion: "The design process." Introduce second activity, "Communication Design."
Day 5	Work on activity. Assign Chapter 11: "Identifying Problems" to be read outside class.
Day 6	Discussion: "Identifying problems." Work on activity.
Day 7	Work on activity. Assign Chapter 12: "Researching Problems" to be read outside class.
Day 8	Discussion: "Researching problems." Work on activity.
Day 9	Work on activity.
Day 10	Complete and summarize activity. Assign Chapter 13: "Creating Solutions" to be read outside class.
Day 11	Discussion: "Creating solutions." Introduce third activity, "House Design."
Day 12–13	Work on activity. Assign Chapter 14: "Selecting and Refining Solutions" to be read outside class.
Day 14	Discussion: "Selecting and refining solutions." Work on activity.
Day 15–17	Work on activity. Assign Chapter 15: "Modeling Solutions" to be read outside class.
Day 18	Discussion: "Modeling solutions." Work on activity.
Day 19–21	Work on activity. Assign Chapter 16: "Testing Solutions" to be read outside class.
Day 22	Discussion: "Testing solutions." Work on activity.
Day 23–26	Work on activity. Assign Chapter 17: "Communicating Solutions" to be read outside class.
Day 27	Discussion: "Communicating solutions." Complete and summarize activity.
Day 28	Assign Chapter 18: "Improving Solutions" to be read outside class. Introduce fourth activity, "Bio-Related Feeder."
Day 29	Discussion: "Improving solutions." Work on activity.
Day 30	Complete and summarize activity. Review and summarize section.

* For a 36-week class, use an additional activity and spread the discussions over the time scheduled for this section. The sketching handouts may also be used with the appropriate chapters.

Preparing to Teach the Section

Before introducing this section, complete the following tasks:

1. Study Section 3 of the Teacher's Resource.
2. Read Chapters 9–18 in the textbook to gather a complete view of the design process.
3. Prepare the following 10–15 minute discussions and presentations:
   A. Invention and innovation. (Textbook, Chapter 9)
   B. The design process. (Textbook, Chapter 10)
   C. Identifying problems. (Textbook, Chapter 11)
   D. Researching problems. (Textbook, Chapter 12)
   E. Creating solutions. (Textbook, Chapter 13)
   F. Selecting and refining solutions. (Textbook, Chapter 14)
   G. Modeling solutions. (Textbook, Chapter 15)
   H. Testing solutions. (Textbook, Chapter 16)
   I. Communicating solutions. (Textbook, Chapter 17)
   J. Improving solutions. (Textbook, Chapter 18)
4. Select appropriate laboratory activities from the following:
   A. Textbook, pages 426–431.
   B. Student Activity Manual, pages 51–61.
   C. Teacher's Resource, pages 105–114.
   D. Other sources of your choosing.
5. Prepare materials and supplies for selected activities and presentations.
6. Work through the activities and student study questions in the textbook, Student Activity Manual, and Teacher's Resource.

Presenting the Section

This section will be taught differently, depending on the length of the course. In a 9-week course, there are two weeks for this section. In two weeks, the students will have time to go through a small design project. There will not be time for an in-depth look at any of the skills needed in the design industry. There is also a problem in the 36-week course, where ten weeks can be spent on this section. In this amount of time, it may seem important to delve deeper into concepts like sketching, rendering, and technical drawing. This course, however, is *not* a course in technical graphics. The development of drawing skills should not be overemphasized. The overall understanding of the process of design is more important than the students' drawing and drafting abilities. The 36-week course simply offers the opportunity to conduct several design experiences.

To remain on schedule throughout this section, do the following:

1. Keep all presentations between 10 and 15 minutes.
2. Demonstrate only fundamental drawing skills.
3. Keep design problems simple. Complex problems will only increase the amount of time it takes students to complete them.
4. Prepare all materials and supplies in advance to ensure that the students have everything they need to complete the activities.
5. Maintain the focus on the design process.

Adjustments for Special Needs and Gifted Students

The activities in this section are both directive and open-ended. Each of the activities will work well as individual or cooperative group work. Using group work is an effective way to adjust instruction for students of varying abilities and interests. Students can be assigned into groups so a mix of abilities, learning styles, and motivation is achieved. A typical group may include a gifted, an average, and a special needs student.

Give special attention to students who may require more structure, especially during open-ended design projects. It is important, however, that the activities are free enough that the design solutions are different. Design, by nature, must be allowed to be creative and unique.

Evaluation Procedures

This section stresses the understanding of the design process. The focus should be on the understanding of the process used to design solutions. There are no right or wrong solutions, but there are solutions that fit the design briefs better than others. This makes evaluation of this section subjective. Therefore, students should be evaluated on the following factors:

1. Contributions to group activities. This evaluation may be done with daily logs (diaries) the students maintain, teacher observations, and student feedback.
2. Creativity and divergent thinking shown during the design activities.
3. Laboratory sheets, sketches, and designs completed during the activities.
4. Participation in discussions conducted during the activities.
5. Scores on quizzes and chapter questions.

Implementing the Laboratory Activities

Eight different activities have been developed for this section. They can be found at the end of Section 3 in the textbook, in Section 3 of the Student Activity Manual, and in the Teacher's Resource. One activity explores the history of inventions. Two short classroom handouts can be used for reviewing concepts, if time permits. Five of the activities are design problems, covering many different contexts. Two of these design problems have been developed as structured activities. The remaining three design problems are open-ended design briefs. When selecting activities, try to choose a sampling of structured and open-ended activities to make sure all students are reached.

While completing the design activities, the students should be encouraged to think "outside the box" and develop new and creative solutions. This does not mean all solutions must be completely original and never seen before. Their solutions can be innovations or adaptations of existing products and ideas.

Procedures for Directing Each Suggested Activity

Invention and Innovation, Textbook, page 426

This activity is a review of the history of inventions. It should be used in the longer courses covering Chapter 9: "Invention and Innovation." This activity asks the students to research an invention and explain not only its history, but also the inventions that led up to it. They will also describe any inventions that have developed from their invention, as well as the impacts of the invention. In order for the students to complete this activity, they must have access to research materials. The Internet and basic invention books, as well as encyclopedias, will all be helpful to the students.

Communication Design, Textbook, pages 427–429, and Student Activity Manual, pages 55–58

This is a prescriptive activity challenging the students to design a television commercial to promote your class. It is organized around the design process outlined in Section 3. At several steps, the students are required to make design decisions. It is important that the students base their decisions on the design problem and design brief, instead of simply selecting the designs they like the most. This activity uses both group and individual work. When the students are designing the storyboards (or possible solutions in other activities), they should be encouraged to work individually. This will help each student gain a better understanding of the design process.

House Design, Teacher's Resource, page 105

This is a structured design activity that still allows for different solutions. It focuses mainly on the design and modeling stages of the design process. Each design step is, however, touched upon. The students will design a floor plan for a single story home with a maximum area of 1800 square feet. This size is large enough to allow the students to add rooms they would enjoy, such as game rooms or exercise rooms. It is small enough, however, to keep the designs and models at a reasonable size. The design brief for this activity (on page 105) includes a client sheet that will help the students define the problem, criteria, and constraints of the house. You should lead the students through the process, making sure the materials are prepared in advance. The students will need the following supplies: paper; 1/4" graph paper; masking or Scotch® tape; rubber cement; corrugated cardboard or foam core board; hobby or utility knives; white, wood, or hot glue; and paint.

Procedure

Identifying a Problem
1. Have the students review the client sheet. This sheet should help them determine the needs (design problem) of the home owner.

Researching the Solution
2. Distribute sample one-story house plans to the students. The plans can be collected from house plan books or local home builders.
3. Discuss simple room layout concepts, such as quiet versus loud zones, service and utility areas, the use of storage space, and typical room sizes.

Creating Solutions

4. Have each student sketch a floor plan on plain paper that would fulfill the requirements of the home owner. At this point, the students should not worry too much about the scales of the drawings. They should just get their basic ideas on paper.

Selecting and Refining Solutions

5. At this point, you should make the decision whether the students will be allowed to work in groups or continue individually.
6. If they are working in groups, each group should gather and select the most promising design.
7. Distribute 1/4" graph paper. The students will need several sheets for each house.
8. Have the students draw their houses at a scale of 1' for every one box on the graph paper.
9. Watch the designs they are creating to make sure the room sizes are appropriate.
10. Have the students mark the doors and windows and label the names of each room.

Modeling the Solution

11. Distribute sheets of cardboard. Local container companies are a good source of cardboard. Because these models are small (1/4" = 1'), thin, corrugated cardboard will work the best.
12. Have the students use rubber cement to glue their plans to the cardboard and cut around the exteriors of their homes.
13. Prepare strips of cardboard that are 2" tall and varied lengths for the students to use as walls.
14. The students should cut their walls to length and then cut out the doors and windows. Doors should be cut 3/4" wide and about 1 5/8" tall. Windows can be cut as needed.
15. Walls should then be painted. If time permits, have the students decorate the walls of each room differently, as opposed to painting every wall the same color.
16. After the walls have dried, the students should assemble them using white, wood, or hot glue.
17. If time permits, have the students build and paint a roof that will lift off of the house.

Testing the Solution

18. Have the students examine their houses to make sure they meet the needs of the client.

Communicate the Solution

19. The students should then draw front and side views of their houses, as the houses would be seen finished from the road. They can add trees and landscaping to the drawings.
20. Have the students present their homes to the rest of the class.

Improving the Solution

21. After viewing the presentations, ask the students how they could have improved their designs.

Locker Design, Student Activity Manual, page 59

This is an open-ended design project challenging the students to design organizers for their lockers. The students should complete each of the design steps in the creation of their solution:

1. Identifying problems.
2. Researching problems.
3. Creating solutions.
4. Selecting and refining solutions.
5. Modeling solutions.
6. Testing solutions.
7. Communicating solutions.
8. Improving solutions.

The Teacher's Resource has a page that can be used for each step of the design process. These generic design sheets can be used for this activity or any other design project.

Design for Space, Textbook, pages 430–431

This is an open-ended design activity challenging the students to develop a model of a space station with living quarters, recreational areas, production facilities, office and managerial areas, mechanical support facilities, and a flight command center. To complete the activity, allow the students to work in groups of two to three students. Give each group a No. 10 tin can (a large can from the school cafeteria or local restaurant) and a supply of poster board, construction paper, a hollow tube or PVC pipe, and dowels. Then, have each group complete the following tasks:

1. Read the design brief in their textbooks.
2. Use the design process to design a space station.
3. Prepare a model of the station.
4. Present the model to the class.

Bio-Related Feeder, Student Activity Manual, page 61

This is an open-ended design brief. It allows students the freedom to create any type of automatic feeder they desire. They may even choose to design an improved model of a product they use to feed a cat or dog at home. Students should be encouraged to use their creativity and imagination. The design sheets in the Teacher's Resource can be used to guide the students' work. In this activity, the students will complete the following steps:

1. Select the type of device they are designing.
2. Create sketches of possible designs.
3. Create displays of their final designs.
4. If time permits, you may choose to have the students build models of their designs.

Section 4

Technological Contexts

Focus: Presentation of agricultural and bio-related, construction, energy conversion, information and communication, manufacturing, medical, and transportation contexts in which technology is applied.

➤ Chapter 19: Agricultural and Related Technology
➤ Chapter 20: Construction Technology
➤ Chapter 21: Energy Conversion Technology
➤ Chapter 22: Information and Communication Technology
➤ Chapter 23: Manufacturing Technology
➤ Chapter 24: Medical Technology
➤ Chapter 25: Transportation Technology

Teaching Materials

Text

Chapter 19, pages 434–464
  Chapter Activities, page 463
  Test Your Knowledge, page 464
Chapter 20, pages 466–490
  Chapter Activities, page 487
  Test Your Knowledge, page 488
Chapter 21, pages 492–513
  Chapter Activities, page 512
  Test Your Knowledge, page 513
Chapter 22, pages 514–554
  Chapter Activities, page 549
  Test Your Knowledge, page 550
Chapter 23, pages 556–581
  Chapter Activities, page 578
  Test Your Knowledge, page 579
Chapter 24, pages 582–603
  Chapter Activities, page 602
  Test Your Knowledge, page 603
Chapter 25, pages 604–627
  Chapter Activities, page 626
  Test Your Knowledge, page 627
Section Activities
  Activity 4A—Agriculture and Bio-Related Technology, page 628
  Activity 4B—Construction Technology, pages 629–634
  Activity 4C—Energy Conversion Technology, pages 635–636
  Activity 4D—Information and Communication Technology, pages 637–640
  Activity 4E—Manufacturing Technology, pages 641–644
  Activity 4F—Medical Technology, pages 645–646
  Activity 4G—Transportation Technology, pages 647–650

Student Activity Manual

Activity 4-1—Sketching and Estimating, pages 83–84
Activity 4-2—Model Bridge Design Contest, page 85
Activity 4-3—Energy Conversion Technology, pages 87–88
Activity 4-4—Information and Communication Technology, pages 89–93
Activity 4-5—Billboard Message, page 95
Activity 4-6—Process Demonstration, pages 97–98
Activity 4-7—Recipe Holder, pages 99–100
Activity 4-8—Transportation Technology, pages 101–102
Activity 4-9—Hovercraft Design, pages 103–104
Chapter 19 Study Questions, pages 105–106
Chapter 20 Study Questions, pages 107–108
Chapter 21 Study Questions, page 109
Chapter 22 Study Questions, pages 111–112
Chapter 23 Study Questions, pages 113–114
Chapter 24 Study Questions, pages 115–116
Chapter 25 Study Questions, pages 117–118

Teacher's Resources

Alternative Activity—Changing Potential Energy to Kinetic Energy, pages 147–150
Alternative Activity—Communicating without Words, page 151
Alternative Activity—Power Boat Design, page 153
Reproducible Masters
19-1—Biotechnology's Earmarks, page 155
20-1—Types of Civil Structures, page 156
20-2—Steps in Constructing a Structure, page 157
20-3—Planning and Management of Construction Projects, page 158
20-4—The Many Tasks in Building a Structure, page 159
21-1—Energy Conversion, page 160
22-1—Types of Communication Systems, page 161
22-2—The Communication Process, page 162
22-3—Creating Media Messages, page 163
22-4—Steps Used to Produce a Message for Print Communication, page 164
22-5—Designing a Message, page 165
22-6—Preparing to Produce a Message, page 166
22-7—Producing a Message, page 167
22-8—Delivering a Message, page 168
22-9—Parts of a Computer, page 169
23-1—Secondary Manufacturing Processes, page 170
24-1—Technology and Medicine, page 171
24-2—Wellness, page 172
25-1—Vehicle Systems, page 173
25-2—Vehicle Systems, page 174
25-3—Vehicle Systems, page 175
25-4—Vehicle Systems, page 176
25-5—Vehicles and Their Systems, page 177
Transparencies (Binder/CD only)
21-1—Energy Converters
22-1—Graphic Communication
22-2—Roughs
22-3—Comprehensives
22-4—Layout
23-1—Producing Finished Products

23-2—Casting and Molding
23-3—Forming
23-4—Separating
23-5—Conditioning
23-6—Assembling
23-7—Finishing
25-1—Transportation
25-2—Transportation Processes
25-3—Transportation Systems
25-4—Transportation Systems
25-5—Transportation System Parts
25-6—Vehicles

Introduction

Technology is used in almost every human activity. It is used to grow and harvest crops, process foods, communicate information, power machines and vehicles, build homes and offices, make products, transport people and goods, and treat injuries and illnesses. These areas in which technology is used are called *contexts*. They are the areas in society in which people use technology to extend their potentials.

Objectives of the Section

At the end of Section 4, each student should be able to do the following:

➤ List and define seven common technological contexts.

➤ Explain how people use technology to grow and harvest crops.

➤ Describe and summarize the uses of the major types of constructed structures.

➤ Discuss the major processes used to select and prepare sites, construct structures, and finish constructed projects.

➤ Give examples of how technology is used to convert and use energy.

➤ Identify and describe several ways to use communication technology.

➤ Name and paraphrase the major steps in producing and delivering a message.

➤ Summarize the major types of manufacturing systems.

➤ Cite and explain the major steps in developing and using a manufacturing system.

➤ Explain how technology is used to maintain wellness and treat illness.

➤ Describe how technology is used to transport people and goods.

➤ Give examples of the parts of a transportation system.

Suggested Schedule and Length of the Section (Three to Eighteen Weeks)

During the study of technological contexts, ask the students to complete the following assignments:

1. Read Chapters 19–25 in their textbooks and answer associated study questions.
2. Participate in presentations and discussions about the seven major technological contexts (agricultural, construction, energy conversion, information and communication, manufacturing, medical, and transportation).
3. Complete the scheduled laboratory activities and follow-up sheets reinforcing the study of the technological contexts.

Schedule for Section 4—9-Week and 12-Week* Classes

Day 1	Discussion: "Technological contexts." Assign Chapter 19: "Agricultural and Related Technology" to be read outside class. Introduce first activity (agricultural, construction, energy conversion, or information and communication technology activity).
Day 2	Discussion: "Agricultural and bio-related technology." Work on activity.
Day 3	Work on activity. Assign Chapter 20: "Construction Technology" to be read outside class.
Day 4	Discussion: "Construction technology." Work on activity.
Day 5	Work on activity. Assign Chapter 21: "Energy Conversion Technology" to be read outside class.
Day 6	Discussion: "Energy conversion technology." Work on activity.
Day 7	Complete and summarize activity. Assign Chapter 22: "Information and Communication Technology" to be read outside class.
Day 8	Discussion: "Information and technology." Introduce second activity (manufacturing, medical, or transportation technology activity).
Day 9	Assign Chapter 23: "Manufacturing Technology" to be read outside class. Work on activity.
Day 10	Discussion: "Manufacturing technology." Work on activity.
Day 11	Assign Chapter 24: "Medical Technology" to be read outside class. Work on activity.
Day 12	Discussion: "Medical technology." Work on activity.
Day 13	Assign Chapter 25: "Transportation Technology" to be read outside class. Work on activity.
Day 14	Discussion: "Transportation technology." Work on activity.
Day 15	Complete and summarize activity. Review and summarize section.

* For a 12-week class, add another activity and spread the discussions over the time scheduled for this section.

Schedule for Section 4—18-Week and 36-Week* Classes

Day	
Day 1	Discussion: "Technological contexts." Assign Chapter 19: "Agricultural and Related Technology" to be read outside class. Introduce first activity (agricultural or construction technology activity).
Day 2–4	Discussion: "Agricultural and bio-related technology." Work on activity.
Day 5	Work on activity. Assign Chapter 20: "Construction Technology" to be read outside class.
Day 6–9	Discussion: "Construction technology." Work on activity.
Day 10	Complete and summarize activity. Assign Chapter 21: "Energy Conversion Technology" to be read outside class.
Day 11–12	Introduce second activity (energy conversion or information and communication technology activity). Discussion: "Energy conversion technology."
Day 13–15	Work on activity. Assign Chapter 22: "Information and Communication Technology" to be read outside class.
Day 16–17	Discussion: "Information and technology." Work on activity.
Day 18	Complete and summarize activity.
Day 19	Introduce third activity (manufacturing, medical, or transportation technology activity).
Day 20–21	Assign Chapter 23: "Manufacturing Technology" to be read outside class. Work on activity.
Day 22–23	Discussion: "Manufacturing technology." Assign Chapter 24: "Medical Technology" to be read outside class. Work on activity.
Day 24–25	Discussion: "Medical technology." Work on activity.
Day 26–27	Assign Chapter 25: "Transportation Technology" to be read outside class. Work on activity.
Day 28–29	Discussion: "Transportation technology." Work on activity.
Day 30	Complete and summarize activity. Review and summarize section.

* For a 36-week class, use an activity for each context and spread the discussions over the time scheduled for this section.

Preparing to Teach the Section

Complete the following tasks before introducing this section to the students:

1. Study Section 4 of the Teacher's Resource.
2. Read Chapters 19–25 in the textbook to establish the scope and sequence of the content.
3. Prepare the following 10–15 minute discussions and presentations:
   A. Technological contexts. (Textbook, Section 4)
   B. Agricultural and bio-related technology. (Textbook, Chapter 19)
   C. Construction technology. (Textbook, Chapter 20)
   D. Energy conversion technology. (Textbook, Chapter 21)
   E. Information and technology. (Textbook, Chapter 22)
   F. Manufacturing technology. (Textbook, Chapter 23)
   G. Medical technology. (Textbook, Chapter 24)
   H. Transportation technology. (Textbook, Chapter 25)
4. Select appropriate laboratory activities from the following options:
   A. Agriculture and related technology.
      (1) Textbook, page 628.
      (2) Other sources of your choosing.
   B. Construction technology.
      (1) Textbook, pages 629–634.
      (2) Student Activity Manual, pages 83–85.
      (3) Other sources of your choosing.
   C. Energy conversion technology.
      (1) Textbook, pages 635–636.
      (2) Student Activity Manual, pages 87–88.
      (3) Teacher's Resource, pages 147–150.
      (4) Other sources of your choosing.
   D. Information and communication technology.
      (1) Textbook, pages 637–640.
      (2) Student Activity Manual, pages 89–95.
      (3) Teacher's Resource, page 151.
      (4) Other sources of your choosing.
   E. Manufacturing technology.
      (1) Textbook, pages 641–644.
      (2) Student Activity Manual, pages 97–100.
      (3) Other sources of your choosing.
   F. Medical technology.
      (1) Textbook, pages 645–646.
      (2) Other sources of your choosing.
   G. Transportation technology.
      (1) Textbook, pages 647–650.
      (2) Student Activity Manual, pages 101–104.
      (3) Teacher's Resource, page 153.
      (4) Other sources of your choosing.
5. Prepare the materials and supplies for the laboratory activities.
6. Identify and construct the necessary tooling to produce the manufactured products using line production techniques.
7. Build a suitable waterway to be used with the transportation activity (textbook, pages 647–650).
8. Work through the laboratory activities and student study questions presented in the textbook, Student Activity Manual, and Teacher's Resource.

Presenting the Section

The goal of this section is to present the societal contexts in which technology is applied. Through the laboratory work, the students will complete hands-on activities designed to present several of these contexts. Skill development is *not* a goal of these activities. To maintain a balance between content presentations and laboratory reinforcement, the following suggestions are made:

➤ Limit all discussions and presentations to 15 minutes. Plan each presentation carefully and use adequate teaching materials (such as transparencies).

➤ Prepare materials for the laboratory activities in advance of implementing the activities. In most cases, it will not be possible to have the students complete a laboratory activity for each context. The activities chosen should be broad and, whenever possible, apply more than one context. For example, a transportation activity applies energy conversion technologies, and constructed works use manufactured components.

To maintain the schedule, the following points should be considered:

➤ Process the materials for the activities before introducing them.

➤ Build all support devices and tooling before introducing an activity.

➤ Produce sample products before introducing an activity.

Adjustments for Special Needs and Gifted Students

Seventeen suggested activities have been developed for this section. Several of them are structured activities, in which the students are given a prescribed procedure to follow. Other activities are open-ended experiences developed around design briefs. Still other activities can be presented as either prescriptive or open-ended activities. All the activities lend themselves to cooperative group work. Using groups is a very effective way to adjust instruction for the differing abilities and varying interests of the students in a class. Each group can be developed to represent a mix of students including gifted, average, and special needs students. In all activities, special attention and individual help should be given to those students who require more structure. You are encouraged, however, to resist developing a high degree of structure for every activity. Generally, highly structured activities lack the level of challenge needed for many gifted students.

Evaluation Procedures

This section emphasizes using technology to produce products and structures. The focus is on the concepts of production, not the skills production workers use. Therefore, the students should be evaluated on the following criteria:

1. Contributions to group activity. This may be done with examination of daily logs (diaries) the students maintain, teacher observations, and student feedback.

2. Creativity and divergent thinking shown during the design and problem solving activities.

3. Laboratory sheets, sketches, and designs completed during each activity.

4. Participation in discussions conducted during the laboratory activities.

5. Scores on quizzes and chapter questions.

Implementing the Laboratory Activities

Time has been allocated to this section for at least two activities. Classes with more time available may include more activities so the students can directly experience more technological contexts. Each activity presented is only suggested, and different activities may be substituted. The goal is to have the students apply technology in two or more societal contexts. It may be wise to consider using a structured activity for one context and a design brief activity for another.

Procedures for Directing Each Suggested Activity

Agriculture and Bio-Related Technology, Textbook, page 628

This is a structured activity introducing students to food processing. They will see how an agricultural product is changed to an everyday food product. The students will make peanut butter from roasted peanuts, peanut oil, and salt. Note: If salted peanuts are used, omit the salt from the ingredients. At the end of the activity, the students may be treated to peanut butter and crackers.

Construction Technology, Textbook, pages 629–634

This is a structured activity introducing students to common residential construction practices. The activity is designed for group work. Each group follows a prescribed procedure to construct a major part of the structure. The parts are then assembled into a finished structure. At the start of the activity, divide the class into groups, as described in the activity. Show the class a completed model. Then, have each group carefully read the instructions for completing their part of the activity. The students should complete these steps, exercising care to follow the safety practices demonstrated during the class. The "input" material should be prepared before the activity is introduced. This includes scale $2 \times 4s$, $1 \times 4s$, and plywood sheets. The sizes of these materials are listed on page 629 of the textbook and are as follows:

➤ $8'\ 2 \times 4s = 1/4'' \times 5/8'' \times 16''$
➤ $8'\ 2 \times 6s = 1/4'' \times 1'' \times 16''$
➤ $8'\ 1 \times 4s = 1/8'' \times 5/8'' \times 16''$
➤ $1/4'' \times 4' \times 8'$ plywood = $8'' \times 16''$ x 6 ply poster board
➤ $1/2'' \times 4' \times 8'$ plywood = $8'' \times 16''$ mat board

Student worksheets to support this activity can be found on pages 83–84 in the Student Activity Manual.

Model Bridge Design Contest, Student Activity Manual, page 85

This is an open-ended activity based on a design brief. The students should be presented with the design problem and the overall procedures. Both of these elements are provided on page 85 of the Student Activity Manual. The bridge tester (Student Activity Manual, Figure 4-4) that will be used should be shown to the class. Have the students complete the following tasks:

1. Review the various truss construction patterns.
2. Select patterns they think will be the strongest.
3. Construct models of a bridge using these patterns.
4. Test and evaluate the finished models.

Bridge Tester

10"

6"

12"

Use 1" x 6" lumber for all parts

Be sure the students use only the prescribed type and quantity of materials so all members of the class have an equal opportunity to design the strongest bridge.

Energy Conversion Technology, Textbook, pages 635–636, and Student Activity Manual, pages 87–88

This is a structured experiment presenting one of the many energy conversion processes. The students should be given the appropriate supplies and equipment. The procedure for the experiment should be demonstrated. Then, the class should be divided into groups. Each group should complete the experiment with one color of can. The groups will share their results with the class so a comparison between color and light-to-heat energy conversion can be made.

Supplies and Equipment

➤ Heat lamp
➤ Large (42 oz.) juice cans or No. 10 cans (from school cafeteria)
➤ Spray cans of various colors of paint—black, red, blue, green, yellow, brown, and white
➤ Thermometer reading up to 200°F
➤ Rubber grommets with a hole the diameter of the thermometer (Corks with a hole in the center can be substituted.)
➤ Data recording sheet (included)
➤ Laboratory summary sheet (included)

Procedure

Preparing for the Activity

Before the activity is introduced, the following steps must be completed:

1. Obtain a heat lamp for each group.
2. Construct a lamp holder (socket, switch, and bracket) for each group. This lamp holder should hold the lamp parallel to the cans and be movable to the three positions specified for the experiment.
3. Obtain juice or vegetable cans.
4. Establish the distances and times to be used in the experiment. These will vary with the heat lamp used. You will want to use times and distances showing significant results. *Experiment* to establish these times and distances. (Start with 12″, 18″, and 24″, for 5, 10, and 15 minutes).

Conducting the Activity

1. Divide the class into groups of two to three students.
2. Assign each group a color and task, such as the following:

   "Determine the heat absorption of your color, in terms of the intensity of light (distance from the can to the light) and the time the light is exposed to the surface."

3. Have each group complete the following steps:
   A. Select an appropriate can.
   B. Locate the laboratory sheet in the Student Activity Manual.
   C. Set up the experiment as demonstrated. Note: Demonstrate the proper, safe way to conduct the experiment.
   D. Take the reading at distance #1 for each can.
   E. Repeat steps 3C and 3D for distances #2 and #3.
   F. Compare results of the experiment by reporting their results.
   G. Record the results from other groups on the laboratory sheet.
   H. Complete a laboratory summary sheet.

Student worksheets to support this activity can be found on pages 87–88 in the Student Activity Manual.

Converting Potential Energy to Kinetic Energy, Teacher's Resource, pages 147–150

This activity allows groups of students to explore how potential energy is converted into kinetic energy. To conduct this activity, you will need to build a test device (pile driver) for each group of students. (See page 148 of the Teacher's Resource.) For each device, you will need the following materials:

➤ Base: One piece of 2″ × 7″ × 12″ wood

➤ Top: One piece of 3/4″ × 2 3/4″ × 12″ wood

➤ Hammer head: One piece of 2 1/4″ × 4″ × 7″ wood

➤ Hammer head face: One piece of 1/8″ × 3/4″ × 5 3/4″ steel

➤ Guide columns: Two 1″ D × 50″ conduits

You will also need four sizes of nails and a 2 × 4 × 6″ block for each group. The procedure for conducting the activity and a data recording sheet are included in the Teacher's Resource.

3/4" x 2 3/4" x 12" wood

1" D x 50" conduit

2 1/4" x 4" x 7" wood

2" x 7" x 12" wood

Radius to fit conduit

Hammer head
2 1/4" x 4" x 7" wood

Hammer head face
1/8" x 3/4" x 5 3/4" steel

Information and Communication Technology, Textbook, pages 637–640, and Student Activity Manual, pages 89–93

This is a structured activity that can be used to study communication technology and custom manufacturing. It challenges the students to first construct simple telegraphs and then communicate messages using their communication devices. The activity contains a specific set of procedures the students are to follow. The students can, then, develop codes to communicate messages. Students may work individually or in groups of two or three students. Half of the individuals or groups should make sending units, and the other half should make receivers. To implement the activity, you will need the equipment and supplies listed on page 637 of the textbook. Each student or group of students should be given a set of supplies. The students are then challenged to complete the following steps:

1. Study the plans for the sender or receiver.
2. Follow the procedures and plans to construct the device.
3. Join another student or group of students who made the other type of device to develop a complete telegraph (hard-wired) communication system.
4. Develop a code to complete the message assignment on page 89 of the Student Activity Manual.

Student worksheets to support this activity can be found on pages 89–93 in the Student Activity Manual.

Billboard Message, Student Activity Manual, page 95

This activity is an open-ended problem solving activity challenging the students to develop one of two different messages:

➤ A billboard to promote safe bicycle riding.
➤ A poster to promote litter control.

To complete the activity, allow the students to work alone or in groups of two to three students. Each student or group should complete the following steps:

1. Develop a theme for the communication campaign.
2. Develop a message for the communication device supporting the campaign theme.
3. Select or develop graphics supporting the message and campaign theme.
4. Prepare a layout for the poster or billboard.
5. Produce the poster or a miniature billboard.

Communicating without Words, Teacher's Resource, page 151

This activity is an open-ended problem solving activity challenging the students to develop a system of communication other than the spoken word. To complete the activity, have the students work alone or in groups of two to three students. Each student or group should complete the following steps:

1. Develop a message.
2. Prepare sketches of at least three systems.
3. Select or develop graphics supporting the message.
4. Prepare a layout for the design (such as a poster or billboard).
5. Produce the system.

Manufacturing Technology, Textbook, pages 641–644

This is a structured activity challenging the students to use drawings and operation sheets to plan for and produce a manufactured product. The activity is organized around a specific set of procedures the students are to follow. Each student can produce a product for himself or herself. This, however, will be very time-consuming. A better approach is to use a teacher-designed production line to produce the product. The line should have the same three branches shown on the operation process chart on page 642 of the textbook. Each of these branches should feed an assembly line. Another approach is to divide the class into groups. Each group should work like a job shop. Group 1 will produce the ends, Group 2 will produce the side strips, and Group 3 will produce the end strips. The parts from the "job shops" will be assembled into products at the end of the activity. In all cases, students should follow appropriate tool and machine procedures as demonstrated in the class. Before introducing the activity, it is necessary to prepare the "input" material for the line. This will require pine strips as follows: $3/4" \times 5" \times$ random length (end strips), $1/4" \times 3/4" \times$ random length (side strips), and $1/4" \times 1\,1/2" \times$ random length (bottom strips). Remember, the goal is to introduce the student to a manufacturing system. Specific skills are not important, and rotating students through jobs is not necessary. Also, tool and material nomenclature is an inappropriate focus for this section.

This activity, like the rubber band car activity in Section 2, is designed to be less structured than the game activity provided in Section 2 of the textbook. The students should be provided the drawings for the CD holder and challenged to complete the following steps:

1. Develop a bill of materials for the product.
2. Prepare a procedure sheet for building the product.
3. List any restrictions they face, such as CD case size, number of CDs the holder will hold, and cost of the solution.
4. Build a prototype of the holder.
5. Discuss the procedures they used, ways to build the product more efficiently, and other similar topics.

Recipe Holder, Student Activity Manual, pages 99–100

This activity is an open-ended problem solving activity challenging the students to develop a manufacturing system for a simple product—a recipe holder. To complete the activity, divide the class into groups of three or four students to plan the system. Each group should complete the following steps:

1. Prepare a bill of materials for the product.
2. Prepare a procedure sheet for each of the two manufactured parts—the base and upright.
3. Highlight steps requiring tooling to maintain tolerance.
4. Check places where the students think inspection (quality control) will be needed. Select the best manufacturing system by choosing one group's work or integrating several groups' ideas. Build the necessary tooling and inspection gages; then, manufacture the product.

Attach with two
3/4" wire nails

Decorate
to please

3/8"

3/8"

1/4"

1/4"

2 pcs. 1/2"x 5"x 5"

4 pcs. 1/4"x 1"x 10"

2 pcs. 1/4"x 1 1/2"x 10"

Medical Technology, Textbook, pages 645–646

This is an open-ended activity based on a design brief. The students should be presented with the design problem and the overall procedures. Both of these elements are provided on pages 645–646 of the textbook. The students will be challenged to design a device allowing a person to pick objects up without bending over. This device will help people with arthritis, back injuries, and other conditions that make bending over difficult. The students should use the design process presented in Section 3 to complete this activity. They can work individually or in small groups to complete the activity.

Transportation Technology, Textbook, pages 647–650, and Student Activity Manual, pages 101–102

This is a structured activity introducing students to vehicle design. They will develop and test boat hulls for effectiveness. To implement this activity, divide the class into groups of two to four students. Provide each group with the supplies listed with the activity on pages 647–648 of the textbook. Have each group carefully follow the procedure listed in their textbooks to build boat hulls, sails, and keels. The groups should test their hulls in a waterway.

Hovercraft Design, Student Activity Manual, pages 103–104

This is a structured activity allowing students to construct and test an air-cushion vehicle (ACV). The students should be divided into groups of two to three students. Each group should be provided the supplies listed with the activity. The groups should then be challenged to complete the following steps:

1. Read the procedures for constructing the ACV.
2. Construct their vehicles.
3. Test the vehicles.
4. Suggest ways to modify the vehicles to improve their performances.

Power Boat Design, Teacher's Resource, page 153

This activity is based on a design brief. The students should be presented with the design problem. To meet the design challenge, they should be provided with the following supplies:

➤ One piece of 1 1/2″ × 5″ × 9″ foam block
➤ One rubber band
➤ Assorted dowels, sheet metal, balsa wood, nails, and pins

The students must solve the problem by completing the following steps:

1. Prepare sketches of at least three designs.
2. Select, refine, and dimension the most promising design.
3. Build a prototype of the design.
4. Test the design.
5. Modify the design to make it more efficient.

Student worksheets to support this activity can be found on pages 107–114 in the Teacher's Resource.

Section 5

Technology and Society

Focus: Exploration of the impacts of technology on resources, people, and the environment and how change will affect the future of technology.

➤ Chapter 26: Technological Impacts

➤ Chapter 27: Technology and the Future

Teaching Materials

Text

Chapter 26, pages 654–677
  Chapter Activities, page 674
  Test Your Knowledge, page 675
Chapter 27, pages 678–691
  Chapter Activities, page 690
  Test Your Knowledge, page 691
Section Activities
  Activity 5A—Technological Impacts, pages 692–693

Student Activity Manual

Activity 5-1—Future Telling, page 119
Activity 5-2—Products of the Future, pages 121–122
Chapter 26 Study Questions, page 123
Chapter 27 Study Questions, page 125

Teacher's Resources

Reproducible Masters
  26-1—Global Warming, page 185
  26-2—Technology Assessment, page 186
  27-1—Modeling a Major Event, page 187
Transparencies (Binder/CD only)
  26-1—Technology Shapes and Is Shaped by the World
  26-2—Energy Resources

Introduction

This section is placed at the end of the course, not as an afterthought, but as a planned conclusion to the study of technology. Knowledge of the impacts and future of technology is vital to the understanding of how society uses and controls technology. Technological resources, designed products, and applications are only useful if their impacts on society are understood. To be technologically literate, one must understand how technology functions in society. It is equally important to understand that we, as a society, control the directions and outcomes of technology. By understanding technology's impacts and recognizing where it is headed, we choose what technological changes to make. Only a technologically literate society has this power.

Objectives of the Section

At the end of Section 5, each student should be able to do the following:

➤ List examples of technological change.
➤ Define *negative impact*.
➤ Explain impacts on people, resources, and the environment.
➤ Describe ways in which the future can be predicted.
➤ Give examples of future technology.

Suggested Schedule and Length of the Section (Five to Ten Days)

Section 5 introduces the students to the topic of technology and society. You should have them complete the following assignments:

1. Read Chapters 26 and 27 in their textbooks and answer associated study questions.
2. Participate in several discussions and presentations.
3. Complete the scheduled activities and follow-up sheets.

Schedule for Section 5—9-Week, 12-Week, 18-Week, and 36-Week* Classes	
Day 1	Discussion: "Technological impacts." Assign Chapter 26: "Technological Impacts" to be read outside class. Introduce first activity, "Technological Impacts."
Day 2	Complete and summarize activity. Discussion: "Forecasting the future." Assign Chapter 27: "Technology and the Future" to be read outside class.
Day 3	Discussion: "Future technologies." Introduce second activity, "Products of the Future."
Day 4	Work on activity.
Day 5	Complete and summarize activity. Summarize course.
* For a 36-week class, use an additional activity and spread the discussions over the two weeks scheduled for this session.	

Preparing to Teach the Section

Before introducing this section, complete the following tasks:

1. Study Section 5 of the Teacher's Resource.
2. Read Chapters 26–27 in the textbook to review the content.
3. Prepare three 10–15 minute discussions and presentations on the following topics:
   A. Technological impacts. (Textbook, Chapter 26)
   B. Forecasting the future. (Textbook, pages 680–685)
   C. Future technologies. (Textbook, pages 685–689)
4. Select appropriate laboratory activities from the following options:
   A. Textbook, pages 692–693.
   B. Student Activity Manual, pages 119–122.
   C. Other sources of your choosing.
5. Prepare materials and supplies for the selected activities and presentations.
6. Work through the activities and student study questions presented in the textbook and Student Activity Manual.

Presenting the Section

This section should be taught by using the three short discussions to reinforce the concepts from the textbook. The activities should be used to support the content from the discussions and readings. The students' understanding of the concepts can be evaluated using the Test Your Knowledge questions in the textbook and study questions in the Student Activity Manual. At the end of the section, it is important that the section is summarized to highlight the importance of understanding the role of technology in society. The last day of this section should also be used to review and summarize the entire course. Prepare a presentation emphasizing the key points of each of the five sections the students have studied.

Adjustments for Special Needs and Gifted Students

The activities in this section are open-ended. They will be easier for the gifted students to handle than the prescriptive activities found in some of the other sections. Give special attention to students who may require assistance in generating ideas and creative solutions. All these activities lend themselves to either individual or group work. Consider using one or more of these activities to combine students of varied abilities. For example, if the "Future Telling" activity is done in a group, it may be beneficial to pair a student with good writing skills and a very imaginative student with low writing ability.

Evaluation Procedures

This section stresses the understanding that technology impacts society and that technology's future can be forecasted. The focus should be on the general understanding of the different types of impacts and forecasts. There are no right answers about the impacts and future of technology. Each student may have different opinions and feelings about how technology has affected and will continue to affect society. Students should be made to feel that their views are not the only views of technology and that even opposing views can be correct. This makes evaluation of this section subjective. Therefore, students should be evaluated on the following factors:

1. Contributions to group activities. This may be done with daily logs (diaries) the students maintain, teacher observations, and student feedback.
2. Laboratory sheets, sketches, and designs completed during the activities.
3. Participation in discussions conducted during the activities.
4. Scores on quizzes and chapter questions.

Implementing the Laboratory Activities

Three different activities have been developed for this section. They can be found at the end of Section 5 in the textbook and in Section 5 of the Student Activity Manual. These activities are suggested activities introducing the students to concepts relating to technology and society. Any activity on the topic of technological impacts or the future of technology can be used in place of any of these.

Procedures for Directing Each Suggested Activity

Technological Impacts, Textbook, pages 692–693

This is an activity allowing students to explore how the technology they use affects society. It is a one to two day activity that will work best if the students work individually or in pairs. They will select everyday objects from around the classroom and explore what happens to the objects when they are discarded. The students will list the ways the objects impact society. To conclude the activity, they will select one of their objects and create a drawing of changes that can be made to decrease the negative impacts.

Future Telling, Student Activity Manual, page 119

This activity is an alternative activity to be used if time permits. It is designed to be creative and open-ended. The students will look into the future and create a scenario of a future event. They should be encouraged to be imaginative and creative. This activity is an individual project. Groups can be used, however, to help the students generate ideas.

Products of the Future, Student Activity Manual, pages 121–122

This is a group activity in which the students forecast the technology that will be available in a certain area of technology in the future. The students will be assigned one context of technology. They will then examine the products and systems currently available in that context. The students will then create drawings of a technology that may exist in the future.

Conducting the Activity

1. Divide the class into seven groups.
2. Give each group one of the seven contexts of technology: agricultural and bio-related, construction, energy conversion, information and communication, manufacturing, medical, or transportation.
3. Have the students generate a list of current technologies in their areas. They should record them on the activity sheet on page 121 of the Student Activity Manual. The students should use the textbook, library, and Internet for ideas.
4. The students should then brainstorm to come up with future technologies in their areas. Space is provided for students to write down their ideas during brainstorming.
5. Have each group choose one of the ideas and develop it further. They are to complete the drawing and explanation portions of the worksheet in the Student Activity Manual.
6. If time permits, in a 36-week course, supply the students with the materials and supplies to build a model of their technologies.
7. To summarize the activity, have the students present their technologies to the rest of the class.

Standards for Technological Literacy Correlation Chart

The International Technology Education Association (ITEA) and its Technology for All Americans Project developed *Standards for Technological Literacy: Content for the Study of Technology* to identify the essential core of technological knowledge and skills for students in grades K–12. This work defined twenty separate standards, divided into five broad categories. Within each standard, benchmark topics are defined for four different grade levels:

➤ Grades K–2
➤ Grades 3–5
➤ Grades 6–8
➤ Grades 9–12

The following chart lists the standards and the benchmark topics for grades 6–8. Adjacent to each standard and benchmark topic are the chapter and page references identifying material in *Technology: Design and Applications* relating to the item.

Standards for Technological Literacy	
1. Students will develop an understanding of the characteristics and scope of technology.	Chapter 1: pages 24–26, 29–30 Chapter 3: pages 56–63 Chapter 9: pages 218–221, 224–225 Chapter 23: pages 576–577
New products and systems can be developed to solve problems or to help do things that could not be done without the help of technology.	Chapter 1: pages 24–26 Chapter 9: pages 218–219
The development of technology is a human activity and is the result of individual or collective needs and the ability to be creative.	Chapter 1: pages 25–26 Chapter 3: pages 56–63 Chapter 9: pages 220–221
Technology is closely linked to creativity, which has resulted in innovation.	Chapter 3: pages 56–63 Chapter 9: pages 224–225
Corporations can often create demand for a product by bringing it onto the market and advertising it.	Chapter 1: pages 29–30 Chapter 23: pages 576–577
2. Students will develop an understanding of the core concepts of technology.	Chapter 1: pages 24–31 Chapter 2: pages 36–39, 43–45, 48–49 Chapter 10: pages 234–247 Chapter 11: pages 262–265 Chapter 14: pages 312–314 Chapter 16: pages 375–376 Chapter 18: pages 413–416

Standards for Technological Literacy

Technological systems include input, processes, output, and, at times, feedback.	Chapter 2: pages 37–39
Systems thinking involves considering how every part relates to others.	Chapter 2: pages 36–37
An open-loop system has no feedback path and requires human intervention, while a closed-loop system uses feedback.	Chapter 2: pages 48–49
Technological systems can be connected to one another.	Chapter 2: pages 36–37
Malfunctions of any part of a system may affect the function and quality of the system.	Chapter 2: pages 37–39
Requirements are the parameters placed on the development of a product or system.	Chapter 11: pages 262–265
Trade-off is a decision process recognizing the need for careful compromises among competing factors.	Chapter 16: pages 375–376
Different technologies involve different sets of processes.	Chapter 2: pages 43–45
Maintenance is the process of inspecting and servicing a product or system on a regular basis in order for it to continue functioning properly, to extend its life, or to upgrade its capability.	Chapter 2: pages 37–39
Controls are mechanisms or particular steps that people perform using information about the system that causes systems to change.	Chapter 2: pages 48–49
3. Students will develop an understanding of the relationships among technologies and the connections between technology and other fields of study.	Chapter 1: page 32 Chapter 2: pages 36–37, 52 Chapter 3: pages 68–72 Chapter 4: page 104 Chapter 5: page 126 Chapter 6: page 147 Chapter 7: pages 152–154, 161 Chapter 8: page 184 Chapter 9: page 228 Chapter 10: page 249 Chapter 11: page 266 Chapter 12: page 282 Chapter 13: page 308 Chapter 14: page 328 Chapter 15: page 358 Chapter 16: page 377 Chapter 17: page 405 Chapter 18: page 423 Chapter 19: page 463 Chapter 20: page 487 Chapter 21: page 512 Chapter 22: page 549 Chapter 23: page 578 Chapter 24: page 602 Chapter 25: page 626 Chapter 26: page 674 Chapter 27: page 690

(Continued)

Standards for Technological Literacy

Technological systems often interact with one another.	Chapter 2: pages 36–71
A product, system, or environment developed for one setting may be applied to another setting.	Chapter 3: pages 68–71
Knowledge gained from other fields of study has a direct effect on the development of technological products and systems.	Chapter 7: pages 152–154
4. Students will develop an understanding of the cultural, social, economic, and political effects of technology.	Chapter 1: page 24 Chapter 3: pages 65–67 Chapter 8: pages 168–176 Chapter 26: pages 664–672
The use of technology affects humans in various ways, including their safety, comfort, choices, and attitudes about technology's development and use.	Chapter 3: pages 65–66 Chapter 8: pages 168–169 Chapter 26: pages 664–672
Technology, by itself, is neither good nor bad, but decisions about the use of products and systems can result in desirable or undesirable consequences.	Chapter 3: pages 66–67 Chapter 26: pages 664–672
The development and use of technology poses ethical issues.	Chapter 26: pages 664–672
Economic, political, and cultural issues are influenced by the development and use of technology.	Chapter 8: pages 168–176 Chapter 26: pages 671–672
5. Students will develop an understanding of the effects of technology on the environment.	Chapter 2: pages 45–48 Chapter 26: pages 666–669 Chapter 27: pages 685–689
The management of waste produced by technological systems is an important societal issue.	Chapter 2: pages 45–48 Chapter 26: pages 666–669
Technologies can be used to repair damage caused by natural disasters and to break down waste from the use of various products and systems.	Chapter 26: pages 666–669
Decisions to develop and use technologies often put environmental and economic concerns in direct competition with one another.	Chapter 26: pages 666–669
6. Students will develop an understanding of the role of society in the development and use of technology.	Chapter 2: pages 40–41, 49 Chapter 3: pages 63–67 Chapter 8: pages 168–169 Chapter 9: page 214–217, 222–223 Chapter 26: pages 657–664
Throughout history, new technologies have resulted from the demands, values, and interests of individuals, businesses, industries, and societies.	Chapter 2: pages 40–41 Chapter 3: pages 63–67 Chapter 9: pages 214–217

Standards for Technological Literacy

The use of inventions and innovations has led to changes in society and the creation of new needs and wants.	Chapter 2: page 49 Chapter 3: pages 65–66 Chapter 9: pages 222–223
Social and cultural priorities and values are reflected in technological devices.	Chapter 3: pages 63–67 Chapter 26: pages 657–664
Meeting societal expectations is the driving force behind the acceptance and use of products and systems.	Chapter 3: pages 63–67
7. **Students will develop an understanding of the influence of technology on history.**	Chapter 1: pages 25–26 Chapter 2: page 45 Chapter 3: pages 57–58, 68–71 Chapter 4: pages 85–88 Chapter 9: pages 214–217 Chapter 10: page 249 Chapter 20: pages 468–469
Many inventions and innovations have evolved by using slow and methodical processes of tests and refinements.	Chapter 9: pages 214–217
The specialization of function has been at the heart of many technological improvements.	Chapter 3: pages 68–71
The design and construction of structures for service or convenience have evolved from the development of techniques for measurement, controlling systems, and the understanding of spatial relationships.	Chapter 2: page 45 Chapter 4: pages 85–88 Chapter 20: pages 468–469
In the past, an invention or innovation was not usually developed with the knowledge of science.	Chapter 3: pages 57–58
8. **Students will develop an understanding of the attributes of design.**	Chapter 10: pages 234–235, 238–240, 248 Chapter 11: pages 261–263 Chapter 12: pages 275–278 Chapter 13: pages 291–292 Chapter 14: pages 314–317, 320–324 Chapter 15: pages 354–356 Chapter 16: pages 364–369 Chapter 17: page 384 Chapter 18: pages 413–416
Design is a creative planning process that leads to useful products and systems.	Chapter 10: pages 234–235
There is no perfect design.	Chapter 10: page 248
Requirements for a design are made up of criteria and constraints.	Chapter 10: pages 238–240 Chapter 11: pages 261–263

(Continued)

Standards for Technological Literacy

9. Students will develop an understanding of engineering design.	Chapter 10: pages 237–248 Chapter 11: pages 258–261 Chapter 12: pages 275–280 Chapter 13: pages 288–306 Chapter 14: pages 312–314, 319–324 Chapter 15: pages 332–334, 354–356 Chapter 16: pages 362–363, 371–376 Chapter 17: pages 402–404 Chapter 18: pages 421–422
Design involves a set of steps, which can be performed in different sequences and repeated as needed.	Chapter 10: pages 237–248 Chapter 11: pages 258–261 Chapter 12: pages 275–280 Chapter 13: pages 288–306 Chapter 14: pages 312–314, 319–324 Chapter 15: pages 354–356 Chapter 16: pages 371–375 Chapter 17: pages 402–404 Chapter 18: pages 421–422
Brainstorming is a group problem-solving process in which each person in the group presents his or her ideas in an open forum.	Chapter 10: pages 240–242 Chapter 13: pages 288–289
Modeling, testing, evaluating, and modifying are used to transform ideas into practical solutions.	Chapter 10: pages 242–246, 248 Chapter 14: pages 319–324 Chapter 15: pages 332–334 Chapter 16: pages 362–363, 375–376 Chapter 18: pages 421–422
10. Students will develop an understanding of the role of troubleshooting, research and development, invention and innovation, and experimentation in problem solving.	Chapter 9: pages 214, 218–219, 224–225 Chapter 12: pages 274–275 Chapter 13: page 308 Chapter 16: pages 370–371
Troubleshooting is a problem-solving method used to identify the cause of a malfunction in a technological system.	Chapter 9: pages 218–219
Invention is a process of turning ideas and imagination into devices and systems. Innovation is the process of modifying an existing product or system to improve it.	Chapter 9: pages 214, 224–225
Some technological problems are best solved through experimentation.	Chapter 9: pages 218–219 Chapter 16: pages 370–371
11. Students will develop abilities to apply the design process.	Chapter 10: pages 237–248 Chapter 11: pages 261–265 Chapter 12: page 282 Chapter 13: pages 291–306 Chapter 14: pages 319–325 Chapter 15: pages 354–356 Chapter 16: pages 363–364 Chapter 17: page 405 Chapter 18: pages 413–416

Standards for Technological Literacy

Apply a design process to solve problems in and beyond the laboratory-classroom.	Chapter 10: pages 237–248
Specify criteria and constraints for the design.	Chapter 11: pages 261–265
Make two-dimensional and three-dimensional representations of the designed solution.	Chapter 10: pages 242–244 Chapter 13: pages 291–306 Chapter 14: pages 324–325 Chapter 15: pages 354–356
Test and evaluate the design in relation to pre-established requirements, such as criteria and constraints, and refine as needed.	Chapter 10: pages 244–246 Chapter 11: pages 261–263 Chapter 14: pages 319–324 Chapter 16: pages 363–364 Chapter 18: pages 413–416
Make a product or system and document the solution.	Chapter 10: pages 246–248 Chapter 17: page 405
12. **Students will develop the abilities to use and maintain technological products and systems.**	Chapter 2: pages 43–45 Chapter 4: pages 84–94 Chapter 5: page 126 Chapter 6: page 147 Chapter 7: page 161 Chapter 8: page 184 Section 2 Activities: pages 194–197 Chapter 10: page 249 Chapter 11: page 266 Chapter 12: pages 271–273, 282 Chapter 15: pages 345–349, 358 Chapter 18: page 423 Section 3 Activities: page 426 Chapter 19: page 463 Chapter 21: page 512 Chapter 22: page 549 Section 4 Activities: pages 630–640, 647–648 Chapter 26: page 674
Use information provided in manuals, protocols, or by experienced people to see and understand how things work.	Chapter 7: page 161 Chapter 10: page 249 Chapter 11: page 266 Chapter 12: pages 271–273 Chapter 15: page 358 Chapter 18: page 423 Chapter 19: page 463 Chapter 22: page 549 Chapter 26: page 674
Use tools, materials, and machines safely to diagnose, adjust, and repair systems.	Chapter 4: pages 84–94 Section 2 Activities: pages 194–197 Chapter 15: pages 345–349 Section 4 Activities: pages 630–634, 647–648

(Continued)

Standards for Technological Literacy

Use computers and calculators in various applications.	Chapter 5: page 126 Chapter 6: page 147 Chapter 8: page 184 Chapter 12: page 282 Section 3 Activities: page 426 Chapter 19: page 463 Chapter 21: page 512 Chapter 22: page 549 Chapter 26: page 674
Operate and maintain systems in order to achieve a given purpose.	Chapter 2: pages 43–45 Section 4 Activities: pages 635–640
13. Students will develop the abilities to assess the impact of products and systems.	Section 2 Activities: pages 194–197 Chapter 10: pages 244–246 Chapter 12: pages 275–278, 280, 282 Chapter 19: page 463 Chapter 20: page 487 Chapter 24: page 602 Section 4 Activities: pages 635–640, 647–650 Chapter 26: pages 664–672, 674 Chapter 27: pages 682–683 Section 5 Activities: pages 692–693
Design and use instruments to gather data.	Section 2 Activities: pages 194–197 Chapter 12: pages 275–278, 282 Chapter 19: page 463 Chapter 20: page 487 Chapter 24: page 602 Section 4 Activities: pages 635–640, 647–650 Chapter 26: page 674
Use data collected to analyze and interpret trends in order to identify the positive or negative effects of a technology.	Chapter 26: pages 664–672, 674 Section 5 Activities: pages 692–693
Identify trends and monitor potential consequences of technological development.	Chapter 26: pages 664–672 Chapter 27: pages 682–683 Section 5 Activities: pages 692–693
Interpret and evaluate the accuracy of the information obtained and determine if it is useful.	Chapter 12: page 280
14. Students will develop an understanding of and be able to select and use medical technologies.	Chapter 19: pages 452–455 Chapter 24: pages 584, 590–591, 596–601
Advances and innovations in medical technologies are used to improve healthcare.	Chapter 24: page 584
Sanitation processes used in the disposal of medical products help to protect people from harmful organisms and disease, and shape the ethics of medical safety.	Chapter 24: pages 596–599

Standards for Technological Literacy

The vaccines developed for use in immunization require specialized technologies to support environments in which a sufficient amount of vaccines are produced.	Chapter 24: pages 590–591
Genetic engineering involves modifying the structure of DNA to produce novel genetic make-ups.	Chapter 19: pages 452–455 Chapter 24: pages 599–601
15. Students will develop an understanding of and be able to select and use agricultural and related biotechnologies.	Chapter 19: pages 440–462
Technological advances in agriculture directly affect the time and number of people required to produce food for a large population.	Chapter 19: pages 440–450
A wide range of specialized equipment and practices is used to improve the production of food, fiber, fuel, and other useful products and in the care of animals.	Chapter 19: pages 440–452
Biotechnology applies the principles of biology to create commercial products or processes.	Chapter 19: pages 452–455
Artificial ecosystems are human-made complexes that replicate some aspects of the natural environment.	Chapter 19: pages 459–462
The development of refrigeration, freezing, dehydration, preservation, and irradiation provide long-term storage of food and reduce the health risks caused by tainted food.	Chapter 19: pages 455–459
16. Students will develop an understanding of and be able to select and use energy and power technologies.	Chapter 6: pages 130–133 Chapter 21: pages 494–507, 509–511 Chapter 26: pages 665–666 Chapter 27: page 687
Energy is the capacity to do work.	Chapter 6: pages 131–133 Chapter 21: page 494
Energy can be used to do work, using many processes.	Chapter 6: pages 130–131 Chapter 21: pages 494–507
Power is the rate at which energy is converted from one form to another or transferred from one place to another, or the rate at which work is done.	Chapter 6: page 133
Power systems are used to drive and provide propulsion to other technological products and systems.	Chapter 21: pages 494–507

(Continued)

Standards for Technological Literacy

Much of the energy used in our environment is not used efficiently.	Chapter 21: pages 509–511 Chapter 26: pages 665–666 Chapter 27: page 687
17. Students will develop an understanding of and be able to select and use information and communication technologies.	Chapter 7: page 157 Chapter 17: pages 385–399 Chapter 22: pages 517, 521–539
Information and communication systems allow information to be transferred from human to human, human to machine, and machine to human.	Chapter 22: pages 521–536
Communication systems are made up of a source, encoder, transmitter, receiver, decoder, and destination.	Chapter 22: pages 528–529, 534–537
The design of a message is influenced by such factors as the intended audience, medium, purpose, and nature of the message.	Chapter 22: pages 538–539
The use of symbols, measurements, and drawings promotes clear communication by providing a common language to express ideas.	Chapter 17: pages 385–399 Chapter 22: pages 517, 532
18. Students will develop an understanding of and be able to select and use transportation technologies.	Chapter 25: pages 607–608, 615–624
Transporting people and goods involves a combination of individuals and vehicles.	Chapter 25: pages 607–608
Transportation vehicles are made up of subsystems, such as structural, propulsion, suspension, guidance, control, and support, that must function together for a system to work effectively.	Chapter 25: pages 615–624
Governmental regulations often influence the design and operation of transportation systems.	Chapter 25: pages 621–624
Processes, such as receiving, holding, storing, loading, moving, unloading, delivering, evaluating, marketing, managing, communicating, and using conventions are necessary for the entire transportation system to operate efficiently.	Chapter 25: pages 608–610
19. Students will develop an understanding of and be able to select and use manufacturing technologies.	Chapter 5: pages 113–118, 120 Chapter 23: pages 558–568, 571–577
Manufacturing systems use mechanical processes that change the form of materials through the processes of separating, forming, combining, and conditioning them.	Chapter 23: pages 558–568

Standards for Technological Literacy

Manufactured products may be classified as durable and non-durable.	Chapter 23: pages 558–568
The manufacturing process includes the designing, development, making, and servicing of products and systems.	Chapter 22: pages 538–539
Chemical technologies are used to modify or alter chemical substances.	Chapter 5: page 120
Materials must first be located before they can be extracted from the earth through such processes as harvesting, drilling, and mining.	Chapter 5: pages 113–118
Marketing a product involves informing the public about it as well as assisting in selling and distributing it.	Chapter 23: pages 576–577
20. Students will develop an understanding of and be able to select and use construction technologies.	Chapter 20: pages 474–479, 481–483
The selection of designs for structures is based on factors such as building laws and codes, style, convenience, cost, climate, and function.	Chapter 20: pages 474–477
Structures rest on a foundation.	Chapter 20: pages 477–479
Some structures are temporary, while others are permanent.	Chapter 20: pages 477–479, 481–482
Buildings generally contain a variety of subsystems.	Chapter 20: pages 482–483

Scope and Sequence Chart

In planning your program, you may want to use this Scope and Sequence Chart to identify the major concepts presented in each chapter of the *Technology: Design and Applications* text. Refer to the chart to select topics that meet your curriculum needs. The chart is divided into five sections. Within these sections, bold numbers indicate the chapters in which the concepts are found. Topics and their corresponding page numbers follow the chapter numbers for easy reference to the text.

Section 1

■ Core Concepts of Technology

1: Technology and Knowledge (26); Technology and Action (29)
2: Technology and Knowledge (42); Technology and Time (42); Technology and Goals (49)

■ Connections to Other Fields of Study

1: Technology and History (25); Technology and Language (30)
3: The Stone Age—Up to about 3500 B.C. (57); The Bronze Age—About 3500 to 1000 B.C. (58); The Iron Age—1000 to 500 B.C. (59); The Industrial Age—1750 to the Late 1900s (60); The Information Age—Late 1900s to Present Day (62)

■ Technology and Its Impacts

1: Assessing Actions (30)
2: Technology and People (40)
3: Historical Context (56); Societal Context (63); Scope of Society (65); Technology and the Individual (65); Technology and Society (66)

■ Design and Problem Solving in Technology

1: Designing Actions (29)

■ Technological Tools and Materials

2: Technology and Natural Resources (41); Technology and Machines (41)

■ Technological Systems and Processes

1: Producing Actions (30)
2: Parts of a System (37); System Inputs (39); System Processes (43); Production Processes (44); System Outputs (45); System Feedback (48)

■ Applications of Technology

1: Using Actions (30)
2: Technology and Energy (42)
3: Functional Context (68)

■ Technological Enterprises

2: Technology and Financial Resources (42); Management Processes (45)

Section 2

■ **Core Concepts of Technology**

6: Energy and Work (131); Power and Energy (133); The Nature of Energy (134); Forms of Energy (134); The Sun As the Primary Energy Source (135); Energy Sources (137); Common Energy Supplies (138)

7: Data, Information, and Knowledge (154); Scientific Knowledge (155); Technological Knowledge (157); Gaining Information and Knowledge (158)

8: Time (180)

■ **Connections to Other Fields of Study**

7: Scientific Knowledge (155); Information Age (155)

■ **Technology and Its Impacts**

6: Alternate Energy Sources (143)

8: Work and a Changing Society (168); Jobs and Technology (170); Jobs in the Information Age (172); Jobs in Technology (173); People and Jobs (176)

■ **Technological Tools and Materials**

4: Toolmakers (83); Tools (84); Measuring Tools (85); Cutting Tools (89); Drilling Tools (90); Gripping Tools (91); Pounding Tools (92); Polishing Tools (93); Mechanisms (94); Lever (94); Wheel and Axle (96); Pulley (96); Inclined Plane (97); Wedge (98); Screw (98); Machines (98); Turning Machines (100); Drilling Machines (100); Milling and Sawing Machines (101); Planing and Shaping Machines (101); Grinding and Sanding Machines (101); Shearing Machines (102)

5: Types of Materials (110); Engineering Materials (111); Metals (111); Ceramics (112); Polymers (112); Composites (113); Obtaining Material Resources (113); Renewable Resources (114); Exhaustible Resources (116); Processing Raw Materials (118); Properties of Materials (120); Physical Properties (120); Mechanical Properties (121); Chemical Properties (122); Thermal Properties (122); Electrical and Magnetic Properties (123); Optical Properties (123); Acoustical Properties (123)

6: Fossil Fuels (138); Waterpower (141); Nuclear Energy (141); Solar Energy (143); Wind Energy (143); Tidal and Wave Energy (143); Geothermal Energy (144); Biomass Energy (145)

■ **Technological Systems and Processes**

5: Processing Raw Materials (118); Mechanical Processing (119); Thermal Processing (119); Chemical Processes (120); Products of Processing (120)

■ **Applications of Technology**

6: Waterpower (141); Nuclear Energy (141); Solar Energy (143); Wind Energy (143); Tidal and Wave Energy (143); Geothermal Energy (144); Biomass Energy (145)

■ **Technological Enterprises**

8: Jobs and Technology (170); Jobs in the Information Age (172); Jobs in Technology (173); People and Jobs (176); Goals, Interests, and Traits (176); Getting Career Information (178); Advancing in a Career (178); Finances (Money) (181)

(Continued)

Section 3

■ Core Concepts of Technology

10: What Is Design? (234)
11: Needs and Wants (256)
12: Types of Research (270)
15: What Is a Model? (332); Why Are Models Used? (332)

■ Connections to Other Fields of Study

15: Mathematical Models (336)

■ Design and Problem Solving in Technology

9: History of Invention (214); Discovery (217); Invention (217); Problem Solving (218); The Invention Process (219); Identify the Problem (219); Collect Information (220); Think Creatively (220); Experiment (221); Review (221); Inventors (221); Why Invent? (222); Invention Teams (223); Innovations (224); Adaptation (225)
10: What Is Design? (234); Types of Design (235); Artistic Design (235); Engineering Design (235); Steps of the Design Process (237); Identifying the Problem (238); Researching the Problem (240); Creating Solutions (240); Selecting and Refining Solutions (242); Modeling the Solution (242); Testing the Solution (244); Communicating Designs (246); Improving Solutions (248)
11: Problems and Opportunities (256); Types of Problems (258); Recognizing Problems (258); Problem Statements (261); Criteria (261); Establishing Constraints (262); Design Briefs (264)
13: Exploring Ideas (286); Ideation (287); Brainstorming (288); Graphic Organizers (289); Questioning (289); Sketching (291); Rough Sketches (291); Refined Sketches (293); Refining Techniques (300); Types of Sketches (302)
14: Selecting Solutions (312); Aspects of Design (314); Appearance (314); Function (315); Human Factors (315); Design Sheets (317); Evaluation Grids (318); Refining Solutions (319); Elements of Design (320); Principles of Design (323); Renderings (324); Detailed Sketches (324)
15: Communication (332); Experimentation (333); Organization (334); Types of Models (335); Graphic Models (335); Computer Models (337); Physical Models (339)
16: Criteria (363); Characteristics (364); Function (364); Human Engineering (366); Appearance (368); Safety (368); Testing Environments (370)
17: Drawings (387); Working Drawings (389); Detailed Drawings (389); Assembly Drawings (390); Schematic Drawings (391); Drawing Classifications (392); Orthographic Drawings (392); Pictorial Drawings (396)
18: Reasons for Improvement (413); Function (415); Safety (416); Sources of Improvement (416)

■ Technological Tools and Materials

15: Modeling Materials (342); Sheet Materials (343); Found Objects (349); Sculpting and Molding Materials (349); Adhesives (352); Modeling Tools (353)
16: Collect Materials and Equipment (374)
17: Communication Documents (384); Bills of Materials (385); Drawings (385); Specification Sheets (386); Mechanical Drawing Tools (399); Traditional Tools (399); Computer-Aided Design (CAD) Tools (401)

■ **Technological Systems and Processes**

9: The Invention Process (219)
10: Steps of the Design Process (237)
12: Library Research (271); Survey Research (273); Experimental Research (274); The Research Process (275); List the Problem (275); Gather Data (275); Analyze Data (278); Make a Conclusion (280)
13: The Sketching Process (294)
14: External Selection (312); Internal Selection (313); Production (317)
15: Making a Model (354)
16: Testing and Assessing (362); The Testing Procedure (371); Determine the Purpose (372); Choose the Test (372); Conduct the Test and Gather Data (375); Analyze the Data (375); Evaluate the Results (375); Assessing the Solution (375)
17: Making Mechanical Drawings (402); Presenting the Solution (404)
18: Making Improvements (421)

■ **Applications of Technology**

9: Social Inventions (222); Financial Inventions (223); Leisure Inventions (223)

■ **Technological Enterprises**

9: Financial Inventions (223)
14: Finances (317)
16: Economics (367)
18: Profit (414); Competition (414); Consumers (417); Competitors (419); Government Organizations (419); Internal Sources (421)

Section 4

■ **Core Concepts of Technology**

25: What Is Transportation? (607)

■ **Connections to Other Fields of Study**

19: Agriculture and Biotechnology (452)

■ **Technology and Its Impacts**

21: Using and Conserving Energy (509)

■ **Design and Problem Solving in Technology**

23: Designing Tooling (573)

■ **Technological Tools and Materials**

19: Power, or Pulling, Equipment (441); Tillage Equipment (442); Planting Equipment (443); Pest Control Equipment (444); Irrigation Equipment (445); Harvesting Equipment (447); Storage Equipment (450)
22: Communication and Information (516); Computer Hardware (545); Computer Software (547)
23: Selecting Equipment (572); Arranging Equipment (573); Designing Tooling (573)

(Continued)

■ Technological Systems and Processes

20: Constructing a Structure (473); Preparing to Build (474); Preparing the Site (477); Setting Foundations (477); Building the Framework (479); Enclosing the Structure (481); Finishing the Interior (483); Finishing the Site (485)

21: Conversion and Converters (494); Energy and Power Transmission (507); Mechanical Transmission (508); Fluid Transmission (508); Electrical Transmission (509); Using and Conserving Energy (509)

22: How We Communicate (517); Types of Communication Systems (521); The Communication Process (534); Encoding (534); Transmitting (535); Receiving (536); Decoding (536); Storage and Retrieval (537); Producing Communication Messages (537); Producing Published Messages (537); Producing Film Messages (541); Producing Broadcast Messages (543)

23: Primary Processes (560); Secondary Processes (560); Manufacturing Systems (569); Custom Manufacturing (569); Intermittent Manufacturing (569); Continuous Manufacturing (570); Flexible Manufacturing (570); Developing Manufacturing Systems (571); Selecting Operations (571); Sequencing Operations (571); Controlling Quality (573); Testing the System (574); Producing Products (575)

25: Transportation Processes (608); Transportation Systems (615); Vehicle Systems (615); Support Systems (621)

■ Applications of Technology

19: Types of Agriculture (437); Agricultural Crops (437); Technology in Agriculture (440); Raising Livestock (450); Special Types of Agriculture (452); Food Production (455); Food Processing (456); Food Preservation (456); Artificial Ecosystems (459)

20: Buildings (469); Residential Buildings (469); Commercial Buildings (471); Industrial Buildings (471); Public Buildings (472); Religious Buildings (472); Civil Structures (472)

21: Conversion and Converters (494); Mechanical Converters (496); Thermal Converters (497); Chemical Converters (501); Solar Converters (503); Fluid Converters (506)

22: Communication Technology (518); Graphic Communication (518); Wave (Electronic) Communication (518); Graphic vs. Electronic Communication (520); People-to-People Communication (522); People-to-Machine Communication (532); Machine-to-People Communication (532); Machine-to-Machine Communication (533); Information Technologies (545)

23: Manufacturing (558)

24: Goals of Health and Medicine (584); Prevention Programs (585); Dealing with Illness and Injury (591); Using Technology in Medicine (591)

25: Types of Transportation (608); Modes of Transportation (610); Land Transportation (610); Water Transportation (612); Air Transportation (614); Space Transportation (614); Intermodal Transportation (614)

■ Technological Enterprises

23: Marketing Products (576)

Section 5

■ **Technology and Its Impacts**

26: The World of Change (657); Technology and Change (659); Negative Impacts (664); Impacts on Resources (665); Impacts on the Environment (666); Impacts on People (669)
27: Futurists (681); The Future of Technology (685)

■ **Technological Tools and Materials**

26: Materials and Change (659); Tools and Change (660); Impacts on Resources (665)
27: Energy and the Future (687)

■ **Technological Systems and Processes**

27: Forecasting (681); Trending (682); Surveying (683); Creating Scenarios (683); Modeling (684)

■ **Applications of Technology**

26: Communication and Change (661); Transportation and Change (662); Construction and Change (663)
27: Agriculture and the Future (685); Construction and the Future (686); Communication and the Future (687); Medicine and the Future (688); Manufacturing and the Future (688); Transportation and the Future (689)

■ **Technological Enterprises**

26: Production and Change (663)

Basic Skills Chart

The Basic Skills Chart has been designed to identify those activities in the *Technology: Design and Applications* text and Student Activity Manual that specifically encourage the development of basic skills.

Academic areas addressed in the chart include reading, writing, verbal (other than reading and writing), math, science, and analytical.

➤ **Reading** activities include assignments designed to improve comprehension of information presented in the chapter. Some are designed to improve understanding of vocabulary terms.

➤ **Writing** activities allow students to practice composition skills, such as letter writing, informative writing, and creative writing.

➤ **Verbal** activities encourage students to organize ideas, develop interpersonal and group speaking skills, and respond appropriately to verbal messages. Activities include oral reports and interviews.

➤ **Math** activities require students to use basic principles of math, as well as computation skills, to solve typical problems.

➤ **Science** activities call for students to use fundamental principles of science to solve typical problems.

➤ **Analytical** activities involve the higher-order skills needed for thinking creatively, making decisions, solving problems, visualizing information, reasoning, and knowing how to learn.

Activities are broken down by chapter, and a page number is given to locate the activity.

Section 1

Reading	**Text:**
	1: Technology and History (25); Technology and Knowledge (26); Technology and Action (29); All Subjects Connection (32)
	2: Parts of a System (37); System Inputs (39); System Processes (43); System Outputs (45); System Feedback (48); Technology and Goals (49); Container Shipping (50)
	3: Historical Context (56); Societal Context (63); Making Peanut Butter (68); Functional Context (68)

Verbal	**Text:**
	1: Technology and Language (30)

Math	**Text:**
	3: Mathematics Connection (72)

Science	**Text:**
	1: Science Connection (32); Activity 1 (32)
	2: Science Connection (52)
	3: Science Connection (72)

Analytical

Text:
1: All Subjects Connection (32); Social Studies Connection (32); Activity 2 (32); Activity 3 (32)
2: All Subjects Connection (52); Social Studies Connection (52); Activity 1 (52); Activity 2 (52); Activity 3 (52)
3: Social Studies Connections (72); Activity 1 (72); Activity 2 (72); Activity 3 (72)
Activity 1A: Using Technology to Make a Product (74)
Activity 1B: What Is Technology? (77)
Student Activity Manual:
1-1: Using Technology to Make a Product (7)
1-2: Marble Counter (11)
1-3: Technological Systems (13)

Section 2

Reading

Text:
4: Toolmakers (83); Tools (84); Mechanisms (94); Machines (98)
5: Types of Materials (110); Engineering Materials (111); Obtaining Material Resources (113); Processing Raw Materials (118)
7: Data, Information, and Knowledge (154); Technological Knowledge (157); Information Age (157); Gaining Information and Knowledge (158); Social Studies Connection (161); Activity 1 (161); Activity 2 (161)
8: Work and a Changing Society (168); Jobs and Technology (170); Jobs in the Information Age (172); Jobs in Technology (173); People and Jobs (176); Advancing in a Career (178); Time (181); Flight Controls (182)

Verbal

Text:
6: Language Arts Connection (147)
7: Social Studies Connection (161); Activity 2 (161)

Math

Text:
4: Mathematics Connection (104)
5: Mathematics Connection (126)
6: Mathematics Connection (147)
7: Mathematics Connection (161)
8: Mathematics Connection (184)

Science

Text:
4: Lasers (102); Science Connection (104)
5: Properties of Materials (120); Geothermal Energy (124); Science Connection (126); Activity 2 (126); Activity 3 (126)
6: Energy and Work (131); Power and Energy (133); The Nature of Energy (134); Biogas (136); Energy Sources (137); Common Energy Supplies (138); Alternate Energy Sources (143); Social Studies Connection (147); Science Connection (147); Activity 1 (147); Activity 2 (147); Activity 3 (147)
7: Scientific Knowledge (155); Science Connection (161)
8: Finances (Money) (181); Science Connection (184)
Activity 2D: Energy As a Resource (198)
Activity 2E: Energy and Machines As Resources (204)
Student Activity Manual:
2-2: Tools and Energy As Resources (25)
2-4: Materials As a Resource (29)
2-6: Energy As a Resource (33)
2-7: Energy As a Resource (35)

(Continued)

Analytical

Text:
4: Social Studies Connection (104); Activity 1 (104); Activity 2 (104); Activity 3 (104)
5: Social Studies Connection (126); Activity 1 (126)
6: Social Studies Connection (147)
7: Social Studies Connection (161); Activity 1 (161); Computer Applications (163)
8: Social Studies Connection (184); Activity 1 (184); Activity 2 (184); Activity 3 (184)
Activity 2A: Tools and Materials As Resources (186)
Activity 2B: Tools As a Resource (193)
Activity 2C: Materials As a Resource (194)
Activity 2F: Information As a Resource (205)
Student Activity Manual:
2-1: Tools and Materials As Resources (21)
2-3: Procedure and Tool Use (27)
2-5: Product Analysis (31)
2-8: Information As a Resource (37)
2-9: Developing an Information Chart (39)

Section 3

Reading

Text:
9: History of Invention (214); Discovery (217); Invention (217); Problem Solving (218); The Invention Process (219); Inventors (221); Why Invent? (222); Invention Teams (223); Innovations (224); Adaptation (225); High-Rise Buildings (226)
10: What Is Design? (234); Types of Design (235); Steps of the Design Process (237); Propellers (244)
11: Needs and Wants (256); Problems and Opportunities (256); Types of Problems (256); Problem Statements (261); Criteria (261); Establishing Constraints (262); Design Briefs (264)
12: Types of Research (270); The Research Process (275); Language Arts Connection (282)
13: Exploring Ideas (286); Ideation (287); Sketching (291); Elegant Solutions (306)
14: Selecting Solutions (312); Aspects of Design (314); Design Sheets (317); Evaluation Grids (318); Refining Solutions (319); Renderings (324); Detailed Sketches (324); Bar Code Readers (326)
15: What Is a Model? (332); Why Are Models Used? (332); Types of Models (335); Modeling Materials (342); Modeling Tools (353); Making a Model (354)
16: Criteria (363); Characteristics (364); Testing Environments (370); The Testing Procedure (371); Canal Locks (372)
17: Communication Documents (384); Drawings (387); Working Drawings (389); Drawing Classifications (392); Mechanical Drawing Tools (399); Making Mechanical Drawings (402)
18: Reasons for Improvement (413); Sources of Improvement (416); Cornflakes (418); Making Improvements (421); Activity 1 (423)

Writing

Text:
9: Language Arts Connection (228)
10: Language Arts Connection (249); Digital Photography (251)
11: Language Arts Connection (266); Activity 3 (266)
13: Language Arts Connection (308)
15: Activity 2 (358)
16: Language Arts Connection (377); System Control Technology (379)
17: Language Arts Connection (405); Activity 1 (405); Activity 3 (405); Video Production (407)
18: Language Arts Connection (423)
Student Activity Manual:
3-3: Communication Design (55)

Verbal

Text:
16: Activity 2 (377)
17: Presenting the Solution (404); Activity 3 (405)
18: Activity 3 (423)
Activity 3B: Communication Design (427)
Student Activity Manual:
3-3: Communication Design (55)

Math

Text:
12: Mathematics Connection (282); Activity 2 (282)
13: Mathematics Connection (308)
14: Mathematics Connection (328)
15: Mathematics Connection (358)
17: Mathematics Connection (405)

Science

Text:
9: Science Connection (228)
10: Science Connection (249)
11: Science Connection (266)
12: Science Connection (282)
13: Science Connection (308)
14: Science Connection (328)
15: Science Connection (358)
16: Science Connection (377)
17: Science Connection (405)
18: Science Connection (423)

Analytical

Text:
9: Social Studies Connection (228); Activity 1 (228); Activity 2 (228);
   Activity 3 (228); Inventions and Innovations (230)
10: Social Studies Connection (249); Activity 1 (249); Activity 2 (249);
   Activity 3 (249); Digital Photography (251)
11: Recognizing Problems (256); Social Studies Connection (266); Activity 1 (266);
   Activity 2 (266)
12: Activity 1 (282); Activity 3 (282)
13: Activity 1 (308); Activity 2 (308); Activity 3 (308)
14: Activity 1 (328); Activity 2 (328); Activity 3 (328)
15: Social Studies Connection (358); Activity 1 (358); Activity 3 (358)
16: Testing and Assessing (362); Assessing the Solution (375); Social Studies
   Connection (377); Activity 1 (377); Activity 3 (377); System Control
   Technology (379)
17: Activity 2 (405); Video Production (407)
18: Activity 2 (423)
Activity 3A: Invention and Innovation (426)
Activity 3B: Communication Design (427)
Activity 3C: Design for Space (430)
Student Activity Manual:
3-1: Shading and Rendering (51)
3-2: Orthographic Sketching (53)
3-3: Communication Design (55)
3-4: Locker Design (59)
3-5: Bio-Related Feeder (61)

(Continued)

Section 4

Reading

Text:
19: Types of Agriculture (437); Agricultural Crops (437); Technology in Agriculture (440); Raising Livestock (450); Special Types of Agriculture (452); Food Production (455); Artificial Ecosystems (459); Landscape Plants (460)
20: Buildings (469); Civil Structures (472); Constructing a Structure (473); Tower Cranes (474); Social Studies Connection 1 (487)
21: Conversion and Converters (494); Energy and Power Transmission (507); Using and Conserving Energy (509); Turbines (510)
22: Communication and Information (516); How We Communicate (517); Communication Technology (518); Types of Communication Systems (521); Instrument Landing Systems (526); The Communication Process (534); Producing Communication Messages (537); Information Technologies (545); All Subjects Connection (549); Social Studies Connection (549)
23: Manufacturing (558); Manufacturing Systems (569); Developing Manufacturing Systems (571); Producing Products (575); Marketing Products (576)
24: Goals of Health and Medicine (584)
25: What Is Transportation? (607); Types of Transportation (608); Transportation Processes (608); Modes of Transportation (610); Transportation Systems (615); Flight Simulators (622)

Writing

Text:
19: Social Studies Connection (463)
21: Activity 1 (512)
22: Social Studies Connection (549); Activity 1 (549)
23: Manufacturing Challenge (580)
25: Activity 3 (626)

Verbal

Text:
22: Communication and Information (516); How We Communicate (517); The Communication Process (534); Producing Communication Messages (537); Activity 1 (549); Activity 2 (549)
25: Activity 3 (626)

Math

Text:
19: Mathematics Connections (463)
20: Mathematics Connection (487)
21: Mathematics Connections (512)
22: Mathematics Connection (549)
23: Mathematics Connection (578)
24: Mathematics Connection (602)
25: Mathematics Connection (626)
Student Activity Manual:
4-1: Sketching and Estimating (83)

Science

Text:
19: Agriculture and Biotechnology (452); Artificial Ecosystems (459); Science Connections (463)
20: Science Connections (487)
21: Energy and Power Transmission (507); Using and Conserving Energy (509); Science Connection (512)
22: Science Connection (549)
23: Science Connection (578)
24: Science Connection (602)
25: Science Connection (626)
Activity 4A: Agriculture and Bio-Related Technology (628)
Activity 4C: Energy Conversion Technology (635)
Student Activity Manual:
4-3: Energy Conversion Technology (87)

Analytical

Text:
19: Activity 1 (463); Activity 2 (463); Activity 3 (463)
20: Social Studies Connection 2 (487); Activity 1 (487); Activity 2 (487); Activity 3 (487); Structural Challenge (489)
21: Social Studies Connection (512); Activity 2 (512); Activity 3 (512); Activity 4 (512)
22: All Subjects Connection (549); Activity 2 (549); Activity 3 (549); Graphic Design (551); Web Page Design (553)
23: Social Studies Connections (578); Activity 1 (578); Activity 2 (578); Activity 3 (578); Manufacturing Challenge (580)
24: Social Studies Connection (602); Activity 1 (602); Activity 2 (602); Activity 3 (602)
25: Social Studies Connection (626); Activity 1 (626); Activity 2 (626)
Activity 4B: Construction Technology (629)
Activity 4D: Information and Communication Technology (637)
Activity 4E: Manufacturing Technology (641)
Activity 4F: Medical Technology (645)
Activity 4G: Transportation Technology (647)
Student Activity Manual:
4-1: Sketching and Estimating (83)
4-2: Model Bridge Design Contest (85)
4-4: Information and Communication Technology (89)
4-5: Billboard Message (95)
4-6: Process Demonstration (97)
4-7: Recipe Holder (99)
4-8: Transportation Technology (101)
4-9: Hovercraft Design (103)

Section 5

Reading

Text:
26: The World of Change (657); Technology and Change (659); Negative Impacts (664); Earthquake Protection (672); Activity 2 (674); Activity 3 (674)
27: Futurists (681); Forecasting (681); The Future of Technology (685)

Writing

Text:
26: Environmental Challenge (676)
27: Language Arts Connection (690); Social Studies Connection (690); Activity 2 (690)
Activity 5A: Technological Impacts (692)
Student Activity Manual:
5-1: Future Telling (119)

Verbal

Text:
26: Activity 4 (674)

Math

Text:
26: Mathematics Connection (674)

Science

Text:
26: Science Connection (674)

Analytical

Text:
26: Social Studies Connection (674); Activity 1 (674); Activity 2 (674); Activity 3 (674); Environmental Challenge (676)
27: Activity 1 (690); Activity 3 (690)
Activity 5A: Technological Impacts (692)
Student Activity Manual:
5-1: Future Telling (119)
5-2: Products of the Future (121)

Technology

Design and Applications

by

R. Thomas Wright

Professor Emeritus, Industry and Technology

Ball State University

Muncie, Indiana

and

Ryan A. Brown

Teacher, Technology Education

Indianapolis, Indiana

Publisher

The Goodheart-Willcox Company, Inc.

Tinley Park, Illinois

Library of Congress Catalog Card Number 2002035435
International Standard Book Number 1-59070-165-8

2 3 4 5 6 7 8 9 — 04 — 08 07 06 05 04 03

Cover image courtesy of NASA

Library of Congress Cataloging-in-Publication Data

Wright, R. Thomas.
    Technology : design and applications / by R. Thomas Wright and Ryan A. Brown.
        p. cm.
    Includes index.
    ISBN 1-59070-165-8
    1. Technology.   I. Brown, Ryan A.   II. Title.
T47 .W738   2003
600--dc21   2002035435

Foreword

Technology: Design and Applications, by R. Thomas Wright and Ryan A. Brown, presents the core of technological knowledge and skills demanded to actively participate in our ever improving society. Studying and applying the lessons included in this textbook will provide you with the solid content and the hands-on/minds-on experiences that will enable you to understand contemporary technology, while preparing you to make good decisions regarding future technology opportunities and options.

This book was written and illustrated to support the growing importance of technology in our democratic society, and it is entirely based upon the *Standards for Technological Literacy: Content for the Study of Technology.* The organization and content of *Technology: Design and Applications* align with the objectives described in the *Standards for Technological Literacy,* regarding what the content of technology education should be.

Technology: Design and Applications presents curriculum content in five easily understood sections, so the reader masters the essential core of technological knowledge. The first section defines technology and explains technology as a system. The second section describes the input resources of tools, materials, energy, information, people, time, and capital. The third section covers creating technology, the invention and design processes, and problem solving in a technological world. The fourth section explains technology contexts, utilizing comprehensive coverage, while providing hands-on applications for the essential technology core areas of agriculture, construction, energy, information and communication, manufacturing, medicine, and transportation. The fifth section relates technology to our society and our future opportunities and challenges.

Additionally, *Technology: Design and Applications* provides exceptional hands-on learning opportunities supporting the Technology Student Association (TSA) competitive events. Active participation with TSA provides not only exciting student learning experiences, but also a framework for individual growth and leadership opportunities.

You are encouraged to study *Technology: Design and Applications* to expand your understanding of the role technology plays in our contemporary society and the role it will play in our future. By applying the lessons, skills, and content of this book, you will be well equipped to make positive decisions to build our technological world for a better tomorrow.

The Publisher

Introduction

What is technology? Why is it important in our lives? We need only look around to see how technology changes our lives. Tools and machines make our work easier. We have automobiles, cellular telephones, computers, and many other products to save us time and improve the quality of our lives.

Construction of all types provides shelter and convenience for all our activities. Houses and apartments keep us comfortable and protect us from the elements. Bridges allow us to cross rivers. We can choose many different kinds of vehicles for travel. Radios, telephones, computers, satellites, and television all help us keep in touch with each other and the rest of the world. All of these advantages are the result of advances in technology.

Technology is the knowledge of doing. It is a means of extending human abilities. Technology allows us to make useful products better and more easily. It enables us to build structures on Earth and, eventually, in space. Technology lets us move people and goods more easily.

This book, *Technology: Design and Applications,* introduces you to the various technologies. It explains the technologies as systems. These systems have inputs (such as people and materials), processes, outputs, goals, and constraints. You will be able to learn about the effects of technology and see that, while most effects of technology are good, some are not. As a result, you will be able to form opinions and make decisions about how to use technology wisely.

Technology: Design and Applications does more than *tell* you about technology. At the end of each section, you will have a chance to apply what you have learned through carefully designed activities. In some activities, you will build and test products. You may even use the products in competition with other students. In other activities, you may be introduced firsthand to mass production or the use of tools. Another activity may ask you to investigate careers of your own choice and rate them against your own interests and expectations. We hope this combination of information and "doing" activities will be a worthwhile experience for you.

R. Thomas Wright
Ryan A. Brown

Brief Contents

Section 1
Scope of Technology 20
  1 What Is Technology? 22
  2 Technology As a System 34
  3 Contexts of Technology 54
Section 1 Activities 74

Section 2
Resources and Technology 78
  4 Tools and Technology 80
  5 Materials and Technology 106
  6 Energy and Technology 128
  7 Information and Technology 150
    TSA Modular Activity:
    Computer Applications 163
  8 People, Time, Money, and
    Technology 166
Section 2 Activities 186

Section 3
Creating Technology 210
  9 Invention and Innovation 212
    TSA Modular Activity:
    Inventions and Innovations 230
10 The Design Process 232
    TSA Modular Activity:
    Digital Photography 251
11 Identifying Problems 254
12 Researching Problems 268
13 Creating Solutions 284
14 Selecting and Refining Solutions 310
15 Modeling Solutions 330
16 Testing Solutions 360
    TSA Modular Activity:
    System Control Technology 379
17 Communicating Solutions 382
    TSA Modular Activity:
    Video Production 407
18 Improving Solutions 410
Section 3 Activities 426

Section 4
Technological Contexts 432
19 Agricultural and Related
    Technology 434
20 Construction Technology 466
    TSA Modular Activity:
    Structural Challenge 489
21 Energy Conversion Technology 492
22 Information and Communication
    Technology 514
    TSA Modular Activity:
    Graphic Design 551
    TSA Modular Activity:
    Web Page Design 553
23 Manufacturing Technology 556
    TSA Modular Activity:
    Manufacturing Challenge 580
24 Medical Technology 582
25 Transportation Technology 604
Section 4 Activities 628

Section 5
Technology and Society 652
26 Technological Impacts 654
    TSA Modular Activity:
    Environmental Challenge 676
27 Technology and the Future 679
Section 5 Activities 692

TSA Modular Activities

The Technology Student Association (TSA) is a non-profit national student organization devoted to teaching technology education to young people. TSA's mission is to inspire its student members to prepare for careers in a technology-driven economy and culture. The demand for technological expertise is escalating in American industry. Therefore, TSA's teachers strive to promote technological literacy, leadership, and problem solving to their students membership.

TSA Modular Activities are based on the Technology Student Association competitive events current at the time of writing. Please refer to the *Official TSA Competitive Events Guide,* which is periodically updated, for actual regulations for current TSA competitive events. TSA publishes two *Official TSA Competitive Events Guides:* one for middle school events and one for high school events.

To obtain additional information about starting a TSA chapter at your school, to order the Official TSA Competitive Events Guide, or to learn more about the Technology Student Association and technology education, contact TSA:

Technology Student Association
1914 Association Drive
Reston, VA 20191-1540
www.tsawww.org

Table of Contents

Section 1
Scope of Technology 20
Technology Headline: Electronic Ink 21

Chapter 1
What Is Technology? 22

Technology and History 25
Technology and Knowledge 26
Technology and Action 29
Technology and Language 30
Curricular Connections 32

Chapter 2
Technology As a System 34

Parts of a System 37
System Inputs 39
System Processes 43
System Outputs 45
System Feedback 48
Technology and Goals 49
Technology Explained: Container Shipping 50
Curricular Connections 52

Chapter 3
Contexts of Technology 54

Historical Context 56
Societal Context 63
Technology Explained: Making Peanut Butter 68
Functional Context 68
Curricular Connections 72

Activity 1A
Using Technology to Make a Product 74

Activity 1B
What Is Technology? 77

Section 2
Resources and Technology 78
Technology Headline: Floating Cities 79

Chapter 4
Tools and Technology 80

Toolmakers 83
Tools 84
Mechanisms 94
Machines 98
Technology Explained: Lasers 102
Curricular Connections 104

Chapter 5
Materials and Technology 106

Types of Materials 110
Engineering Materials 111
Obtaining Material Resources 113
Processing Raw Materials 118
Properties of Materials 120
Technology Explained: Geothermal Energy 124
Curricular Connections 126

Chapter 6
Energy and Technology 128

Energy and Work 131
Power and Energy 133
The Nature of Energy 134
Technology Explained: Biogas 136
Energy Sources 137
Common Energy Supplies 138
Alternate Energy Sources 143
Curricular Connections 147

Chapter 7
Information and Technology   150

Data, Information, and Knowledge   154
Scientific Knowledge   155
Technological Knowledge   157
Information Age   157
Gaining Information and Knowledge   158
Curricular Connections   161
TSA Modular Activity: Computer Applications   163

Chapter 8
People, Time, Money, and Technology   166

Work and a Changing Society   168
Jobs and Technology   170
Jobs in the Information Age   172
Jobs in Technology   173
People and Jobs   176
Advancing in a Career   178
Time   178
Technology Explained: Flight Controls   182
Finances (Money)   182
Curricular Connections   184

Activity 2A
Tools and Materials As Resources   186

Activity 2B
Tools As a Resource   193

Activity 2C
Materials As a Resource   194

Activity 2D
Energy As a Resource   198

Activity 2E
Energy and Machines As Resources   204

Activity 2F
Information As a Resource   205

Section 3
Creating Technology 210
Technology Headline: Interactive Television 211

Chapter 9
Invention and Innovation 212

History of Invention 214
Discovery 217
Invention 217
Problem Solving 218
The Invention Process 219
Inventors 221
Why Invent? 222
Invention Teams 223
Innovations 224
Adaptation 225
Technology Explained: High-Rise Buildings 226
Curricular Connections 228
TSA Modular Activity: Inventions and Innovations 230

Chapter 10
The Design Process 232

What Is Design? 234
Types of Design 235
Steps of the Design Process 237
Technology Explained: Propellers 244
Curricular Connections 249
TSA Modular Activity: Digital Photography 251

Chapter 11
Identifying Problems 254

Needs and Wants 256
Problems and Opportunities 256
Types of Problems 258
Recognizing Problems 258
Problem Statements 261
Criteria 261
Establishing Constraints 262
Design Briefs 264
Curricular Connections 266

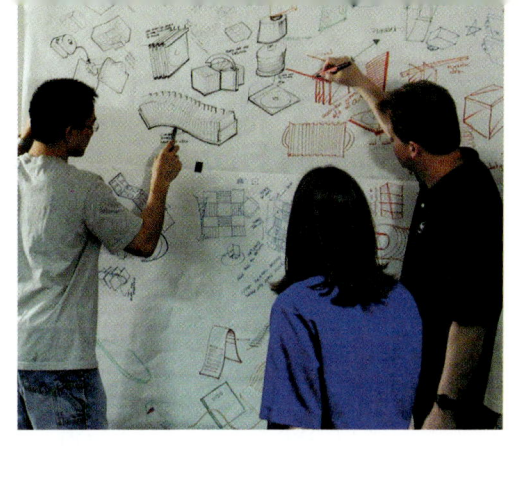

Chapter 12
Researching Problems   268

Types of Research   270
The Research Process   275
Curricular Connections   282

Chapter 13
Creating Solutions   284

Exploring Ideas   286
Ideation   287
Sketching   291
Technology Explained: Elegant Solutions   306
Curricular Connections   308

Chapter 14
Selecting and Refining Solutions   310

Selecting Solutions   312
Aspects of Design   314
Design Sheets   317
Evaluation Grids   318
Refining Solutions   319
Renderings   324
Detailed Sketches   324
Technology Explained: Bar Code Readers   326
Curricular Connections   328

Chapter 15
Modeling Solutions   330

What Is a Model?   332
Why Are Models Used?   332
Types of Models   335
Modeling Materials   342
Modeling Tools   353
Making a Model   354
Curricular Connections   358

Chapter 16
Testing Solutions   360

Testing and Assessing   362
Criteria   363
Characteristics   364
Testing Environments   370
The Testing Procedure   371
Technology Explained: Canal Locks   372
Assessing the Solution   375
Curricular Connections   377
TSA Modular Activity: System Control Technology   379

Chapter 17
Communicating Solutions   382

Communication Documents   384
Drawings   387
Working Drawings   389
Drawing Classifications   392
Mechanical Drawing Tools   399
Making Mechanical Drawings   402
Presenting the Solution   403
Curricular Connections   405
TSA Modular Activity: Video Production   407

Chapter 18
Improving Solutions   410

Reasons for Improvements   413
Sources of Improvement   416
Technology Explained: Cornflakes   418
Making Improvements   421
Curricular Connections   423

Activity 3A
Invention and Innovation   426

Activity 3B
Communication Design   427

Activity 3C
Design for Space   430

Section 4
Technological Contexts 432
Technology Headline: Facial Recognition Systems 433

Chapter 19
Agricultural and Related Technology 434

Types of Agriculture 437
Agricultural Crops 437
Technology in Agriculture 440
Raising Livestock 450
Special Types of Agriculture 452
Agriculture and Biotechnology 452
Food Production 455
Artificial Ecosystems 459
Technology Explained: Landscape Plants 460
Quality Control in Agriculture and Biotechnology 462
Curricular Connections 464

Chapter 20
Construction Technology 466

Buildings 469
Civil Structures 472
Constructing a Structure 473
Technology Explained: Tower Cranes 474
Curricular Connections 487
TSA Modular Activity: Structural Challenge 489

Chapter 21
Energy Conversion Technology 492

Conversion and Converters 494
Energy and Power Transmission 507
Using and Conserving Energy 509
Technology Explained: Turbines 510
Curricular Connections 512

Chapter 22
Information and Communication Technology 515

Communication and Information 516
How We Communicate 517
Communication Technology 518

Types of Communication Systems   521
Technology Explained: Instrument Landing Systems   526
The Communication Process   534
Producing Communication Messages   537
Information Technologies   545
Quality Control in Communication Systems   548
Curricular Connections   549
TSA Modular Activity: Graphic Design   551
TSA Modular Activity: Web Page Design   553

Chapter 23
Manufacturing Technology   556

Manufacturing   558
Manufacturing Systems   569
Developing Manufacturing Systems   571
Producing Products   575
Marketing Products   576
Curricular Connections   578
TSA Modular Activity: Manufacturing Challenge   580

Chapter 24
Medical Technology   582

Goals of Health and Medicine   584
Curricular Connections   602

Chapter 25
Transportation Technology   604

What Is Transportation?   607
Types of Transportation   608
Transportation Processes   608
Modes of Transportation   610
Transportation Systems   615
Technology Explained: Flight Simulators   622
Quality Control in Transportation Systems   624
Curricular Connections   626

Activity 4A
Agriculture and Bio-Related Technology   628

Activity 4B
Construction Technology   629

Activity 4C
Energy Conversion Technology   635

Activity 4D
Information and Communication Technology   637

Activity 4E
Manufacturing Technology   641

Activity 4F
Medical Technology   645

Activity 4G
Transportation Technology   647

Section 5
Technology and Society   652
Technology Headline: Speech Recognition Software   653

Chapter 26
Technological Impacts   654

The World of Change   657
Technology and Change   659
Negative Impacts   664
Technology Explained: Earthquake Protection   672
Curricular Connections   674
TSA Modular Activity: Environmental Challenge   676

Chapter 27
Technology and the Future   678

Futurists   681
Forecasting   681
The Future of Technology   685
Curricular Connections   690

Activity 5A
Technological Impacts   692

Technical Terms   694

Index   713

Features

TSA Modular Activities

Computer Applications   163
Inventions and Innovations   230
Digital Photography   251
System Control Technology   379
Video Production   407
Structural Challenge   489
Graphic Design   551
Web Page Design   553
Manufacturing Challenge   580
Environmental Challenge   676

Technology Explained

Container Shipping   50
Making Peanut Butter   68
Lasers   102
Geothermal Energy   124
Biogas   136
Flight Controls   182
High-Rise Buildings   226
Propellers   244
Elegant Solutions   306
Bar Code Readers   326
Canal Locks   372
Cornflakes   418
Landscape Plants   460
Tower Cranes   474
Turbines   510
Instrument Landing Systems   526
Flight Simulators   622
Earthquake Protection   672

Technology Headlines

Electronic Ink   21
Floating Cities   79
Interactive Television   211
Facial Recognition Systems   433
Speech Recognition Software   653

Curricular Connections

Curricular Connections are found at the end of each chapter. These activities tie the technological topic to a related language arts, mathematics, science, or social studies skill. The following chart identifies the types of activities contained in each chapter.

Chapter	All Subjects	Language Arts	Mathematics	Science	Social Studies
1	●			●	●
2	●			●	●
3			●	●	●
4			●	●	●
5			●	●	●
6		●	●	●	●
7			●	●	●
8			●	●	●
9		●		●	●
10		●		●	●
11		●		●	●
12		●	●	●	
13		●	●	●	
14			●	●	
15			●	●	●
16		●		●	●
17		●	●	●	
18		●		●	
19			●	●	●
20			●	●	●
21			●	●	●
22	●			●	●
23			●	●	●
24			●	●	●
25			●	●	●
26			●	●	●
27		●			●

Acknowledgments

The authors and publisher wish to thank the following companies, organizations, and individuals for their generous contributions of photographic images, artwork, and resource material.

Agricultural Research Service, U.S. Department of Agriculture
Alaska Airlines
Alcoa
All-Glass Aquarium
Amax
AMEC Wind
American Baseball Company
American Greeting
American Petroleum Institute
American Red Cross
Aminoil UAS, Inc.
Amoco Corp.
AMP, Inc.
AMS Phoenix/Jim Dore
Andersen Corp.
Army ROTC
Art Design International
Artromick International
Arvin Industries
Asphalt Roofing Manufacturers Association
AT&T Network Systems
Bayer Corp.
Bethlehem Steel Corp.
Biomet, Inc.
Boeing
Bren Instruments Inc.
Buick
Carolina Power and Light
Case IH
Caterpillar, Inc.
Cessna Aircraft
Cherry Marketing Institute
The Chicago Tribune Co.
Chrysler Corp.
Cincinnati Milacron
Clorox Co.
Coachman Industries
Colorado State University
Conoco, Inc.
CSX Corp.
Custom Modeling & Graphics Studio, Inc.
DaimlerChrysler
Dana Corporation
Deere and Company
Dell
Delta Air Lines, Inc.
Design Central
Design Edge
Duracell
Eastman Chemical Company

EIS Div., Parker-Hannifin
Eli Lilly and Company
Faro Technologies, Inc.
Federal Emergency Management Agency
Fine Art Lamps
Fischertech, A Division of Fischer Gauge Limited
FMC Corporation
Ford Motor Company
Forrest M. Mimms III
Frist Center for the Visual Arts/Timothy Hursley
Gannett Co.
GE Medical Systems
GE Plastics
General Motors
Goodyear Tire and Rubber Company
Graphic Arts Technical Foundation
Habitat for Humanity International
Harris Corp.
Hewlett Packard
Hibbing Taconite Company
Hirdinge, Inc.
Honda
Honeywell
Hunt Corportation
IDEO/Rick Engish
IDEO/Steven Moeder
IKEA Home Furnishings
Ingenico
Inland Steel Company
Intel
Jack Klasey
Jerry E. Howell
Kalb
Keith Nelson
Lakeside Equipment Corporation
Laura Trimble
The LEGO Group
Lexington Homes
LTV Steel Company
Lucent Technologies, Inc./Bell Labs
Makita
Mark Chrapla
McDonnell Douglas Company
Medlife
Microsoft Corp.
Monster Cable Products, Inc.
Motoman
Motorman
Motorola, Inc.
MyCelticRings.Com

Napa Valley Balloons, Inc.
NASA
NASA/JPL/Caltech
NASA's Marshall Space Flight Center and Science@NASA
Natchez Trace Parkway, National Park System
Natural Gas Supply Assoc.
Norfolk Southern Corp.
Northern Telecom
Occupational Safety and Health Administration
Ohio Art Co.
Owens Corning
Owens-Brockway
Piaggio Aviation
Product Development Technologies, PDT
R. R. Donnelley & Sons Company
Rollerblade
Sauder
SportsArt
Staedtler Mars
Standard Oil of California
Standard Oil of Ohio
StarTrac
Stratasys, Inc.
Tandy Co.
Technomagnete
Tillamook County Creamery Assoc.
Tonka Toys
Tony Gothard
Toolbox
U.S. Department of Agriculture
U.S. Department of Labor
U.S. Park Service, Natchez Trace Parkway
U.S. Patent and Trademark Office
Underwriter's Laboratory
Vapo Oy
Velux-America
Viacom International, Inc.
Visteon
VX Corporation
WB Automotive
Werner Co.
Westinghouse Electric Corp.
Weyerhaeuser Co.
Whirlpool Corporation
Xerox

About the Authors

Dr. R. Thomas Wright is one of the leading figures in technology education curriculum development in the United States. He is the author or coauthor of many Goodheart-Willcox technology textbooks. Dr. Wright is the author of *Manufacturing and Automation Technology, Processes of Manufacturing, Exploring Manufacturing,* and *Technology.* He is the coauthor of *Exploring Production* with Richard M. Henak.

He has served the profession through many professional offices, including President of the International Technology Education Association (ITEA) and President of the Council on Technology Teacher Education (CTTE). His work has been recognized through the ITEA Academy of Fellows Award and Award of Distinction, the CTTE Technology Teacher Educator of the Year, the Epsilon Pi Tau Laureate Citation and Distinguished Service Citation, the Sagamore of the Wabash Award from the Governor of Indiana, the Bellringer Award from the Indiana Superintendent of Public Instruction, the Ball State University Faculty of the Year Award and George and Frances Ball Distinguished Professorship, and the EEA-Ship Citation.

Dr. Wright's educational background includes a bachelor's degree from Stout State University, a master of science degree from Ball State University, and a doctoral degree from the University of Maryland. His teaching experience consists of 3 years as a junior high instructor in California and 37 years as a university instructor at Ball State University. In addition, he has also been a visiting professor at Colorado State University, Oregon State University, and Edith Cowan University in Perth, Australia.

Mr. Ryan A. Brown is currently a technology teacher at Lawrence Central High School in Indianapolis, Indiana. He teaches a variety of technology courses, including Design Processes, Fundamentals of Engineering, and Transportation Systems. Mr. Brown has written titles in both the *HITS* and *KITS* series for the ITEA, as well as in the *Activity!* series for the Center for Implementing Technology Education.

Mr. Brown's educational background includes a bachelor's degree and master's degree from Ball State University. He has been recognized with the Young Professionals Award from the Technology Educators of Indiana. The Foundation of Technology Education also named him a Maley Spirit of Excellence Outstanding Graduate Student.

Section 1
Scope of Technology

This book is about the way we control our environment. If we are cold, we can provide heat or heavier clothing. If we wish to travel, we can provide fast and efficient methods of transportation. This ability to create change in our environment is the result of applying knowledge to solve problems. As a result of these problem-solving abilities, we are always changing the way we live, work, and move. New inventions and discoveries are a part of our everyday lives.

What is new today soon becomes commonplace. One million years ago, fire was a new tool. It was used to fashion crude stone tools, cook food, and keep warm. Fifty years ago, computers took up whole rooms. They were big, loud, and slow. Today, computers are operated by chips that can fit on your fingertip. These chips operate everything from our telephone systems to our automobiles.

As you study the following three chapters, you will begin to see how important technology is to you. You will begin to see technology as a series of carefully organized efforts. Also, you will begin to see there are right ways and wrong ways to use technology.

1 What Is Technology?

2 Technology As a System

3 Contexts of Technology

Technology Headline

Electronic Ink

The invention of paper and ink, almost 2,000 years ago, greatly impacted the way the world communicates. Think about how many times a day you use paper! Perhaps you read a newspaper in the morning, use textbooks at school, and flip through a magazine in the evening. In the future, you may be able to read all of this information from a single sheet of electronic paper.

Scientists are now working on developing electronic paper and ink. This revolutionary technology may soon replace paper as we know it. If they are successful, in a few years, we will be able to carry around an entire library in one book!

Electronic ink (e-ink) is a thin film of microscopic capsules. Each capsule can appear as black or white, depending on an electric signal. Electronic paper (e-paper) is a thin, paper-like plastic sheet coated with e-ink. The display on the e-paper is controlled by electronic signals, similar to a computer monitor. Therefore, the display can be changed by sending a new signal. Unlike a computer monitor's display, however, the display of type on e-paper will be similar to the display on a printed page.

The day could arrive when the daily newspaper is downloaded automatically to your sheet of e-paper. This same sheet of e-paper may also have all your school textbook pages in its memory. Magazines could also be downloaded to your e-paper automatically. Several sheets of e-paper could be bound together to create an electronic book. Eventually, a single electronic book could allow instance access to thousands of books.

Currently, e-ink is in the early stages of development and can produce only black and white images. As time goes on, scientists will develop a color version of e-ink.

You probably won't need to throw away all of your paper or stop using your computer any time soon. Electronic ink will slowly be incorporated into daily communication. There will always be a need to have printed pages and computer monitors, but e-ink is an exciting technology that will greatly improve the way we are able to communicate!

Discussion Starter

Ask the students to carefully look through all the pages of this section, from page 22 to page 77. As they are looking through the pages, they should choose two illustrations from each chapter that they find the most interesting. Then they should choose one of those six illustrations to study more closely and report its technological significance to the rest of the class.

Group Activity

Try to get the students thinking about *technology*. Ask them what they think the "scope of technology" might include. Break the students into groups of three or four students and ask them to brainstorm ways we use technology. The students should focus on the technology they use in their daily lives. Have them spend about ten minutes listing as many uses of technology as they can. After they have finished, have them prioritize their lists. Ask the students, "If you could only use technology in one way, which one would it be?" Have each group rank its top three uses of technology. Bring the groups back together and have one member of each group present its list, giving reasons why their three uses of technology are the most important. Discuss the lists and supporting statements, then tell the students that in this section, they will be learning some of the most important ways we all use technology in our everyday lives.

Writing Assignment

Have the students read the section introduction on page 20. Then have them write two or three paragraphs on what they think they will learn and what they would like to learn in this section. Ask volunteers to share their essays with the class.

Chapter 1
What Is Technology?

Did You Know?

> The word *technology* comes from a combination of two Greek words. The first is *techne,* which means art or craft. The second is *logos,* which means word or speech. The combination means the discussion of the practical and fine arts.

Objectives

The information given in this chapter will help you do the following:

➤ Define and explain *technology*.

➤ Describe some positive and negative aspects of technology.

➤ Explain the focus of technological history.

➤ List and describe the four major types of knowledge.

➤ Identify and explain the seven types of technological knowledge.

➤ Name and describe the four major technological actions.

➤ Give examples of the major ways to communicate technological information.

Key Words

These words are used in this chapter. Do you know what they mean?

agricultural knowledge

assess

construction knowledge

descriptive knowledge

design

energy knowledge

humanities knowledge

information and communication knowledge

language

manufacturing knowledge

medical knowledge

produce

scientific knowledge

technical drawing

technological knowledge

technology

transportation knowledge

use

vocabulary

Discussion
Ask the students to think of five technological devices they use. Ask them what they would do if the devices had not been invented.

Example
Make a transparency of or photocopy a newspaper article with scientific and technological content. Present the article to the students and ask them to identify examples of technological, scientific, humanities, and descriptive knowledge and terms used in the article.

Chapter Outline
Did You Know?
Objectives
Key Words
Technology and History
Technology and Knowledge
Technology and Action
  Designing Actions
  Producing Actions
  Using Actions
  Assessing Actions
Technology and Language
Summary
Curricular Connections
Activities
Test Your Knowledge

Resource
Transparency 1-1: Technology Is Knowledge

Brainstorm
Divide students into small groups and have them explore some examples of tools and equipment (technology) they may have used.

Research
Have the students explore ways they think technology directly helps and harms their way of life.

Tech Terms
Technology—Using tools, materials, and processes to make things that help us live better.

Almost everyone has heard and used the word *technology,* yet the word means different things to different people. See **Figure 1-1.** Technology, to many people, means equipment and tools. People often think of things like robots, spaceships,

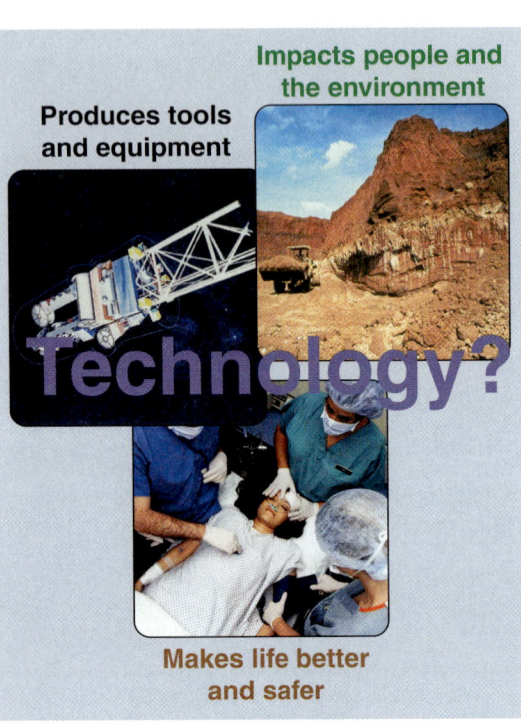

Figure 1-1. Technology means many things to different people. What does it mean to you?

and computers when they use the word *technology*. In this context, technology is the hardware that has changed life for better or worse.

To other people, technology causes many of the world's problems. These people see technology as the origin of inventions that have polluted the air and water around us. These opponents of technology believe it has caused problems such as crime, unemployment, and global warming.

To still other people, technology is the hope for a better life. This group of people sees new technological products and systems as the solutions to many of society's immediate and future problems. These products and systems can help do things that could not be done without the help of technology. This group of people points out that technology can improve personal lives by providing efficient transportation, rapid communication, comfortable housing, and plentiful food. People can travel faster with the use of engines. Water pumps can move water to distant locations where it is needed. Cures for diseases can be provided to decrease health risks. Machines can be used to help with hard labor.

To some extent, all three of these views are correct. See **Figure 1-2.** Technology involves tools and processes that help us live better lives. It can also cause changes that make life more complicated and stressful. Technology can provide more food and clothing, but it can be used to create a society in which some people live in great comfort and others live in poverty. Used well, technology can be an agent of good. Used poorly, it can cause great harm to people and the environment. To fully understand technology, you must understand that it involves knowledge and actions and that it has a history and vocabulary of its own.

TechnoFact: The most commonly used means for assessing the effects of technological advancements on people and the environment are government agencies, the courts, and the insurance industry.

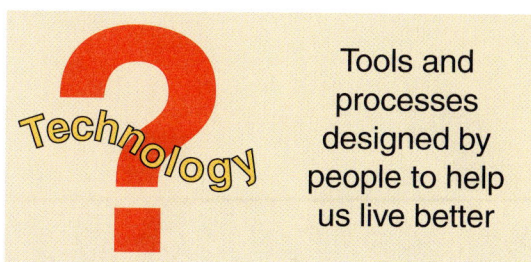

Technology? Tools and processes designed by people to help us live better

Figure 1-2. Here is a simple definition for *technology*.

Technology and History

Often the study of history is one of wars and kings, as history is usually presented from a political viewpoint. The information in most history textbooks focuses on who was in power and the wars fought to solidify that power. Another way to view history, however, is through technological history. This type of history describes how people have lived and attempted to cope with the natural world throughout the course of time. See **Figure 1-3.** Technological history looks at the types of homes in which people have lived. It also describes how humans have transported themselves and their belongings. This type of history explains how people have grown and preserved food. It also presents the communication systems they have used and the ways in which they have made common utensils and tools.

This view of history accepts that people have always wanted to live better lives. Early humans lived in caves and crude shelters. They gathered fruits and berries or killed game for food. These people had harsh and unpredictable lives. They had to move whenever food became scarce. To help live more efficiently, they developed crude tools. To make weapons, these early humans tied sharp stones to sticks. These new tools allowed them to become better hunters. These people tanned animal hides and made crude clothing from them. They sewed these garments together using bones for needles and tendons for thread. These early humans also developed stone axes and hammers, allowing them to make better houses.

Figure 1-3. Early humans used simple technologies to improve their lives.

As civilization advanced, people started to grow and harvest plants. This allowed them to have more reliable sources of food, so they no longer had to move constantly. They selected tree limbs with certain shapes that could be used as digging sticks. This early plow allowed them to till the

TechnoFact: The earliest known human artifacts have been found in eastern Africa and are over two million years old. They were flaked stones, known as Oldowan tools, and were used for chopping and scraping.

Figure 1-4. The invention of the plow was the start of efficient farming. (Deere and Company)

soil, and agriculture was born. See **Figure 1-4.** With the plow and other early farming tools, the foundation for modern civilization was set. People could gather in villages and cultivate nearby land. Without the ability to grow large amounts of food in one place, concentrated population centers would have been impossible. The invention of the plow was followed by the development of tools to gather and harvest crops. Later, irrigation was developed in ancient Egypt and other areas along large rivers. Dams were built, and ditches were dug. Simple devices to raise the water from one level to another were developed. Irrigation projects were some of the first "engineered" works built by people. The knowledge used in these projects was applied to other building projects, such as the pyramids.

The skills and knowledge developed for one project were applied to other projects. Each new project allowed people to develop more skills and knowledge. The process of applying old knowledge and skills while developing new knowledge and skills is a characteristic of technology. This is true because technology is a product of the human mind. Technology involves the engineering spirit and is the result of human ambition, vision, and control. People have developed all forms of technology to meet the needs and wants of other people. These developments are the result of human needs and the ability to be creative. Creativity and innovation allow us to generate new products and systems to make life easier. See **Figure 1-5.**

Technology and Knowledge

People use knowledge every day. They use it to make decisions, complete work, and entertain themselves. There are several types of knowledge.

TechnoFact: Early technologies were primarily focused on practical tools. However, other uses of technology were developed. For example, colorful minerals were pulverized to make pigments used to adorn the human body, baskets, clay utensils, and clothing.

Figure 1-5. People develop technology for other people to use.

Resource
Transparency 1-4: Types of Knowledge

Tech Terms
Scientific Knowledge— Knowledge about the laws and principles governing the natural world.
Technological Knowledge— Knowledge about using tools, materials, and processes to make and operate systems.
Humanities Knowledge— Knowledge about the development and functioning of social and political systems and ways to express creative thought.
Descriptive Knowledge— Knowledge of ways to use words and numbers to describe relationships, events, and ideas.

Research
Instruct the students to explore how the knowledge of surveying developed by the Egyptians to plot farmland was used to build the pyramids.

First, there is knowledge about the world around us. It describes the plant and animal kingdoms and the land, water, and air we depend on for life. This knowledge explains the laws and principles governing the universe. We call this type of knowledge *scientific knowledge.* See **Figure 1-6.**

Second, people have knowledge of the society around them. They usually

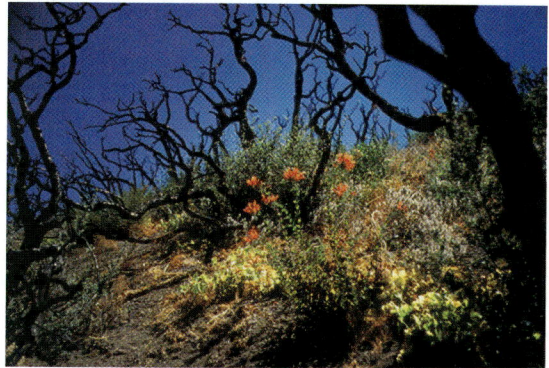

Figure 1-6. Scientific knowledge describes the natural world.

have religious ideals and beliefs and know what is considered right and wrong. Most people know about the histories of their families, communities, and world. They understand how their societies were formed and how they are organized and governed. This type of knowledge is called *humanities knowledge.*

Also, people have a way to use signs and symbols to communicate with each other. They use an alphabet to form words that can be spoken or written. We call these symbols *language*. People can also use numbers to show relationships and values. They can use lines to show shapes and forms in drawings. We call this type of knowledge *descriptive knowledge.* This knowledge helps people describe objects and events.

Finally, people have knowledge of how to use tools and materials. They

TechnoFact: Technology has been described as the process people use to create and use tools and machines to increase their understanding and control of the natural environment. Technology is the foundation for advanced civilizations.

Standards for Technological Knowledge

2

Extend
Instruct the students to explore ways they have used each type of technological knowledge in the last week.

Looking Ahead
The types of technological knowledge will be explored in depth in Chapters 19–25.

Figure 1-7. Technological knowledge is used to develop the human-built world. (Northern Telecom)

use this knowledge to make things and construct buildings. This knowledge helps them process information, grow and harvest crops, convert and use energy, and heal diseases. This knowledge of actions is used to design, produce, and use things other people create. It is the knowledge of the human-built world and is called *technological knowledge.* See **Figure 1-7.**

There are many types of technological knowledge. They can be categorized in a number of ways. See **Figure 1-8.** One example would divide technological knowledge into these categories:

➤ *Agricultural knowledge.* This is the knowledge about using machines and systems to raise and process foods.

➤ *Construction knowledge.* This is the knowledge about using machines

and systems to erect buildings and other structures.

➤ *Energy knowledge.* This is the knowledge about using machines and systems to convert, transmit, and apply energy.

➤ *Information and communication knowledge.* This is the knowledge about using machines and systems to collect, process, and exchange information and ideas.

➤ *Medical knowledge.* This is the knowledge about using machines and systems to treat diseases and maintain the health of living beings.

➤ *Manufacturing knowledge.* This is the knowledge about using machines and systems to convert natural materials into products.

➤ *Transportation knowledge.* This is the knowledge about using machines and systems to move people and cargo.

Types of Technology

Agricultural

Construction

Energy and power

Information and communication

Medical

Manufacturing

Transportation

Figure 1-8. These are some types of technology.

TechnoFact: Both science and technology use thinking processes. Science uses a process that has been called *the scientific method.* Technology uses the design process, experimentation, invention, and innovation to create new products and systems and develop new knowledge.

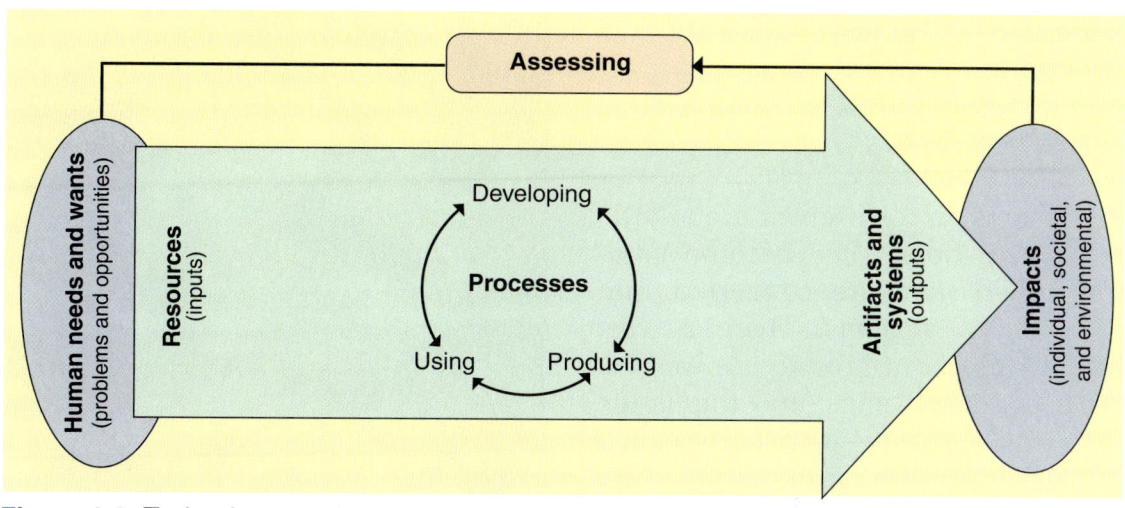

Figure 1-9. Technology involves designing, producing, using, and assessing actions.

Standards for Technological Knowledge

Tech Terms
Technological Actions—The actions people take to develop, produce, use, and assess technological products and systems.

Extension
Ask the students what types of "using" information they would need to be a good consumer of a product.

Technology and Action

People apply technological knowledge and skills in different activities and settings. Consumers use technological knowledge to select devices to fit their needs. They use technological skills to operate and maintain these products. Designers use technological skills and knowledge to create or improve products and systems. Workers use technological abilities and knowledge to make products to meet our needs. Some products, however, are developed before a need is identified. Corporations can use technological knowledge and abilities to create demand for these unnecessary products. They do this by bringing the products onto the market and advertising them. These and hundreds of other activities fall into four categories of *technological actions*. See **Figure 1-9.** These actions change resources into outputs we want and need:

➤ *Designing* and engineering technological products and systems.

➤ Using tools and machines in *producing* products and systems.

➤ *Using* the products and systems to meet human needs and wants.

➤ *Assessing* the impacts of products and systems on people, society, and the environment.

Designing Actions

Technology is created by the purposeful actions of people. All technology started in the minds of people. These people identified problems needing solutions or opportunities that could be met. Products and systems were then developed. They were designed to meet the problems or address the opportunities. A common technique used to create products or

TechnoFact: A major reason why the United States became a technological leader in the twentieth century was its system of technical education. The foundation for this type of education was the mechanical arts schools of the eighteenth century; the land-grant colleges with their emphasis on agriculture, mining, and engineering of the nineteenth century, and a system of vocational education providing training in basic technical skills developed in the twentieth century.

Extension
Ask the students what types of information they would need to build a simple product, like a book holder.

Tech Terms
Technological Assessment—
Evaluating the positive and negative impacts caused by technology on people, society, and the environment.

Discussion
Technology is a major means for creating new physical and human environments. Ask the students if it is possible that technology will also destroy global civilization.

systems is *design.* A common design process is presented in Chapters 9 through 18.

Producing Actions

Designs, in themselves, are of little value to people. They become valuable when they are converted into products or systems. There is vast knowledge about how to use tools to make or grow things, and this is the basis for a variety of actions called *producing.* Production actions are used to build products, erect buildings, grow and harvest crops, communicate information, and transport cargo. Information about these and other production actions is included in Chapters 19 through 25.

Using Actions

People *use* technological knowledge and products to meet many demands. See **Figure 1-10.** They use products in their roles as workers, consumers, family members, and citizens. For most people, using technology is a daily action. It involves selecting an appropriate product and determining which technological service to use. Product service and repair decisions are also using actions. These and other using actions will be explored in Chapter 26.

Assessing Actions

Using technology is more than simply meeting human wants. Technology must fit within the social and

Figure 1-10. People use technological devices to meet their needs. (Deere and Company)

political systems of a community or nation. Also, it must be in harmony with the environment. This means people must *assess* the impacts and importance of each form of technology they use. They must decide what is effective or appropriate technology. The description of what is "appropriate," however, changes from country to country and over time. The process of assessing technology and its impacts will be explored in Chapter 26.

Technology and Language

Each major type of knowledge has its own language. Scientists use words many of us find hard to understand. Likewise, historians have their own terms and ways of communication.

TechnoFact: E. F. Schumacher wrote a book entitled *Small Is Beautiful* in 1973. His followers believe people must recognize that Earth's resources are limited and that human life must be structured around a commitment to control the growth of industry, the size of cities, and the use of energy.

Figure 1-11. Drawings are part of the language of technology. (Ford Motor Company)

Technology has two communication tools unique to it. First, it has a *vocabulary* that has been developed to describe actions with tools and the products of these actions. Suppose you had a dictionary from the 1800s. You would not find words like microchip, airplane, television, and photocopy in it. These are new words developed to describe new technological devices. Other new words were coined to describe tools and actions.

New technological words are developed every year. First, they are part of the language of technology, and later, some of them become part of the common language. For example, *debug* was a technical term in computer technology. Early computers had exposed wires and vacuum tubes that were hot and attracted bugs. When the bug got close to the hot tube, the heat killed it. The dead bugs built up on the circuits and shorted them out. People had to open the computer cabinets and physically remove the dead bugs. This process was called *debugging*. Later, the term was used to describe correcting circuit and software problems in computers. Today, some people use the term to describe diagnosing problems in any system.

Second, technology uses *technical drawings* as a form of communication. These drawings are tools of engineers and architects, showing how products are to be made or buildings are to be built. They also can show how communities are to be developed or parks are to be landscaped.

Drawings may be as simple as a sketch of a designer's ideas. They may be so complex that computer systems are used to help in their development. See **Figure 1-11.** Chapter 17 will discuss technical or engineering drawings more completely.

Summary

Technology plays a part is everyone's life. People use it to meet their needs and wants. It is a series of actions in which products and systems are designed, produced, used, and assessed. Technology is the knowledge of tools and materials and the actions that can be accomplished with them. It has its own history and ways of communication. Technology is unique and important to each of us.

Standards for Technological Knowledge

Research
Have your students use the dictionary to find words that are used to describe technological actions and products.

Tech Terms
Working Drawing—
An accurately measured and detailed drawing used as a guide to workers in constructing a structure or machine or of any part of one.

Curricular Connections

All Subjects

Read a current events magazine or newspaper article. Highlight examples of descriptive, humanities, scientific, and technological knowledge. Use a different colored pencil or marker for each type of knowledge.

Social Studies

Ask several older people what they think technology is. Record their answers. Later, arrange the comments under three headings: Computers and Hardware, Source of Society's Problems, Hope for a Better Life. Note: A person's responses may have entries under more than one heading.

Science

Research a technological device that has helped scientists learn more about the universe. Describe the device, its use, and when it was developed.

Standards for Technological Knowledge

3

Activities

1. Science is the knowledge of the natural world. Technology is the knowledge and actions used to create the designed world. Divide a sheet of paper into two columns. Write *science* at the top of the right-hand column and *technology* at the top of the left-hand column. On your way to school, list things you see explained by each of these:

   A. Scientific laws or theories. In the right-hand column, for example, you might list an apple falling to the ground, a flower dying, or a building casting shade on the sidewalk.

   B. Technological principles or applications. In the left-hand column, for example, you might list a pothole being fixed, a car being towed away, a building being painted, or a billboard being installed.

2. Develop a simple information sheet that would help people understand the four major actions of technology.

3. Develop an assortment of photos and other pictures showing the various types of technological knowledge.

TechnoFact: Humans have the unique ability to change materials found in the natural world into tools and machines that can be used to improve their quality of life.

Test Your Knowledge

Do not write in this book. Place your answers to this test on a separate sheet of paper.

1. Select the statement best defining *technology:*
   A. Knowledge of the laws and principles of the natural world.
   B. Knowledge about developing and making products.
   C. Knowledge about social systems people have developed.
   D. Knowledge about numerical relationships.

2. What are some negative effects technology can cause?

3. Why is the study of technological history important?

4. Identify each of the following types of knowledge:
   A. Knowledge about the world around us.
   B. Knowledge about society.
   C. Knowledge about communicating with other people.
   D. Knowledge of how to use tools and materials.

5. Describe each of the following terms:
   A. Agricultural knowledge.
   B. Construction knowledge.
   C. Energy knowledge.
   D. Information and communication knowledge.
   E. Medical knowledge.
   F. Manufacturing knowledge.
   G. Transportation knowledge.

6. Actions used to develop new or improved products or systems are called _____ actions.

7. Actions used to build products, erect buildings, grow and harvest crops, communicate information, and transport cargo are called _____ actions.

8. Actions involved in selecting, operating, and servicing products are called _____ actions.

9. Actions used to determine the value or impact of a technology are called _____ actions.

10. The language of technology involves _____ and _____.

F. Knowledge about using machines and systems to convert natural materials into products.
G. Knowledge about using machines and systems to move people and cargo.
6. designing
7. producing
8. using
9. assessing
10. vocabulary or words, technical drawings

Chapter 2
Technology As a System

Did You Know?

➤ The solar system is a very large natural system. It consists of the Sun, nine planets, and sixty-eight satellites of the planets.

➤ A computer's operating system is the first software seen when a computer is turned on. It is the last software seen when the computer is turned off. The operating system directs all the programs used on a computer. It manages the hardware and software elements of the computer system. This system provides a way for applications to deal with the hardware.

➤ Not all computers have operating systems. Computers controlling simple objects, like microwave ovens, do not need operating systems. The computer in a microwave oven runs a single program all the time.

Objectives

The information given in the chapter will help you do the following:

➤ Define and explain a *system*.

➤ Explain the differences between a natural system and a technological system.

➤ Summarize how a system works.

➤ List and describe the seven major inputs to technological systems.

➤ Recognize and give examples of the major types of processes used in technological systems.

➤ Identify and describe the major types of outputs of a technological system.

➤ Name the two types of feedback systems.

➤ Describe some major goals for technological systems.

Key Words

These words are used in this chapter. Do you know what they mean?

closed-loop system
desired output
energy
feedback
feedback system
goal
input
intended output
knowledge
machine
material
natural system
open-loop system
output
pollution
process
system
technological system
time
undesirable output

Career Connection
A *systems engineer* is an engineer who specializes in the development and implementation of production systems.

Discussion
Ask the students to describe the difference between a part and a system.

Chapter Outline
Did You Know?
Objectives
Key Words
Parts of a System
System Inputs
  Technology and People
  Technology and Natural Resources
  Technology and Machines
  Technology and Knowledge
  Technology and Energy
  Technology and Financial Resources
  Technology and Time
System Processes
  Production Processes
  Management Processes
System Outputs
System Feedback
Technology and Goals
Summary
Curricular Connections
Activities
Test Your Knowledge

A system is ...

... a group of parts that work together to do a task.

Figure 2-1. This machine is a technological system people developed. (Cincinnati Milacron)

You hear about systems all the time. The weather forecaster talks about a new weather system moving into your region. A doctor tells you your digestive system is upset. The auto mechanic explains that the ignition system of your family's car needs work. So what is a system?

Simply put, a *system* is *a group of parts working together to complete a task.* See **Figure 2-1.** There are both natural and human-made systems. *Natural systems* appear in nature without human interference. They are part of the world around us. These systems include the universe. Natural systems also include all the living things populating Earth. People study natural systems as they explore science. Typical natural systems are the transpiration systems of plants, the circulatory systems of animals, the solar system, and ecosystems found in lakes and ponds. See **Figure 2-2.**

Figure 2-2. The universe is made up of natural systems.

TechnoFact: The term *system* can have several definitions: 1. A combination of parts forming a complex whole (a railroad system). 2. A set of correlated members (currency system). 3. A comprehensive group of facts, principles, or doctrines in a field of knowledge (system of philosophy). 4. A method or procedure (a system of government).

Figure 2-3. People build technological systems to meet their needs. (Amax)

Human-made or designed systems are used throughout society. People create these systems to meet their needs. There are many types of human-made systems. Political systems are designed to develop laws and regulations governing human actions. Judicial systems are created to enforce these laws and regulations. Economic systems are used to exchange items of value. Social systems are created to help people interact with one another. *Technological systems* are used to make the artifacts and services people want or need. See **Figure 2-3.**

Technological systems include agricultural, communication, transportation, manufacturing, construction, and energy conversion systems. Computers and information processing systems are technological systems. Irrigation systems found on farms and radiation treatment systems found in hospitals are technological systems

as well. Technological systems can be linked to one another and often interact with each other. Sometimes the output of one system is the input to another system, and sometimes one system controls another system. Often, several systems work together to produce the technology we use.

Systems thinking requires considering how each part connects to the others. Analyzing a system can be done by looking at its individual parts. It can also be done by looking at how the system as a whole interacts with other systems.

Parts of a System

All systems have a basic structure. See **Figure 2-4.** Generally, they are made up of five basic elements:

➤ Goals. These are the reasons for developing and operating the system.

➤ Inputs. These are the resources the system uses to meet the identified goals.

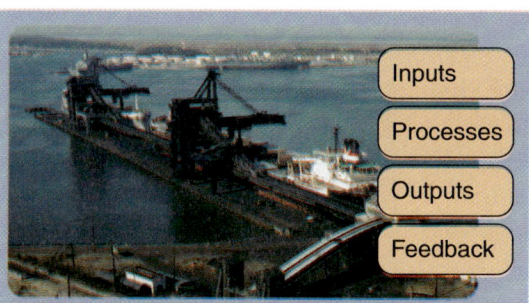

Figure 2-4. This system to transport coal has inputs, processes, outputs, and feedback. (Norfolk Southern Corp.)

Standards for Technological Knowledge

Resource
Transparency 2-1: Technological Systems

Extend
Ask the students to identify four technological systems and list reasons people developed them.

Demonstrate
Disassemble a ballpoint pen. Have the students name and describe the function of each part in the system.

Tech Terms
Technological System—A human-designed and human-made system.

TechnoFact: For most of human history, technology has fulfilled simple human needs in simple ways. In the last few centuries, complex technological systems have been developed to meet increasing human expectations.

➤ Processes. These are the actions taken to use the inputs to meet the goals.

➤ Outputs. These are the results obtained by operating the system.

➤ Feedback. This includes the adjustments made to the system to control the outputs.

All five parts of a system are extremely important to the proper operation of the system. Malfunctions of any one part may affect the function and quality of the entire system. When part of a system fails or works incorrectly, the results can range from a nuisance to a catastrophe. Because the elements of a system are so interrelated, it is vital to understand how they all work together. This can be done by examining a common system.

Let's consider a typical heating system found in a home. See

Fan

Fan motor relay

Hot air (output)

Heat exchanger

Burning fuel (process)

Fuel (input)

Wire conductors

Thermostat (feedback)

Fuel control valve

Figure 2-5. A heating system is a technological system developed to keep buildings warm.

Figure 2-5. The primary *goal* of the system is to keep the building comfortable during cold weather. The system is designed to maintain an even, livable temperature in the house. The main *input* to a heating system is an energy source. A fuel, such as natural gas, propane, coal, wood, or fuel oil, provides the energy. The *process* involves converting the fuel's energy into heat. This conversion uses a natural process called *combustion*. The fuel is burned to convert its stored energy into heat energy. This heat energy is the *desired output* of the system. Other outputs could include carbon dioxide gas, nitrogen oxide, water, or ash. These outputs are not needed. They are, however, unavoidable.

An important part of most heating systems is a temperature regulator. This regulator is called a *thermostat*. It keeps the temperature within a set range so the house is neither too hot nor too cold. A thermostat has a part that senses the room's temperature. When the temperature reaches the lower limit of the thermostat's setting, it turns the heating unit on. When the room has reached the upper limit of the thermostat's setting, it turns the heater off. The thermostat is part of the *feedback* loop of the system. It uses data (room temperature) to control the output of the system (heat). Feedback loops adjust and control the operations of systems.

Although it is not considered one of the main elements of a system, a successful system also must include

TechnoFact: Systems analysis is the evaluation of a system to identify its desired objectives and determine how the system attains them. Computer system analysis involves the study of the data-processing needs of a business to determine its hardware and software requirements.

some form of maintenance. Maintenance is the process of inspecting and servicing a product or system on a regular basis to ensure it continues to work correctly, to prolong its life, or to improve its capability. All technological systems will ultimately break down, but maintenance decreases the risk of early failure. If maintenance is not done, failure is guaranteed. The rate of failure depends on such issues as how complex the system is, what kinds of environment it must function in, and how well it was initially made.

System Inputs

Technology consists of human-built systems. Technology includes all the material things people have created over the years. Creating these products and systems takes resources. People use resources in creative ways to meet their needs and wants. The resources for creating technology are called *system inputs*. See **Figure 2-6.** These inputs can be grouped into seven major categories:

➤ People.

➤ Natural resources (materials).

Standards for Technological Knowledge

2

Resource
Transparency 2-2: Inputs to Technological Systems

Resource
Reproducible Master 2-3: Systems

Tech Terms
Technology—
Includes all the things people have designed and built over the years.

Research
Have your students use the Internet to determine the rate of failure for a specific type of product.

Figure 2-6. These are the inputs to technological systems.

TechnoFact: One of the most complex systems humans have invented is the city. This form of technological, social, and political structure was developed about 5,000 years ago.

Figure 2-7. People develop, operate, control, and use technological products and systems. (Inland Steel Company)

➤ Machines.
➤ Knowledge (information).
➤ Energy.
➤ Financial resources (money).
➤ Time.

Each of these inputs is essential for developing new or improved technological devices. These inputs are used to create and produce the things people design and use.

Technology and People

People have developed and produced all forms of technology. Technological devices and systems are the results of people's ideas and work. Without people's minds and efforts, there would be no technology. Likewise, people use technology. If people did not want a device or technological system, there would be no need to develop it. In short, people develop and produce technology for other people to use.

People bring many skills and roles to technology. See **Figure 2-7.** Designers and engineers develop technological devices and systems. Machine operators and laborers make devices and systems. Advertisers develop commercials to promote the use of products. Managers organize and supervise development and production processes. Retailers operate stores and e-commerce sites. Customers buy and use products

TechnoFact: There are many types of people who work with technology. One is the engineer, a person who solves technological problems. Engineers may deal with everything from designing an automobile brake, to developing and testing an artificial hand, to developing computer hardware and software.

Figure 2-8. Material resources start with natural resources found on Earth and in the air. (American Petroleum Institute)

and systems. Politicians and public officials regulate production and use of technology. Without people, there would be no technology.

Technology and Natural Resources

Technology includes the devices and systems people design, build, and use. These things are made out of *materials* found on Earth and in the air. See **Figure 2-8.** Materials may be classified as gases, liquids, or solids.

Gases may be the fuels used to make devices. Also, gases may become part of products. Typical gases used in technology are natural gas fuels and inert gases. Liquids are also used as fuels. They can be used as finished

products as well. Liquids may be lubricants used during production. The products themselves may use them. Typical liquids used in technology are gasoline and water. Solid materials are shaped and formed to become structures and products. They may become frames for buildings. Solids may become structures of products. Typical solids used in technology are metal sheets and bars, glass sheets, lumber, and plastic pellets and sheets.

Technology and Machines

Technology requires *machines* to shape materials, process information, convert energy, and transport cargo. See **Figure 2-9.** Machines are used to construct roads, produce bread, and broadcast television programs. They are used to carry people and cargo, make furniture, and fabricate computers. Machines extend humans' abilities to do work. They help us grow and harvest crops, heal illnesses, and manufacture products. Machines help us

Figure 2-9. This ship is an example of a technological machine.

TechnoFact: The first toolmakers were traveling groups of people who utilized the pointed edges of rocks as tools.

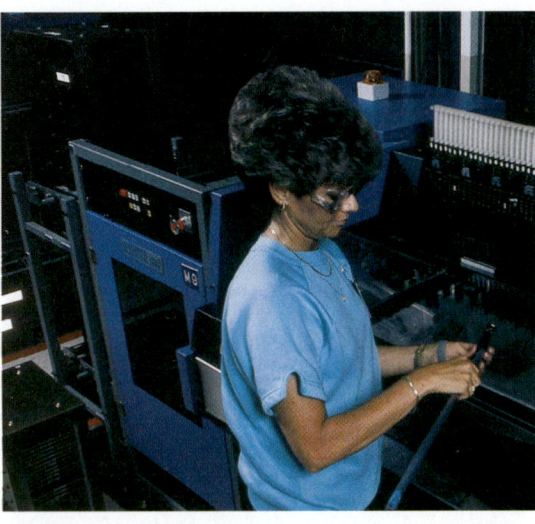
Figure 2-10. This person is applying production knowledge to operate an injection molding machine. (AT&T Network Systems)

transport goods, convey messages, and build buildings.

Technology and Knowledge

Without *knowledge,* we would not have technology. People apply their knowledge about materials and processes to design and make things. They use knowledge about using tools to create technology. People use this knowledge to use technology. They use design knowledge to create products and systems. People apply production knowledge to use machines to make these products from materials. See **Figure 2-10.** They use knowledge to select, use, service, and repair devices. People use knowledge to measure or predict the impacts technology has on other people and the environment.

Technology and Energy

All technological systems are designed to do a task. Most of these applications require *energy.* Energy is needed to make furniture from lumber. It is needed to broadcast television programs. Energy is required to harvest grain, build a road, and haul gravel to a building site. Technological systems could not operate without energy.

This energy may come from inexhaustible sources, such as wind and water. It is also available from renewable sources, such as wood and corn. Energy is found in exhaustible sources, such as coal, petroleum, and natural gas, as well.

Technology and Financial Resources

People design technological systems. They also build these systems. To do this, materials must be located and purchased. Machines and buildings must be bought or leased. Energy supplies must be obtained. People must be employed. Patents must be sought. All of these activities require finances. It takes money before a technological system can be created, operated, and maintained. See **Figure 2-11.**

Technology and Time

It takes *time* to move a creative idea from inside a designer's mind into a product. Time is needed to design and engineer the product or system and to develop a production facility to make the device. It is needed

TechnoFact: The idea that science provides the ideas for technological innovations is a myth. Many great technological developments cannot be traced back to the science laboratory. Important inventions were made before the science behind them was understood. For example, the steam engine was in widespread use before the science of thermodynamics could explain its basic principles.

Standards for Technological Knowledge

Resource
Transparency 2-3:
Processes of
Technological
Systems

Tech Terms
Production—
Changing the form
of resources to
increase their value.
Management—
Planning, directing,
and controlling pro-
ductive activities.

Figure 2-11. Labor, materials, and machines cost money to obtain. They also require financial resources to use. (FMC Corporation)

to build and test the product or system, to ship products to stores, and to install new products. Time is a key resource in developing and operating technological systems. It is measured in machine time, work hours, product life, or service intervals.

System Processes

The understanding of how a system works is vital if one is to operate and maintain it successfully. Technological systems are designed to be operated and maintained in order to achieve a given purpose and create a desired output. Producing these outputs involves processes. See **Figure 2-12.** Different technologies entail different sets of processes. Typically, technological systems use two types:

➤ Production processes.
➤ Management processes.

These two types of processes work together to meet the system's goals, such as transporting cargo, growing crops, communicating information, or manufacturing products.

TechnoFact: An expert system is a computer system or program that makes use of artificial intelligence systems to resolve problems usually demanding an experienced person. Artificial intelligence employs computers to develop human logic and knowledge.

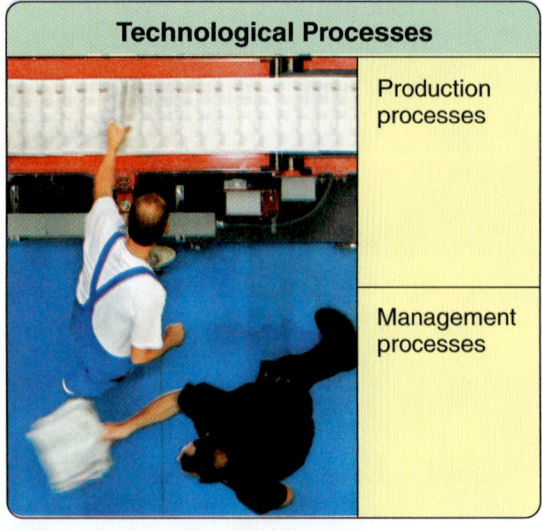

Technological Processes

Production processes

Management processes

Figure 2-12. Technological systems involve production and management processes. This state-of-the art printing press requires both types to function smoothly. (Gannett Co.)

Production Processes

Production processes are actions changing inputs into outputs. See **Figure 2-13.** These processes include a number of actions:

➤ Growing and harvesting crops (agricultural technology).

➤ Changing materials into products (manufacturing technology).

➤ Changing materials and manufactured products into constructed works (construction).

➤ Converting information and ideas into printed or broadcast messages (communication technology).

➤ Using devices to improve health and fight diseases (medical technology).

➤ Converting energy and applying it to do work (power and energy technology).

➤ Using technical means to move people and cargo (transportation technology).

Production processes are at the core of all technological systems. They are the activities in which people use

Inputs

Processes

Outputs

Figure 2-13. Production processes can change materials into products and buildings.

TechnoFact: The farmer who produces wheat, the miller who crushes the wheat into flour, and the baker who changes flour into bread are illustrations of manufacturers who add value through their production processes.

Figure 2-14. Technological systems produce products we need and want (left). They also produce other outputs, such as pollution (right). (Goodyear Tire and Rubber Company, American Petroleum Institute)

Standards for Technological Knowledge

2 **5** **7**

12

Resource
Transparency 2-4: Outputs of Technological Systems

knowledge, machines, and materials to create objects and systems.

Management Processes

Production processes do not run by themselves. People must oversee their operations. This involves four key actions:

➤ *Planning* for the use of the production processes and a course of action; setting goals.

➤ *Organizing* the resources: material, labor, energy, information, and machines; seeing that the right resources are available when the processes are operated.

➤ *Directing* and motivating the workers operating the systems; seeing that they know how to do their jobs and want to do them well.

➤ *Controlling* the operations of the systems; seeing that the outputs meet the goals of the operations.

These four actions are called *management*. The people carrying them out are called *managers*.

System Outputs

Nearly all technological systems have two types of *outputs*. See **Figure 2-14.** The first type of outputs is the one for which the system was designed. These are the *intended outputs.* Typical outputs include prepared or preserved food, constructed structures, and communicated messages. They also include transported goods and people, manufactured products, and medicines. Each of these outputs is considered good. It is intended to help people live better or longer lives, be more informed, or have more comfort. These outputs were produced in response to human needs and wants.

Tech Terms
Industrial Management—The highly organized methods of planning, directing, and controlling industrial operations.

Brainstorm
Ask the students how they would plan, organize, direct, and control candy sales for a club.

TechnoFact: The first major effort in improving production efficiency was made by Frederick Taylor. In the 1880s, he conducted a series of studies to determine whether workers used unnecessary motions in completing their jobs. From these studies, he reorganized the job and workplace to reduce the time needed to perform operations at various machines. For this pioneering work, Taylor has been called the father of scientific management.

Brainstorm
Have the students list some intended outputs for each type of technology.

Tech Terms
Intended Output—
The product, structure, or system a technological system was designed to produce.

Figure 2-15. The aluminum cans (left) can be used to make the extruded aluminum shapes (right). This process of using scrap to make new products is called recycling. (Alcoa)

These intended outputs can be grouped into products and services. *Products* are things like refrigerators, sweaters, and houses. Technological services include activities such as moving people on transportation systems and providing information through communication systems.

Most technological systems, however, produce other outputs. These are in addition to the desired or intended outputs. These ***undesirable outputs*** fall into two major categories:

➤ Scrap and waste.

➤ Pollution.

Scrap and waste are the materials left over after the production processes are completed. Many of these materials will not harm the environment. They are things like pieces of metal, wood, plastics, and ceramics. They may include empty shipping containers, shavings and sawdust, unused exposed film, and nutshells and grain hulls.

Scrap is material that can be used for other purposes. Small pieces of metal, wood, and plastics may be used to make other products. See **Figure 2-15.** Sawdust may be compressed into wood products. Wood chips can be used to make paper. Cardboard shipping containers may be recycled into new corrugated containers.

Waste cannot be easily reused. It must be discarded. Waste generally ends up in landfills. Examples of waste are unusable ceramic items, small pieces of wood, and some plastic materials.

Scrap and waste cannot be avoided. Cutting round parts out of sheet material will always produce some scrap or waste. Production problems will create scrap ceramic materials. Machining materials will create chips, shavings, and dust. The challenge is to reduce scrap and waste to the absolute minimum, and this is an important societal issue. Recycling materials has

TechnoFact: Recycling is a process used to recover and reuse waste products from household use, manufacturing, agriculture, and business. It became important during World War I and World War II, as shortages of essential materials led to collection drives for silk, rubber, steel cans, and other commodities. In recent years, concern over damage to the environment caused by solid waste has renewed the focus on recycling.

decreased the waste sent to landfills. This has reduced the need for new dumping grounds. The management of waste requires careful planning and effective production controls.

Pollution is output harming the environment. (The environment is the air, water, and soil around us.) Typically we talk about air, water, soil, and noise pollution. Chemicals from production processes cause air, soil, and water pollution. Smoke from processing activities causes pollution. Power plants also cause pollution. Vehicle exhausts are another major source of pollution. Unwise forestry and farming practices can cause water pollution. Likewise, excessive use of pesticides (bug killers) and herbicides (weed killers) by farmers and homeowners is a major source of soil and water pollution. Unwise actions by consumers and service personnel can also cause pollution. Loud activities, such as landing aircraft at airports and driving vehicles with poor mufflers, cause noise pollution. Loud radios and noisy machines are other sources of noise pollution.

A new type of pollution is visual pollution. Structures reducing the beauty of nature cause visual pollution. Unsightly buildings, highway billboards, and wind farms (wind powered electric generators) are examples of things some people think are unsightly. They can cause visual pollution.

Again, many technological activities create outputs that can cause pollution. The challenge is to reduce these emissions and to control those remaining. See **Figure 2-16.** Installing scrubbers on coal-fired power plants

Standards for Technological Knowledge

5

Tech Terms
Solid Waste— Discarded materials other than fluids.
Scrap—Materials left from production processes that can be used for another purpose.
Waste—Materials left from production processes that cannot be used easily for other purposes.

Brainstorm
Have the students look at a typical packaged product. Ask them how they could redesign the product and its packaging so solid waste is reduced.

Figure 2-16. These workers are testing refinery wastewater to ensure it does not pollute local water supplies. (American Petroleum Institute)

TechnoFact: In the United States, over 200 million tons of solid waste is generated each year. This amounts to over four pounds per person per day. Municipal garbage includes over 12 million tons of glass and 80 million tons of paper and paperboard each year.

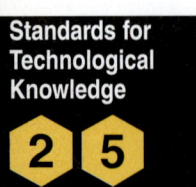
reduces the nitrogen oxide emissions that cause acid rain. Better exhaust systems reduce vehicle emission pollution. Quieter jet engines reduce noise pollution near airports. Laws regulating billboard numbers and placement reduce visual pollution. These are four examples of the many ways to reduce pollution. They represent a start on a challenging path all people must travel. Individuals, households, and businesses must work together to reduce pollution.

System Feedback

The outputs of technological systems should meet the intended uses of the systems. To ensure this happens, systems have control features that cause them to change if necessary. Controls are mechanisms or specific actions people take using information about the system. Control mechanisms compare information about what is happening to what is desired, and then they adjust the systems to make the desired outcomes more probable. Control systems are often also called *feedback systems.*

There are two types of feedback systems. The first is called a *closed-loop system.* This type of control uses feedback and is a built-in part of the device. Components within the device monitor conditions. They also adjust the device to produce the desired output. The heating system discussed earlier is an example of a closed-loop system. See **Figure 2-17.** The thermostat measures the room temperature. It turns on the furnace when the temperature is too low. When the proper temperature is reached, it turns off the heating system.

The second type of system is called an *open-loop system.* See **Figure 2-18.** This system does not have a built-in feedback or control path. Instead, external control—human intervention—is needed. An example of an open-loop system is the speed control system in an automobile. The driver presses on the accelerator pedal. This causes the engine to produce power to move the

Figure 2-17. This is an example of a closed-loop control or feedback system.

TechnoFact: Over one-fourth of solid waste in the United States is recycled. Products that are recycled in large quantities include paper and paperboard, ferrous and nonferrous metals, plastics, glass, and yard waste.

Figure 2-18. This is an example of an open-loop control system.

Standards for Technological Knowledge

Brainstorm
Ask the students what positive contributions they would make if they started a solid-waste recycling activity in your school.

Extend
Ask the students to describe how container shipping promotes global trade.

vehicle. As the vehicle moves, its speed is shown on the speedometer. The driver adjusts the speed by changing the throttle setting.

Technology and Goals

Technological systems have one primary goal. This is to *meet the needs and wants of people.* Without this goal, any technological activity is doomed to failure.

Many technological actions, however, are carried out as part of the economic system. Businesses are involved in performing the actions. Companies and individual proprietors are involved in the systems. They transport people, build buildings, deliver medical services, raise crops, generate electricity, and make products. These companies and individuals expect to be paid for their efforts. They expect to *make a profit.* Profits, the second goal of technology, are the rewards for creative ideas, personal labor, and financial risks.

A third goal of technological activity is to *make a positive contribution to society.* This means we expect that all technological activities will help make life better. We use them to help people in their various life pursuits. The systems should contribute to a clean environment. They should also promote community and international goals.

Summary

Technology is the knowledge of efficient actions involving tools and materials. It is using technical means to extend the human potential. It allows us to do more with less effort. With technology, we can see well. We can hear more clearly. Technology helps us maintain our health. It helps us grow more food. Technological systems directly impact every person.

Technological systems are comprised of parts working together to complete a task. They have inputs that are processed into intended outputs. Also, technological systems produce undesirable and unintended outputs. These outputs include scrap, waste, and pollution. Many technological systems use feedback for control.

These systems are developed to meet personal and societal goals. They are intended to make life better and more comfortable. These systems should improve our lives, while protecting the environment.

TechnoFact: Technological systems are human-built systems that allow people to do many things. A typical system is a transportation system that moves people on land and water and through the air. One example is a subway system. Below is a list of major city subway systems, the year completed, and the size in miles.

London, 1863, 258
New York City, 1904, 230
Moscow, 1935, 211
Tokyo, 1927, 175

Seoul, 1974, 173
Paris, 1900, 131
Mexico City, 1969, 125

Technology Explained

container shipping: Using sealed containers to group and contain items for bulk shipment.

Shipping cargo across oceans has become a commonplace activity. In earlier times, each crate or box was lifted individually onto a ship and stored in the hold. This technique was time-consuming and expensive. It tied the ship up in port for a number of days at the end of each trip.

To increase the efficiency of ocean shipping, a new type of ship was needed. To meet this need, the container ship was developed in the 1960s. This type of ship uses steel containers 20´ or 40´ (6 or 12 m) long, 8.5´ (2.6 m) high, and 8´ (2.4 m) wide. Special high cube containers are the same length and width. They are, however, 9.5´ tall. Each container can hold up to 30 tons (27 metric tons) of cargo.

Container shipping works by grouping a number of small shipments going to one place into one large load. The grouping of shipments by destination is called *consolidation*. Each item is handled only twice: once when it is placed in the container and again

1. Loading the containers

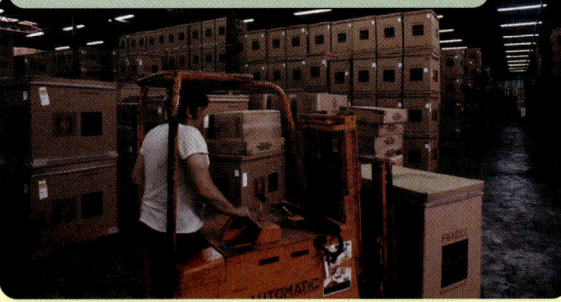

3. The containers arriving at the port

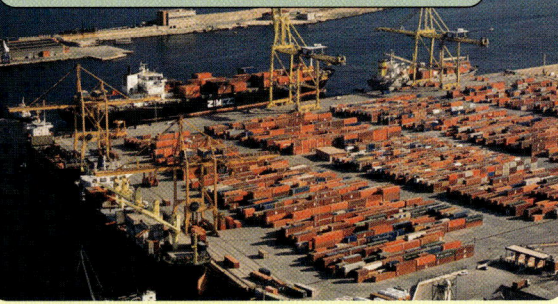

2. Moving the containers to the port

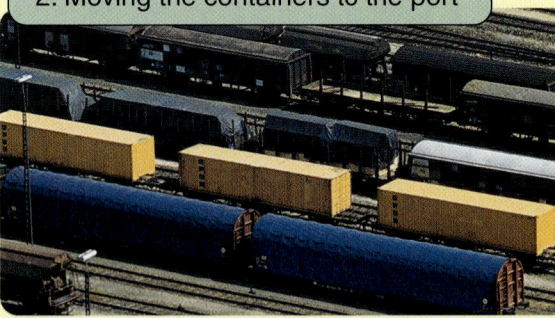

4. Loading the containers onto a ship

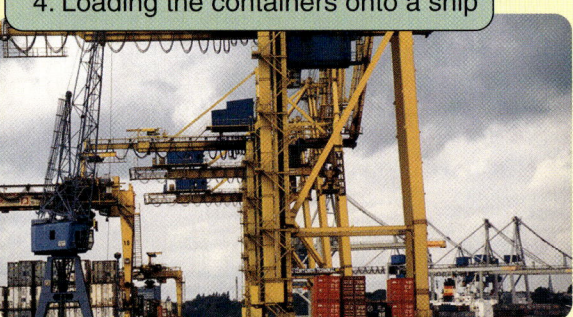

Figure A. These are the steps used in container shipping by sea. Once the ship reaches its destination port, these steps are reversed. The containers are unloaded from the ship and then moved to their individual destinations.

TechnoFact: Profit is the monetary return on capital invested in a company. It is determined by subtracting the costs of obtaining and managing land, materials, labor, and capital from the income generated by the company's operations.

when it is removed from the container.

Container shipping has several steps, **Figure A.** First, the cargo is picked up from the shipper. A number of different shipments destined for the same location are put together. These shipments are loaded into a container. The container is then transported to the dock. Containers can be placed on a set of wheels to form a semitrailer, which is then pulled by a truck. The truck can pull the container to the dock. Trucks sometimes haul containers to rail yards.

There, the containers are lifted off the set of wheels and set onto railcars. They make the rest of their journey by rail.

Often, entire trains are made up of railcars loaded with containers. The train moves the containers to the port, where they are stored in large stacks. When the ship comes to port, the containers are moved to the dock. Large cranes lift them and place them on the ship. Using this process, dockworkers can load an entire ship in less than a day. Using the older loading techniques, it could take up to five days to unload a ship and another five days to load it.

In addition to being faster, container shipments do not require warehousing at dock sites. Each container is a small, weatherproof "warehouse." Container shipping is now being adapted for air transportation, **Figure B.** The development of wide-body airplanes allows the use of large containers. Container shipping is popular with overnight parcel companies.

Tech Terms
Technology—The knowledge of efficient actions involving tools and materials.
System—Parts working together to complete a task.

Figure B. Wide-body aircraft, such as this 747, allow for containers to be shipped by air.

Curricular Connections

All Subjects

Use current magazines or newspapers to identify as many systems as you can. Your list can include natural and human-designed systems.

Social Studies

Describe how systems are used in governing society. Explain how these systems have changed over time. Describe how technology (such as global communication, military weapons, and rapid transportation) has impacted these systems.

Science

Select a natural system, such as weather, lake ecology, or the human body. Describe the system in terms of its inputs, processes, outputs, and feedback.

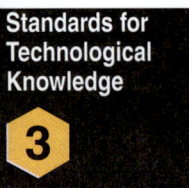

Standards for
Technological
Knowledge

3

Activities

1. List all the ways pollution is created in your (a) school and (b) home.
2. Prepare a poster that would help elementary school students understand the goals, inputs, processes, outputs, and feedback of technological systems.
3. Collect a number of pictures showing technological processes. Label the types of processes shown.

TechnoFact: A trend of the late twentieth century was the greater use of computerized production process in integrated machinery and to replace human labor.

Test Your Knowledge

Do not write in this book. Place your answers to this test on a separate sheet of paper.

1. Select the statement best defining a *system:*
   A. Natural events working without human intervention.
   B. Products people designed for their use.
   C. A groups of parts working together to complete a task.
   D. Tools and machines doing a task.
2. Summarize the differences between natural and technological systems.
3. List the five parts of a system. Describe them.
4. List the inputs to a technological system.
5. What are the two types of processes used in technological systems?
6. People who plan, organize, direct, and control production processes are called _____.
7. Prepared foods, buildings, and manufactured goods are examples of _____ outputs.
8. Scrap and pollution are examples of _____ outputs.
9. How is a closed-loop system different from an open-loop system?
10. Paraphrase the three major goals of a technological system.

Answers to *Test Your Knowledge* Questions

1. C. A group of parts working together to complete a task.

2. Natural systems appear in nature without human interference, while technological systems are human-made.

3. Goals, inputs, processes, outputs, and feedback. Goals are the reasons for developing and operating the system. Inputs are the resources the system uses. Processes are the actions taken. Outputs are the results obtained. Feedback includes the adjustments made.

4. People, natural resources or materials, tools or machines, knowledge or information, energy, financial resources or money, and time.

5. Production and management.

6. managers

7. Intended

8. undesirable

9. A closed-loop system is a built-in part of the device, while external control is needed in an open-loop system.

10. Meet human needs and wants, make a profit, and positively contribute to society.

Chapter 3
Contexts of Technology

Did You Know?

Early humans built the Seven Wonders of the World:

➤ Pyramids of Egypt. The largest, near Giza, is thought to have been built from 2200 to 1100 B.C. The actual dates are unknown.

➤ Babylon. This city was built about 600 B.C. and had hanging gardens, high walls, and the Temple of Jupiter Belus. It covered about 225 square miles. Babylon had walls 350 feet tall and over 85 feet thick.

➤ Temple of Diana at Ephesus. This temple was built in 544 B.C. It was about 425 feet long and 200 feet high. The temple had over 125 marble columns that were 60 feet tall. It burned in 356 B.C., was rebuilt over two hundred years, and burned again in 260 A.D.

➤ Statue of Jupiter Olympus. This gold and ivory statue was built in 433 B.C. The seated figure was nearly 60 feet tall.

➤ Mausoleum at Halicarnassus. Artemisia built this burial place for her dead husband about 350 B.C.

➤ The Colossus at Rhodes. The brass statue of Apollo was built about 290 B.C. It was about 105 feet tall. An earthquake collapsed this statue.

➤ Pharos at Alexandria. This was a 550-foot tall white marble lighthouse built about 280 B.C. Fires constantly burning at its top produced the light.

Objectives

The information given in the chapter will help you do the following:

➤ Identify the major contexts in which technology can be viewed.

➤ Recognize the major ages in human history.

➤ Give examples of the positive and negative impacts technology has on society.

➤ List the seven common functional areas of technology.

Key Words

These words are used in this chapter. Do you know what they mean?

agricultural technology
Bronze Age
communication and information technology
construction technology
energy and power technology
First World country
Industrial Age
Information Age
Iron Age
manufacturing technology
medical technology
Stone Age
Third World country
transportation technology

Chapter Outline
Did You Know?
Objectives
Key Words
Historical Context
　The Stone Age—Up to about 3500 B.C.
　The Bronze Age—About 3500 to 1000 B.C.
　The Iron Age—1000 to 500 B.C.
　The Industrial Age—1750 to the Late 1900s
　The Information Age—Late 1900s to Present Day
Societal Context
　Scope of Society
　Technology and the Individual
　Technology and Society
Functional Context
Summary
Curricular Connections
Activities
Test Your Knowledge

Technology, as a word, first appeared in English in the 1600s. It was used to describe discussions about the applied arts. In the twentieth century, the term was expanded to cover a broad range of tools, ideas, and processes. To understand technology, people study it in a number of contexts.

People study technology in its *historical context.* They explore how people developed and used technology in times past. These scholars may be historians. They may be anthropologists. Other people study technology as it affects people and society. They examine technology's impacts on jobs, communities, and governments. These people look at its *societal context.* They may be political and social scientists. Still other people investigate technology as it is used in various areas of activity, such as transportation, agriculture, or manufacturing. They explore the application of technological knowledge and skills. These people explore technology's *functional contexts.* They could be engineers, economists, or business managers.

Historical Context

Technology has always been with humans, even though it may not have had that name. It is the process of creating new tools and products to meet the needs of people. Other creatures make things. Beavers build dams to create ponds and lakes within streams. Bees build complex hives to store honey and house the queen bee. Birds build nests to lay eggs. These processes are driven by natural instincts.

Humans, however, do not have these instincts. Instead, they have the ability to think systematically. Humans can plan actions before they take them. Also, people have creativity, which has resulted in innovation. See **Figure 3-1.** Technology is directly related to creativity. Perceived needs and wants inspire most inventions, but some inventions are linked to creative ideas and the ways people use them. Inventions can always be improved. The ability to try new ideas is often key to that improvement. People can create, adapt, and

Figure 3-1. People throughout the ages have used their creativity to help them live more successfully in the world. This model shows an early Chinese cantilever bridge built in the 1600s. The bridge was made of wood. It had a span of over 110'.

TechnoFact: History is the story of past events and activities. It is often a recording of the past as found in books, newspapers, printed documents, personal papers, objects (artifacts), and oral accounts. It helps people understand why certain events happened as they did and how they are linked to earlier events.

Figure 3-2. Early humans used stone tools for hunting and other tasks.

explore. They can modify their environments to better meet their needs. These abilities are the basis for technology.

The development of technology has taken place over a long period of time. It was slow at first. The main force driving the development of technology was a realization of basic social needs. People needed better food supplies, shelter, and clothing. As time passed, the development of technology grew more rapidly.

The Stone Age—Up to about 3500 B.C.

Humans have been technologists throughout history. In the past, however, inventions and innovations were not usually developed with the knowledge of science. Humans have always been toolmakers and tool users. Early humans lived in small nomadic groups. They traveled and used tools to gather food, such as berries, fruit, and roots. These humans also traveled and used basic technology to hunt game and fish.

Humans used their creativity to make tools. They used the tools to help in tasks, such as hunting and gathering food. Early in history, humans mastered the ability to make tools from stone. See **Figure 3-2.** They used shaped stones to form arrows, harpoons, and spears. The simplest way humans made tools was by finding properly shaped stones. Another way

TechnoFact: Human ancestors living before the Stone Age most likely used natural objects, such as tree limbs, animal bones, and rocks, as tools. This use of "natural tools" is found today among the higher primates, including the chimpanzees. As history progressed, humans in different parts of the world began making and using stone tools. This activity happened at different times in different places. Also, humans abandoned stone for metal tools at different times in different places. Today, some cultures are not far from the Stone Age, in terms of their toolmaking and tool usage.

Standards for Technological Knowledge

1   7

Tech Terms
Bronze Age—A period in the development of technology when metals were first used regularly in making tools and weapons.

Demonstrate
Hammer a piece of copper foil into the shape of a bowl.

Figure 3-3. This arrowhead was found in central Indiana. Early Native Americans produced it.

was by breaking stones to get usable pieces. More advanced toolmakers chipped away parts of a stone to make the desired shape. See **Figure 3-3.** This process was called *flaking*. The use of stones as the main material for tools led historians to call this stage of human history the *Stone Age.* It began roughly 2.5 million years ago. The Stone Age ended in some parts of the world about fifty-five hundred years ago.

During this period, people may have lived in caves. Most likely, however, they built simple shelters. Their bodies, like ours, were not well protected from the weather. People made simple clothing from animal skins and other natural materials. Also, they mastered the use of fire, which was a major step in human development.

The Bronze Age—About 3500 to 1000 B.C.

As humans progressed, they learned that some plants grow from seeds. Humans started to plant these seeds. They harvested the resulting crops. With these actions, farming was developed. We are not sure why people needed to make this advancement. Some scientists suggest population growth outgrew natural sources of food. Others suggest environmental changes created the need to develop agriculture.

Regardless of the cause, farming allowed people to remain in one place. Villages and towns started to appear. The first urban revolution was born. The roots of the modern city were developed. This achievement took place in several locations, like Egypt, Central America, and Syria.

During this time, people learned to grind grain, spin yarn, and weave fabric. They started making clay storage and cooking vessels. Hammering shaped soft metals, such as copper, silver, and gold. Later, ores (metal-bearing rocks) were heated (smelted) to obtain the pure metals. Copper and tin or lead were combined to make bronze (an alloy). Metal tools replaced stone ones. This stage of human history is called the *Bronze Age.* It started in an area near Thailand as early as 4500 B.C. The Bronze Age was widely seen in the Western world by 3000 B.C.

Irrigation systems were developed during this period. This advancement allowed farming to increase the food supply. Cities grew. Building techniques

TechnoFact: The mining, smelting, and casting operations needed to make copper and bronze tools required the specialization of labor. This, in turn, required the production of surplus food to support people to search for raw materials and artisans to make the products.

Figure 3-4. Many construction techniques were developed during the Bronze Age. The Egyptian pyramids are examples of these techniques.

Standards for Technological Knowledge

1

Tech Terms
Iron Age—A period in the development of industry that began with the general use of iron and continues into modern times.

Research
Have the students research how steel is made.

developed. See **Figure 3-4.** People moved goods on water, using simple boats with oars or sails. Pack animals were used on land. Carts with wheels appeared for the first time. Fired clay bricks were developed. They supplemented building stones.

The Iron Age—1000 to 500 B.C.

Iron ore had been available to the developing cultures for a long period of time. People could not, however, get the metal from the rocks. They could not develop the required temperature of about 2,800°F. The development of better furnaces changed this. People learned to smelt iron around 1000 B.C. By 500 B.C., the metal was widely available in the Western world. It introduced a period called the *Iron Age.* The actual time frame for this age varies by culture. The use of iron as a primary toolmaking material did not impact all areas at the same time.

The new metal led to better agriculture, with the development of the iron-tipped plowshare. Cooking ware and weapons improved. During this period, new building techniques developed. See **Figure 3-5.** Fired clay bricks and tiles were widely produced. They came into widespread use. The traditional building materials, such as limestone and marble, also became

TechnoFact: Iron beads, made from meteoric iron, were produced in Egypt about six thousand years ago. However, the first known iron object shaped by hammering is an Egyptian dagger made more than thirty-three hundred years ago.

Figure 3-5. The Roman Colosseum was first used in 80 A.D., after eight years of construction. It rises 165' above the streets. The Colosseum is 610' long and 515' wide.

widely used. Large buildings, temples, and monuments appeared.

Also, ships were improved with new types of sails and a keel. Boats became oceangoing vessels. Carefully laid out road systems were developed to connect major cities. See **Figure 3-6.**

The Industrial Age—1750 to the Late 1900s

The use of iron tools led to rapid advancements in technology. Village crafts grew in number. Many villages had craftsmen who worked with leather and metal. Other trades included making rope, barrels, candles, and soap. Demand for the products of these workers grew.

New machines and sources of power were developed to support the industry of the day. The steam engine was invented. With time, it was perfected. Metal cutting and shaping machines were developed. These and other developments led to a period called the *Industrial Revolution* or *Industrial Age.* It was a period in

which most Western countries changed from rural to urban. The primary focus moved from farming to manufacturing.

The Industrial Revolution impacted different cultures at different times. It is generally believed to have started in Great Britain about 1750. The Industrial Age began in the United States around 1850. It traveled to Japan and Russia during the first half of the twentieth

Figure 3-6. This is a view of a street in Pompeii. An eruption of the volcano Mount Vesuvius destroyed it in 79 A.D. Notice the sidewalks, cobblestone streets, and remains of the building.

TechnoFact: Mass production involves assembling products from standardized, interchangeable parts by a number of workers, each doing a portion of the total assembly job.

century. Many developing (Third World) countries are just now having their industrial revolutions.

The Industrial Revolution was primarily born in the textile industry of England. Government support helped this process. The first textile factories started to appear in 1740. Over the next one hundred years, factory-made cotton garments replaced the woolen garments the British wore. The factory system in England aided this change. Cheap cotton from the American colonies also played a part in this shift. The invention of the cotton gin by Eli Whitney in 1793 greatly aided this development.

The factory concept quickly spread to other activities. See **Figure 3-7.** Steel mills, clay tile factories, and other plants soon appeared. Along with new factories came new transportation and communication technologies. The railroad and automobile were developed. Telegraph, telephone, and radio (wireless) communications were developed. Printing became common with the advent of moveable type and the printing press.

A key to the Industrial Revolution was the variety of new power sources. Reliance on waterpower disappeared with the inventions of the steam engine and electric motor. Later, the steam engine gave way to gasoline and diesel engines. Homes were wired for electricity. Natural gas and heating oil provided fuel for heating systems.

Standards for Technological Knowledge

1

Research
Have the students research an invention that contributed to the Industrial Revolution.

Tech Terms
Industrial Revolution—
The changes that marked the transition from an agricultural/handicraft society, based on simple tools, to an industrial society, based on complex machines.

Brainstorm
Ask the students how life changed as society moved into the Industrial Age.

Figure 3-7. The factory system was developed to produce textiles. It soon spread to many other industries. This shows a continuous production process producing glass containers. (Owens-Brockway)

TechnoFact: Many historians trace the start of mass production to Eli Whitney, who in the 1800s, organized a firearms factory using the labor process of modern industrial production.

Figure 3-8. Modern farm machinery allows more food to be raised by fewer people. People who were once needed to raise food can now move to the cities. There they can work in factories or do other work. (Deere and Company)

Cities grew. Farms became mechanized. See **Figure 3-8.** Tractors, combines, row cultivators, and other implements made farming more efficient. Skyscrapers were developed. Also, for the first time, large numbers of people worked for other people. The factory system developed division of labor. Owners were not laborers like in the older village crafts. A system of managers and workers developed. With it, labor unions were born.

The Information Age— Late 1900s to Present Day

The twentieth century saw rapid changes in technology. Many countries became industrial powers with the development of assembly lines and mass production. Companies made conscious efforts to innovate. They worked hard to improve their production processes. A new material called *plastic* came into use. Synthetic fibers were developed. Products became plentiful. A large middle-class with money to spend developed.

The computer was developed in 1944. The transistor was invented in 1947. It was late in the century, however, that the real impacts of these technologies were felt. Then, the microchip and personal computer were developed. With these devices came a revolution as important as the Industrial Revolution. It was the *Information Revolution.* With it, the *Information Age* was born.

TechnoFact: With the establishment of the Bank of England in 1694, the foundation for capitalism was laid. With it, a new type of commercial entrepreneur was developed to supplement the old merchant class. These new business owners developed factories to produce products, using machines already in existence.

Again, the focus of societies changed. Economies built on manufacturing started to focus on information processing. Manufacturing did not disappear. The ways products were manufactured, however, changed. Automation entered the workplace in many areas. See **Figure 3-9.** Computers, rather than people, directly controlled many machines.

New and larger farm machines were developed. They allowed farmers to work greater acreage. The number of people involved in agriculture continued to drop. Travel increased with the jet aircraft and high-speed rail systems. The use of the automobile for personal transportation increased. Instant communication appeared with the communication satellite and cellular (cell) phone. People became more closely tied to technology. Therefore, some people call the period we are living in the *Technological Age.*

Societal Context

Technology has always had an impact on society. Likewise, society and its expectations have impacted technology. People have different views because of the technology they use. Also, social and cultural priorities

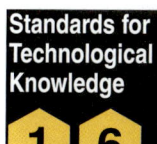

Research
Have the students explore important inventions that contributed to the Information Age.

Tech Terms
Information—
(1) The knowledge communicated or received concerning a particular fact, event, or circumstance. (2) The knowledge gained through observation, research, study, communication, investigation, or experimentation.

Figure 3-9. Automation and computer process control are parts of the Information Age. This woman is working in a concentrator control room. (Hibbing Taconite Company)

TechnoFact: A major icon of the Information Age is Bill Gates, who was born in 1955. In 1974, he and Paul Allen founded the Microsoft Corporation as a computer software firm. In 1980, they developed the operating system, called MS-DOS (Microsoft Disk Operating System), for the personal computer developed by International Business Machines (IBM). Later, Microsoft was responsible for many new software programs, including the Windows operating systems and the Office Suite (Word, Excel, and PowerPoint). These programs have made Microsoft the world's largest producer of microcomputer software.

Tech Terms
Low Technology— Any technology that uses relatively simple equipment and production techniques.
High Technology— Any technology that uses the most advanced equipment and engineering techniques.

and values dictate which technology devices are developed and used.

Three major societal factors affect technology. First, there must be a *societal need* for the technology. See **Figure 3-10.** People develop technologies to serve their needs. All through history, the demands, values, and interests of individuals, businesses, industries, and societies have resulted in new technologies. Without the need, there would be no reason for developing the device or system.

Second, the proper *resources* must be available. The society must have capital (money), labor (workers), and

materials available. The people must possess the knowledge needed to develop and use the technology. Proper energy resources must be available. Creative ideas for new technologies will die without proper resources.

Third, the society must *value* the technology. Meeting societal expectations is the major influence behind the acceptance and use of products and systems. The people must feel the technology will benefit them and agree with their economic, political, cultural, and environmental concerns. They must be receptive to new ideas, devices, and ways of doing things.

Figure 3-10. Technology meets societal needs, such as housing, transportation, and useful products.

TechnoFact: Not all technology is developed in response to a human need or want. Often inventors create things for the fun of it—they like to tinker. The products of their efforts may not have any apparent immediate use. For example, the ideas for lasers were developed in 1917, but were not applied to successful products until 1960.

Scope of Society

In ancient times, the society meant the tribe. People lived in close-knit groups. They traveled little. People seldom received news from other groups. In more recent times, people have developed strong national identities. They consider themselves citizens of a nation. Present times require a broader view. Rapid communication and transportation systems have made us citizens of the world. The events in one nation or region impact all other regions. Pollution, hunger, and war ignore national boundaries. We now live in a *global society*, due in a large part, to technology. Understanding technology in terms of its impacts on people and societies requires several views. It requires looking at technology as it relates to individuals, communities, nations, and the world.

Technology and the Individual

Technology impacts people in several ways. It changes the way they live and view their world. The use of technology influences people in many aspects, affecting things such as their well-being, comfort, choices, and opinions about technology's development and use. People's knowledge about and attitudes toward the uses of technology vary greatly. Their moral, social, and political beliefs often influence how they feel about technological developments.

In early history, people were satisfied with some simple technology. They were happy when they had a few tools to hunt with, a small shelter, and simple clothes. The use of inventions and innovations led to changes in society. Later generations needed more material things to be happy. They wanted more products of technology to help them. As these more recent generations acquired new technology, their needs and wants changed even more. The development of technology sometimes creates the demand for new technology.

Let's consider communication as a means of exchanging information and ideas. Once, simple writing was enough. Hieroglyphics (ancient Egyptian writing) on walls and buildings were considered adequate. Then paper, ink, and pens were invented (technological advancements). People could carry their written message with them. They started to expect more information. The few educated people started to exchange information freely. Knowledge passed from one generation to another more easily. People could exchange information with those in other areas.

In the mid-1400s moveable type and the printing press were developed. Printed books and newspapers became available. More people could receive printed messages. Information could be the property of all people. This encouraged more people to learn to read. Technology changed their views of basic education. It also changed the way they thought about their world. Knowledge started to become power.

Later, the telegraph, telephone, and radio were developed. The age of telecommunications was born. Television,

Tech Terms
Value—The relative worth, importance, or merit of an idea, object, or belief.

Brainstorm
Ask the students why technology is worth something to, or valued by, people.

TechnoFact: People have banded together in communities of like interest, which may be for patriotic, cultural, religious, political, economical, or security purposes. When these communities become fairly large, they are called *societies*.

Brainstorm
Ask the students how the development of electronic communication media has changed people's expectations about information.

Research
Have the students research a major development in communication.

Figure 3-11. Communication has changed from simple writing and drawings (left) to computer-based systems (right).

communication satellites, the facsimile (fax) machine, and the cell phone followed. People today are not satisfied with only the written word. In fact, large numbers of people read very little. Technology has allowed them to avoid reading. Many people have become passive information gatherers. They depend on electronic media (radio and television) as sources of information. Many people feel they must be in constant contact with their friends and business associates. They use the cell phone, electronic mail (e-mail), and the Internet many hours a day.

Technology has changed our views about communication. See **Figure 3-11.** We no longer want occasional sources of information. People want immediate and constant contact with others. Likewise, technology has changed our views about transportation. We want quick, flexible, and inexpensive systems. Our views of products have also changed. We expect a wide selection of easy-to-use devices. People expect buildings to be heated in the winter and cooled in the summer. We want comfortable levels of light. These and other examples indicate that our expectations have changed. Because of these expectations, new technological products and systems are developed.

Technology and Society

Society can be viewed from several perspectives. Young children view their families as society. As they grow, their perspectives widen. Children start seeing their communities as society. Later, they may see the nation as society. The most developed view sees us as part of a global society. See **Figure 3-12.**

Global society means everyone on Earth is part of a large society. Everyone's needs and wants are considered, as leaders shape policies. These leaders understand that the actions of one country affect all people. They guide policy. Leaders control technological development with a global view.

The type of society people live in is partly established by technology.

TechnoFact: Early communication technology depended on carving or painting symbols on a surface for others to see and interpret. *Hieroglyphics* (a word derived from the Greek term meaning "sacred carving") is an example of this type of communication. It uses symbols to represent objects and ideas.

The regions having the resources (land, knowledge, people, materials, and machines) and desire (values and acceptance) for technology have developed more rapidly. These are called the *industrialized countries* or **First World countries.** Towns, states, or sections within a country having resources and desire have also prospered.

Where the resources are poor and human will is absent, technological advancement has been slower. Such countries are called **Third World countries.** Lack of education and poor political leadership have limited their growth. Resistant personal or religious beliefs, low personal incomes, and overpopulation also slow growth. Likewise, towns and states having poor resources and leadership that has resisted change have failed to prosper.

Having all the new technology is not, however, always good. In fact, technology, by itself, is neither good nor bad. How people use it in society determines its worth. Choices about the use of products and systems can result in desirable or undesirable outcomes. Nuclear technology can be used for good in treating diseases. It can be used for destruction in war. Coal-fired generating plants provide electric power needed for homes and industry. They also, however, produce pollution, a source of acid rain. This has negative impacts on forests. The impacts of technology on individuals and people will be discussed more fully in Chapter 26.

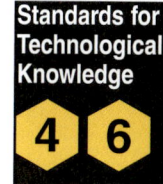

Standards for Technological Knowledge

4 **6**

Tech Terms
Global—Pertaining to the entire world.
Globalize—To extend influence or operations to all parts of the globe.

Brainstorm
Ask the students what impact globalization has on technological development.

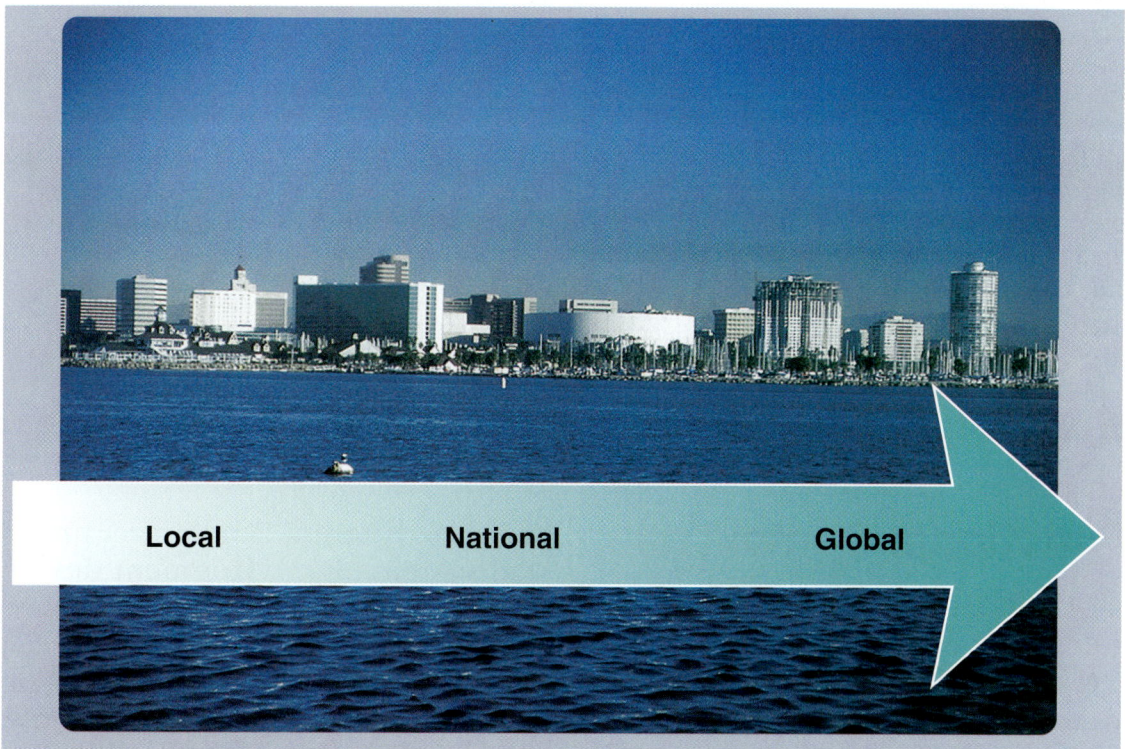

Local   National   Global

Figure 3-12. Over time, society's focus has changed from local to national to global.

TechnoFact: The technological advancements of recent times have changed the ways cultures interact. People from around the world can transmit and receive information almost instantaneously. Global commerce has reached all parts of the globe. Governments and corporations have immense political power through military and economic influence. Multinational corporations have created a global culture based on worldwide markets.

Technology Explained

making peanut butter: Peanuts are rich in many vitamins and minerals. The nuts can be processed into a spread called *peanut butter*.

Peanuts are not nuts, but legumes that ripen in the ground. They are native to South America, but are grown in many countries with warm climates. The grinding of peanuts into an edible paste goes back several centuries in South America. In the United States, however, making refined peanut butter started only about 100 years ago, **Figure A.**

Making peanut butter begins with *blending* peanuts from various growing areas, **Figure B.** This is done to ensure proper taste and texture for the end product. Then, the peanuts enter a series of cleaning operations. A process called *aspiration* uses blasts of air to remove soil, stones, and other

Figure A. Peanut butter is a favorite food for sandwiches and snacks.

Functional Context

There are many types of technology developed for specific uses. The specialization of function has been at the core of many technological advancements. Each type of technology serves a different function. A product, system, or environment developed for one situation, however, may be useful to another situation as well. In this book, the functional uses of technology will center on seven technologies. See **Figure 3-13.** Each of the

TechnoFact: *Third World* is a term used to describe countries that are technologically less developed and generally poor. They have economies that depend on the export of raw materials to the developed countries. They import most of their finished products.

foreign objects. Then, a *gravity table* is used to remove lighter objects at one end and heavier ones at the other.

The cleaned peanuts are *roasted* until they have an even color. The temperature of this process reaches about 300°F (150°C) and takes about 20 minutes. Cold air cools the hot nuts. This stops the nuts from cooking and makes it easier to remove their skins.

A process called *blanching* removes the skins of the cool peanuts. This is done using stiff bristles or rubber rollers. The skinless peanuts are then *ground* into a paste. The type of grinding plates used determines if the product will be smooth or crunchy. Finally, a stabilizer is added to stop the oil from separating from the peanut butter. The final product is placed in jars, sealed, and labeled.

Figure B. This is the peanut butter manufacturing process.

Standards for Technological Knowledge

3 **7**

Tech Terms

Communication and Information Technology— Developing and using devices and systems to gather, process, and share information and ideas.

Construction Technology—Using systems and processes to erect structures on the sites where they will be used.

Energy and Power Technology— Developing and using devices and systems to convert, transmit, and use energy.

Manufacturing Technology— Developing and using systems and processes to convert materials into products in a factory.

Medical Technology— Developing and using devices and systems promoting health and curing illness.

Transportation Technology— Developing and using devices and systems to move people and cargo from an origin point to a destination.

common functional areas (or contexts) for technology will be discussed in a chapter in Section 4 of this book:

➤ *Agricultural technology.* This area involves developing and using *devices and systems to plant, grow, and harvest crops.* Also, it includes *raising livestock* for food and other useful products. Broadly defined, this area can include farming, fishing, and forestry activities. Agriculture includes related biotechnology

TechnoFact: All technologies have unique processes. These are the systematic series of actions directed to the goal for which the technology was designed to meet. For example, communication processes use a series of actions to allow people to share information and ideas.

Standards for Technological Knowledge

3 7

Tech Terms

Technological Literacy—A person's ability to understand and apply the knowledge of technology to meet personal and societal needs.

Agricultural technology

Construction technology

Medical technology

Energy and power technology

Manufacturing technology

Communication and information technology

Transportation technology

Figure 3-13. These are the major types of technologies.

activities. These pursuits use living organisms to produce technological products.

➤ *Communication and information technology.* This context involves developing and using *devices and systems to gather, process, and share information and ideas.* These systems may involve communication using print or electronic (radio, television, and Internet) media. They may involve information processing techniques used in computer applications and e-mail systems.

➤ *Construction technology.* This area involves using *systems and processes to erect structures* on the sites where they will be used. These structures may be buildings or civil works, such as dams, roadways, and power transmission lines. They

may be residential, commercial, governmental, or industrial structures.

➤ *Energy and power technology.* This context involves developing and using *devices and systems to convert, transmit, and use energy.* These devices can be used to provide power, heat, light, or sound. Energy and power technology is used in residential, commercial, industrial, and governmental settings.

➤ *Manufacturing technology.* This area involves developing and using *systems and processes to convert materials into products* in a factory. It includes finding and extracting natural resources. This technology involves converting these resources into industrial materials. Finally, manufacturing

TechnoFact: Technological advancements have both positive and negative impacts on humans and the environment. Throughout history, technological advances have made life easier and more enjoyable. However, they have produced unwanted and often unexpected long-term effects. For example, agricultural chemicals increase crop yields, but pollute streams and drinking water. Automobiles make transportation flexible, but exhaust gases contaminate the air we breathe.

technology includes converting these materials into products.

➤ *Medical technology.* This context involves developing and using *devices and systems promoting health and curing illnesses.* This area includes devices used to study the human body. Also, devices replacing parts of the body are part of this area. Finally, devices used to treat illnesses and medical conditions are part of this technology. The area of workplace health and well-being can be included in this area.

➤ *Transportation technology.* This area involves developing and using *devices and systems to move people and cargo* from an origin point to a destination. This technology includes the means of transportation (such as vehicles or pipelines) and the support systems (terminal or routes, for example). This area includes land, water, air, and space transportation systems.

Summary

The study of technology is valuable in many ways. Studying its history lets us see how and why things were invented and used. Studying the relationship between technology and society allows us to see how technology impacts people and cultures. The functions or areas of technology show us how people are using technology today. With this understanding, people can decide which technology is good and which is not appropriate. This knowledge can help them direct technological development. It can also reduce negative impacts. Knowledge can give people the power to control the direction technology is going.

Standards for Technological Knowledge

Extend
Ask the students why they think people should study technology.

Tech Terms
Technological Impact—Influence or effect developing, producing, and using technology has on people, society, and the environment.

TechnoFact: Henry Ford was born on a farm near Dearborn, Michigan, on July 30, 1863, and died in 1947. During his early industrial career, he was a machinist's apprentice, mechanical engineer, and chief engineer. In 1893, he built his first automobile, and in 1903, he founded the Ford Motor Company. In 1913, he began using standardized interchangeable parts and assembly line techniques in his factory. Ford did not originate, nor was he the first person to use, such practices. However, he was responsible for their general use in American industry. These practices caused a great expansion of mass production and provided an increased standard of living for industrial workers.

Curricular Connections

Social Studies

Make a display or poster presenting the major technologies shaping each age. Include the Stone Age, Bronze Age, Iron Age, Industrial Age, and Information Age.

Science

Select a major type of technology, such as agricultural, medical, or transportation. Identify and explain a major scientific discovery aiding in the development of the technology you selected.

Mathematics

Select a type of technology and explain how people using that technology use mathematics. For example, in construction technology, surveyors use trigonometry to measure the distances across rivers.

Social Studies

Trace the development of a major type of technology over recorded history. You may select a technology such as housing, transportation, or food preservation.

Standards for Technological Knowledge

3

Activities

1. Prepare a chart with the five major ages (Stone, Bronze, Iron, Industrial, and Information) across the top. List the major types of technologies (agricultural, communication and information, construction, energy and power, medical, manufacturing, and transportation) along the side. List the major technological advancements that have happened for each age in each technology. You may want to work in groups to complete this activity.

2. Build a model of a technological advancement contributing to the growth of one of the technology areas (agricultural, communication and information, construction, energy and power, medical, manufacturing, or transportation).

3. Build a diorama showing life for the people in one of the ages (Stone, Bronze, Iron, Industrial, or Information). Show aspects such as a typical home and transportation devices.

TechnoFact: Humans have developed new technologies that raise serious ethical questions. Developments in genetic engineering and cloning give people, for the first time, the unprecedented power to bring about new life or to devise new forms of living things. The assessment of the right and wrong uses for technology is becoming more difficult each day.

Test Your Knowledge

Do not write in this book. Place your answers to this test on a separate sheet of paper.

1. Name the three major contexts in which technology can be viewed.
2. How did the Stone Age get its name?
3. Copper and tin were major metals used in the _____ Age.
4. Select the statement best describing the Iron Age:
   A. The period of history when the focus moved from farming to manufacturing.
   B. The period of history when the focus moved from manufacturing to information processing.
   C. The period of history when villages and towns started to appear and metal tools replaced stone ones.
   D. The period of history when building techniques improved, road systems were developed, and new cooking ware and agricultural tools were created.
5. The steam engine was a major development in the _____ Age.
6. The computer is the driving technology in the _____ Age.
7. We live in a (local, national, global) society.
8. Summarize some ways in which technology has changed our expectations of things, such as communication and transportation.
9. Technology, in itself, is neither good nor bad. True or False.
10. Define each of the following:
    A. *Agricultural technology.*
    B. *Communication and information technology.*
    C. *Construction technology.*
    D. *Energy and power technology.*
    E. *Manufacturing technology.*
    F. *Medical technology.*
    G. *Transportation technology.*

F. Developing and using devices and systems promoting health and curing illnesses.
G. Developing and using devices and systems to move people and cargo.

Answers to *Test Your Knowledge* Questions
1. Historical, societal, and functional.
2. During this stage of human history, stone was used as the main material for tools.
3. Bronze
4. D. The period of history when building techniques improved, road systems were developed, and new cooking ware and agricultural tools were created.
5. Industrial
6. Information
7. global
8. Answers will vary. (example: technology has led people to expect systems and devices that are immediate, constant, flexible, inexpensive, and easy-to-use)
9. True.
10. A. Developing and using devices and systems to plant, grow, and harvest crops and raise livestock.
B. Developing and using devices and systems to gather, process, and share information and ideas.
C. Using systems and processes to erect structures on the sites where they will be used.
D. Developing and using devices and systems to convert, transmit, and use energy.
E. Developing and using systems and processes to convert materials into products.

Using Technology to Make a Product, Textbook, page 74, and Student Activity Manual, pages 7–10

This activity will allow students to explore technology by, first, producing a product without tools and, then, making the same product using technological aides. The activity is designed to be pre-scriptive in nature, with the procedure provided for the students.

During *phase one* of the activity, each student forms and deco-rates a 1/4" × 4" × 4" clay tile using his or her hands. Have students complete the following steps:

1. Read the direc-tions for Activity 1A-1 on pages 74–75 of the textbook.

2. Form as many tiles as he or she can in a set time period, such as ten minutes.

3. Place his or her tiles with the other ones the class mem-bers produce.

4. Complete the first section of the Activity 1-1 follow-up sheet, found on page 7 of the Student Activity Manual.

5. Discuss the prob-lems of producing a product without technology.

6. Recommend ways to develop tech-nology (tools) to increase the speed of producing the tiles.

During *phase two*, divide the class into groups of five students.

Demonstrate the process the students will use to make tiles

(continued)

Using Technology to Make a Product

Introduction

You have learned that technology is the use of tools, machines, or systems to make a job easier to do. The tool may be as simple as a shaped stick or as complex as a computer system. Each tool, however, lets people do more work with less effort. In this activity, you will first make a simple product with only your hands. Then, you will use some special tools to do the same task. You will be able to see how tools extend your ability to do work.

Equipment and Supplies

Activity 1

➤ Tote tray or pan

➤ Rags or paper towels

➤ 25 lbs. of red clay (potter's clay)

➤ Water

Activity 2

For each group of five students:

➤ Materials from Activity 1

➤ Rolling sheets (approxi-mately 30—⅛˝ x 12˝ x 12˝ tempered hardboard)

➤ Rolling pin

➤ Thickness strips (¼˝ x ¾˝ 12˝ hardboard)

➤ Cutting jig. See **Figure 1A-1.**

   ➤ 1 pc. 26–28 ga. x 2˝ x 16˝ galvanized steel

   ➤ 1 pc. 26–28 ga. x 1˝ x 2˝ galvanized steel

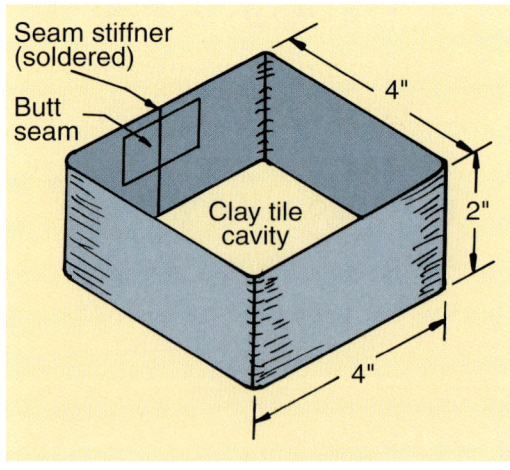

Figure 1A-1. This is a tile cutter.

➤ Grooving jig. See **Figure 1A-2.**

    ➤ 4 pc. ¼ʺ x ½ʺ x 4½ʺ pine

    ➤ 1 pc. ⅛ʺ x 4ʺ x 4ʺ hardboard

➤ Groove forming tool (¼ʺ dowel with one end rounded)

➤ Curing boards (approximately 50—¼ʺ x 5ʺ x 5ʺ hardboard)

Figure 1A-2. Here is the template for forming the decorative grooves.

Procedure

Activity 1

1. Select a ball of clay approximately 3ʺ in diameter.

2. Form a decorative wall tile using only your hands. See **Figure 1A-3.** The tile is ¼ʺ thick and 4ʺ square. It has a decorative groove across each corner. The grooves are 1ʺ from each corner.

3. Compare your tile with those other students make.

Activity 2

1. Divide into groups of five students.

2. Assign a number to each member of the group from 1 to 5.

3. Each student will work with the tools or supplies listed:

    Student 1—About 5 lbs. of red clay and six rolling sheets.

    Student 2—A rolling pin and two thickness strips.

    Student 3—A tile cutter.

    Student 4—A grooving jig.

    Student 5—Ten curing boards.

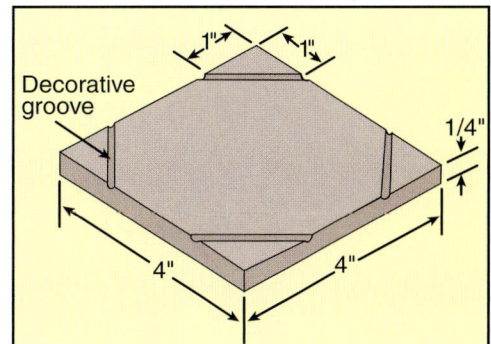

Figure 1A-3. This is the decorative tile.

4. Arrange yourselves around a workbench in numerical order.

Using Technology to Make a Product *(continued)*

using technology. Have each student record the steps on the activity procedure sheet found on page 8 of the Student Activity Manual.

Each group should complete the following steps:

1. Read the directions for Activity 1A-2 on pages 74–76 of the textbook.

2. Collect the supplies:

• Two 1/4ʺ × 3/4ʺ × 12ʺ wood strips

(continued)

Using Technology to Make a Product *(continued)*

- Rolling pin
- Tile cutter
- Tile decorating press
- Potter's clay
- Sponge
- Can of water

3. Create groups of students by counting off numbers from one to five.
4. Produce twelve tiles using these procedures.
5. Analyze the tiles.
6. Complete the activity follow-up sheet, found on page 9 of the Student Activity Manual.
7. Discuss the problems of producing the products using technology.
8. Complete the activity summary, found on page 10 of the Student Activity Manual.
9. Recommend ways to improve the technology (tools) used to produce the product.

5. Repeat the following tasks until you have made ten tiles.

Student 1:

a. Wedge the clay into 2½″–3″ balls. Wedging means pounding and rolling the ball on the bench top. This removes air bubbles.

b. Place the ball on a rolling sheet and slide it to student 2.

Student 2:

a. Space thickness guides about 5″ apart. Place clay ball between strips.

b. Roll the clay to proper thickness. Make sure you produce a sheet about the same size in width and length. Use a rolling pin over the thickness guides. See **Figure 1A-4.**

Figure 1A-4. Here, clay is being rolled.

c. Remove the thickness strips. Slide formed clay and the rolling sheet to student 3.

Student 3:

a. Using the tile cutter, cut a 4″ square from the clay sheet. SAFETY: Edges of cutter may be sharp. Wear gloves to prevent cuts.

b. Remove the excess clay and give it to student 1.

c. Slide the clay square on the rolling sheet to student 4.

Student 4:

a. Place the grooving jig over the clay square. Use the dowel to form the grooves. You may need to dip the dowel in water before each forming pass.

b. Remove the grooving jig.

c. Slide the clay tile and rolling sheet to student 5.

Student 5:

a. Carefully remove the tile from the rolling sheet.

b. Place the tile on a curing sheet.

c. Return the rolling sheet to student 1.

6. Write a short report discussing how technology helped you make a product an easier way.

Activity 1B

What Is Technology?

The Challenge

Lift a 4 oz. weight 36″ vertically, using the energy the air blowing from a standard box fan produces.

Use wind energy to lift a 4 oz. weight

Energy source   Weight

Background

Technology is used to help people do work, live better, or in some other way, control and modify the environment. A common task facing all people is moving weight. We lift, pull, shove, or slide objects from one location to another. Early humans used their physical strength to complete this task. People have limited strength, however, and are limited to the size and weight of the object they can move. Modern life uses a number of technological devices to lift objects from one level to another. Each of these devices has an energy source and a unique technique for lifting the weight

Materials

Develop a technological device using any or all of the following materials:

- ¼″ diameter dowel
- Bond paper
- Copper wire
- Index paper
- Masking tape
- Paper clips
- Poster board
- Soda straws
- String
- Thread spool
- Tongue depressors
- White glue

What Is Technology?, Textbook, page 77
This activity is designed to allow the students to use their creativity to design an energy conversion system. They will use the mechanical energy from wind to lift a weight. There is no right answer. To use this activity, divide the class into groups of two to four students. Have each group complete the following steps:
1. Define the problem they are attempting to solve.
2. List any restrictions they face, such as materials that can be used (a list is included with the activity description in the textbook) and the strength of the wind.
3. Develop sketches of two to five ways to solve the problem.
4. Select and refine the best solution.
5. Build a prototype of the solution using only the materials listed in the textbook. (Limiting the materials introduces a common restriction all product designers face.)
6. Demonstrate the solution.
7. List ways to improve the design.

Section 2
Resources and Technology

Resources are the many kinds of supplies, materials, and services you use to get something done. Whether you are writing a letter, cooking dinner, or building a house, you need resources. Consider the simple process of writing a letter. What resources do you need?

First, there is the matter of tools or machines. At the very least, you will need a pencil. If you use a pencil, you may need a sharpener. You may even decide to use a machine to write your letter—a typewriter or computer.

You will need a chair for sitting and a surface on which to write or type. Also, you will need paper and an envelope. Mailing your letter requires help from people at the post office. If you use your electric typewriter or computer, you will need energy for power.

Anything else? What about information? There is a good chance you will write about something you have heard, read, or experienced.

Technology is the system that puts the resources to use. No matter what we are doing, we use the same basic resources. You will see how resources are tapped and tied into one another as you read the next five chapters.

4 Tools and Technology

5 Materials and Technology

6 Energy and Technology

7 Information and Technology

8 People, Time, Money, and Technology

Technology Headline

Floating Cities

There is growing concern around the world that within the next several decades, the world's population will be so big, there will not be enough living space for all the people. Some scientists and technologists have suggested building colonies in space. This may be one solution to this problem, but now there is another solution, which may prove to be more realistic for the near future. Plans are already underway for a very large ocean vessel called *Freedom Ship,* which will serve as a floating city!

This nearly mile-long ship will be home to more than 60,000 people, and visitors will be welcome to vacation there. Its 25 floors will contain everything a city needs, including a library, a hospital, schools, an airport, a golf driving range, a marina, bicycle paths, one of the world's largest shopping malls, offices, warehouses, light manufacturing and assembly enterprises, and 200 open acres for recreation. The city will travel, circling the globe every two years, and it will make stopovers at various islands and shores.

Residents of this city will be able to participate in nearly any activity available in traditional cities. They will be able to go to restaurants, nightclubs, casinos, and movie theaters. Tennis courts, bowling alleys, swimming pools, gyms, ice skating rinks, and fishing areas will be available for all residents. In addition, each home will have satellite television and Internet access.

Besides being comfortable and exciting, this self-contained floating city will also be environmentally friendly. All waste created on the ship will be reused in some way. There will be no need for a sewage treatment plant or any sewage to spill into the ocean, as all sewage will be burned and put in the flower beds. Instead of being dumped into the ocean, waste oil will be used to generate electricity. Finally, all used paper, metal, and glass will be recycled and sold on the ship. It is estimated that residents of this city will produce 80 percent less waste than people living on land.

Although *Freedom Ship* might sound like a dream come true, it will most likely be a home only to the world's wealthiest people, at least at first. Suites on this ship currently range from $121,000 to $11 billion! The people who live here, however, will not have to pay any local taxes, including income tax, real estate tax, sales tax, business tax, and import duties.

This enormous ship is expected to begin sailing the world's oceans within the next few years. If this floating city proves to be successful, you can anticipate more in the coming decades. This is only the beginning, and it might not be very long before you are living on the ocean!

Discussion Starter
Ask the students to describe how their lives would change if various resources were no longer available to them.

Group Activity
Refer the students to the chapter titles in this section. Organize the class into five groups and assign one of the chapter titles to each group. Ask the students in each group to draw a picture or cartoon illustrating the title and reflecting what they think the content of the chapter will be. Display the students' work. After the students complete the section, discuss how closely the pictures reflect the chapters' contents.

Writing Assignment
Ask the students to imagine they are time travelers who have interrupted the development of resources used in modern technology. Have them describe the differences they think will exist in their home, school, and community.

Chapter 4
Tools and Technology

Did You Know?

➤ The metric system was developed in France during the reign of Napoleon in the 1790s. It missed being adopted by the United States by one vote in the Continental Congress in the late 1700s or early 1800s.

Objectives

The information given in the chapter will help you do the following:

➤ Describe the unique abilities of humans.

➤ Explain how humans are toolmakers and tool users.

➤ Summarize the differences between tools, mechanisms, and machines.

➤ Describe and give examples of the six major types of primary tools.

➤ Explain and give examples of the six mechanisms, or simple machines.

➤ Summarize the use of mechanisms as force and distance multipliers.

➤ List and explain the major parts of a machine tool.

➤ Give examples of the major types of machine tools.

Chapter Outline
Did You Know?
Objectives
Key Words
Toolmakers
Tools
  Measuring Tools
  Cutting Tools
  Drilling Tools
  Gripping Tools
  Pounding Tools
  Polishing Tools
Mechanisms
  Lever
  Wheel and Axle
  Pulley
  Inclined Plane
  Wedge
  Screw
Machines
  Turning Machines
  Drilling Machines
  Milling and Sawing Machines
  Planing and Shaping Machines
  Grinding and Sanding Machines
  Shearing Machines
Summary
Curricular Connections
Activities
Test Your Knowledge

Key Words

These words are used in this chapter. Do you know what they mean?

distance multiplier

drilling machine

force multiplier

grinding and sanding machine

inclined plane

lever

machine

machine tool

mechanism

milling and sawing machine

planing and shaping machine

pulley

screw

shearing machine

simple machine

tool

turning machine

wedge

wheel and axle

Humans are unique in the world of living things. People have the ability to reason. Using this ability, we can predict events and plan actions. For example, you can say, "If *this* is true, then *that* must also be true," "If I do *this,* I can expect *that* to happen," or "I must do *this* first, then do *that* next."

People also can think about the future. They can plan activities over a period of time and decide to do something at a later date. In addition, humans can think of different ways to do the same task. They can offer alternatives for action. People can adjust their actions to meet different situations and events. Their minds connect things, actions, and relationships. All of these traits combine to produce an ability only humans have. This ability is called *rational thought.*

A second unique ability humans have is *complex language.* People use this language to describe their thoughts to other people. Early language was based on sounds that became the spoken language. Later, symbols were used to represent these sounds. These symbols evolved into an alphabet, the basis for written language. See **Figure 4-1.**

Humans also have value systems. They have a sense of what is right and what is wrong. They can think, "It is okay to do *this,* but I shouldn't do *that.*" This reasoning is called *making moral judgments.*

The abilities to reason, use complex language, and make moral judgments are positive differences between humans and other mammals. Humans, however, lack some basic

<table>
<tr><td>α</td><td>β</td><td>γ</td><td>δ</td><td>ε</td><td>ζ</td><td>η</td></tr>
<tr><td>ϑ</td><td>ι</td><td>κ</td><td>λ</td><td>μ</td><td>ν</td><td>ξ</td></tr>
<tr><td>ο</td><td>π</td><td>ρ</td><td>σ</td><td>τ</td><td>υ</td><td>φ</td></tr>
<tr><td>χ</td><td>ψ</td><td>ω</td><td></td><td></td><td></td><td></td></tr>
</table>

<table>
<tr><td>Α</td><td>Β</td><td>Γ</td><td>Δ</td><td>Ε</td><td>Ζ</td><td>Η</td></tr>
<tr><td>Θ</td><td>Ι</td><td>Κ</td><td>Λ</td><td>Μ</td><td>Ν</td><td>Ξ</td></tr>
<tr><td>Ο</td><td>Π</td><td>Ρ</td><td>Σ</td><td>Τ</td><td>Υ</td><td>Φ</td></tr>
<tr><td>Χ</td><td>Ψ</td><td>Ω</td><td></td><td></td><td></td><td></td></tr>
</table>

Figure 4-1. This is an example of the Greek alphabet.

survival traits and innate knowledge other animals have. People lack an inherited "blueprint" of action on how to lead our lives. Animals can survive with their built-in knowledge, called *instincts.* Beavers instinctively know how to build dams. Bees use their instincts to collect pollen. Salmon know how to return to their spawning ground after several years in the ocean. Birds know the path from their summer homes to winter habitats. Humans, however, have to learn most of these things. We cannot instinctively build homes, grow food, or make clothing.

In fact, humans are poorly equipped physically to meet the challenges of the world. People cannot lift much weight, run very fast, withstand cold temperatures very well, or see very far. Other animals are better equipped to survive in the natural environment. Eagles can see movement of prey from hundreds of feet in the air. Lions have great speed to catch food. Geese have the strength to fly

Research
Have the students explore how individual languages name the same technological objects.

Tech Terms
Language—A cultural system allowing people to share information and ideas through vocal and written symbols.
Judgment—Forming an opinion after considering information or ideas.

TechnoFact: Two ways of thinking are induction and deduction. Observing events (induction) and using principles already known (deduction) are used to gain insight into events and situations. From these mental tasks, new hypotheses are formulated. These hypotheses are tested in real-life applications, and a new understanding is gained.

hundreds of miles south for warmer climates in the winter. Ants can lift several times their weight. Bears have heavy fur to fight the cold. In contrast, modern humans cannot survive with only their natural equipment. They would starve or freeze to death without help.

Toolmakers

To overcome physical weaknesses, humans have another special ability. They can design, make, and use tools. See **Figure 4-2.** Tools are devices people develop and use to do specific tasks.

Over the history of humankind, people have developed many types of tools. Early humans used things they could find to make their tools. They used sticks, rocks, bones, and other natural elements for tools. These humans used pointed sticks to hunt animals and spear fish, tree limbs were used as clubs, and animal bones could be used as needles to make clothing.

As civilization advanced, tree limbs were used to plow the soil. Weapons were made from copper and other metals. Clay vessels were shaped and fired. Hammers were made from the newly developed steel.

Today, people have tools for every job. Dentists have a set of tools to work on their patients' teeth. Surgeons have special tools to perform operations. Auto mechanics use different tools than plumbers use. Have you ever seen a carpenter at work? Carpenters use many different types of tools. They use a heavy hammer to drive stakes into the ground. A smaller hammer or an air-nailer is used to nail 2 x 4s together.

Demonstrate
Shape a simple clay vessel storage tool. Present the act as toolmaking.

Research
Ask the students to research and list some typical tools developed during the Stone Age.

Figure 4-2. Humans have designed, built, and used many tools and machines. (Caterpillar, Inc.)

TechnoFact: Stone Age humans made tools from a wide range of materials other than stone. They probably used available natural items, such as seashells, animal horns, and tree bark to make containers. Most likely, they used available limbs (wood) for spears, clubs, and digging sticks and shell, hide, bark, or horn to fashion containers.

Extend
Ask the students to list some examples of machines that include tools and mechanisms.

Example
Additional Language Tools: Word processor, typewriter, and photocopier
Additional Educational Tools: Charts, software programs, games

Tech Terms
Tool—A human-designed device used to cut, shape, or otherwise handle a material.
Flake Tool—A tool consisting of a stone flake that was generally modified by additional human action (chipping or flaking).

| Tool | + | Mechanism | = | Machine |

| Saw | + | Wheel and axle | = | Circular saw |

Figure 4-3. People use tools, mechanisms, and machines to do work. (Combining a tool and a mechanism produces a machine.)

Still another hammer or nailer is used to attach the trim around windows and doors. Each profession has its own set of tools. See **Figure 4-3.** These tools can be understood better by looking at the major differences between a tool, a mechanism, and a machine:

➤ A *tool* is a device people design and use to complete a task.

➤ A *mechanism* is a device people design and use to adjust or power a tool.

➤ A *machine* is a combination of tools and devices people design and use to complete complex tasks.

Tools

Throughout history, humans have survived by using their hands, arms, and brains. In earliest times, the hand held a rock to form a crude hammer. The brain directed the arm to move in a controlled manner. Later, a stick was attached to the rock to make the first hammer. Exchanging the rock with a sharp stone produced a hatchet. Each

of these examples is a human-made device called a *tool*. See **Figure 4-4.** Over time, humans have developed many different types of tools:

➤ Tools used with language, such as pens, pencils, and printing presses.

➤ Tools used in education, such as projectors, models, and maps.

Figure 4-4. Over time, humans have developed many types of tools. This offshore drilling platform is a special tool of the petroleum industry. (American Petroleum Institute)

TechnoFact: The Clovis people were early human inhabitants of North America. They are named after a site near Clovis, New Mexico, where early remains of this society have been found. They used handmade stone projectile points fluted at the base for mounting to a wooden shaft. Clovis artifacts have been found in many parts of North and Central America.

➤ Tools used in business and trade, such as calculators, scales, and money.

➤ Tools used in religion, such as vessels and special clothing.

➤ Tools used in art, such as easels, paintbrushes, modeling tools, and sculpture chisels.

➤ Tools used by the government, such as military weapons and police equipment.

➤ Tools used in games and sports, such as basketball goals, tennis rackets, soccer balls, and baseball bats.

➤ Tools used in the pure sciences, such as telescopes, microscopes, and chemical apparatus.

➤ Tools used in technological systems, such as agricultural, communication, construction, energy conversion, manufacturing, medical, and transportation equipment.

➤ Tools used in managing companies, such as computers and word processors.

Each of these groups of tools makes human action more efficient. It extends the human potential to do jobs.

All tools can be traced back to another type of tool. See **Figure 4-5.** We call these older tools *primary tools.* They are the tools people use to make new tools. Without them, humans could not develop new tools and machines to make lives easier.

Standards for Technological Knowledge

7 **12**

Resource
Reproducible Master 4-1: Measurement

Research
Have the students research and list some early historical measuring units, such as a cubit.

Tech Terms
Measuring—
Determining the size or extent of something by comparing it to an established standard.

Types of Tools

Measuring tools

Gripping tools

Cutting tools

Drilling tools

Pounding tools

Polishing tools

Figure 4-5. These are types of tools. We can use them to build other tools and mechanisms.

Measuring Tools

Early humans lived in small groups that moved from place to place searching for food. These people needed only a few tools to hunt game and gather food. As the population grew, however, the nomadic life gave way to permanent settlements. With this growth of civilization, people needed new tools to measure things. They had to plot fields to farm. These people had to measure material as they made houses. Grain placed in the village storehouses had to be weighed.

Measurement became an essential part of life. Over the years, humans have developed several different types of measurements. These measurements include volume and weight measurements used in commerce. For

TechnoFact: Time, length, area, volume, pressure, mass, force, and energy are frequently measured elements. To perform measurements, an instrument measuring the desired units is used to determine a quantity.

Extend
Have the students discuss how people relate to distance. For example, shouting distance means people are close enough together to hear each other's shouts.

Tech Terms
Measurement System—An established system used to determine the size (dimensions or area), volume, or weight of something.

Figure 4-6. There are many kinds of common measuring tools (left). A steel tape measures a board to length (right).

example, many items sold in a grocery store use these measures. Milk is sold by the gallon (volume). Fruit and vegetables can be sold by the pound (weight).

Two other types of measurement are also important. People measure distances and relationships. We measure the size of objects (distance) and how one surface relates to another (for example, squareness).

Measuring Distances

Measuring distance tells us how far it is from one point to another.

Humans have developed a number of different tools to measure distances. See **Figure 4-6.** Some of these tools are considered precision measurement devices. They provide very accurate measurements. Most precision measurement devices measure distances in thousandths (1/1000) of an inch or smaller. Other measurement devices are nonprecision devices. They give us measurements in *fractions* (such as $\frac{1}{8}$, $\frac{1}{16}$, and $\frac{1}{32}$) of an inch. See **Figure 4-7.**

A *ruler* can be used to measure the size of a sheet of paper. An automobile *odometer* can be used to measure the

Figure 4-7. The dyes in this stamping press were developed using precision measurement (left). The package designer here is using standard measurements (right). (Dana Corporation)

TechnoFact: Measurement accuracy is affected by the precision of the individual making the measurement and the condition of the measuring tool used.

distance between cities. Both of these devices are non-precision. They provide you with a general idea of the size or distance you are measuring. You would not, however, use an odometer to measure the racetrack or a ruler to measure the diameter of a ball bearing. These devices are not accurate enough. More precise devices are needed.

A ruler is a common measuring tool. It is used to measure distances along a flat surface. A ruler can be used to measure length, width, and thickness. This tool is a strip of material with measurement marks along its length. These marks will be in one of two systems:

➤ U.S. Conventional. This is a measurement system in which the inch is the standard length. One foot equals 12 inches, and one yard equals 36 inches. Longer measurements are made in miles. A mile is equal to 5,280 feet or 1,760 yards.

➤ SI Metric. This is a measurement system using the meter as its basic unit. The meter is divided into decimal parts. Common measuring units are the *centimeter* (1/100 meter), *millimeter* (1/1000 meter), and *kilometer* (1000 meters).

Most countries in the world use the metric system of measurement. The metric system has several advantages over the American system:

➤ It is based on a decimal system (powers of ten).

➤ It simplifies calculations by using a set of prefixes.

➤ It has commercial and trade advantages because most other nations of the world use it. An American manufacturer having domestic and international trade must absorb the added cost of dealing with two measurement systems.

Some common prefixes used in the metric system include the following:

Prefix Unit	Common Use
milli	1/1000 milligram, millimeter
centi	1/100 centimeter
kilo	1000 kilogram, kilobyte, kilometer
mega	1,000,000 megahertz, megabyte
giga	1,000,000,000 gigabyte

A ruler is not effective in measuring diameters. For this task, a caliper can be used. The caliper is used in one of two ways. For inside diameters, the caliper is fitted to touch the inside surfaces of a hole. For outside surfaces, the tool is fitted to touch the outside surfaces of a shaft or rod. The measurement can then be read on a scale or digital screen. This tells us the diameter of the object or hole. See **Figure 4-8.**

The ruler gives us standard measurements. *Precision* measurements are often made with a micrometer. To use a micrometer, the part to be measured is placed between the spindle and the anvil. The spindle is brought into contact with the part. The measurement is read on the micrometer barrel. See **Figure 4-9.**

Standards for Technological Knowledge

7 12

Research
Have the students use the Internet to research the historical development of the U.S. Customary and SI Metric systems of measurement.

Tech Terms
Measurement Standard—A recognized standard used for determining the size (dimensions or area), volume, or weight of something.
Unit of Measurement—A unit in a measurement system used to determine the size (dimensions or area), volume, or weight of something.

Brainstorm
Ask the students why they think one system of measurement might be better than another for technological activities.

TechnoFact: English units of measurement consist of two associated systems: the U.S. Customary System of units and the British Imperial System. The terms for the units of measure are, by and large, the same in both systems, but their sizes vary. The United States is the only major industrialized country using English units of measurement.

Figure Discussion
Figure 4-9 shows a micrometer. Micrometers are used to measure extremely small distances. The standard micrometer can measure the thickness or diameter of objects within an accuracy of one ten-thousandth of an inch.

Research
Have your students use a catalog or the Internet to research the different types of squares.

Figure 4-8. Internal and external measurements can be made with a caliper. The measurements are read on a scale on traditional calipers (top). Digital calipers show the measurements on a digital screen (bottom).

Figure 4-9. A micrometer is used for precision measurement.

Measuring Relationships

The relationship between two surfaces is often important. People want to know if a part is square. They need to know that a side is 90° (at a right angle) to an end. In another case, it is very important that the ends of parts of a picture frame are at 45°. These measurements are made with tools called *squares.* See **Figure 4-10.**

A square is an "L" shaped tool. Often, it has a head and a blade. The two parts are exactly 90° to each other. The common types of squares are rafter, try, and combination squares. The combination square can measure both 90° and 45° angles.

Figure 4-10. Squares are tools measuring angles. Note how a try square is used to mark a 90° angle on a board. Squares are often combined with rules.

TechnoFact: The relationship between two surfaces is an important concept in technology. The most common relationship is "square." This means two adjacent surfaces are at right angles to one another. The sides of a bookcase need to be square with its top. *Plumb,* another important relationship term, means an object is perfectly vertical. Walls in a home need to be plumb.

Cutting Tools

Cutting tools remove material to size and shape parts. Each tool cuts away unwanted material until the part has a desired shaped. Cutting tools include three major types. These are sawing, slicing, and shearing tools. See **Figure 4-11.**

Sawing Tools

A saw uses a set of teeth to cut the material. The tooth is a sharp projection (point) on a body. The teeth can be arranged along a strip. Examples are hand, coping, scroll, or band saws. Other saws have the teeth on a disc.

The saw must move to cause a cutting action. All handsaws are strips of metal with teeth along their edges.

Figure 4-11. These are the different families of cutting tools.

The saw moves back and forth over the material to make the cut. Handsaws usually cut on the forward stroke. Typical handsaws are the woodworking crosscut, rip, and backsaws. See **Figure 4-12.** Those designed for metal cutting are called *hacksaws*.

Disc (circular) saws are used with cutting machines. Circular saw blades are commonly found on table, radial,

Standards for Technological Knowledge

12

Research
Have the students use a catalog or the Internet to research the different types of saws.

Tech Terms
Saw—A hand-operated or power tool with a thin metal blade or disk and toothed edge used for cutting hard materials, such as wood, metal, ceramics, or plastic. The word *saw* is derived from the Middle English word *sawe* and from the Old English word *sagu*.
Sawyer—A person employed in sawing wood, usually in a lumber sawmill.

Figure 4-12. These are different kinds of handsaws.

TechnoFact: When people use the word *cut,* it can mean a number of things. In technology, to cut means to separate material into parts with a tool (manufacturing and construction), to mow, reap, or harvest a crop (agriculture), or to stop filming, edit a film, or record a performance on a medium (communication).

Figure 4-13. Typical cutting tools include planes, chisels, and files. Hand planes smooth board surfaces.

and cutoff saws used in woodworking. Also, cutters used on metalworking milling machines have teeth on a disc or shaft. These and other machines will be discussed later.

Figure 4-14. Shearing tools use two sharp opposing edges to cut the materials. There are several varieties of tin snips (top). Tin snips cut sheet metal (bottom).

Slicing Tools

Slicing tools use a sharp, wedge-shaped edge to separate the material. The wedge cuts away unwanted material in the form of shavings. Typical slicing tools are knives, chisels, carving tools, and woodworking planes. See **Figure 4-13.**

Shearing Tools

Shearing tools fracture material between two opposing edges. The workpiece is placed between the edges. The knives coming together cause the material to separate. Common shearing tools are tin snips and scissors. See **Figure 4-14.**

Drilling Tools

One of the first tools ancient humans developed was the drill. The first one was no more than a sharp, pointed stone attached to a wooden shaft. The tool user rotated the shaft in the palm of the hands. The stone tip was pointed downward and rested on the piece to be drilled. When the hands were rubbed back and forth, the shaft would rotate. As the shaft

TechnoFact: Acheulean refers to remnants of a prehistoric industry found near the French town of Saint-Acheul. The site contained many stone hand axes that were more advanced than those found in other sites called *Oldowan sites*. The Acheulean hand ax had a large, pointed or oval-shaped head made by removing flakes around the outside of a stone. These axes were used to butcher (cut and slice) animals.

Figure 4-15. Here is a sample of drilling tools (left). Drills must be rotated to make them cut (right).

turned, downward pressure forced the stone point into the work.

Today, drilling tools and machines use the same action. A steel shaft is formed with cutters on its end. It is rotated, while downward pressure forces it into the workpiece. This action cuts a series of chips, producing a hole.

Common drilling tools are twist drills, spade bits, and auger bits. These are often held and rotated in a hand drill, brace, portable electric drill, or drill press. See **Figure 4-15.**

Gripping Tools

The human hand is a natural gripping device. People hold and twist things with their hands. The hand, however, has serious limitations. It can only exert a limited amount of force. Also, the human hand is easily damaged when it holds rough materials. Therefore, holding and gripping tools have been developed to extend the power of the hand. Pliers are familiar gripping tools.

Holding Tools

Holding tools replace the gripping power of the hand. They are used to hold an object in place. These tools squeeze the part and keep it from moving. Typical holding tools are vises and clamps. See **Figure 4-16.**

Turning Tools

People often need to turn objects to position them or tighten fasteners. Parts are rotated to fit into slots. Screws are turned into wood, plastic, or metal. Nuts are turned onto bolts. Shafts are turned to align them properly.

Turning tools are used to perform these tasks. Typical turning tools are wrenches and screwdrivers. Wrenches are made adjustable or in fixed sizes. Fixed-size wrenches usually come in sets to fit common bolts and nuts. They are either open-end or box styles.

TechnoFact: The human hand consists of the wrist, palm, four fingers, and thumb. The thumb can be repositioned opposite to the other four digits (opposable). Opposable thumbs make gripping tiny items feasible. Many tools, such as pliers, use the same principle to hold things.

Tech Terms
Hammer—A hand tool with a handle and an attached head used for striking or pounding an object or material.

Research
Have the students use a catalog or the Internet to explore the different types of hammers.

Figure 4-16. Holding tools multiply the squeezing force of the hand. There are many types.

Commonly used adjustable wrenches are open-end wrenches and pipe wrenches. See **Figure 4-17.**

Screwdrivers are commonly used to set or tighten wood, sheet metal, and machine screws. They are available in several lengths and blade sizes. They also are made to fit different slots in screw heads. There are screwdrivers fitting standard screws and Phillips head screws. See **Figure 4-18.**

Pounding Tools

From early times, people needed to apply a striking force to objects. They needed to pound things into the ground, break them, or shape them. This need to pound objects gave rise to the development of pounding tools. Typically, these tools consist of a heavy head attached to a handle.

Today, we call these tools hammers. Hammers come in many types and styles:

➤ Claw hammers for driving all types of nails.

➤ Sledge hammers for driving stakes.

➤ Shingling hammers for attaching roof shingles.

➤ Mallets for driving woodworking chisels.

Figure 4-17. These tools are useful for assembling parts using bolts and nuts.

TechnoFact: Many people use the verb *hammer*. This means to hit or pound something into a shape with a special tool called a *hammer*. The word *hammer* comes from the Middle English word *hamer*, which was derived from the Old English word *hamor*.

Figure 4-18. These are standard and Phillips screwdrivers. They turn screw fasteners into parts.

➤ Ball peen hammers for striking cold chisels.

➤ Rubber or plastic mallets for striking parts to align them.

Several types of hammers are illustrated in **Figure 4-19.**

Polishing Tools

Many parts and products require smooth surfaces. This type of surface finish can be produced using scrapers and abrasive grits. These tools remove small amounts of material to improve the surface of the material. Hand and cabinet scrapers use a *curled edge* (burr) to scrape away the unwanted material.

Sanding and grinding tools use abrasives (mineral grit) to create uniform scratches on the material. The action replaces dents and large scratches with small, straight scratches. As these scratches become very small, they cannot be seen. There is a degree of roughness, but the eye and the hand tell us the material is smooth. Typical

Standards for Technological Knowledge

12

Demonstrate
Show how rotten-stone, pumice, or emery can be used to polish a metallic surface.

Brainstorm
Ask the students why materials are polished.

Tech Terms
Brightwork—Metal parts or products made shiny by polishing.

Figure 4-19. Here are some of the many hammers we use. They pound and drive certain types of fasteners.

TechnoFact: Many materials can be used as polishing compounds. These include rotten-stone (soft decomposed limestone), emery (fine-grained impure corundum), and pumice (powdered, lightweight, and porous lava).

Resource
Reproducible Master 4-2: Basic Mechanisms

Extend
Look at a mechanical product and identify the simple machines used in it.

Tech Terms
Simple Machine—A physics term describing any of six elementary mechanisms: lever, wheel and axle, pulley, screw, wedge, and inclined plane.
Complex Machine—A physics term describing a combination of simple machines.

abrasive tools are loose and sheet abrasives, sharpening stones, and buffing compounds.

Mechanisms

Hand tools served early humans well, but as civilization grew, demand for food and products also grew. People using hand tools could not meet the increasing demand. Machines to produce more goods were needed. These improvements required technological advancements.

There is not, however, a direct step from a hand tool to a machine. The hand tool must be combined with a mechanism to produce a machine. What is a mechanism? *It is a basic device controlling or adding power to a tool.* Science calls these mechanisms *simple machines.* These simple machines, or basic mechanisms, multiply the force applied or distance traveled. An understanding of them helps us understand more complex machines. There are six simple machines that

Principle	Simple Machine
Lever	Lever
	Wheel and axle
	Pulley
Inclined plane	Inclined plane
	Wedge
	Screw

Figure 4-20. The six simple machines are based upon two basic principles.

work off two basic principles. Levers, wheels and axles, and pulleys use the lever principle. Inclined planes, wedges, and screws use the inclined plane principle. See **Figure 4-20.**

Lever

Have you ever seen someone pry open a crate using a crowbar? If you have, you have seen a *lever* in action. See **Figure 4-21.** A lever, like all simple machines, is a force

 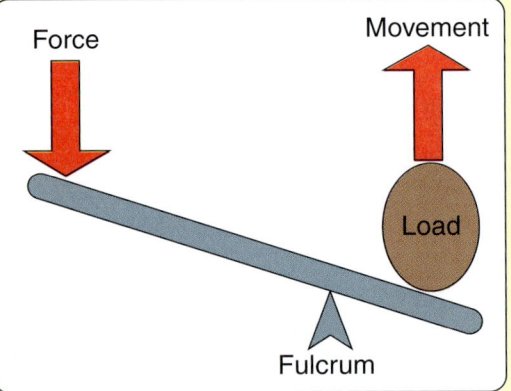

Figure 4-21. Levers multiply force. The bar on the left is an application of the lever illustrated on the right.

TechnoFact: Early machines were developed for many reasons, such as hunting, weapons, food production, and construction.

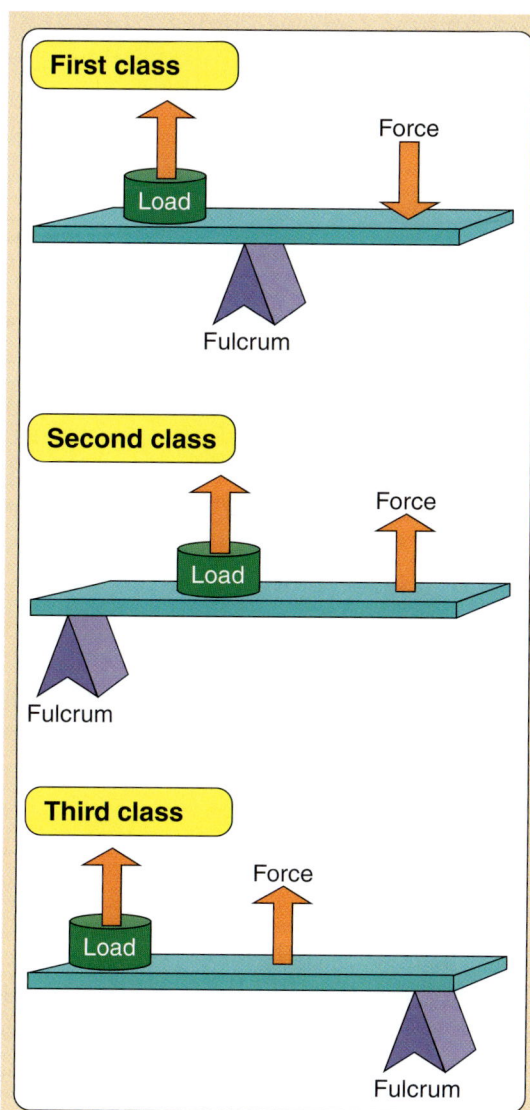

Figure 4-22. These are the three classes of levers. Try to think of tools you use fitting these classes.

produce a large amount of movement at the other end. This is using a lever as a *distance multiplier.*

A lever is a device consisting of a lever arm and a fulcrum. The fulcrum is a pivot point upon which the lever arm rotates. A lever consists of the lever arm, the fulcrum, the load to be moved, and the force to be applied. There are three basic arrangements for these components of a lever. These arrangements are called *classes of levers.* See **Figure 4-22.** Scissors and pry-bars use the principle of the *first-class* lever. Wheelbarrows and hand trucks are devices using *second-class* levers. When you use brooms or baseball bats, you are using *third-class* levers.

Levers can be force or distance multipliers:

➤ Force multiplier. See **Figure 4-23** (top). The fulcrum of this first-class lever is located near the load to be moved. Thus, a small amount of force, moving a greater distance, will move a load a shorter distance than the force movement.

➤ Distance multiplier. See **Figure 4-23** (bottom). The fulcrum is located near the force applied to the lever. Thus, a force moving a short distance can cause a load to be moved a much greater distance. Of course, a much larger force must be applied to multiply the movement of the load.

Look around the school laboratory for examples of levers in use. Did you identify pliers, tin snips, or claw hammers (when pulling a nail)?

or distance multiplier. It can increase the force applied to the work. A lever makes us stronger than we really are. This is using it as a *force multiplier.*

A lever can also let us change the amount of movement created. A small amount of movement at one end will

TechnoFact: In some machines, a small force is applied to move a much greater load. To do this, the applied force must move a greater distance than if the load was moved directly.

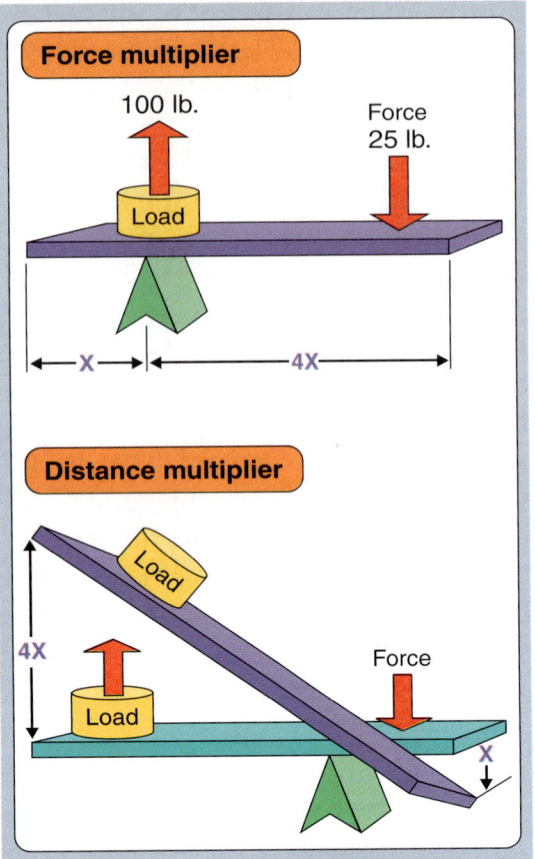

Figure 4-23. We use levers to multiply force and distance. A twenty-five pound force is multiplied four times (top). Great force applied at right moves a short distance to move loads four times further (bottom).

Wheel and Axle

The *wheel and axle* is another mechanism, or simple machine. It consists of a shaft attached to the center of a disc. See **Figure 4-24.** This mechanism operates a second-class lever. The axle is the fulcrum, and the load is applied to the wheel or axle.

If the force is applied to the axle, the mechanism becomes a distance multiplier. One revolution of the axle will cause the wheel to rotate one time.

The *circumference* (distance around) of the wheel, however, is greater than the axle. Therefore, the mechanism will move a greater distance. Bicycle drives and automotive *differentials* (gears turning the axle) use this action.

If the force is applied to the wheel, the mechanism is a force multiplier. A screwdriver is a good example of this action. Try to drive a screw by gripping the shaft (axle) of a screwdriver. Then repeat the task gripping its handle (wheel). You will find it is much easier to turn the mechanism by applying the force to the wheel. This principle is used for automobile steering wheels, other control knobs and wheels, and woodworking braces.

See **Figure 4-25.** Note how wheels and axles can be used as both force and distance multipliers. Can you think of other examples?

Pulley

A *pulley* is a third type of simple machine, or mechanism. It is a wheel with a grooved rim attached to a loose axle. People use pulleys for many tasks,

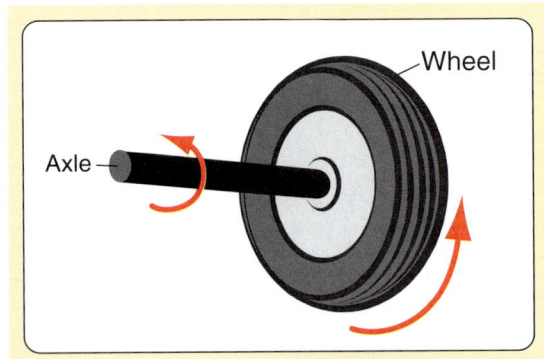

Figure 4-24. The wheel and axle is an application of the lever.

TechnoFact: In nearly all machines, a force is amplified, its direction is altered, or one type of energy or motion is changed into another type.

Figure 4-25. Like the lever, the wheel and axle can be either a force or distance multiplier.

including raising sails on ships, hoisting cargo, and pulling parts together.

Pulleys are used by themselves or in sets. They can do three things:

➤ Change the direction of force.
➤ Multiply force.
➤ Multiply distance.

See **Figure 4-26.** Notice how pulleys are used to do each of these jobs.

Inclined Plane

An *inclined plane* is the fourth type of mechanism, or simple machine. It is a mechanism using a sloped surface. See **Figure 4-27.** This mechanism operates on the principle that moving up a slope is easier than lifting straight up.

A simple experiment will test this principle. Pull a smooth weight up a

Extend
Have the students give an example of using a pulley to do each of the following: 1. Change the direction of the force applied. 2. Multiply the force applied. 3. Multiply the distance a load is moved.

Figure Discussion
Explain why the steering wheel shown in Figure 4-25 can be classified as a wheel and axle.

Tech Terms
Pulley—A simple machine consisting of a grooved wheel within a support housing and a rope threaded around the wheel.

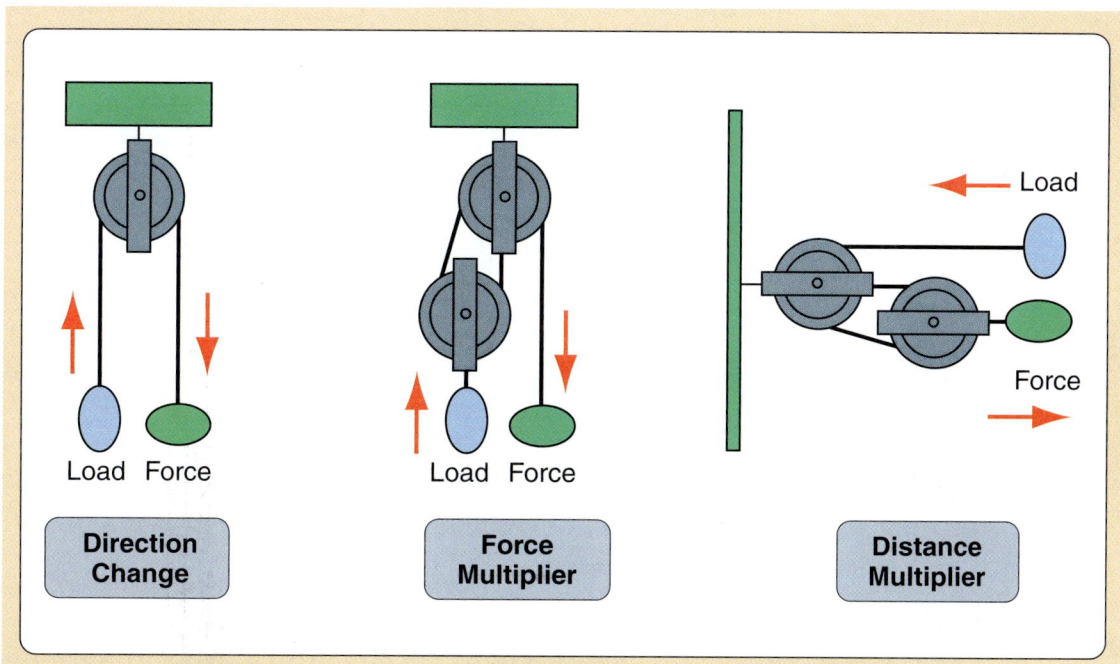

Figure 4-26. Pulleys change direction, multiply force, and multiply distance.

TechnoFact: Pulleys can be used to raise things by fastening one end of a cord strung through the pulley to the object and pulling on the other end of the cord. A single fixed pulley, like the one of a flagpole, changes the direction of the force applied to the end of the cord. The use of several interconnected pulleys can alter both the direction of the force and the amount of force required to raise something. Using multiple pulley systems lessens the amount of force needed to raise heavy equipment.

Figure 4-27. Here is an inclined plane. Loads can be lifted with less force. The lower the slope, the easier the load is to move.

Resource
Reproducible Master 4-3: From Tool to Mechanism to Machine

Figure Discussion
Explain why the wedge shown in Figure 4-28 splits the log.

Brainstorm
Ask the students to list some of the uses for a wedge.

Tech Terms
Screw—A simple machine consisting of a cylindrical shaft with one or more helical or advancing coiled threads.
Wedge—A simple machine consisting of a piece of material that is wide at one end and tapered to a thin edge at the other end.

slope. Use a scale to measure the force. Then lift it. You will find that the longer the slope, the easier it is to move the object.

Wedge

A *wedge* is the fifth type of simple machine, or mechanism. It is a set of two inclined planes. See **Figure 4-28.**

Figure 4-28. People use wedges to split logs. Do you see that the wedge is two inclined planes put together?

This mechanism is used in many simple hand tools. The wood chisel, knife, ax, splitting wedge (for firewood), and cold chisel operate on the wedge principle. Also, nails are wedges.

Screw

A *screw* is the last type of mechanism, or simple machine. It is actually an inclined plane wrapped on a round shaft. The threads move slowly up the shaft as they go around it. The screw is a great force multiplier. It takes a great deal of rotating motion to move a nut a short distance onto a bolt.

Consider a ½″ x 12 bolt (½″ diameter with 12 threads per inch). To move the nut 1″, the bolt must be turned 12 times. A point on the circumference of the bolt would move almost 19″.

Machines

Every type of technology uses machines of one type or another. There are machines used in agriculture. People use machines to till soil,

TechnoFact: The mechanical advantage of a machine is the factor by which it multiplies an applied force. The mechanical advantage is the relative amount of the input force to the output force or the distances through which they are applied.

plant and cultivate crops, and harvest output. There are machines to irrigate plants, apply insecticides and herbicides, and store harvests.

Machines are used in communications. Printing presses produce newspapers and magazines. Transmitters send signals through the airways to radios and television sets. Switching gears interconnects our telephones.

There are energy conversion machines. They change energy from one form to another. These machines change the energy in coal into electricity. They change the energy in petroleum into mechanical energy to power vehicles. Energy conversion machines change electrical energy into heat and light.

Machines are used in construction. Bulldozers prepare construction sites. Cranes lift structural steel into place. Cement mixers prepare concrete for use. Power trowels smooth the concrete, while saws cut joints into the concrete.

There are machines used in manufacturing. They change the forms and shapes of materials. These machines make products. Presses stamp out parts. Welders fuse metal. Robots move parts from place to place.

Machines are used in medicine and health care. They restore or improve the health of people and animals. Machines in hospitals monitor patients' vital signs. X-ray machines help doctors diagnose illnesses. Pharmaceutical (drug manufacturing) companies use machines to develop drugs to treat illnesses. Veterinarians use machines to treat diseases in pets and farm animals.

There are machines used in transportation systems. Trucks, trains, ships, and airplanes move people and cargo. Conveyors load and unload cargo from vehicles. Computers maintain reservation information.

Many of these machines will be presented later in this book, but all of these machines can be traced back to other types of machines. They are a special type of manufacturing machine. These machines are called *machine tools.* Machine tools are the machines that make other machines. Without them, there would be no agricultural, communication, construction, energy conversion, general manufacturing, medical, or transportation machines. Machine tools change raw materials into parts. These parts later become machines and other products. See **Figure 4-29.** Machine tools have four major elements:

➤ A basic structure (such as a frame, bed, or table).

Figure 4-29. Every machine tool has these three basic parts.

Extend
Have the students look at several pictures of machines and identify their power units, tools, control units, and structures.

Brainstorm
Have the students compare the parts of a bicycle to the four major elements in machines. Ask the students if the bicycle is a machine.

Tech Terms
Machine Tool—A power tool used for shaping and finishing metal parts.
Machine Shop—An area where machine tools are housed.

TechnoFact: The Machine Age began with the Industrial Revolution and continues to this day. It grew out of a series of inventions that applied energy from fossil fuels, applied metallurgical knowledge and processes, used advancements in electricity and electronics, and used the internal combustion engine.

➤ A power unit (such as an electric motor or hydraulic drive).

➤ A control unit (such as a feed, speed, or depth of cut control).

➤ A tool (a device to produce a cut).

Machine tools can be grouped into six major classes:

➤ Turning machines.

➤ Drilling machines.

➤ Milling and sawing machines.

➤ Shaping and planing machines.

➤ Grinding and sanding machines.

➤ Shearing machines.

See **Figure 4-30.** Each machine tool has its own way of operating. It cuts materials into shapes using different motions and tools.

Turning Machines

Turning machines were some of the first machines to be developed. They are almost as old as civilization itself. The potter's wheel is an example of an early turning machine. Later turning machines are lathes.

Turning machines use a stationary tool. The material to be shaped is rotated around an axis. The stationary tool is fed into the work. The cut is created by slowly moving the tool along or into the rotating work.

Drilling Machines

The first drill was probably invented over forty thousand years ago. You learned earlier that it was no

Turning   Drilling   Milling and sawing

Shaping and planing   Grinding and sanding   Shearing

Figure 4-30. All machine tools can be classified as one of these six basic types.

TechnoFact: Machine tools operate by removing material from a piece of material called the *workpiece.* Five major machine tool operations are turning (shaping of a piece that has a round external contour), facing (shaping of a flat face on a round part), milling (shaping of a flat or contoured surface), drilling (forming a round hole), and boring (finishing an existing round hole, such as one formed by drilling). In addition, there are operations such as sawing, grinding, gear cutting, broaching, threading and tapping, polishing, buffing, and honing.

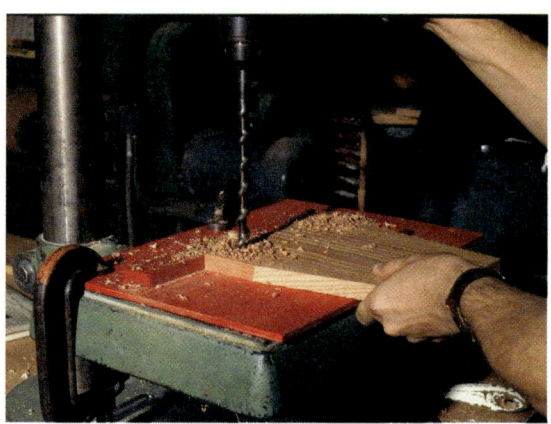

Figure 4-31. A drill press is the most common drilling machine.

more than a pointed stone on a shaft. Today, clamping a drill bit in a chuck (tool holder) is a more common way to produce holes. The chuck is rotated and pushed downward. This causes the rotating drill to feed into the work and cut a hole. The most common *drilling machine* is the drill press. See **Figure 4-31.**

Milling and Sawing Machines

Milling and sawing machines use either the straight or circular saw blades discussed earlier. A motor makes the blade move. The work is fed into the moving blade, creating a cut.

Many of these machines use rotating circular blades or cutters. The most common machines of this type are the milling machine, table saw, and radial saw. The woodworking surfacer, jointer, shaper, and router use the same cutting action.

Power hacksaws, scroll (jig) saws, and saber saws use reciprocating straight blades. The band saw uses a straight blade that has been welded into a loop. The blade travels around two wheels. This produces a *linear* (straight-line) cutting motion at the workpiece.

Planing and Shaping Machines

Planing and shaping machines are generally limited to cutting metals. Both machines use a single-point tool. The shaper moves the tool into the work to produce the cut. The metal planer moves the work into the tool. Both machines produce a flat cut on the surface of the work.

Grinding and Sanding Machines

Grinding and sanding machines use abrasives to cut materials from the workpiece. The abrasives can be bonded into wheels or onto a backing for sheets, discs, and belts. Generally, the work is moved against the moving abrasive. Grinders and sheet, disc, and belt sanders are the common machines in this group.

Shearing Machines

Shearing machines slice materials into parts. They use opposed edges to cut the workpiece. The material is placed between the cutting edges (knives or blades). One edge is moved down, forcing the material against the second edge. As more force is applied, the material is cut. A pair of scissors uses

Discussion
Compare the cutting actions of a table saw, band saw, and scroll saw.

Example
Sawing is used to prepare many standard materials in the wood products industry. For example, a sawmill is both a plant where timber is sawed into boards and a large machine used for sawing lumber.

TechnoFact: Machines used to process materials vary in size from handheld devices to large stationary tools. Some tools can do only one job, while others can do many tasks. For example, a boring machine is used primarily to drill holes, while a lathe can be used for turning, facing, drilling, threading, boring, and other operations.

Technology Explained

laser: A device that emits a beam of coherent, mono-chromatic light.

We are all probably familiar with lasers in one form or another. The word *laser* stands for light amplification by stimulated emission of radiation. A laser is a device that *amplifies* (strengthens) light. The laser is based on the findings of Albert Einstein and Niels Bohr. When atoms are exposed to an outside source of energy, such as electricity or light, the electrons become excited. This raises the electrons to a higher energy level within the atom. When the electrons fall back to their original energy level, they give off

Figure A. Lasers produce coherent light, meaning the light travels in the same direction as the triggering light. It, thus, amplifies the triggering light. In incoherent light systems, such as the Sun and the electric lightbulb, light travels in different directions, thus reducing its power.

light. When this light hits another atom, it causes more light to be emitted.

Scientists discovered that all the atoms of a material give off light that is the same wavelength. Because the light has one wavelength, it has one color. Thus, laser light is *mono-chromatic*. Laser light is also *coherent,* meaning that all the wave crests line up, **Figure A.** Beams of laser light are very intense and do not spread out as they pass through the air. The heat from the light can become so strong it can burn a hole in a rock.

The first practical laser used a ruby rod, **Figure B.** Ruby is an aluminum oxide crystal. Some chromium atoms that replace a few aluminum atoms in the crystal cause its deep red color. The ruby rod is

---

Resource
Reproducible Master
4-4: Tool
Relationships

Resource
Reproducible Master
4-5: Tool Usage
Chart

Resource
Reproducible Master
4-6: Tool Safety

Figure Discussion
Explain how sanding, shown in Figure 4-33, is a cutting operation. Discuss the types of sanders.

Tech Terms
Shearing—A process in which a material is shaped or finished by fracturing the unwanted material away from the workpiece.

a shearing action. Common shearing machines are sheet metal shears, punch presses, and paper cutters.

Tool Storage

Tools and equipment must be regularly inspected and maintained so they remain safe and reliable. Always follow the guidelines in the owner's manuals for inspecting, maintaining, storing, and servicing tools and materials.

Care should always be taken to properly store tools and materials. They should be kept in an organized fashion in a place where they are easily accessible. Some tools have sharp points that can injure people if not stored carefully. Other tools and materials must be protected from the environment. For example, computer disks should be protected from dust and dirt when stored. If they are not, there is a risk of losing important data.

TechnoFact: Many machine tools have a name indicating their principal function. Common machines used in working with metals and wood are the drill press, jointer, band saw, broach machine, milling machine, and shaper.

Figure B. The first laser used a ruby rod excited by a flash tube. Both ends of the rod are reflectors, but one end is a partial reflector. This allows the laser light to escape the rod when it has enough energy.

polished at each end. The ends are coated with a reflective material. One of the ends has a thinner coating than the other. To start the laser, outside energy excites the chromium atoms. Some of the light waves the chromium atoms emit strike the reflective ends. The light waves hitting the ends bounce back through the ruby rod. The waves strike other atoms, which become excited. The light is amplified as it bounces back and forth between the reflective ends of the rod. Once the light becomes strong enough, some will pass through the end of the rod with the thin coating.

We use many types of lasers today. Lasers are grouped by the *amplifying medium* they use. For example, the ruby laser used a ruby rod as the amplifying medium.

Other lasers use gases, dyes, or semiconductor materials to amplify the light. All lasers operate on the same basic principle as the ruby laser. Each type of laser gives off light that has a different wavelength and, therefore, a different color.

We use lasers in more applications every day. For example, we use them for measurement. We also use them to send messages through fiber optic cables. Compact disc players and some computer disk drives use lasers to store and retrieve data. People also use lasers for surgery, welding, cutting, and in bar code readers.

action is the use of tools. Tools are devices humans have developed to do jobs. Tools may be classified as measuring, cutting, drilling, gripping, pounding, and polishing tools.

Tools, however, are used by hand. Handwork may be fun, but it is slow. The demand for more efficient farming, products, communication media, energy, medical care, buildings, and transportation devices gave rise to machines. Humans found that some basic mechanisms could be combined with tools to create these machines.

They joined the lever, wheel and axle, pulley, inclined plane, wedge, and screw with the basic tools. From these advancements came machine tools. People developed turning, drilling, milling and sawing, shaping and planing, grinding and sanding, and shearing machines. These machines can be used to make all other machines. From these come agricultural, communication, construction, energy conversion, manufacturing, medical, and transportation machines.

Extend
Describe how machine tools are used to make machines for all technologies.

Tech Terms
Tool—Any device people have developed to help them do a task.

TechnoFact: Tools are the very foundation of all civilizations, past and present. Archaeologists find human artifacts in many forms as they explore early civilization. Tools are commonly found because stone tools and clay vessels (storage "tools") can remain buried for extended periods of time, while wood and bone implements often decompose in the soil. Archaeologists have developed special methods for classifying artifacts so these artifacts can help reveal the type of people that developed them. These classifications are based on distinctive features of the items, such as the shape of clay vessels, types of decoration on the artifact, and methods use to produce the items (for example, hand forming or flaking).

Curricular Connections

Mathematics

Examine a complex tool or simple woodworking or metalworking machine. Identify a place where mechanical advantage is used. Calculate and graph the mechanical advantage using several different forces applied.

Social Studies

Trace the development of a tool or machine over time. Start with early developments in places like Egypt and Rome.

Science

Examine a complex tool or simple woodworking or metalworking machine. Identify as many uses of simple machines (mechanisms) as possible. List and describe them.

Standards for Technological Knowledge

3

Activities

1. List five common tools you use at home.
   A. Identify the categories of tools in which they belong.
   B. Identify any mechanisms (simple machines) they use.
   C. Describe how they work.
2. List three machines you have seen or used.
   A. Classify them as one of the six types of machines.
   B. Describe the task each machine performs.
3. Build a simple item using tools.
   A. List each tool you use.
   B. Group the tools by the family of tools to which they belong.
4. Select a tool from your classroom lab. Prepare a presentation for the class reviewing the proper safety, inspection, maintenance, and storage procedures for the tool. Include a handout with your presentation.

TechnoFact: Tools and machines are present in all technological activities. There are agricultural, communication, construction, energy conversion, manufacturing, medical, and transportation tools and machines.

Test Your Knowledge

Do not write in this book. Place your answers to this test on a separate sheet of paper.

1. Why are humans different from other species of living things?
    A. They can adjust their behaviors to different situations.
    B. They can design, make, and use tools.
    C. In order to control their environments, they have to depend on tools.
    D. All of the above.
2. Devices people develop and use to complete tasks are called _____.
3. A _____ is a basic device controlling or adding power to a tool.
4. Give an example of each of the six primary tools.
5. Which type of measurement is most widely used in the United States?
    A. Conventional.
    B. Metric.
6. Explain each of the six mechanisms.
7. A lever producing large movement at one end when a force is applied to the other end is being used as a _____.
8. A pulley can be used to change the direction of force, multiply force, and multiply distance. True or false.
9. Which of the following are major elements of a machine tool? You may choose more than one answer.
    A. Frame.
    B. Tool.
    C. Electric motor.
    D. Feed control.
    E. The part being made.
10. Summarize the qualities of each of the six major classes of machine tools.

10. Turning machines use a stationary tool to cut a material rotating around an axis. Drilling machines rotate and push downward to cut a hole into a work. Milling and sawing machines have motors that cause blades to feed into a work in order to create a cut. Planing and shaping machines use single-point tools to produce flat cuts on the surface of a work. Grinding and sanding machines have moving abrasives, which cut materials from the workpiece. Shearing machines use opposed edges to slice materials into parts.

Answers to *Test Your Knowledge* Questions

1. D. All of the above.
2. tools
3. mechanism
4. Answers will vary. [examples: ruler (measuring), hand saw (cutting), spade bit (drilling), vise (gripping), claw hammer (pounding), sharpening stone (polishing)]
5. A. Conventional.
6. A lever is a device consisting of a lever arm and a fulcrum, and it can increase the force applied to the work or the amount of movement created. A wheel and axle consists of a shaft attached to the center of a disc, and it operates as a second-class lever. A pulley is a grooved rim attached to a loose axle, and it can act to change the direction of force, multiply force, or multiply distance. An inclined plane is a sloped surface, and it operates on the principle that moving up a slope is easier than lifting straight up. A wedge is a set of two inclined planes. A screw is an inclined plane wrapped on a round shaft.
7. distance multiplier
8. True
9. A. Frame., B. Tool., C. Electric motor., and D. Feed control.

Chapter 5
Materials and Technology

This chapter covers the benchmark topics for the following Standards for Technological Literacy:

3 12 19

Did You Know?

➤ The word *plastic* refers to new substances chemical means produce. The first such material to be manufactured was Parkesine. The British inventor Alexander Parkes developed it. He won a bronze medal for his invention at the International Exhibition of 1862 in London.

➤ Celluloid was developed in the late 1800s. The solid form was used as imitation horn for billiard balls. Celluloid sheets were used for men's collars and photographic film.

➤ Leo H. Baekeland, a U.S. inventor, introduced a new class of large molecule plastics in 1909. He developed Bakelite® plastic. This plastic is a hard, chemically resistant thermosetting plastic.

Objectives

The information given in this chapter will help you do the following:

➤ Describe natural materials.

➤ Name and give examples of the three major types of materials used in technological systems.

➤ List and give examples of the four major types of engineering materials.

➤ Summarize the differences between renewable and exhaustible materials.

➤ Explain the three major ways to process raw materials.

➤ Identify and describe the major properties of materials.

Key Words

These words are used in this chapter. Do you know what they mean?

acoustical property
alloy
ceramic
chemical process
chemical property
composite
ductility
electrical property
engineering material
exhaustible resource
magnetic property
material
mechanical process
mechanical property
metal
natural resource
optical property
physical property
plastic
polymer
process
property
renewable resource
thermal conductivity
thermal expansion
thermal process
thermal property

Chapter Outline
Did You Know?
Objectives
Key Words
Types of Materials
Engineering Materials
  Metals
  Ceramics
  Polymers
  Composites
Obtaining Material Resources
  Renewable Resources
  Exhaustible Resources
Processing Raw Materials
  Mechanical Processing
  Thermal Processing
  Chemical Processes
  Products of Processing
Properties of Materials
  Physical Properties
  Mechanical Properties
  Chemical Properties
  Thermal Properties
  Electrical and Magnetic Properties
  Optical Properties
  Acoustical Properties
Summary
Curricular Connections
Activities
Test Your Knowledge

Extend
Have the students explore how, over time, trees have been used as feedstock for natural materials (digging sticks and club handles), modified natural materials (lumber), processed natural materials (waxed paper), and human-made materials (cellulose).

Tech Terms
Synthetic Material—A material produced by chemical, rather than natural, processes.

Natural materials → Modified natural materials → Processed natural materials → Human-made materials

Figure 5-1. The development of materials has gone through four stages.

Humans have always used materials as they have adapted to their world. Over time, there have been four basic phases of material use. See **Figure 5-1.** First, people used the materials they could find. They searched for rocks having the right shape for spearheads. Carefully selected tree branches became digging sticks (the early plow). Rocks, limbs, and tree bark became houses. During the next phase, people started forming natural materials to fit their needs. Stones were chipped to make arrowheads. See **Figure 5-2.** Natural copper was shaped to make metal tools and utensils. Trees were cut into boards. As human technological knowledge advanced, a third phase of material use appeared. Combining old materials developed new materials. Iron, coke (a coal product), and limestone were converted into steel. Copper and zinc were combined to form the alloy we call *brass*. Thin strips of wood were glued together to form plywood. Finally, people created synthetic materials. In the late 1800s and thereafter, plastics were developed. These new materials were created to fit special product needs. Plastics were developed for a wide range of products.

Figure 5-2. Early Native Americans, like these shown in this diorama, were masters at using natural materials.

TechnoFact: Early agricultural implements were constructed with basic materials. These implements included the digging stick, hoe, scythe, and plow. The use of human-designed implements developed slowly over the centuries, as new materials were developed.

Special plastics became fibers for new textiles.

Today, as in past years, we live in a world of materials. Everything around us is made of materials. See **Figure 5-3.** Some of these materials appear in nature. We call these *natural materials* or ***natural resources.***

Trees, grass, soil, minerals, coal, petroleum, and many other materials are found naturally on Earth. See **Figure 5-4.** Many of these materials have limited use to us in their natural states. Rocks can be used to make retaining walls. Plants can form windbreaks. Trees give us shade on a hot day. Grass and other plants make parks look nicer. Stones can be used for walkways.

Most objects we use, however, are made of natural materials that have been modified. People have processed them into new forms and compositions.

Figure 5-3. Things made from materials surround us. (Lexington Homes)

The desks, chairs, and tables in your school are made from materials that were once in their natural states. The tabletops may be made from wood. The tables may have metal legs that were once an ore. The chairs may have plastic seats made from petroleum. In this

Figure Discussion
Ask the students what materials are shown in Figure 5-3. Ask them if the materials are natural or synthetic.

Research
Ask the students to find out what natural resources are produced in your state.

Figure 5-4. Natural materials include things growing or found on Earth (left). Trees grow in soil and provide wood (right). Minerals are found underground. (Weyerhaeuser Co. and Amoco Corp.)

TechnoFact: A mineral is an inorganic substance occurring in nature. Each mineral has a unique and homogeneous chemical composition, definite physical properties, and, generally, a distinctive crystalline form. Several common minerals, such as carbon, gold, silver, copper, lead, mercury, platinum, and iron are elements. However, the vast majority of all minerals are chemical compounds.

chapter, you will explore the roles materials play in technology.

Types of Materials

A *material* is any substance out of which a useful item can be made. Not all materials are alike, however, nor do they fulfill the same purposes. For this discussion, three major types of materials used in technological systems will be explored.

The first type of material provides energy to operate technological systems. This type is the input to energy conversion systems. It powers engines, turbines, and other energy converters. Some of these materials are petroleum products like fuel oil, gasoline, kerosene, or diesel fuel. Falling water and wind are also sources of energy, as are uranium, wood, grain converted into ethyl alcohol, and coal. These materials will be discussed in Chapter 6.

The second type of material includes process-supporting materials. These are the liquids and gases supporting natural and technological processes. They include the air we breathe and water we drink. These materials also include the water and fertilizers promoting plant growth on farms and forests. This type of material also cools industrial processes, lubricates moving parts, and provides the raw materials for products and thousands of other uses.

A third type of material is called *industrial material* or *engineering material.* This group will be the main focus of this chapter. Engineering materials have rigid structures. They withstand outside forces, and their shapes are hard to change. Most of us know these materials as solids. Engineering materials form the primary structures for all products having set forms. See **Figure 5-5**.

Figure 5-5. The steel on the left and the lumber on the right are examples of solids, or engineering materials.

TechnoFact: Natural resources conservation embraces the broad concept of conserving Earth by protecting its capacity for self-renewal. The science of ecology deals with the complex interrelationships among humans, other living things, and the environment. This is important at a time when growing population and industrial activity have shown how easily ecological relationships can be disrupted.

Typical products made of engineering materials are bicycles, tennis rackets, cola bottles, garbage bags, microwave ovens, and T-shirts. These products all share one characteristic. They hold their shapes over time. Products made of engineering materials have basic structures that remain constant. Soap powders, toothpaste, lipstick, cherry cola, and pizza are different. All of these products are made from materials without structure. They must be placed in or on containers to hold their shapes.

Engineering Materials

Engineering materials are the building blocks for many products. They can be grouped into four major categories:

➤ Metals.

➤ Ceramics.

➤ Polymers.

➤ Composites.

Industries and individuals use each of these materials. Also, each type of material is very different from the others.

Metals

Metals are inorganic materials. This means they were never living things. Metals, as a group, have similar internal structures. Their molecules are arranged in a box-like framework.

Figure 5-6. This is what a section of metal looks like when magnified one thousand times. (Bethlehem Steel Corp.)

These material structures are called *crystals*. They combine to form the grains of metal. See **Figure 5-6.** The crystal framework produces a rigid, uniform material.

Metals are seldom used in their pure forms. We do not often use pure gold, silver, copper, aluminum, or iron. Most metals we use are mixtures of two or more different base materials. Iron is combined with carbon and other elements to make steel. Aluminum is often combined with copper, silicone, and magnesium to make more useful materials. These mixtures of two or more elements are called *alloys.* Bronze is an alloy of copper and tin. Sterling silver is an alloy of silver and copper. Stainless steel is a nickel-chromium alloy of steel.

Demonstrate
Use simple tools to show some properties of metals by bending, abrading, and pounding them.

Tech Terms
Metals—A group of chemical elements exhibiting most or all of the following properties: are solid at room temperatures, are opaque, are good electrical and thermal conductors, are lustrous when polished, and have a crystalline structure in the solid state.

TechnoFact: Metals are usually durable and resilient to most kinds of stresses. Most metals are solid, flexible, pliable, and elastic, and have a high tensile strength. They are unaffected by constant strain, abrasion, exterior deformation (damage), breakage, and fatigue.

Tech Terms

Ceramic Materials—Any hard, brittle, heat-resistant, and corrosion-resistant material made from a nonmetallic mineral, such as clay, that is fired at a high temperature.

Ceramic Product—An object, such as earthenware, porcelain, or tile, made of ceramic materials.

Ceramics—The art and techniques used to make objects of ceramic, especially from fired clay.

Figure 5-7. A Pueblo Indian from the southwest United States made this ceramic pot.

Ceramics

Ceramics are probably the oldest materials humans have used to make products. See **Figure 5-7.** Ceramics, like metals, are inorganic materials. They are, also, made of crystals.

Ceramic materials are widely used because they are very stable. Heat, moisture, and chemicals do not affect them. Ceramic materials are stiff and brittle. They do not bend, and they break under stress. There are three major types of ceramic materials:

➤ Clay-based materials. These materials are substances made up of crystals held in place by a glass matrix (binder). Typically, clay-based ceramics are used to make bricks, decorative and drain tiles, dinnerware (earthenware and china), and sanitary ware (sinks and toilets).

➤ Refractories. These are crystalline materials held together without binders. They can withstand high temperatures. Firebrick (used to line furnaces and fireplaces) and the space shuttle reentry tiles are made of refractory materials.

➤ Glass. This is an amorphous material (without a regular structural pattern) commonly made from silica sand. Glass is a very thick liquid appearing to be a solid under normal conditions. Glass is used for windows, containers, and many other products.

Polymers

Polymers are organic materials. This means they are formed from materials that were once living. The molecules in the material form chain-like structures. Most polymers are made from natural gas and petroleum. Nylon and polyethylene are examples of polymers made from these base materials. See **Figure 5-8.** Wood and other cellulose fibers can also be used to make polymer materials. Rayon is a cellulose-based polymer.

Figure 5-8. These hot air balloon are made from nylon, a plastic. (Jack Klasey)

TechnoFact: Glass is typically made by fusing silicates with a metallic oxide. The material solidifies from the molten state without crystallization. Glass is generally hard, brittle, and transparent or translucent.

Polymers may be natural or human-made. A typical natural polymer is natural rubber. Synthetic (human-made) polymers are often called *plastics.* They are materials designed and produced to meet specific needs. Polymers can be grouped into three categories:

➤ Thermosets. These are materials that form rigid shapes under heat or pressure. After a part or product is made, the structure remains rigid. Once set, the material will not soften when heat or water is applied. Cases for television sets, furniture components, and many automotive parts are made from thermoset plastics.

➤ Thermoplastics. These are materials that soften under heat and become rigid when cooled. They can be heated and reshaped a number of times. Product packages are commonly made from thermoplastic materials.

➤ Elastomers. These are materials that can be stretched, but rapidly return to their original shape. To be called an *elastomer,* the material must withstand being stretched at least twice its length. Rubber bands and balloons are common products made from elastomers.

Composites

Composites are combinations of materials. Each material, however, retains its original properties. Mixing cement, sand, gravel, and water forms a composite. We call this composite *concrete.* The cement binds the sand and gravel together. Water starts the curing action of the cement. The sand and gravel, however, have not changed. If you look at a piece of broken concrete, you can see the sand and gravel. This is different from an alloy. You cannot see the copper and zinc particles in a piece of brass.

Likewise, wood chips are glued and pressed together to form particleboard and wafer board. Thin wood sheets, called *veneer,* are glued and pressed together to form plywood. Wood fibers can be heated and pressed under high pressure to make hardboard. All of these wood products are composites.

Still another common composite is made with glass fibers and plastic binders. The fibers are coated with a liquid plastic. When the plastic cures, we have a product called *fiberglass.* It is widely used for automobile, aircraft, boat, and truck parts; bathroom shower enclosures and bathtubs; fast-food serving trays; and sporting goods.

Obtaining Material Resources

All materials we use can be traced back to Earth. As you have learned, they may be living things, or they may be minerals and other elements found in the ground, sea, or air. These materials must be located before they can be removed from the earth through such processes as harvesting, drilling,

TechnoFact: The first polymer, Bakelite, was discovered by Leo H. Baekeland in 1909. This led to additional development of a number of polymers, including neoprene, nylon, Orlon® polymers, and Dacron® fabric.

Figure 5-9. Young trees grow next to a stand of mature trees. Why should mature trees be cut, rather than younger ones? (Weyerhaeuser Co.)

and mining. Few materials occur in nature in usable forms. Most must be changed into new forms before they can be used as inputs in the manufacturing process. All materials may be grouped into two categories:

➤ Renewable material resources.

➤ Exhaustible material resources.

Renewable Resources

Renewable resources are living things. They sprout from seeds or shoots, or they are born. Over time, they grow and mature. Finally, they die. See **Figure 5-9.** These materials may go through their life cycles without human care. They may grow in a forest, on a mountainside, or on a plain. Renewable resources may be

part of what we call *nature.* People, however, grow many of these materials. Humans engaged in *farming* (growing food and fibers) and *forestry* (growing trees) may produce these materials. Commonly used renewable material resources are trees, grains, animals, fish, fruits, and vegetables.

Locating and Harvesting Natural Resources

Some renewable materials must be located before they can be harvested. These are the materials growing wild in nature. Foresters search forests to locate mature trees. They also select the correct species. See **Figure 5-10.** Some trees have value for lumber, while others are used to make paper and particleboard. Still others are good only for firewood. In fact, some trees have little or no commercial value. The western juniper and the pin oak are two examples. You may want them in a yard for shade and appearance. These trees, however, produce little valuable lumber or wood fibers. Fish, also, must be found before they are harvested. Many people will tell you that finding a tree is easier than finding fish.

Growing and Harvesting Natural Resources

Most people are familiar with farming. Farmers plant seeds in the spring. They care for the crop during the growing season. Fertilizer applied to the soil increases growth. Weeds and pests are controlled. At the end of the growing season, the crop is harvested.

TechnoFact: Natural resources include plant life, animals, mineral sediment, fossil fuels (coal, oil, and natural gas), earth, unpolluted water, and uncontaminated air. Natural resources are sorted into two groups, renewable and nonrenewable.

Figure 5-10. Foresters carefully select the trees to be harvested. (Weyerhaeuser Co.)

Standards for Technological Knowledge

Research
Have the students research the common life cycle of one type of crop and one type of livestock.

Tech Terms
Forestry—The management of forests from tree planting to harvesting. A main emphasis is on planting, growing, cutting, and replenishment.

See **Figure 5-11.** The bounty is placed in storage or used immediately.

In a similar manner, ranchers raise cattle and sheep. Adult livestock give birth to their young each year. The young animals are fed and cared for. They grow to be young adults. Some are saved to raise additional young animals. The majority is "harvested." They are butchered for their meat, hides, and other parts. Additionally, shearing adult sheep harvests wool.

Farming and ranching, however, are not the only ways people grow renewable resources. Many wood product companies operate large tree nurseries. They grow trees to be planted in forests or on tree farms. The newly planted trees are carefully tended. They receive fertilizer and often are pruned to increase growth. The goal is to grow the maximum amount of wood fiber in the least time. Similarly, fish are now being raised on farms. A large industry has emerged to grow catfish for sale.

The key for all renewable resources is to harvest them when they are *ripe*

Figure 5-11. Farmers harvest mature crops, like this grain, before they spoil. (Deere and Company)

TechnoFact: The three major methods of tree cutting are clear cutting, selective cutting, and seed tree. In clear-cutting all trees in an area are cut, and new trees are planted to reforest the area. Selective cutting harvests large, mature trees, while leaving younger trees to grow. In the seed tree method, about 10 percent of the trees in the cutting area are used as a natural seed source.

Brainstorm
Ask the students to list some of the ways they could reduce their use of exhaustible resources.

Tech Terms
Exhaustible or Nonrenewable Resource—A resource having a limited supply that cannot be replaced or can be replaced only over extremely long periods of time.

Figure 5-12. This machine harvests mature trees that will be used to make paper.

(ready to be used). See **Figure 5-12.** Cutting trees that are too small is wasteful. Also, letting a tree mature and die wastes valuable resources.

Likewise, catching young fish breaks nature's cycle. Later, we pay a price. There will be too few mature fish to provide food. Entire species of fish may become extinct.

Exhaustible Resources

Many resources are exhaustible. There are only so many of these resources on Earth. When **exhaustible resources** are used, they will be gone and can never be replaced. This type of resource provides a unique challenge. People must use each resource very wisely. Wasted resources are gone forever. Changes in habits and attitudes will not bring them back. New resources cannot be grown.

Common exhaustible resources include petroleum, natural gas, coal, *mineral ores* (such as iron, copper, and aluminum), farm soil, clay, and *chemical deposits* (such as salt and sulfur). See **Figure 5-13.** Obtaining exhaustible materials involves two major steps:

➤ Finding deposits of the resource.
➤ Extracting the raw material.

Figure 5-13. Modern farming practices help protect soil from erosion. (Deere and Company)

TechnoFact: Conservation activities for nonrenewable resources concentrate on sustaining an ample amount of these resources well into the future. Natural resources are conserved for their biological, financial, and recreational values. In addition, the resources are conserved for their environmental splendor and significance to native societies.

Finding the resource requires special skill, training, and a certain degree of luck. The challenge is often the job of geologists. See **Figure 5-14.** Geologists study Earth to determine possible locations of the materials. They may use aerial maps and satellite pictures. Seismic studies are also used. These involve firing explosives buried in the ground. The resulting shock waves are read on special equipment. "Pictures" of the rock layers below the surface are obtained. These pictures are charts and graphs the equipment makes. They are read to locate promising sites for exploration.

Once a body of ores, minerals, or petroleum is located, it must be extracted. Holes are often drilled to get core samples. Core samples are samples of the rock below the surface. The samples provide additional evidence of the presence of the resource. If the samples show promise that the resource is there, extraction begins.

Two major methods of extraction are used. See **Figure 5-15.** Wells are

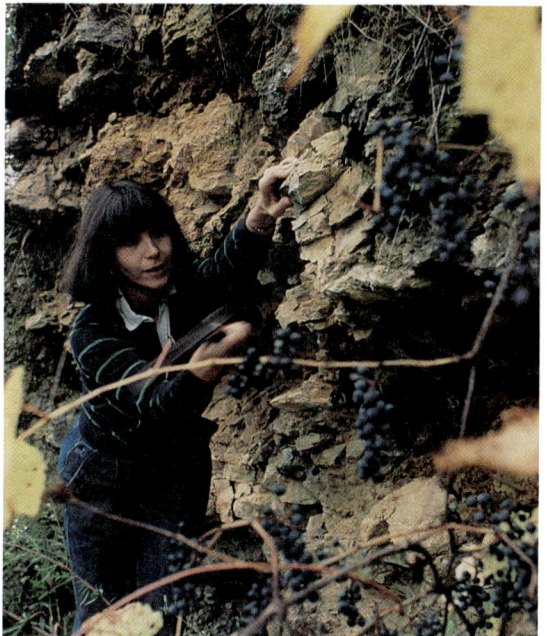

Figure 5-14. This geologist searches for natural resources located in the earth. (Amoco Corp.)

drilled to remove liquid resources. Mines are dug to remove solid resources. Some may be big holes dug into Earth's surface. These holes are called *open pit mines.* Others are tunnels extending into the sides of mountains

Standards for Technological Knowledge

19

Research
Have the students research the way seismology is used to discover petroleum and other resources deep beneath the surface of the earth.

Tech Terms
Geology—The study of the rock, soil, and mineral structure of the earth.
Mining—The processes used to remove resources from the earth.

Brainstorm
Ask the students to list the advantages and disadvantages of open-pit mining.

Extracting Natural Resources

Mining

Drilling

Figure 5-15. Mining and drilling extract natural resources. (Caterpillar, Inc. and Amoco Corp.)

TechnoFact: Mining is a major process used to extract resources from the earth. It includes both underground mines and surface (open-pit and strip) mines. Advances in technology may soon allow the mining of the seafloor for metallic ores.

or vertically into Earth. See **Figure 5-16.** These tunnels are called *underground mines.*

Processing Raw Materials

Most raw materials have little value to people in their own rights. We cannot use them as they come from the ground. These materials must be *processed* into something more useful. Often the resource found is mixed with other materials. In this condition, it has limited value. Iron ore is of little use to us in its natural state. Likewise, a pile of harvested trees cannot be used easily. See **Figure 5-17.** Processing raw material falls into three general categories:

Figure 5-16. This is the entrance of an old gold mine in eastern Oregon.

➤ Mechanical processing.
➤ Thermal processing.
➤ Chemical processing.

Mechanical

Thermal

Chemical

Figure 5-17. There are three types of processes used to convert natural resources into usable materials. (Inland Steel Company)

TechnoFact: Examples of production activities include the people who grow trees, the sawmill that changes logs into lumber, and the contractor who converts lumber and other products into homes.

Mechanical Processing

Mechanical processing uses machines to cut, slice, crush, or grind material into new forms. The size and shape of the natural resources are changed. Grain is ground into flour. Rocks are crushed into gravel. Animal carcasses are cut into meat. Trees are cut into lumber, veneer, and chips. See **Figure 5-18.**

Thermal Processing

Often, we rely on heat to process materials. The materials are melted or softened so they can be changed into more useful materials. This is called *thermal processing.* Iron ore, coke, and limestone are heated together in a blast furnace. This action produces pig iron. Later, this material is thermal processed into steel. See **Figure 5-19.**

Figure 5-18. Saws cut logs into lumber. Sawing is a mechanical means for changing material. (Weyerhaeuser Co.)

Most metal ores are processed using heat. They are first crushed into manageable pieces. Heat is then used to melt the metal away from the impurities.

Likewise, heat can be used to fuse materials. Silica sand is heated to high

Demonstrate
Use a tool to crush a sugar cube to show the concept of mechanical processing.

Brainstorm
Have the students list the reasons metals are processed using thermal processes.

Tech Terms
Mechanical Processing—Using mechanical force to change a resource into a new form. Examples are sawing lumber from logs or crushing rocks to make gravel.
Thermal Processing—Using heat to change the form of a resource. Examples are smelting steel or making glass.

Figure 5-19. This worker is controlling a steelmaking (thermal) process.

TechnoFact: The open-hearth steelmaking process was developed in France by the Martin brothers in 1863. They used a furnace developed by Frederick Siemens to melt the ingredients for the steel.

Resource
Reproducible Master
5-1: Properties of
Materials

Discussion
Discuss some
common types of
standard stock, such
as lumber, plastic
pellets, and sheet
metal.

Tech Terms
Property—A
characteristic that a
material possesses.

Figure 5-20. Each material used to build and equip this kitchen was chosen to meet special requirements. (Velux-America)

temperatures. It is melted and fused to make glass.

Chemical Processes

Chemical materials are altered using chemical action. Gold is removed from its ore using an arsenic *chemical process.* Natural gas changes into plastic when chemicals react with it. Aluminum is produced using a combination of electricity and chemicals. This is called an *electrochemical process.*

Products of Processing

The products of processing actions are industrial materials. This type of material comes in a standard form or size. It is called *standard or industrial stock.* Each of these materials needs further processing to be useful. Flour mixes with other materials to make bread. Lumber must be cut and shaped into useful products. This further

processing is called *manufacturing.* It will be studied in Chapter 24.

Properties of Materials

There are hundreds of materials available. They can be used in various ways to meet human needs and wants. For each job (product or use), a material must be selected. See **Figure 5-20.** It would be unwise to use wood for a heat shield in a welding booth. Even at a low temperature, it would burn. Steel would be a poor choice for a baseball bat. It is too heavy (dense).

Understanding material *properties* helps people select the right material for the job. *A material property is a characteristic the material has.* See **Figure 5-21.** These properties can be divided into seven groups:

➤ Physical.
➤ Mechanical.
➤ Chemical.
➤ Thermal.
➤ Electrical and magnetic.
➤ Optical.
➤ Acoustical.

Physical Properties

Physical properties describe the basic features of a material. One of these features is density. This is a comparison of weight to size. Density is often given in pounds per square foot.

Another physical property is moisture content. It tells us the amount of

TechnoFact: Standard stock or industrial materials are materials that were produced in primary processing, but must be further processed to make a product. For example, lumber (a standard stock) is produced in a lumber mill. The lumber needs to be processed further into products, like furniture, decorative items, and homes, before it is useful to most people.

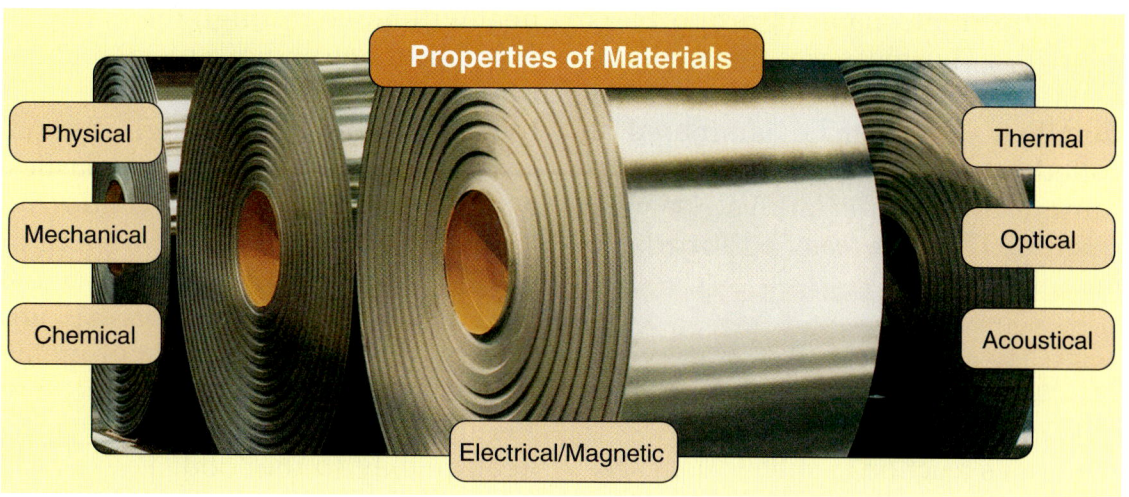

Properties of Materials

Physical

Mechanical

Chemical

Thermal

Optical

Acoustical

Electrical/Magnetic

Figure 5-21. These are the properties of materials. (LTV Steel Company)

Brainstorm
Ask the students which mechanical properties are most important in choosing a flooring material.

Tech Terms
Mechanical Properties—The properties determining how a material will react to mechanical forces or loads.

water in the material. Lumber used for furniture should have a moisture content of 6 to 8 percent. Construction lumber will have a moisture content in the 12 to 16 percent range.

A third physical property is surface smoothness. This property will affect a material's use. A very smooth material makes a poor floor. It becomes too slippery when wet. Similarly, a rough material is a poor bearing surface. The parts will not slide easily. Abrasive materials have a rough surface, while glass has a smooth surface.

Mechanical Properties

Mechanical properties affect how a material reacts to mechanical force and loads. See **Figure 5-22.** They describe how a material will react to twisting, squeezing, and pulling forces.

Strength is a common mechanical property. It measures the amount of force a material can withstand before breaking. A strong material can hold heavier loads than weak materials.

Steel is very strong, while writing paper is fairly weak.

Hardness is another mechanical property. This property measures the resistance to denting and scratching. Often, hard materials are also brittle. They break easier than softer materials. Diamonds and glass are both hard materials. They will *shatter* (break) if they receive a sharp blow.

Figure 5-22. These windows are manufactured to withstand many mechanical forces. (Andersen Corp.)

TechnoFact: Strength is a common material property. It is described as the capacity of metals, woods, plastics, ceramics, and composites to withstand stress and strain. Stress is the internal force exerted by one portion of the material upon the adjoining portion. Strain is the deformation or change in dimension caused by stress (applied load).

Extend
Ask the students if all metals rust. Ask them to explain their answers.

Figure Discussion
Have the students discuss some ways a steel part could be protected from rusting, aside from the method in Figure 5-23.

Tech Terms
Corrosion—
Atmospheric oxidation of metals caused by chemical reaction between the material and its surroundings (air or water).

A third mechanical property is *ductility.* This measures the ability of a material to be shaped with force. A ductile material can be hammered or rolled into a new shape. Many aluminum and copper alloys are very ductile. They are easy to form into complex shapes.

Chemical Properties

Chemical properties describe a material's reaction to chemicals. These chemicals may be of any kind. They may be water, some other liquid, or acids in the air.

These chemicals may cause corrosion. They may react with the material, causing unwanted by-products. Water reacts with steel to form rust. When copper reacts with chemicals, it begins to tarnish. Most ceramic materials, however, resist chemical actions. That is why ceramics are used for dishes, bathroom fixtures, and food containers.

Thermal Properties

Thermal properties are a material's reaction to heat. Most materials *expand* (become longer and wider) when heated. This property is called *thermal expansion.* The property can be used in many ways. For example, gently heating a metal jar lid may make it easier to remove. The metal lid expands more than the glass jar.

Many materials allow heat to move through them. The measure of this property is called *thermal conductivity.* The higher the conductivity, the easier

heat will move through the material. If you heat one end of a metal rod, the other end will quickly become hot too. Most metals have high thermal conductivity. Wood and most plastics, however, have low thermal conductivity. If you heat one end of a piece of wood or plastic, it will catch fire before the other end gets warm. This explains why most kettles are made of metal, while their handles are plastic. See **Figure 5-23.** The metal conducts the heat quickly to the food. The handle insulates the heat from the cook's hand.

Electrical and Magnetic Properties

Electrical properties describe the material's reaction to electrical current. Materials that will easily carry the current are called *conductors.* Other materials resist the flow of electricity. They are called *insulators.* Metals are generally good conductors. Most polymers and ceramics are good insulators. Plastic insulated copper wire is used to

Figure 5-23. A material's thermal properties can be advantageous when designing a product.

TechnoFact: The most notable kind of corrosion is the rusting of iron. Rusting is basically a process of oxidation in which iron mixes with water and oxygen to create a reddish-brown coating. Rust is a chemical compound (ferric oxide) that is brittle and flakes off the outside of the iron to constantly expose a clean surface.

carry electricity in homes and other buildings.

Similarly, *magnetic properties* describe the material's reaction to magnetic forces. Some materials will become magnetized easily. Other materials will never become magnets. Most iron alloys (steel) can be magnetized. Almost all other materials are nonmagnetic.

Optical Properties

Optical properties govern the material's reaction to light. Some materials let light pass through them easily. They are called *transparent.* Other materials stop all light. They are said to be *opaque.* Still other materials let some, but not all, light through. They are called *translucent.*

Another optical property relates to reflecting light. Each material absorbs part of the light striking it. Other parts of the light spectrum are reflected. The part reflected reacts with the human eye. We see the material as having color. Light absorbing and reflecting properties are used in photographic filters. These filters allow certain colors of light to pass through and reflect the other colors.

Acoustical Properties

The *acoustical properties* of a material determine its reaction to sound waves. Some materials absorb the waves. They are sound insulators or deadeners. Others reflect sound. Some materials carry sound. They are sound transmitters. Smooth hard materials generally reflect sound. Softer, uneven material will absorb sound. A room having a lot of hard material in it causes sound to echo. That is why most auditoriums have a lot of soft material in them. Curtains, insulating tiles, and curved walls improve the sound quality of a room.

Material Storage

All materials must be handled and stored properly so they remain safe and effective. Both raw and processed materials must be appropriately handled and stored to avoid damaging them or wasting the effort it takes to process them. During the manufacturing process, materials usually need to be transported several times and often need to be stored between processing steps.

Prior to being processed, raw materials are sometimes sent to storage. Here, they must be stored correctly so they are not exposed to contaminants. Care must be taken to prevent the materials from rotting, rusting, or deforming in any way.

Many products are processed to meet quality specifications. Storage facilities have a responsibility to make sure that the quality is maintained throughout the storage period. Processed materials usually must be handled and stored even more carefully than raw materials. Finished products are often packaged with protective material, which helps protect them from damage while they are being handled. It is especially important to use care when handling finished products.

Extend
Ask the students for which applications they would want glass to be transparent, translucent, and opaque.

Figure Discussion
Ask the students which materials, beside plastics, could be used for the handle of the kettle shown in Figure 5-24.

Tech Terms
Insulation—A material or device that reduces or prevents the conduction of heat or electricity to undesirable areas or surfaces.

TechnoFact: Insulators control the flow of heat and electricity. Common heat insulators are fur, feathers, fiberglass, cellulose fibers, stone, firebricks, wood, and wool. Common electrical insulators are dry air, glass, porcelain, rubber, and plastics.

Technology Explained

geothermal energy:
Energy derived from the natural heat present in the earth.

People are actively seeking new sources of energy. One of these sources is actually very old. It is the natural heat present in the earth. The earth's magma, heating the earth's crust and water below, develops this heat. The effect of this heating can be seen in geothermal areas, such as Yellowstone National Park, **Figure A.** Eruptions of volcanoes, such as

Figure A. Here is a geyser at Yellowstone National Park.

Geothermal heat pump

Figure B. This is a diagram of a residential geothermal heating unit.

Mount Saint Helen's, are another example of the earth's geothermal energy.

One of the ways to use this energy is with a geothermal heat pump in a home, **Figure B.**

Summary

Humans have used materials throughout history. Materials are key to our existence. They are used for almost everything we need and want. Materials are found in our clothing and shelters. They become the products and fuels we use. Materials are the food and fibers sustaining our lives.

All materials are either renewable or exhaustible. Renewable resources can be grown and harvested. Proper management will give us a constant supply of these resources. Exhaustible resources have a limited supply. Once used, they cannot be replaced. They require careful use if future generations are to have them, too.

All materials must be grown or located. They then must be harvested or extracted. Finally, they are processed into industrial or standard materials.

Extend
The term *conservation* came into use in the late nineteenth century and referred to the management of such valuable natural resources as timber, wildlife, topsoil, and minerals. Ask the students why conservation is important today.

TechnoFact: Static tests are performed to reveal a material's rigidity, ductility, elastic threshold, thermal conductivity, and other properties.

To use this type of system, two wells are drilled. Warm water is brought to the heat pump. The unit extracts heat from the water in the winter. It can add heat to the water (cool the home) in the summer. The altered water is then pumped back into the ground. This type of unit will work anywhere. It uses the difference in temperature between the water and outside air.

Another use of geothermal energy is in power plants, **Figure C.** These pumps work much differently than residential heat pumps do. First, the earth's magma heats water to at least 350°F. A well and pipes bring the hot water to the power plant. Some of the water is allowed to enter a set of steam separators. Here, a part of the water is allowed to flash

Figure C. Here is a diagram showing the parts and operation of a geothermal power plant.

into steam. The remaining hot water is flashed into steam in a second separator.

The pressurized steam is moved to a turbine through pipes. Here, it turns the turbine connected to an electric generator. The output of the generator

is fed into the power grid through wires.

A condenser pulls the cooled steam from the turbines and allows it to change back into water. A gas emissions treatment unit removes sulfide gas. The hot water is allowed to cool in a cooling tower. Then, the water is returned to the geothermal reservoir through an injection well.

Like most technological systems, there are some drawbacks to geothermal power. The water contains chemicals that can deposit on the walls of the pipes. This will restrict the flow of the hot and cold water. Also, the plants are more expensive to build than conventional power plants are.

All materials must be grown or located. They then must be harvested or extracted. Finally, they are processed into industrial or standard materials.

Engineering materials are an important group of materials. They have set structures. These materials are the rigid materials used in many durable products.

Metals, ceramics, polymers, and composites are engineering materials. Each type of material has its own set of properties. These properties include the physical, mechanical, thermal, chemical, electrical and magnetic, optical, and acoustical qualities of the material. The combination of properties in a material dictates its range of use. Materials are a basic input to technological actions as systems. They are used in agricultural, energy conversion, information processing and communication, construction, manufacturing, medical, and transportation technologies.

Extend
Ask the students why they should know about materials and their properties.

TechnoFact: Dynamic tests expose a material to expected operating conditions, such as vibration, changing loads, shock, and impact.

Curricular Connections

Mathematics

Research the properties of several materials in reference books or on the Internet. Develop a set of graphs comparing the properties of these materials.

Social Studies

Select a type of material, such as steel, plywood, or nylon. Trace its historical development:

➤ List the important events in its development.

➤ Indicate where and when these developments took place.

➤ List any inventors contributing to the development.

Science

Look at a pencil. List the properties the materials in it have. Describe why these properties are important.

Standards for
Technological
Knowledge

3 **12**

Activities

1. Use magazines to develop a picture set showing at least three uses of each of the following engineering materials:

   A. Metals.

   B. Ceramics.

   C. Plastics.

   D. Composites.

2. Complete an experiment comparing a property, such as strength, for two different materials.

3. Gather a group of five to ten different materials. Arrange the materials in the kit first by hardness, then by smoothness, and finally, by density.

4. Select a material from your classroom lab. Prepare a presentation for the class reviewing the proper handling and storage procedures for the material. Include a handout with your presentation.

TechnoFact: Materials can be tested using a number of techniques, including stress and strain testers, hardness testers, polarized light, X rays, ultrasonic waves, and microscopic examination.

Test Your Knowledge

Do not write in this text. Place answers to test questions on a separate sheet.

1. Paraphrase the definition of *natural material.*

2. Which of the following materials are natural? You may choose more than one answer.

    A. Plastic.

    B. Trees.

    C. Rocks.

    D. Petroleum.

3. Give an example of a material used to power engines, turbines, and other types of converters.

4. What is the name of materials having rigid structures?

5. Indicate which of the following are engineering materials. You may choose more than one answer.

    A. Modeling clay.

    B. Metal.

    C. Glass.

    D. Plastic.

    E. Concrete.

    F. Petroleum.

6. Trees and wheat are examples of _____ resources.

7. Resources that cannot be replaced once they are used are called _____ resources.

8. Match the processes on the left with the descriptions on the right. Give one more example of each type of process.

    _____ Producing plastic   A. Mechanical processing.

    _____ Making lumber   B. Chemical processing.

    _____ Producing steel   C. Thermal processing.

9. Explain what the term *property* means in the context of this chapter.

10. List the seven properties of materials and define each.

Chapter 6
Energy and Technology

This chapter covers the benchmark topics for the following
Standards for Technological Literacy:

3 **12** **16**

Did You Know?

➤ *Geo* means earth, and *thermal* means heat. *Geothermal* means earth heat.

➤ *Hydro* means water, and *hydro-electric* means making electricity using waterpower.

➤ Some two thousand years ago, the Greeks used mirrors to focus the Sun's rays on Roman ships, causing them to catch fire.

Objectives

The information given in this chapter will help you do the following:

➤ Explain the differences between natural and technological uses of energy.

➤ Define *energy* and *work.*

➤ Recognize the difference between energy and power.

➤ List and summarize the different forms of energy.

➤ Identify and describe several sources of energy.

➤ Name the three most common types of energy supplies.

➤ Give examples of alternate energy sources.

Key Words

These words are used in this chapter. Do you know what they mean?

bioenergy
biomass
chemical energy
coal
electrical energy
energy
exhaustible
fission
fossil fuel
fusion
geothermal
heat energy
hydroelectric
kinetic energy
mechanical energy
natural gas
nonrenewable
nuclear energy
peat
petroleum
potential energy
power
radiant energy
solar energy
thermal energy
tidal energy
wave energy
wind energy
work

Chapter Outline
Did You Know?
Objectives
Key Words
Energy and Work
Power and Energy
The Nature of Energy
  Forms of Energy
  The Sun As the Primary Energy Source
Energy Sources
Common Energy Supplies
  Fossil Fuels
  Waterpower
  Nuclear Energy
Alternate Energy Sources
  Solar Energy
  Wind Energy
  Tidal and Wave Energy
  Geothermal Energy
  Biomass Energy
Summary
Curricular Connections
Activities
Test Your Knowledge

Standards for Technological Knowledge

Research
Ask the students why fire is one of the basic tools of human culture.

Tech Terms
Fire—The natural phenomenon of combustion that emits light and heat.

Figure 6-1. Human activity requires energy we get from food.

About a half million years ago, early humans learned to make fire. They collected and burned wood. With the heat generated, they cooked food, warmed their homes, and made tools. Much later, ancient Egyptians developed the first sails. They used these sails to power their boats on the Nile River and Mediterranean Sea. Still later, the waterwheel was developed. It raised water into irrigation ditches and powered crude machines. In recent history, the steam engine was developed. It sparked the Industrial Revolution. This engine powered factories and transportation vehicles. Today, we are connected into a worldwide communication system. We are developing new ways to move people and cargo around the world and into space. New manufacturing processes are being used. Cutting edge medical treatments are available. Energy, in one form or another, was a key part of all these technological developments.

People use the word *energy* almost every day. They talk about the energy it takes to get out of bed. People say they don't have the energy to run a mile. They worry about the energy crisis the cost of foreign oil caused. Not all people, however, understand the role of energy in their lives. Energy is one of the most fundamental parts of our universe.

Energy is used in natural and technological systems. In natural settings, our bodies transform food into energy to do work. See **Figure 6-1.** We use energy as we walk and run. As we think, read, and write, we use energy. Likewise, animals use energy as they go about their daily activities. They use energy to gallop, fly, build burrows,

TechnoFact: The controlled use of fire was a major event in human history. It enabled people to protect themselves against predators. Fire allowed people to camp in the open or in caves where hostile animals existed. It also allowed them to cook food and cope with cold temperatures in harsh climates. The mastery of fire allowed the first settlements of *Homo erectus* in Europe.

Figure 6-2. These waves crashing on rocks contain a great amount of energy.

and do other natural activities. Also, energy is released as snow melts, waves crash onto rocks, and earthquakes shake the ground around us. See **Figure 6-2.**

Technological activities also use energy. Energy can be used to do work, using several processes. It powers our cars, trains, and planes and lights our homes, stores, and offices. Medical systems use energy as illnesses are treated. Energy is used to plant, cultivate, and harvest crops. It heats buildings and cooks food. Energy provides entertainment through sound systems and television sets. It powers machines in factories and offices. There could not be technology or industry without energy. See **Figure 6-3.**

Energy and Work

Energy is essential for meeting human needs and wants. It is the foundation for raising standards of living. Energy contributes to increased life expectancies. Over time, people have used more energy. Today, each person uses more than one hundred times the energy early humans used. In fact,

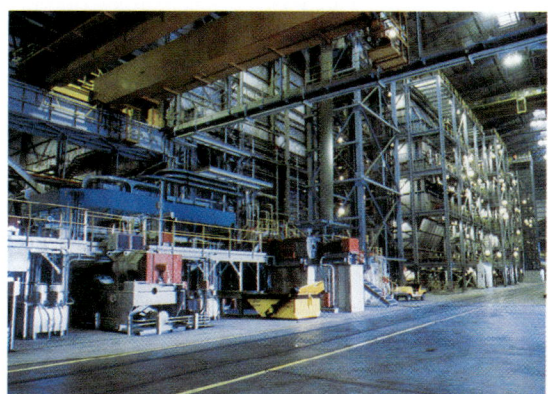

Figure 6-3. Industrial processes require large amounts of energy. (Inland Steel Company)

TechnoFact: The development of technology and civilization is closely linked to the development of sources of useful forms of energy.

over the last one hundred years, our energy use has more than tripled. Some people estimate that the demand for energy will double in the next fifty years. See **Figure 6-4.**

So what is energy? Simply defined, *energy* is the ability to do work. It provides a force. All forms of energy are associated with some type of motion. *Work* is the use of force to create movement. It can be calculated using a simple formula: Work equals the force required, times the distance traveled.

Energy makes things happen. It is either moving or stored. Stored energy is called *potential energy.* It is waiting to be used when needed. Moving energy is called *kinetic energy.* It is the energy of action. To see the difference between potential and kinetic energy, think of a pencil. If it is placed on a desk, it has potential energy. The energy is stored because of its position above the floor. If the pencil falls to the floor, it has kinetic energy. It has the energy of motion. Picking it up and placing it on the desk restores the potential energy. The kinetic energy the person moving the pencil uses is stored as potential energy.

Likewise, think of a rubber band. If you stretch it, your kinetic energy (moving hands) is changed into potential (stored) energy in the rubber band. As it slips from your grip, the potential energy is changed into kinetic energy. The rubber band moves through the air.

Figure 6-4. Our energy use has changed over time.

Ancient humans 1,000,000 B.C.
Hunter-gatherer 100,000 B.C.
Ancient farmer 5000 B.C.
Developed farmer 1500 A.D.
Industrial worker 1850 A.D.
Technological worker 1950 A.D.

TechnoFact: The movement of a pendulum demonstrates the difference between kinetic energy and potential energy. As the suspended body (weight at the end of the pendulum) moves upward in its swing, its kinetic energy (energy of motion) is continuously changed into potential energy. The higher it moves, the greater its potential energy becomes. This is due to its higher position. At the top of the swing, the change from kinetic to potential energy is complete. During its downward motion, the potential energy is converted back into kinetic energy.

Figure 6-5. The water in the reservoir on the left has potential energy. The water rushing through this power plant has kinetic energy.

Standards for Technological Knowledge

16

Research
Have the students find out which units of measurement are used to express power.

Therefore, we can say kinetic energy is the energy an object has because of its motion. Potential energy is stored energy that depends upon the relative position of an object. See **Figure 6-5.**

Power and Energy

Power and *energy* are two words people often confuse. They are not the same thing. As we said, energy is the ability to do work. *Power,* on the other hand, is the amount of work done in a set period of time, or the rate at which work is done. It is the result of using energy for a definite time span. The rate at which energy is changed from one form to another or moved from one place to another is also called *power.*

Suppose you pedal your bicycle for one mile. That is work. You have used energy to create this movement. The energy you used can be calculated. Typically, one of two measurements is used to measure the energy consumed. The first unit of measure is called a *Btu,* or *British thermal unit.*

A *Btu* is defined as the amount of heat energy needed to raise the temperature of one pound of water by 1°F, at sea level. One Btu equals the energy in one kitchen match. As you can see, it is a fairly small unit of measurement. Often we measure energy by thousands of Btu. One thousand Btu equal the energy in an average candy bar. The Btu is the measurement unit used in the United States. In countries using the metric system, energy is measured in Newton-meters, or *joules* (pronounced *jewels*). Also, scientists around the world use this unit of measurement. One thousand joules equal one Btu.

When you measure the energy it takes you to ride a bicycle one mile, you are calculating *power.* As you just read, power is energy used over time. See **Figure 6-6.** It can be calculated with a simple formula: Power equals the distance traveled, times the force required, divided by the time.

Tech Terms
Power—The rate at which energy is produced or consumed.
British Thermal Unit (Btu)—The amount of heat energy needed to raise the temperature of one pound of water by one degree Fahrenheit, at sea level.

TechnoFact: Power is a measure of the rate of doing work and is generally expressed in wattage or horsepower. The term *power* also can be used to denote the force or energy used to drive machinery (mechanical power). The term *power* is also used to describe the available electricity (electric power).

Figure 6-6. Can you explain energy, power, and work in terms of these gondola operators?

The Nature of Energy

Energy is everywhere. People and machines use it. Energy is used to move things, light areas, lift weight, and warm spaces.

One form of energy can be changed into another. See **Figure 6-7.** Stored chemical energy in a flashlight battery can be converted into light energy. Your body uses the stored chemical energy in food to do work. A television set changes electrical energy into light and sound energy. An automobile engine changes the chemical energy in gasoline into heat and mechanical energy to power the car.

Energy cannot be created, however, nor can it be destroyed. As we change its form, it can be wasted. For example, the lightbulb is an energy converter. Its purpose is to change electrical energy into light energy. In doing this task, not all the electrical energy ends up as light energy. Some of it becomes heat energy. The electrical energy that becomes heat energy is not destroyed. It could be considered wasted, however. This energy is not in a form that is useful in the situation, nor is it wanted.

Forms of Energy

Not all energy is alike. See **Figure 6-8.** It appears in several different forms:

➤ Light or *radiant energy.* This is the energy in the atomic motion present in sunlight and fire.

➤ *Heat energy.* This is the energy in the increased activity of molecules

Figure 6-7. These wind generators change mechanical power into electrical power.

TechnoFact: Heat energy is related to variations in temperature between a system and its environment or between elements of the same system.

Radiant

Heat

Mechanical

Chemical

Electrical

Nuclear

Figure 6-8. How many forms of energy can you see in this picture? (Caterpillar, Inc.)

found in matter whose temperature has been raised.

➤ *Mechanical energy.* This is the energy in the movements that humans, animals, moving water and air, and machinery produce.

➤ *Chemical energy.* This is the energy in reactions between substances such as carbon (coal) and oxygen.

➤ *Electrical energy.* This is the energy in the movement of electrons in matter, caused by lightning, batteries, and generators.

➤ *Nuclear energy.* This is the energy from the splitting of atoms.

The Sun As the Primary Energy Source

The Sun is key to much of the energy we use. It is a huge energy "factory." Every day, the Sun gives Earth enormous amounts of heat and light. See **Figure 6-9.** In twenty days,

Research
Assign each of the students one form of energy and have them research ways it is applied by technological devices to meet human needs and wants.

Tech Terms
Solar Energy—Any form of energy radiated by the Sun, including light, radio waves, and X rays. Solar energy usually refers to the visible light of the Sun.

Brainstorm
Ask the students to list the forms of energy they used to get ready for school.

Figure 6-9. The Sun's energy acts as a huge water pump for Earth. It heats water that turns to vapor. The vapors rise to form clouds. In the clouds, the water vapor condenses to make rain. (Natural Gas Supply Assoc.)

TechnoFact: Solar energy is required by green vegetation for the process of photosynthesis, which is the fundamental source of all food.

Technology Explained

biogas: A gas produced by processing animal and plant waste.

As petroleum supplies become scarcer, people are searching for new energy sources. One of these is called *biogas.* This is gas produced by processing biological waste.

A dairy cow can produce over 100 lbs. of manure per day. A large dairy farm with a thousand cows has to deal with over 50 tons of manure per day. This waste must be disposed of without harming the environment. One way to do this is to convert it into a gas that can be used to generate electricity, **Figure A.**

Figure A. Livestock waste can be a major source of alternative energy. (Tillamook County Creamery Assoc.)

A process called *anaerobic generation* produces biogas. This process has been used in Europe for a number of years. In the United States, it is starting to be a source of alternate energy, **Figure B.**

Discussion
Describe how solar energy is used to air-dry fruit.

Research
Have the students research how heat from the Sun is used to produce salt from seawater.

Earth receives as much energy from sunlight as is stored in all coal, oil, and natural gas reserves in the world. Even so, only 13 percent of all solar energy leaving the Sun actually reaches Earth.

Some of the sunshine striking Earth creates kinetic energy. Another large share of it gets stored as potential energy. Solar activity accounts for wind power, waterpower, and fossil fuels.

Wind, a type of kinetic energy, comes from uneven heating of the atmosphere by the Sun. Heated air rises and causes cold, heavier air to move in under it. The resulting movement is called *wind.* It is often strong enough to drive windmills and electric generators.

The Sun also works like a giant water pump. Radiant energy heats water until it vaporizes. The water vapors rise into the sky. There, they form clouds. When cooled, the vapors become a liquid again and fall as rain. The rain becomes rivers that can power hydroelectric power plants.

Much of our solar energy becomes stored in plant life. This energy remains stored until something releases it. Animals and humans eat

TechnoFact: The energy in fossil fuels (coal, peat, natural gas, and petroleum) and other organic fuels (such as wood) is derived from solar energy.

The manure from 100 cows can be used to produce enough energy to power an average home.

Collection pit

Digester

Gas engine generator

Heat exchanger

Transformer

Figure B. Here is the biogas production system.

Anaerobic generation starts in the dairy barn. Here, the manure is scraped into a collection pit. A low volume pump moves the manure to an outside tank. From there, the waste is deposited into a digester. This is an airtight, aboveground, concrete tank. A heat exchanger maintains the temperature at about 95°F. This is the temperature at which anaerobic bacteria work best.

Over the course of two to four weeks, the bacteria work on the manure. They break the waste into smaller particles. These particles are primarily organic acids, hydrogen, and sugar. At the same time, another group of organisms is at work. These organisms are archaebacteria, which change the acids and hydrogen into a gas. This biogas is a mixture of about 40 percent carbon dioxide and 60 percent methane.

The gases rise to the top of the tank and are collected. They are fed into a natural gas engine connected to an electric generator. When the gases are burned in the engine, they provide the energy to drive the generator. The electric output is fed into the electric grid (power lines) by using a transformer on a power pole.

plants for energy and growth. Burning releases the energy in wood and other plants. Plants and animals decay and become coal and petroleum. Burning these materials also releases energy.

Energy Sources

Technological systems need available, easy-to-control energy. When the energy is controlled, it can deliver power when and where it is needed. Energy sources were limited for early humans. These humans relied upon natural sources. People burned wood to cook food and keep warm. They used water and wind to do work for them.

Still, these energy sources sometimes failed. Rivers were low during droughts. Waterwheels did not always have enough waterpower to drive them. Sailing ships could not move when winds did not blow. Windmills sometimes stopped turning. Over time, people developed energy sources they could control. They moved from dependence on wind power and waterpower.

Resource
Reproducible Master 6-1: Nature's Energy Storehouse

Research
Have the students trace the types of energy used by humans during important periods of history.

Tech Terms
Wind—The flow of air relative to Earth's surface.

TechnoFact: A windmill is a technological device that harnesses wind power for a variety of uses, including pumping water, grinding grain, and powering electrical generators. Windmills probably appeared in Europe during the eleventh century and became familiar landmarks in Holland, France, Germany, and England.

Research
Have the students research how peat is cut for use as a fuel.

Tech Terms

Fossil Fuel—Any carbon-based fuel (coal, peat, petroleum, and natural gas) derived from the decomposed remains of prehistoric animals and plants.

Peat—Compact, dark brown organic material containing a high carbon content and found in swamps and bogs in various parts of the temperate zone.

Energy resources used today can be considered to be either primary or secondary. Primary energy resources are in the form of natural resources. These include resources such as wood, coal, petroleum, natural gas, uranium, wind, water, and sunlight. Many of these resources are changed into more usable forms. These altered forms are secondary energy resources, such as electricity, gasoline, and propane.

Primary energy sources can be either renewable or nonrenewable resources. Renewable resources are inexhaustible energy sources. Natural or human actions can replace them. These resources include solar energy, wind energy, wave energy, biomass (wood or crops, such as sugar), geothermal energy, and hydropower.

Nonrenewable energy resources have limited quantities. When they are used up, they cannot be replaced. They will be gone forever. Typical nonrenewable energy sources include fossil fuels (coal, oil, and natural gas) and uranium. Today, the fossil fuel resources provide over 80 percent of our energy.

Common Energy Supplies

Developed countries, like those in North America and Western Europe, use several different types of energy supplies. The most used and practical of these are the following:

➤ Fossil fuels.

➤ Waterpower.

➤ Nuclear energy.

Fossil Fuels

All fossil fuels were formed millions of years ago. This was before the time of the dinosaurs. For this reason, these fuels are called *fossil fuels.* The age they were formed is called the *Carboniferous period.* This period gets its name from *carbon,* the basic element in coal and other fossil fuels.

Fossil fuels are the remains of once living matter. They are the products of dead plants or animals. A process called *decay* formed them. When living matter decays, it reacts with oxygen. Fossil fuels, however, were formed under conditions where there was very little oxygen. See **Figure 6-10.** Therefore, the dead matter did not fully decay. The partially decayed matter became a material high in carbon.

The carbon in fossil fuels makes them burn easily. Burning is rapid oxidation and requires large quantities of oxygen. In the process of burning, large amounts of heat energy are released. Fossil fuel deposits can be found in several forms. These include peat, coal, petroleum, and natural gas.

Peat

Peat is decayed plant matter that sank in swamps. It is made up of moss, reeds, and trees. Other plant matter and soil covered and pressed on it. The decaying matter compressed into solid material. It became somewhat like coal. In some parts of the world, it heats homes. Also, gardeners use peat as a soil additive. See **Figure 6-11.**

TechnoFact: Peat is produced over time by the partial decay and carbonization of vegetation in acid water. It is the earliest stage of the transition of compressed plant growth into coal. Peat is still used as fuel in some countries.

Figure 6-10. This is a view of peat that has been cut for fuel. (Vapo Oy)

Coal

Coal is also decayed plant matter. See **Figure 6-11.** It started out just like peat. The weight of soil and rock, however, greatly compacted the coal. Coal became a hard, black, rock-like substance. It is mined in deep mines or in strip mines closer to the surface.

Industry and homes once widely used coal. Coal provided heat for powering steam locomotives. See **Figure 6-12.** Coal-fired stoves and

Figure 6-11. Fossil fuels were formed from plant and animal life that existed millions of years ago. (Standard Oil of California)

TechnoFact: There are a number of carbon fuels. These include carbon, peat, lignite, and the various forms of coal, such as black lignite, bituminous coal, and anthracite.

Figure 6-12. This wood-burning steam locomotive can also burn coal.

Tech Terms
Petroleum—A naturally occurring oily liquid composed of a variety of organic chemicals.

Research
Have the students research how petroleum is refined.

Discussion
The chemical composition of all petroleum is primarily hydrocarbons. In addition, there are a few sulfur-containing and oxygen-containing compounds present. The consistency of petroleum varies from a liquid as thin as gasoline to a liquid so thick it will barely pour.

plants. Also, it is used in steelmaking and as a base for many other products.

Petroleum

Gradually, petroleum products and natural gas have replaced coal. *Petroleum* is a thick liquid. It is thought to be the decayed remains of plant and animal life. Like coal, it became buried and partially decayed. It is pumped from pockets or "wells" deep below ground.

Petroleum requires refining. Refining is a heat process separating the petroleum into more usable products. The products include liquids such as gasoline, diesel fuel, kerosene, and fuel oil. See **Figure 6-13.**

furnaces were common in homes, factories, and public buildings. Today, coal is a major fuel in electricity generating

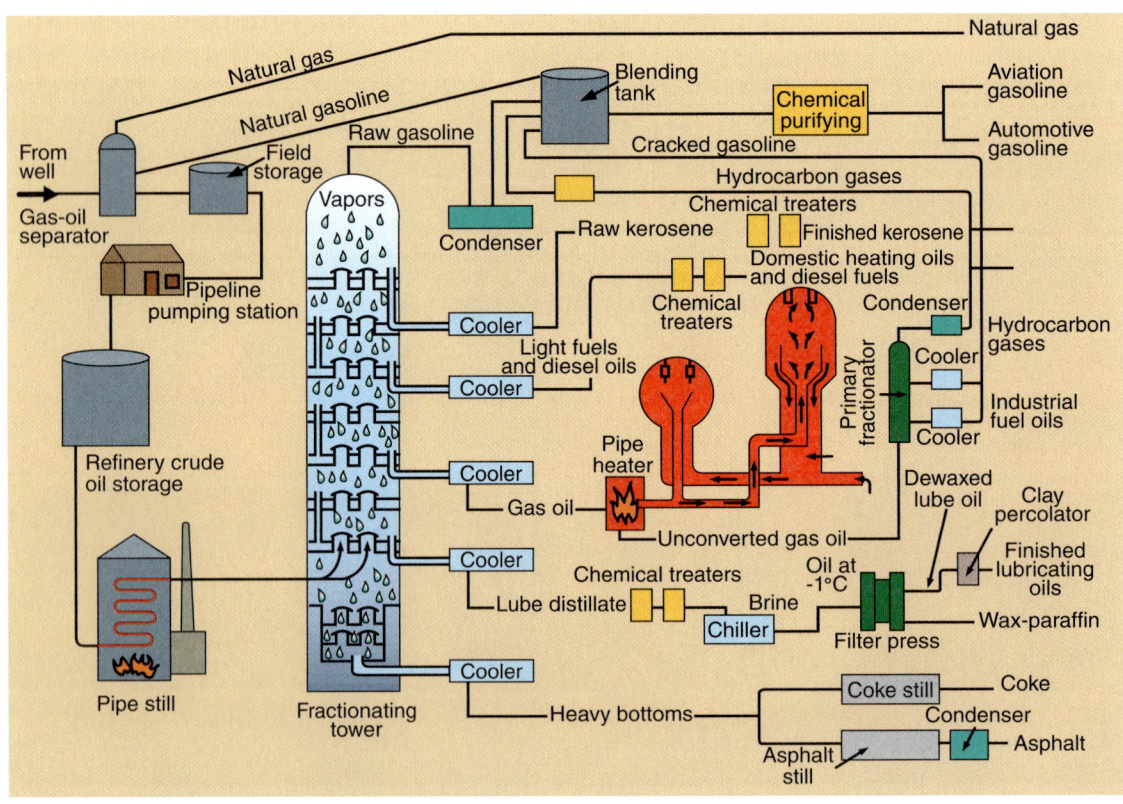

Figure 6-13. When petroleum is refined, many different products are produced. Some of these products are used as fuels for their energy.

TechnoFact: Petroleum is found in large quantities in the ground. It is used, primarily, as a fuel or feedstock (raw material) for the chemical industry in the manufacture of products such as plastics, medicines, fertilizers, building materials, paints, and synthetic fibers.

Figure 6-14. This is a picture of a restored waterwheel used to power a flour mill in the nineteenth century.

Natural Gas

Natural gas is lighter than air. It is mostly made up of a gas called *methane.* This gas is a simple chemical compound made up of carbon and hydrogen atoms.

It is usually found near petroleum underground. Like petroleum, natural gas is pumped from below ground. Natural gas is used for many of the same purposes for which we use petroleum. It is, however, generally used for heating buildings and producing electric power. Natural gas is the key ingredient for making many plastics.

The process producing fossil fuels is called *fossilization.* It still goes on. We, however, are using these fuels one hundred thousand times faster than they are being replenished. One day, fossil fuels will be gone. Therefore, we call them *nonrenewable,* or *exhaustible.*

Waterpower

When it rains, the water becomes streams. The streams run together to form rivers. The rivers flow toward the oceans. This moving or falling water has kinetic energy. It can be used to do work. For early humans, this was a primary source of power. In early times, moving water turned wooden wheels called *waterwheels.* See **Figure 6-14.** The turning wheels created mechanical energy. These wheels turned big stone wheels to grind wheat and corn into flour.

Today, we use moving water to make electricity. This action takes place in hydroelectric power plants. *Hydro* means water, and *hydroelectric* means making electricity from waterpower. To make electricity from water, dams are built across rivers. See **Figure 6-15.** These dams hold river water, forming a reservoir behind the dam. The water in the reservoir flows through pipes into a turbine. The turbine spins and turns electric generators.

Unlike fossil fuels, hydropower is a renewable energy source. This means it will always be available. Water in rivers is constantly replaced by rainfall. Surface water is evaporated by the Sun's heat. It falls back to Earth as snow, rain, and dew. This cycle occurs endlessly.

Nuclear Energy

Nuclear energy is the energy found in atoms. Scientists learned how to unlock this energy in recent times. In the 1940s, they discovered a

TechnoFact: One of the world's largest deposits of natural gas is located in the Prudhoe Bay region of Alaska. Most of the gas is injected back into the ground to add pressure to the oil deposits found in the region. This process helps make oil extraction more efficient.

Figure 6-15. This dam on the Columbia River is used for flood control and electricity generation.

Extend
Discuss the use of nuclear reactors for research and power production.

Tech Terms
Nuclear Energy—
Energy released during the splitting or fusing of atomic nuclei.
Nuclear Reactor—
A technological device designed to produce a controlled release of nuclear energy.

process that causes the atom's nucleus to be split apart. In turn, this causes other atoms to split. Each splitting is called a reaction. Once started, the reactions continue. This process is called *fission.* See **Figure 6-16**. With this process, a tremendous amount of heat and light energy is released. This energy, when let out slowly, can be harnessed to generate electricity. The great heat produced is usually transferred to water. The water vaporizes into steam, which spins a turbine. An electric generator, coupled with the turbine, produces electric power.

A second type of nuclear reaction is called *fusion.* It causes parts of hydrogen atoms to *fuse* (join). Like fission, fusion releases huge amounts of energy. The Sun releases energy through the fusion of hydrogen and helium atoms.

Our technology has been able to start fusion reactions. So far, it has been unable to keep them going.

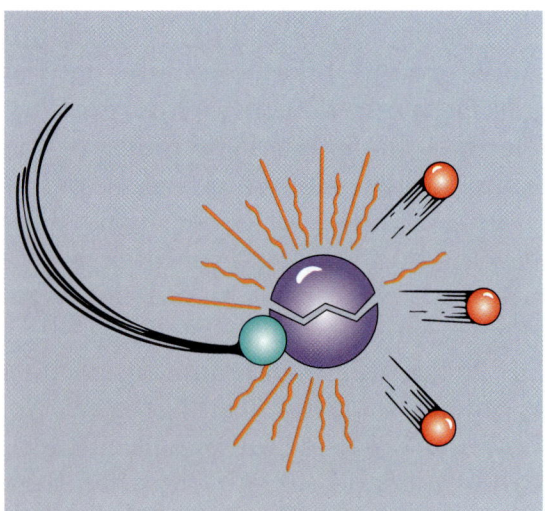

Figure 6-16. Nuclear energy is released when atoms are split. (Westinghouse Electric Corp.)

TechnoFact: Nuclear reactors can be used for power production or research. A power reactor uses the heat produced by the nuclear reaction to power standard steam turbines. A research reactor is intended to emit different beams of radiation for experimental use.

Figure 6-17. The solar cells on the top of this sign convert solar energy into electricity.

Alternate Energy Sources

People are becoming increasingly concerned about the use of nonrenewable energy resources. They are turning to alternate energy sources to reduce our primary dependence on fossil fuels. Individuals, government agencies, and private companies are studying these sources. Large-scale, efficient methods to harness and use alternate energy sources are being explored. Like waterpower, these sources will always be available. We will never run out of them. The more important alternate energy sources are solar, wind, tidal or wave, geothermal, and biomass energy.

Solar Energy

The Sun has been an energy source since the beginning of time. Plants use the Sun's light to make food in a process called *photosynthesis.* Early in history, people learned to use *solar energy.* It was used to dry food and warm buildings. When you hang a wet swimsuit outside to dry, you are using solar energy to do work. Today, the Sun is being used to heat water for homes and businesses. Campers can use solar energy to cook food. The two major commercial uses for solar energy technology are to heat water and produce electricity. See **Figure 6-17.**

Wind Energy

Wind is an indirect use of solar energy. It can be used to do work. A sailboat uses *wind energy* to push it through the water. Ranchers use wind energy to pump water for their cattle. In Holland, windmills have been used for centuries to pump water from low-lying areas. Wind, like the Sun, is free and increasingly harnessed for electricity production. See **Figure 6-18.**

Tidal and Wave Energy

Energy is present in the ocean. The ocean can produce two types of energy: *thermal energy* from the Sun's heat, and *mechanical energy* from the tides and waves.

Oceans cover more than 70 percent of Earth's surface. This makes the oceans huge solar collectors. The Sun's heat primarily warms the surface water. This water is warmer than the water

TechnoFact: The typical Dutch windmill is called a *tower windmill.* It has a large tower made of stone, brick, or wood. At the top of the structure are four to six arms attached to a central hub that can revolve. The 20′ to 40′ long arms have sails constructed of lightweight wood or canvas attached to them. A fan keeps the wheel facing the wind.

Extend
Discuss tidal and
ocean thermal
energy conversion
technologies.

Tech Terms
Ocean—An inter-
connected mass of
saltwater covering
over 70 percent of
the surface of Earth.
Wave—An oscillat-
ing movement up
and down of a body
of water (lake or
ocean) caused
primarily by the fric-
tion between the
wind and water.
Wavelength—The
distance between
two successive
crests of waves.

Figure 6-18. This windmill pumps water for livestock in Australia.

at greater depths. The temperature difference creates thermal energy, which can be used to make electricity.

Tidal energy is another type of ocean energy. See **Figure 6-19.** Unlike thermal energy, it uses the mechanical energy of moving water. Ocean mechanical energy is quite different from ocean thermal energy. The Sun is the ultimate source for thermal energy. The gravitational pull of the Moon, however, primarily drives tides. The wind primarily drives waves.

Ocean tide energy uses the differences between high tide and low tide. *Wave energy* uses the forces present in the coming and going of waves near the shore. Both of these sources can power electricity-generating equipment.

Geothermal Energy

Not all renewable energy resources, however, come from the Sun. *Geothermal* energy uses the natural heat below Earth's surface. Earth is a molten mass of hot material covered

Figure 6-19. The Grand Coulee Dam in Washington is used to harness tidal energy to make electricity.

TechnoFact: The four main components making up the world's ocean are the Atlantic, Indian, Pacific, and Arctic Oceans. The Antarctic Ocean is sometimes considered a fifth, separate ocean. The main oceans are subdivided into smaller areas called *seas, gulfs,* and *bays.* Technological systems utilizing the twice-daily ebb and flow of the ocean's tides have met with limited success.

Figure 6-20. This diagram shows how geothermal energy can be captured. (Aminoil UAS, Inc.)

by a thick crust. Deep under the surface, water sometimes makes its way close to the hot rock and turns into hot water or steam. See **Figure 6-20.** The hot water can reach temperatures of more than 300°F. This water is at a temperature above its boiling point. When this hot water comes up through a crack in Earth, we call it a

geyser, or *hot spring*. Drilling a well can also tap this water. This hot water from below the ground can warm buildings, keep roads clear of ice, and power electric generating plants. See **Figure 6-21.**

Biomass Energy

Biomass includes all the living organisms in an area. It is matter people often treat as garbage. Biomass includes materials such as yard waste, dead trees, and tree branches. It can be paper products and other household waste. Biomass includes animal waste and by-products. Also, it can be housing rubble, leftover crops, and bark and sawdust from lumber mills.

Figure 6-21. Geothermal energy from Earth can be used as an energy source. It is seen as a geyser. (Aminoil UAS, Inc.)

Research
Have the students research New Zealand's efforts to generate electricity from geothermal deposits.

Tech Terms
Geothermal Energy—Energy related to the internal heat of the earth.

TechnoFact: Hot springs are a natural discharge of hot groundwater resulting from the water passing through or near hot, igneous rocks. Yellowstone Park, the North Island of New Zealand, and Iceland are famous for their hot springs. Hot springs, a type of geothermal energy, are being used in California, Italy, and Iceland as an alternate energy source.

Biomass can be used as an energy source (fuel). Also, it can be recycled and made into other products, such as paper and fertilizer. Some biomass is made into compost (decayed plant or food products) that helps plants grow. Biomass can be used for *bioenergy,* which is defined as the energy from organic matter. Bioenergy is used in three ways:

➤ Biochemicals. Biomass can be converted into chemicals to generate electricity.

➤ Biofuels. Biomass can be converted into liquid fuels for transportation.

➤ Biopower. Biomass can be burned directly to generate electricity.

Summary

Energy is key to our modern way of life. It heats our homes, provides us with light, powers our entertainment devices, and makes our vehicles move. We use energy in the form of radiant or light, heat, mechanical, chemical, electrical, and nuclear energy. Fossil fuels—peat, coal, petroleum, and natural gas—primarily provide this energy. These fuels, along with nuclear energy resources, are exhaustible. They will not always be available. To conserve these energy resources, people are turning to inexhaustible sources. These include water (hydro), solar, wind, tide and wave, geothermal, and biomass energy sources.

Extend
Discuss how bioenergy technologies help protect the environment by making use of renewable plant material (sawdust, tree trimmings, and grain straw), industrial waste, and paper found in municipal solid waste.

Tech Terms
Biomass—A fuel consisting of plant material, vegetation, or agricultural waste.

Department of Energy (DOE) Notes
- Inadequate insulation is one of the leading causes of energy waste in most homes.
- The Fuel Economy Guide, produced in partnership with the Environmental Protection Agency, provides fuel economy data, including estimated annual fuel cost, for all new model year vehicles.
- The Center for Transportation Analysis (CTA) conducts innovative, cost-effective research and development on transportation energy and environmental issues, national transportation planning and policy, transportation systems engineering, and military transportation and logistics.
- The DOE sponsors basic scientific research in the fields of high energy and nuclear physics, whose principal goal is to explore the behavior of matter and energy at their deepest and most fundamental levels.
- The DOE National Biofuels Program seeks cost-effective ways to produce ethanol and other fuels and chemicals from biomass resources, such as agricultural and forestry residues or fast-growing trees and grasses. The DOE's Geothermal Energy Program seeks to establish geothermal energy as an economically competitive contributor to the U.S. energy supply by developing technology to enable use of the full range of geothermal resources.

Curricular Connections

Language Arts

Trace the roots of several words used in discussing energy. Such words could include *energy, geothermal, hydroelectric,* and *biomass.*

Social Studies

Draw an illustration describing the geology of petroleum or a coal deposit.

Science

Prepare a display explaining the science behind geothermal or ocean thermal energy.

Mathematics

Use the Internet or other resources to discover the energy use for several countries. Prepare a graph showing the relationship between the types of energy and countries selected.

Social Studies

Develop a map highlighting the distributions of energy reserves (such as fossil fuels and uranium) in the United States or world.

Activities

1. Visit the local electric company and find out how their electricity is generated. Prepare a poster to show the processes it uses to make electric power.
2. Make a model of a petroleum deposit and a well used to extract the resource.
3. List all the ways you use energy. Indicate the type of energy used with each activity. Describe the probable source of that energy.

TechnoFact: The primary fuel for most of the world was wood until about 1800. The energy in wood is derived from solar energy accumulated in plants during their lifetimes. Since the Industrial Revolution, people have depended on fossil fuels (coal, petroleum, and natural gas) for energy. This energy is also ultimately derived from stored solar energy.

Test Your Knowledge

Do not write in this book. Place your answers to this test on a separate sheet of paper.

1. Summarize the different ways energy is used in natural and technological systems.

2. The ability to do work is called _____.

3. Using force to create motion is called _____.

4. The amount of work done in a period of time is called _____.

5. Label the six forms of energy.

   A. The energy present in sunlight.

   B. The energy in reactions between substances.

   C. The energy caused by batteries.

   D. The energy produced by moving water.

   E. The energy found in matter whose temperature has increased.

   F. The energy caused by atoms splitting.

6. Paraphrase the difference between primary and secondary energy resources.

7. List the four major types of fossil fuels.

8. Another name for waterpower is _____.

9. Energy from natural hot water or steam from Earth is called _____ energy.

10. Energy from organic matter is called _____.

Working with Hazardous Materials and Wastes

We are surrounded by hazardous materials at school, at work, and at home. Cleaning materials, finishing materials, art and science supplies, lawn and garden chemicals, and gasoline are only a few of the materials that we encounter. Follow these guidelines to prevent accidents:

- Read the Material Safety Data Sheet (MSDS) before using any chemical or material for the first time.
- Follow the instructions in the MSDS on how to use, store, and dispose of the chemical.
- Store flammables in an approved container or in an approved flammables cabinet.
- Wear the appropriate personal protective equipment (i.e., gloves, goggles, respiratory equipment, etc.) when using a hazardous substance.
- Keep emergency phone numbers posted.

You must be able to identify and classify hazardous materials and wastes. Be familiar with all hazardous materials in the lab or shop area. Review the MSDSs periodically to refresh your memory and also to make sure you are aware of any new materials.

If you encounter a substance and you cannot identify it immediately, assume that it is hazardous. Notify your teacher immediately. Look for nearby containers and then review the MSDSs.

When disposing of hazardous materials wastes, also follow the requirements specified in the MSDS. Use the proper protective equipment (also specified in the MSDS).

Material Safety Data Sheets

The Occupational Safety and Health Act (OSHA) requires employers to keep a list of all hazardous chemicals and materials used on their premises and maintain a file of Material Safety Data Sheets (MSDS) on the chemicals. They also must train their employees to use the chemicals and respond properly should an accident occur. The MSDS includes the following information:

- Chemical name(s) and symbol.
- Manufacturer's contact information.
- Flammability and reactivity data (vapor pressure and flash point information).
- Health hazards.
- Overexposure limits and effects.
- Spill cleanup procedures.
- Personal protection information.

Chapter 7
Information and Technology

Did You Know?

➤ Americans have thousands of sources of information. In the United States alone, there are over 1,500 newspapers; 11,500 periodicals (magazines and journals); 27,000 video rental stores; 360 million television sets; and 400 million radios.

Objectives

The information given in this chapter will help you do the following:

➤ Summarize how people use knowledge.

➤ Explain the differences between data, information, and knowledge.

➤ Paraphrase the three beliefs that are the foundation of the scientific method.

➤ Describe technological knowledge.

➤ Define and explain the *Information Age.*

➤ Give examples of the three types of research.

Key Words

These words are used in this chapter. Do you know what they mean?

data

descriptive research

experimental research

historical research

information

knowledge

Chapter Outline
Did Your Know?
Objectives
Key Words
Data, Information, and Knowledge
Scientific Knowledge
Technological Knowledge
Information Age
Gaining Information and Knowledge
Summary
Curricular Connections
Activities
Test Your Knowledge

Figure Discussion
Ask the students how the geologist shown in Figure 7-2 might use each type of knowledge.

Tech Terms
Scientific Knowledge— Explains the laws and principles governing the universe.
Humanities Knowledge— Explains how people have expressed themselves through art and music, formed societies, and developed values.
Descriptive Knowledge— Explains how people have used words and numbers to describe objects and events.
Technological Knowledge— Explains how people use tools and materials to produce products and systems.

Humans are unique because they have advanced brains. These brains allow them to think, plan, and process information. The ability to know is a basic input to technology. Knowledge lets people develop and use technological products and systems. See **Figure 7-1.** It allows them to be inventive and creative. In Chapter 1, you learned there were four types of knowledge:

➤ Scientific knowledge. This explains the laws and principles governing the universe.

➤ Humanities knowledge. This explains how people have formed societies, developed values, and expressed themselves through art and music.

➤ Descriptive knowledge. This explains how people have used words and numbers to describe objects and events.

➤ Technological knowledge. This describes how people use tools and

Figure 7-1. Knowledge allows people to develop and use technology. (American Petroleum Institute).

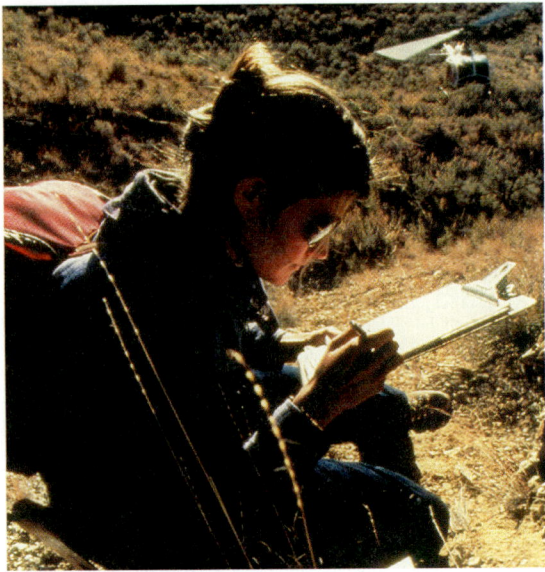

Figure 7-2. This geologist is searching for petroleum resources. (Standard Oil Company of Ohio)

materials to produce products and systems.

People who develop technological products and systems use all four of these. The types of knowledge obtained from various areas of study have direct effects on these developments. Designers use material science knowledge to select materials for specific applications. They use physics knowledge as they design mechanical structures and electrical circuits. These people use chemical knowledge as they develop manufacturing and energy conversion systems. Biological knowledge helps them develop agricultural and medical devices. Geologists use their knowledge of Earth to find resources. See **Figure 7-2.**

Likewise, designers use humanities knowledge in their work. Knowledge about personal, social, and religious values is used. This knowledge allows

TechnoFact: Information technology is the use of computer and electronic technologies to gather, process, and distribute information. This technology has developed over many years, but grew dramatically since 1971, the year the microprocessor (a computer central processing unit on one chip) was developed. This electronic device had the same computing power as the original thirty ton Eniac computer built in the1950s.

Figure 7-3. Historical knowledge was used in recreating this historical western fort.

Standards for Technological Knowledge

3

Extend
Describe how a designer uses the four types of knowledge to develop a consumer product.

designers to develop products people want and will value. Historical knowledge is used in restoring significant buildings. See **Figure 7-3.** Knowledge of gender roles in society is also used. In some societies, certain tasks are considered "men's" work. Other tasks are considered "women's" work. In other societies, males and females may do the same work. Knowing these distinctions allows designers to match products to body sizes, shapes, and strengths. Knowledge of style and design tastes allows designers to create products people think are attractive.

Descriptive knowledge is used to tell other people about design ideas. Words, mathematical formulas, and drawings are used in the process. See **Figure 7-4.** They communicate sizes, shapes, and relationships. Also, descriptive knowledge is used in preparing production and operation directions. Descriptive knowledge is used in preparing documents to tell suppliers of material and energy needs.

Finally, technological knowledge is at the heart of technological enterprises. This knowledge of tools and materials is vital in developing, producing, and using devices. It lets people select appropriate tools and materials. Technological knowledge allows these tools and materials to be used in making products and systems.

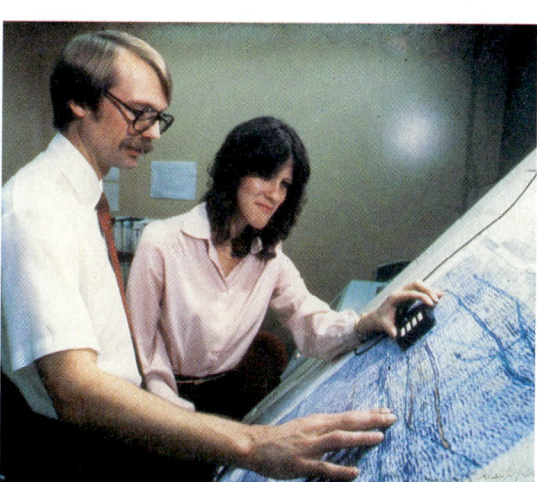

Figure 7-4. The seismic map these people are using is an example of descriptive knowledge. It describes the features of Earth below the surface. (Getty Oil Co.)

TechnoFact: Mathematics is not considered a science, but is closely associated with the sciences. It is frequently called the *language of science* because of its use in communicating the results of science.

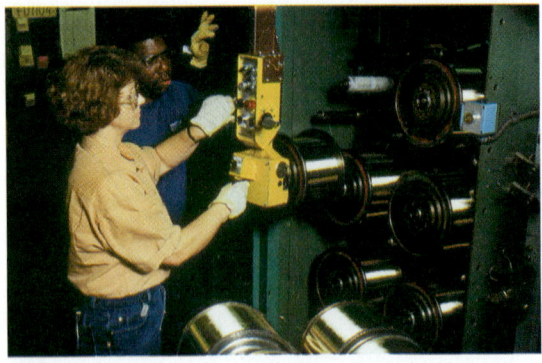

Figure 7-5. Technological knowledge allows people to use machines and materials to make products. (Goodyear Tire and Rubber Company)

See **Figure 7-5.** It lets people select and use technological devices properly. This knowledge allows people to assess and control technological actions.

In actual practice, knowledge is not used in isolated areas. We don't say, "Now, I'm going to use scientific knowledge. Then, I will use technological knowledge as I work on this problem." People may learn things in specific areas, such as science or technology. This approach may help the learning process. In daily life, however, we use knowledge as a whole. We apply scientific, humanities, descriptive, and technological knowledge together. To deal with problems and opportunities, we use knowledge as an integrated mass. Knowledge is a resource in our daily lives.

Data, Information, and Knowledge

Directly related to these types of knowledge are data and information. People often use the three words *data,* *information,* and *knowledge* as if they describe the same thing. This is wrong. The terms describe different things.

Data is individual facts, statistics, and ideas. These facts and ideas are not sorted or arranged in any manner. They are simply an accumulation—a collection of facts, numbers, and ideas.

By contrast, *information* is data that has been sorted and arranged. It is organized facts and opinions people receive during daily life. It may come directly from other people, electronic media (radio and television), the Internet, books and magazines, or many other sources. Information is data that has been put in a condition useful to people or machines (computer systems). See **Figure 7-6.** It is the product of direct human action or computer processing. Using computer systems to change data into information is called *data processing.*

Finally, *knowledge* is information humans can apply to situations. See **Figure 7-7.** When information is organized in a logical way, it is called a *body of knowledge.* This knowledge

Figure 7-6. These workers are viewing information about materials on a computer screen. (Inland Steel Company)

TechnoFact: Data processing or information processing requires operations such as handling, sorting, merging, and computing, which use defined procedures.

can be acquired by systematic study. It is a body of useful information a person has accumulated over time. A body of knowledge is the result of schooling and life experiences. People have general knowledge. They understand facts and principles about many things. People may know how the universe operates or how devices work. They may understand how people react in various situations. This general acquaintance with the world is useful to all people. Also, people have specific knowledge. They can apply information to do a task or job. For example, they may have design knowledge or machine operation knowledge. They can use this knowledge to develop new products or operate machines to make something.

To put these three terms into context, think of going to an airport. There are thousands of flights each day. Each flight goes from an origin to a destination. A list of all these flights represents data. It would be a massive list. The data would be of little use to the average person. Sorting this data into groups by origin city and destination city would change the data into information. Arranging the data further by time of departure would make the information more useful. Being able to use this information to select an appropriate flight is knowledge.

Scientific Knowledge

As you learned earlier, scientific knowledge explains how the natural world operates. This knowledge includes the laws and principles

Extend
Ask the students what knowledge they need to plan an automobile trip and what sources of data and information they could use to develop this knowledge.

Figure 7-7. This worker is using knowledge to control a steelmaking process. (Inland Steel Company).

TechnoFact: *Data* is now both singular and plural, where it was once the plural of the Latin term *datum*. For example, the statements "This data is important in solving the problem," and "These data are important in solving the problem," are correct.

Extend
Ask the students what scientific information they have used today.

Tech Terms
Science—The body of knowledge concerning the physical (natural) world.

governing natural interactions. Scientific information is gathered using several basic methods. One of the most important of these is the scientific method. It starts with a person who believes the following ideas:

➤ Everything that happens in nature can be understood. You have to ask the right questions. Then you must do the right experiments.

➤ Nature is always the same. Time or distance makes no difference. A scientist in one place working on an experiment should get the same results as another scientist anywhere else at any time.

➤ There is a relationship between a cause and its effect. A specific cause will produce a specific effect. Scientists design experiments to produce given effects. Then, they change the conditions one by one. This way, they can find which conditions are producing the effect.

Starting with these ideas, scientists develop hypotheses. They state what they think will happen if certain actions take place. Then, they design and conduct experiments to test this hypothesis. See **Figure 7-8.** They collect and analyze data. Finally, they draw conclusions about the causes and effects shown during the experiments. The experiments are designed to produce useful information. Scientists seek answers to questions puzzling them. Experiments must often be repeated to prove the results are not accidental. There is no end to the kinds of experiments scientists can conduct. For example, scientists might launch

Figure 7-8. This scientist is examining a culture as part of a scientific experiment. (Consulate General of Israel)

TechnoFact: The scientific method has come to be described as a series of steps used to investigate a natural phenomenon. The scientific method is as follows:
Step 1: Information is gathered by careful observation of the phenomenon being studied.
Step 2: A preliminary generalization, or hypothesis, is formed.
Step 3: A number of implications are tested by observations and experiments.
Step 4: If the original hypothesis proves successful, it is accepted as a scientific theory or law.

Figure 7-9. Technological knowledge is used when people make products. (Dana Corporation)

a space probe to photograph a distant star, or they might work in a laboratory studying molds or bacteria.

Technological Knowledge

Technological knowledge is the knowledge of action. It includes how to do or make something. This knowledge is knowing how to use tools and other resources to make products and systems. See **Figure 7-9.** It involves using information to control and modify the natural and human-made environments.

This knowledge is gained by trying something. It is gained through purposeful action. Technological knowledge involves building something, trying it out, evaluating its operation, modifying it, and trying it out again. It is gained through the processes of creating, building, testing, and evaluating devices and systems.

Without technological knowledge, agriculture could not feed the people of the world. Technological knowledge prevents food from spoiling before it can be delivered to people. We also use this knowledge to locate minerals and petroleum. To process these materials into usable products, we use technological knowledge. We need technological knowledge to harness atoms and electrons for power.

Information Age

We are becoming overwhelmed with information. We are living in a period of time some people call the *Information Age.* It started with the development of printing with moveable type by Johannes Gutenberg in 1450. The age was hastened by many inventions (technological advancements). These include lithography (a method of printing) in 1798, Bell's telephone in 1876, Edison's sound recorder in 1877, the first successful movie system in 1891, Marconi's radio in 1896, the first workable television in 1926, and the tape recorder in 1935. The advent of the computer after World War II, however, brought the age into being. Today, we have more information available than anyone can read or use. People may be drowning in information. They risk having information overload. The challenge is to gather the information you need and ignore what is not helpful.

TechnoFact: *Information Age* is a term used to describe the present state of developed nations having widespread electronic access to information. It depends upon the use of computer technology to find, select, process, and report information.

Gathering Technological Information

- Determine type needed
- Identify possible sources
- Gather
- Sort and arrange
- Apply

Figure 7-10. These are the steps in gathering technological information. (AT&T Network Systems)

Gaining Information and Knowledge

You may gain new technological information by reading about it, listening to the radio or television, having someone show it to you, or developing and testing your own designs. The foundation to all these approaches is research. It involves seeking and discovering information.

To gather technological information, a person must have a focused approach. See **Figure 7-10.** He or she should do the following:

- ➤ Determine the type of information needed to address the technological problem or opportunity.
- ➤ Identify possible sources of this information.
- ➤ Gather information about the problem or opportunity.
- ➤ Sort and arrange the information.
- ➤ Apply the information to the problem or opportunity.

Gathering information may require research. This research may be historical, descriptive, or experimental.

Historical research gathers information already existing. It describes how other people have solved similar problems. See **Figure 7-11.** Historical information may include existing designs or processes used in the past. It may come from examining products on the market or reading operating manuals for equipment. Sources of historical information include books, magazines, museums, films, videotapes, photographs, drawings, and models.

Descriptive research gathers information by measuring and describing products and events. It describes something as it is. See **Figure 7-12.** For example, it may describe how people are doing a job. It may describe the physical attributes (such as size and

TechnoFact: Unwritten information gathered from studying the physical remains of past civilizations is used to describe the technologies developed and used by people of early times. Often this information is incomplete because much of the information is lost over the long periods of time.

Figure 7-11. Historical research describes how other people have solved similar problems. (American Association of Blacks in Energy).

Research
Measure the hands and fingers of all your classmates. Summarize (average) these measurements in an ergonomic chart that shows the twenty-fifth, fiftieth, and seventy-fifth percentiles for each measurement.

Tech Terms
Descriptive Research— Gathering information by measuring and describing products and events.

shape) of products. Henry Dryfus conducted a pioneer study of this type. He and his research colleagues measured large numbers of men, women, and children. He analyzed and categorized the data he collected. Then, he published his findings in great detail. For example, he provided information about the size of men's, women's, and children's hands. This information is valuable for glove makers. Likewise, the data on human torso sizes is important to clothing manufacturing companies. The data on the width of people is useful for aircraft and stadium seating designers.

Experimental research is typical of the research scientists conduct. See **Figure 7-13.** This approach structures activities so changes or improvements can be measured. For example, a team may carefully study how people are doing a job. The job may be modified to see if the production improves. The

Figure 7-12. Descriptive research gathers information about existing conditions. This worker is gathering information about inventories. (Datastream)

TechnoFact: Ergonomics is the use of technical information about people for the design of things, systems, and environments for human use. A good ergonomically designed product or system increases performance because it takes advantage of the strengths and weaknesses of its human and machine components.

Figure 7-13. Scientists often conduct experimental research.

quality of the product, the time required to do the job, or the amount of scrap and waste may be compared.

Historical research describes what was possible. Descriptive research describes what is possible. Experimental research describes what can be possible. See **Figure 7-14.**

Summary

Technological information and knowledge have helped people control and modify the environments around them. They have helped people build better ways of life. This information and knowledge have given people better food, clothing, and shelter. They have made movement from one place to another easier. Technological information and knowledge have increased the standard of living and life expectancy of the U.S. population.

Types of Research

| Historical | Descriptive | Experimental |
| What was? | What is? | What can be? |

Figure 7-14. This shows the focus of each type of research.

Conducting a Research Project
The reason many research projects are conducted is to answer the question "What happens if...?" This question explores the effect changing one element (variable) has on another element. Research involves at least five steps:
1. Research Question. This is the "What if?" question written as a simple sentence.
2. Hypotheses. These statements are what you think will happen as a result of the research.
3. Procedure. This is the step-by-step plan used to conduct the research.
4. Results. This is the data or information you collected while conducting the research.
5. Conclusion. This presents what was learned by conducting the research. It explains what happened in the research and whether or not it supported the hypotheses.

Curricular Connections

Social Studies

Research the effects of a specific invention on the Information Age. Use inventions aiding in communication, such as the radio, television, or printing press.

Science

Read an owners' manual for an electrical appliance or tool. Describe the scientific information included in it. Look for examples of physical, chemical, and biological information.

Mathematics

Measure the size of the hands of your classmates—the length of each finger, width of the hand across the knuckles, and length of the hand from the heel to the tip of the third finger. Average these measurements for boys, girls, and the class as a whole. Provide a drawing of the average female and male hand.

Standards for Technological Knowledge

3 12

Activities

1. Read a current events magazine or newspaper article. Indicate examples of descriptive, humanities, scientific, and technological knowledge, using colored pencils or highlighters.
2. Develop a timeline of inventions that have led to the Information Age. You may want to develop a timeline for a specific strand. Typical strands would be the written word, broadcasting, or recording.

TechnoFact: Motivation research is a special type of research used by companies in their advertising and marketing efforts. It applies knowledge and techniques to understanding consumer attitudes and behavior.

Test Your Knowledge

Do not write in this book. Place your answers to this test on a separate sheet of paper.

1. Knowledge communicating size and shape is _____ knowledge.
2. Knowledge about mechanical structures and electricity is _____ knowledge.
3. Knowledge about using products and devices is _____ knowledge.
4. Information that can be used to solve a problem is called _____.
5. Facts arranged in a useful order are called _____.
6. Rewrite, in your own words, the basic beliefs underlying the scientific method.
7. Why is technological knowledge essential to our existence?
8. What are three sources of information?
9. Research answering the question "What can be?" is called _____ research.
10. Research telling us the condition of something is called _____ research.

Modular Activity

This activity develops the skills used in TSA's Computer Applications event.

Computer Applications

Activity Overview

In this activity, you will create grade tracking reports for three of your classes. You will design the reports using word processing software, and will incorporate elements developed with spreadsheet software and graphics software.

Materials

➤ Paper and pencil
➤ Computer with word processing, spreadsheet, and graphics software and clip art
➤ Printer

Background Information

General. Consider the principles of design as you develop your reports. Locate and size elements to achieve balance and proportion. Use contrast to emphasize key elements. Select elements to provide rhythm.

Word processing. Design your reports to be attractive and functional. Using a two-column layout or text wrapping in the section of the report containing the illustration may improve visual appeal. The individual scores on assignments should be readable, but the focus should be on your final grade.

Spreadsheet. When creating your spreadsheet, keep in mind that the information will be inserted into an 8½″ x 11″ report. If you do not know how many entries you will have for a component of your final grade (for example, if your instructor does not have a set number of quizzes), design the spreadsheet to allow you to add as many entries as needed.

Graphics. Use brainstorming techniques to develop a list of items or themes appropriate for each art piece. Select the theme(s) you wish to use. Review illustrations of similar objects (from books, clip art, or the Internet) to identify possible design elements.

Guidelines

➤ Begin by developing rough sketches for several possible report layouts. Develop several rough sketches for the art pieces as well.

➤ Each report must include an original piece of artwork created in a graphics program. The artwork must be related to the specific course.

➤ The report for each class must be one page in length.

➤ Each report includes your name, the course title, instructor, period, and a description of each component included in your final course grade. It also must include an explanation of final grade calculation, and a spreadsheet table that automatically calculates averages for each component and final grade.

Evaluation Criteria

Your reports will be evaluated using the following criteria:

➤ Design elements

➤ Technical functionality

➤ Creativity

Employer Expectations

Many general qualities are needed to be successful in the workplace. Behavior required for professional success and advancement includes the following:

- **Cooperation.** An employee must cooperate with supervisors, other employees, and customers.
- **Dependability.** A dependable employee is timely, completes all assignments, and sets realistic goals for completing projects. A dependable employee is trusted by others.
- **Work ethic.** Good employees put an honest effort into their work.
- **Respectful.** In order to be respected, employees must show respect for others, the company, and themselves.

Today's workplace emphasizes equality. *Equality* is the idea that all employees are treated in the same way. To make the workplace a pleasant place to work, negative behaviors, such as discrimination and harassment, are not tolerated. *Discrimination* is expressing a biased attitude, act, or behavior toward another person. People sometimes discriminate against individuals based on personal characteristics such as race, gender, age, or religion. *Harassment* includes offensive and unwelcome actions against another person. Behaviors such as discrimination and harassment often result in termination of employment.

Teamwork is also important in today's workplace. Solving problems often requires you to work with other people. Teamwork can help get a job done quicker than doing it alone, and it allows for more possible solutions to be considered. Working in teams almost always improves the efficiency of developing a design solution, but it takes organization and cooperation among the team members.

A team consists of team members and one or more team leaders. The team members provide their ideas and abilities. Problems are successfully solved using teamwork when all the members and their contributions are involved. The team leaders are responsible for guiding the direction and setting the goals for the team. They also ensure that all the team members contribute to the team's success.

The team leaders guide the rest of the team. The quality of their leadership determines whether the team will be a success or a failure. Good leaders have a vision of what must be done and look for ways to reach the team's goals. They also are skilled communicators who are able to encourage the team members to assist each other and cooperate. A leader's job is not always easy, but good leaders are willing to work as long as it takes to achieve success. They can organize and direct the team's activities. Good team leaders are fair and honest. They take responsibility for their own actions and give credit to others when it is due. Finally, good leaders know how to delegate authority. They may even assign leadership roles to other team members to get the job done as efficiently as possible.

Chapter 8
People, Time, Money, and Technology

This chapter covers the benchmark topics for the following
Standards for Technological Literacy:

3 **4** **6** **12**

Did You Know?

➤ In a recent survey, company leaders thought listening and interpersonal skills are the most important skills managers need.

➤ Good career decision making involves knowing the following:

  ➤ Your interests, skills, and values.

  ➤ How your personal criteria relate to career options.

  ➤ What those careers involve in terms of education, experience, and skills.

➤ According to the U.S. Department of Labor, these are the ten fastest growing occupations:

  ➤ Computer engineers.

  ➤ Computer support specialists.

  ➤ Systems analysts.

  ➤ Database administrators.

  ➤ Desktop publishing specialists.

  ➤ Paralegals and legal assistants.

  ➤ Personal care and home health aides.

  ➤ Medical assistants.

  ➤ Social and human service assistants.

  ➤ Physician assistants.

Objectives

The information given in this chapter will help you do the following:

➤ Describe the roles of people in technology.

➤ Summarize the difference between white-collar and blue-collar jobs.

➤ Recognize the changes in job opportunities that came with the Information Age.

➤ Paraphrase what a company and an industry are.

➤ Give examples of the tasks that managers, engineers, technologists, technicians, and workers carry out.

➤ Explain the factors to consider in selecting a career path.

➤ Describe what a career ladder means.

➤ Give examples of time as a technological resource.

➤ List the ways money is used in technological activities.

Key Words

These words are used in this chapter. Do you know what they mean?

career ladder

company

engineer

industry

middle manager

operative management

salary

technician

technologist

upper management

wage

worker

Chapter Outline
Did You Know?
Objectives
Key Words
Work and a Changing Society
Jobs and Technology
Jobs in the Information Age
Jobs in Technology
People and Jobs
  Goals, Interests, and Traits
  Getting Career Information
Advancing in a Career
Time
Finances (Money)
Summary
Curricular Connections
Activities
Test Your Knowledge

Standards for Technological Knowledge

4 **6**

Brainstorm
Ask the students why it is not possible to farm effectively using techniques like the one shown in Figure 8-1.

Tech Terms
Factory System—A working environment in which a number of workers work together to build products for human use.
Factory—One or more buildings on a site where a fairly large workforce produces goods.

We are in the greatest growth period in the history of humankind and gaining new knowledge at an unbelievable pace. Compared to past periods of history, we know much more about our universe and how it was formed. We are developing new technological knowledge daily, learning more about using machines effectively, and rapidly creating new products, systems, and structures. All of these advancements are the products of the efforts of people.

Work and a Changing Society

Our country has already passed through several historical stages. Each of these stages saw great changes in what kinds of work people did. During pioneer days, we were an agricultural society. Growing crops and raising livestock were our main industries. See **Figure 8-1.** Most people worked on farms and ranches. Some people worked in shops and stores. Other people worked as doctors, lawyers, teachers, and nurses. A small group of people practiced village trades, like blacksmithing and carpentry. Still other people worked in newly developing industries. Regardless of the type of work they did, most people worked for themselves. They tilled the land, ran the businesses, practiced the trades, and served the professions.

With the Industrial Revolution that started in the late 1800s, work changed. Factories started to sprout up in many parts of the nation. This

Figure 8-1. Early in American history, most people worked on farms.

TechnoFact: Engineering is a vital career dedicated to designing, building, and operating the structures, equipment, and other devices of industry. The main kinds of engineering are chemical, civil, electrical, industrial, and mechanical.

TechnoFact: Implementing the factory system in many countries caused serious social problems, including worker boredom from routine production work, urban congestion, and pollution.

Figure 8-2. This worker is using a computer system to monitor processes in an oil refinery. (Honeywell)

Standards for Technological Knowledge

4 6

Research
Have the students trace the development of a major technological advancement, such as television or computers.

was particularly true in the Northeast and upper Midwest. Manufacturing became an important industry. Work was concentrated in newly developed factories. Machines were invented to help make products efficiently. People moved off the farms and ranches into expanding cities. Many of these people started to work for other people for the first time. They were trained to run the machines making an expanding array of products.

After World War II, a new age started to develop. It started to become more important for people to exchange information. Many inventions helped lay the foundation for this information exchange. The telephone became more important, as did the telegraph and radio. Books, magazines, and newspapers had become plentiful and inexpensive. Television came into general use during the 1940s and early 1950s. Computers became common in the later part of the twentieth century. They became a key part of information processing and manufacturing process control. See **Figure 8-2.**

Experts are predicting even greater advances in the twenty-first century. They think more new discoveries will be made during this period than in all of human history. With these changes will come new demands on people. Jobs and communities will change. Whole societies will be different.

TechnoFact: Starting in the 1960s, factory production has become a global activity. Products are now produced and assembled in more than one country. Production activities have caused companies to move their factories from industrial countries to parts of the world that have cheap labor and low taxes.

Standards for Technological Knowledge

4

Resource
Reproducible Master 8-1: Management

Extend
Discuss the types of jobs involved in producing a product (such as engineers, designers, managers, machine operators, and stock clerks).

Tech Terms
Wage—Payment received by an employee in exchange for hourly work.
Salary—Fixed payment for work paid to a person on a regular basis.
Payroll—The total amount of money paid to employees for a given period of time.
Skilled Worker—A production worker who possesses a broad, recognized skill in a specific area, such as a tool-maker, millwright, or carpenter. The skill is acquired through education and on-the-job experience.

Jobs and Technology

People have always worked. First, they worked to find food and build shelters for themselves and their families. Then communities developed. People gathered together and shared work. Some people raised crops, while others fished or hunted. Simple trades, like shoemaking and candle making, developed. Society continued to progress over the years. Likewise, the work people did changed. Today, there are literally thousands of different jobs. It is impossible to know about all of them. The various jobs, however, can be grouped into several categories. A common breakdown was developed to reflect the Industrial Era of the nineteenth and twentieth centuries. It divided jobs in two categories: management and workers. See **Figure 8-3.**

Management included all people paid salaries. Actually, not all people in the group were managers. They were, however, paid a weekly or monthly *salary.* This group was often referred to as *white-collar* workers. The name implies these jobs were clean, office-type jobs. White-collar jobs allowed workers to wear white shirts and blouses. Typical white-collar jobs were engineering, sales, accounting, and supervising.

Workers were people paid hourly *wages.* They earned money for each hour they worked. This group was often called *blue-collar* workers. This name suggests the work was somewhat dirty. The workers wore colored clothes so the dirt was not as noticeable. Blue-collar jobs were generally divided into three groups:

➤ Skilled workers. These were people having extensive training and work experience for a broad job. See **Figure 8-4.** Often, they served apprenticeships. This training is a combination of classroom and on-the-job instruction. It often takes four or more years to complete an apprenticeship. Typical skilled

Figure 8-3. During the Industrial Age, employees were thought to be either managers (left) or workers (right). (Conoco, Inc. and Motorola, Inc.)

TechnoFact: Blue-collar and white-collar are used to describe workers. Originally, blue-collar described workers who worked with machines and wore colored work shirts that did not show dirt. White-collar described workers who did deskwork and wore white dress shirts.

Figure 8-4. This skilled engraver prepares a printing plate for a greeting card. (American Greeting)

jobs are carpentry, plumbing, tool and die making (making manufacturing tools), and mold making (making casting molds for metal, plastic, and glass).

➤ Semiskilled workers. These were people that received specific training for a specific job. See **Figure 8-5.** Semiskilled workers had a high degree of skill, but in a narrow range. They could be machine operators in a manufacturing plant, roofers on a construction site, or truck drivers.

➤ Unskilled workers. These were people doing routine work requiring little training and experience. They could do cleanup, product packing, or other routine jobs.

The use of the terms *white-collar* and *blue-collar* is less meaningful today. It was closely related to manufacturing. This sector of employment is shrinking. More manufacturing processes are being automated. See

Standards for Technological Knowledge

4

Brainstorm
Have the students categorize the jobs in your school as skilled, semiskilled, or unskilled.

Tech Terms
Semiskilled Worker—People who possess a high degree of skill, but in a narrow range of work.
Unskilled Workers—People who do routine work requiring little training and experience.

Figure 8-5. This machine operator would be considered a semiskilled worker. (Dana Corporation)

TechnoFact: Having skilled, semiskilled, and unskilled labor is related to the concept of *division of labor*. This term means separating the job into its parts and assigning workers to do one or more of these parts. Division of labor is supported by standardization of production, modern machinery, and large-scale industry.

Figure 8-6. Automation is one reason there are fewer manufacturing jobs. (Chrysler Corp.).

Figure 8-6. Fewer workers are needed to do the same amount of work. Also, foreign competition has caused additional manufacturing jobs to be lost.

Jobs in the Information Age

With the Information Age, however, times have changed. The society has changed from one based on hardware (machines and materials) to one based on "thoughtware" (ideas and knowledge). Information processing is more common than material processing. Many production workers work in clean environments. See **Figure 8-7.** They are knowledge workers rather than manual laborers. The jobs are more dependent on mental skills than physical skills.

There are fewer jobs for people failing to complete high school. The industrial jobs requiring little education and

Figure 8-7. Many of today's production workers work in clean environments. (Inland Steel Company)

Figure 8-8. Most technology-related jobs are with companies. (AT&T Network Systems)

training are rapidly disappearing. There is a high demand for people having technical and professional training

after high school. The days of dropping out of school and getting a fairly good job are gone. In the future, even people with only a high school education may have trouble getting a good job.

Jobs in Technology

There are many jobs directly related to technology. Most of these are associated with businesses or companies. See **Figure 8-8.** These businesses are economic organizations that change resources into products and services. They are generally organized to make money by providing something consumers want. See **Figure 8-9.** In brief, a *company* is all of the following:

➤ An economic institution.

➤ A user of resources (labor, material, energy, information, finances, and time).

➤ A producer of a product, structure, or system.

➤ A profit maker.

Standards for Technological Knowledge

4

Research
Have the students use a phone book to identify manufacturing, transportation, and construction companies in your town or city.

Tech Terms
Company—An economic institution that uses resources to produce a product, structure, or system with the intent of making a profit.
Industry—An economic organization that produces goods and services in an economy.

Economic institution

that uses

resources

to produce

products or systems

intending to make a

profit

Figure 8-9. Here is a simple definition of *industry.* (Carolina Power and Light)

TechnoFact: A company is a commercial (dealing with commerce; selling goods and services) organization that can be called by a number of different names. Some of these names are *business, corporation, enterprise, establishment,* or *firm.*

Standards for Technological Knowledge

4

Resource
Reproducible Master 8-2: Functions of Management

Resource
Reproducible Master 8-3: Levels of Authority and Responsibility

Discussion
Discuss the role of management in modern business.

Tech Terms
Management— Organized methods of doing business operations.

Figure 8-10. The managers, at the left and center of this photo, organize the work for the company. (Gannett Co.).

In a free enterprise economic system, production and prices are determined by the principles of supply and demand. The government has little impact on markets. Businesses are free to produce any products they choose. As consumer demands change over time, some businesses are created, some close, and some modify their production to meet current demands. In this system, business plays the important roles of owning production equipment and processes, determining product supply, and reacting to market demands.

The demands of the market are extremely important to businesses. A company's products or services must meet customer expectations if they are going to sell well. Customers want products, services, and systems that meet their needs and wants. For this reason, most businesses have quality control systems. These systems help to ensure customer satisfaction. The image of a company depends on how satisfied the customers are with its products and services. If a business is successful at satisfying its customers,

it will be profitable and will remain stable in the economy. If the customer is not satisfied or disappointed in the company's product or service, the customer develops a negative image of the company.

Companies are present in all technological contexts. They exist in agriculture, communication, construction, energy conversion, manufacturing, medical services, and transportation. The companies within a technology can be grouped into a number of different industries. Farming, food processing, flour milling, and fertilizer industries would be part of agriculture. Each *industry* is composed of a number of companies producing competing products. For example, several companies make cereals or frozen vegetables. Government and universities also provide technology-based employment. Typically, these technology-based careers include managers, engineers, technologists, technicians, and workers.

Managers are people who organize resources to produce products or services efficiently. See **Figure 8-10.** Effective managers have special leadership qualities. They are good at organizing people to get a job done. These managers are not afraid to make decisions. They have great determination and confidence even when things are not going well. Effective managers seem to enjoy the challenges of difficult jobs.

The top managers, or *upper management,* in a company are the officers that can include the *president* and *vice presidents.* Other titles used for top managers are Chief Executive

TechnoFact: Managers organize, direct, and control a company and its parts. For example, production managers supervise the making of products. Marketing managers deal with selling the company's output (goods and services) to customers. Human resource managers help recruit, screen, train, and place new workers.

Officer (CEO) and Chief Operating Officer (COO). Information-based companies may have a Chief Information Officer (CIO). These managers set company goals and plot courses of action.

Working under top management are *middle managers.* These people direct most company operations. Middle managers oversee production, marketing, finance, personnel, and engineering activities.

The lowest level of management is called *operative management* or *first-line supervision.* These managers help plan the day-to-day operations of their businesses. They also set up the work schedules and supervise production workers, salespeople, bookkeepers, and other employees.

Engineers design processes, products, and structures. See **Figure 8-11.** They apply knowledge of physical and mechanical principles and mathematics to do their work. These people use this knowledge to design roads, bridges, and skyscrapers. They design products and machines. They may direct the work of other engineers and technologists.

The technologist is a fairly new type of professional person. As engineers became busier in design, technologists have taken some of their duties. *Technologists* work closely with engineers to implement their work. They may be called *industrial, manufacturing, food, medical,* or *construction technologists.* Technologists are the major link between engineers and the factory floor or construction site. They help managers and workers make the objects the engineers design.

Technicians are people who know how to operate machines. They diagnose and solve problems at the operational level. Generally speaking, engineers know the theories of how and why something works. Technicians know how to make it work.

These people work in many different fields. See **Figure 8-12.** Over sixty different kinds of technicians work in the fields of engineering and other technologies. Some work on computers, robots, and automatic machines. The greatest numbers of technicians work in health care. They operate and repair medical instruments and devices.

Workers are the people who construct and assemble various products. They operate cameras in television studios, printing presses at newspaper publishers, and bulldozers on

Figure 8-11. Engineers, like the one shown here, design products and systems.

Standards for Technological Knowledge

4

Resource
Reproducible Master 8-4: Preparing a Company Organizational Chart

Resource
Reproducible Master 8-5: Preparing an Organizational Chart for Your School

Tech Terms
Manager—A person who plans, organizes, directs, and controls company operations.
Engineer—A person involved in designing, constructing, and operating structures, machines, and systems.

TechnoFact: Industrial management deals with the performance of machines, as well as people. Machines are kept in good working condition to ensure the quality of their production. The flow of materials through a plant is supervised to ensure that neither workers nor machines are idle.

Figure 8-12. This technician works for a large dairy. (Toolbox)

construction sites. See **Figure 8-13.** These workers are the people driving buses, flying airplanes, and making cheese.

Workers are the people who actually produce the products and services that have been designed. These people operate the machines to build products, erect buildings, harvest crops, provide health care, and generate power. They may work with machines and systems that engineers and technicians have designed.

People and Jobs

How do jobs affect you? Most of us will have several roles in life. We will have roles in a family. Most of us will have citizenship responsibilities to our community, nation, and world. We will be workers and consumers of products and services. Most of us will work for or operate businesses.

Goals, Interests, and Traits

You are a special person, and you have special abilities we sometimes call *talents*. For example, you may be very good in math, or you may excel in communication skills, such as languages or grammar. These talents should guide you in selecting a career path or job. Where and how you will work will depend upon the following factors:

➤ Personal goals. Not all of us want the same things out of life. Some want to live and work among friends. For example, you may

Figure 8-13. This worker is a welder on a construction site.

TechnoFact: Thomas Alva Edison was a very talented person. When he died, he left behind an estimated five million pages of technical, business, and personal papers. He held 1,093 patents, plus he filed an estimated 500 to 600 unsuccessful or abandoned patent applications. His first successful patent was filed in 1868. In 1882, at the height of his work on electric light and power, he completed 106 successful applications.

want to live in the community where you grew up. Perhaps being home every night is important to you. On the other hand, you may prefer traveling, working with new ideas, and making new friends.

➤ Interests. If you have held jobs, you may already know what you prefer. You may like working with others. Perhaps you get along well as part of a team. Then you should consider a job where you meet and work with people. You may prefer working alone, using tools, instruments, or computers. This could indicate that you should look for a career working with tools or machines. A job where you will be alone most of the time may be attractive.

➤ Ability. It is good to identify those courses you like and in which you do well. These courses indicate where you will be most successful. An ability to do math is important for accounting or engineering careers. On the other hand, communication careers depend on English and grammar skills. Abilities in several areas may lead to a career in which these skills and interests can be combined. For example, an interest in writing and working with people is important for authors, teachers, and editors. An ability to work with machines and materials is important for auto mechanics, carpenters, computer operators, and printers. See **Figure 8-14.**

➤ Physical traits. Careers make demands on our mental abilities,

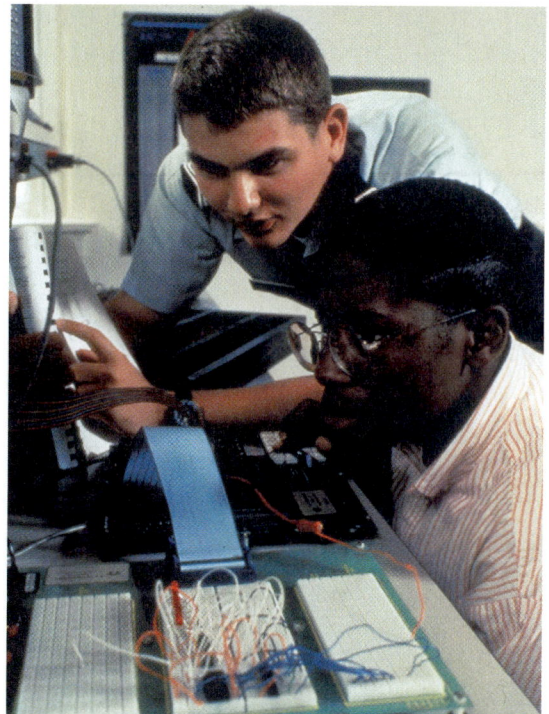

Figure 8-14. These people like to work with tools and machines. (Army ROTC).

but they also call on our physical abilities. Certain jobs may require long hours of physical activity. Others will depend on above average strength or good eyesight. Still others need nimble fingers or quick reaction times.

Getting Career Information

Deciding on a career is important. It is not like any other activity you do in school. Try to learn as much as you can about each kind of work. For each kind, match the job requirements and major responsibilities with your own abilities, interests, and values. Also, pick out the requirements that do not

Research
Have the students compare their goals, interests, and abilities to those needed to perform one or more jobs.

Tech Terms
Goal—What a person intends to do or achieve.
Interest—Curiosity about or attention to some situation, event, or product.
Ability—A natural or acquired competence in a specific task or activity. It is similar to adeptness, expertise, proficiency, skill, technique, and know-how.

Brainstorm
Have the students list how they can develop the abilities required to perform a specific task, such as driving a racecar, designing a product, and running a six-minute mile.

TechnoFact: Goals are important for almost every human undertaking. Other terms related to goal are *aim, intent, objective, purpose,* and *target.*

Tech Terms
Occupation—Jobs and positions employing the skills and knowledge of workers.
Trade—An occupation involving considerable training and physical skill.
Profession—An occupation requiring lengthy education and training, such as architecture, dentistry, law, or engineering.

Figure Discussion
Ask the students what types of interests, abilities, and personal traits would be required to do the work shown in Figure 8-15.

match your characteristics. Decide if there is something you can do to remove the mismatch. If not, then that particular career may not be for you.

There are many ways to learn about jobs. The world of technology offers many satisfying and challenging jobs. One of the best ways is to look for books on careers in your school library, guidance office, or public library. Also, the U.S. Department of Labor website has occupational information. These sources will describe jobs in a number of fields. They also will tell you what is needed to get and hold the job of your choice.

The U.S. Department of Labor groups jobs into a number of categories. Each of these groups contains jobs that are somewhat alike. Each job has its own requirements and duties. The table in **Figure 8-15** lists the groups and typical jobs.

Advancing in a Career

Changes in technology are opening up new jobs every day. At the same time, old occupations are being eliminated. People entering the job field today need to be flexible. This means they must be willing to change jobs. Many people will make career changes several times during their working lives.

The *career ladder* means working your way up to better jobs, as your skills and knowledge improve. For many people, a beginning job is the first step to a better one. It is wise to choose a job that gives you a chance to

move to a better job. See **Figure 8-16.** Usually people move on to related jobs. Each new job gives added responsibility, more pay, and greater satisfaction. Sometimes the job promotion and training come from within the company, while sometimes there are offers from another company. As you decide on a career, you should think ahead to the chance for promotion.

It is important to show your superiors you have developed new abilities. Also, you must show interest in taking on more responsibilities. Being on time for work, doing your job with enthusiasm, and responding to job challenges are important. They show your supervisor you are mature and trustworthy.

Time

Time measures duration of an event. In technology, time is a resource. It measures the duration it takes to make a product, perform a task, or transmit information. See **Figure 8-17.**

To deal with labor and production processes, time is used. In our technological society, people are one of the inputs for producing products and structures. People, however, cannot produce technology without time any more than they can produce technology without knowledge. People are paid by the hour, month, or year for the time spent on the job. In one sense, time is considered renewable. There will always be more time tomorrow, the next day, and the next. In a stricter sense, time wasted or lost is lost forever.

TechnoFact: The aim of the U.S. Department of Labor is to advance and support the well-being of the employment seekers, wage earners, and retirees of the United States. This is done by improving working environments, developing opportunities for gainful employment, defending retirement and health care benefits, helping companies locate employees, strengthening free collective bargaining, and tracing variations in employment, prices, and other nationwide financial measurements.

Job Classification	Typical Jobs or Occupations
Executive, Administrative, and Managerial	Accountant Public relations manager Construction and building inspector Construction manager Farmer Farm manager Financial manager Health services manager Hotel manager Industrial production manager Restaurant and food service manager
Professional and Technical	Aircraft pilot Flight engineer Engineer Engineering technician Drafter Computer programmer Mathematician Food scientist Geologist Lawyer Judicial worker Urban and regional planner Social worker Schoolteacher Librarian Dentist Veterinarian Dental hygienist Broadcast and sound technican Photographer Actor
Marketing and Sales	Cashier Insurance sales agent Manufacturer's sales representative Real estate agent Real estate broker Retail salesperson Securities sales representative Travel agent
Transportation and Material Moving	Bus driver Taxi driver Truck driver

Job Classification	Typical Jobs or Occupations
Administrative Support, Including Clerical	Bank teller Communications equipment operator Computer operator Hotel desk clerk Receptionist Mail clerk Shipping and receiving clerk Mail carrier Bookkeeping clerk Brokerage clerk Statement clerk File clerk Library assistant
Service	Janitor Groundskeeper Pest controller Chef Dental assistant Medical assistant Barber Flight attendant Correctional officer Firefighter
Installations and Repairs, Including Mechanical	Computer repairer Aircraft mechanic Automotive mechanic Musical instrument tuner Bricklayer Stonemason Carpenter Roofer
Production	Assembler Production supervisor Fishing vessel supervisor Butcher Inspector Machinist Welder Water treatment plant operator Printing press operator Texitle machinery operator Upholsterer Dental laboratory technician Photographic process worker

Figure 8-15. This table lists a sample of jobs with each major job classification. (U.S. Department of Labor)

Demonstrate
Show the students how to use the Internet to access the Department of Labor's Occupational Outlook Handbook.

Research
Have the students use the Internet and the Occupational Outlook Handbook to find the educational, training, and skill requirements for a job that interests them.

Brainstorm
Ask the students to list the job categories in Figure 8-16 appealing to them and why these categories do.

Tech Terms
Job—An activity, such as a trade or profession, for which a person is paid.

TechnoFact: During the twentieth century, wages rose, fringe benefits grew, and working conditions improved for American workers. The American workforce was much better off at the end of the century than it was at the beginning.

Figure 8-16. These employees have moved up the career ladder. As they advanced, they were given more authority and responsibilities. (Motorola, Inc.)

Figure 8-17. For an airline, it is important to control the time it takes to load baggage on an airplane. (Alaska Airlines)

Extend
Introduce the concept of an occupation as a combination of the work content and the skills and education a person brings to the job.

Tech Terms
Career Ladder—
The means for a person to advance to better jobs as skills and knowledge improve.
Career Advancement—
A promotion to a job with more responsibilities and better pay.
Job-Hopping—
Changing jobs to attain quick financial gains or career advancement.

TechnoFact: In most countries, working people specialize by offering a unique service or producing a product people want. This leads to people holding different types of jobs requiring differing levels of skill and education or training. Each of these occupations provides its own status that is generally referred to as an *economic class.*

Likewise, time is related to processes. Most tasks have a time limit. Machines can run only so fast or so slow. Materials can be cut only within a range of speed. Plastics take a certain amount of time to melt. A casting must cool for a certain amount of time. Crops have a growing season. X rays must be exposed for a set period of time. Companies must meet time limits in contracts. We call these *deadlines* for completion.

Managing Time

Time is an important resource for all technological processes. It is also an important resource for every activity in our lives. People need time to perform a task, whether it is manufacturing a product or completing math homework. Because time is such an important input to any process, it is essential to learn how to manage your time wisely.

Before completing an activity, it is important to develop a work schedule. This requires an assessment of the task and an estimate of how long it will take to complete. If you are writing a report on an emerging technology, you will need to plan time to research your information, organize it, and write the paper. This work schedule may extend over several weeks. If you are growing crops, you will need to plan time to prepare the soil, plant the seeds, water and fertilize the plants, and harvest the crops. This work schedule will span several months. In planning your work schedule, it helps to analyze your goals and set your priorities. It also may be beneficial to set a routine for yourself in which you work on this task at the same time every day.

Once your work schedule is developed, you must follow it. Maintaining your schedule is necessary if you wish to complete the task on time. If you are having trouble sticking to your work schedule, you may want to try writing down all your time commitments in a day planner. Make sure your work location is conducive to completing your task. For example, when you sit down to do your homework, make sure you are in a quiet place, away from distractions and interruptions. It is also important to assess your progress as you work and to take short breaks often to relax and rejuvenate yourself.

Learning how to manage your time is essential in order to meet deadlines. Most activities need to be completed within a certain time limit. Companies often have contracts that state the deadlines for completion. Your history essay probably has a due date. Developing and maintaining your work schedule will help you meet these deadlines. Prioritizing your tasks, recording your schedule in a planner, and avoiding interruptions can help you finish on time. Depending on the task, you may need to work with others to meet your deadline. This requires additional planning and cooperation. When working with a group, it is often best to divide the work and delegate specific tasks to each group member. Using these time management techniques, you should be able to complete your tasks successfully and meet your deadlines.

Discussion
Discuss what was meant by the statement "It takes money to make money."

Tech Terms
Lead-Time—The elapsed time between initiating a project or activity and achieving results.

Technology Explained

flight controls: A series of movable control surfaces that can be adjusted to cause an airplane to climb, dive, or turn.

An airplane is the only transportation vehicle that can move in these three distinct ways: go forward, turn side to side, and climb and dive. Airplanes can change direction through a system of flight controls the pilot operates. Flight controls change the *lift* acting on the wings and other control surfaces.

Lift is what makes an airplane fly. The engine pushes or pulls the airplane through the air. Air flows over the wings, creating lift. If not enough air is

Right turn

Left turn

Figure A. An airplane moves to the right when the pilot moves the rudder to the right side, lifts the aileron on that side, and lowers the aileron on the other wing. The airplane turns to the left with the opposite movements.

flowing over the wings, no lift is produced, and the plane loses altitude. The airplane must maintain forward motion to fly.

Pilots make turns by using two types of control surfaces, the rudder and the ailerons. The rudder is a movable flap on the vertical portion of the tail assembly. The pilot controls the rudder with two pedals. The ailerons are flaps on the rear of each wing, **Figure A.** The pilot controls the ailerons with a yoke or stick. To make a turn, the pilot, using the pedals, moves the rudder to one side. Using the yoke or stick, the pilot simultaneously lifts the aileron on that side and lowers the one on the other wing. This causes the plane to bank (turn)

Finances (Money)

It has been said "It takes money to make money." Technological activities are no exception. People must be paid for their labor. Buildings and machines must be purchased or leased. Material and energy must be bought. All of this takes money. A simple statement can show this use of money: *Money* is used to *purchase resources* (materials, machines, and labor), which are used to *produce a product or service*, which is then exchanged in the *marketplace* for *money.*

TechnoFact: An important time measure is called a *deadline.* This is a time limit for completing a project or assignment. An important part of managing an operation is to set realistic deadlines and monitor work as it progresses toward those dates.

in the direction the rudder is pointed.

Surfaces on the horizontal stabilizers control climbing and diving. These control surfaces are called *elevators*. When the elevators are tilted downward, the tail is forced upward. The nose is pointed downward, so the plane will lose altitude or dive. When the elevators are tilted upward, the tail will be forced down, and the nose will point up. The plane will then gain altitude or climb, **Figure B.**

Figure B. The elevators control climbing and diving.

Summary

We are in a period of history when knowledge is expanding rapidly. In the next fifteen to twenty years, knowledge will advance more than it has in all of recorded human history. How you will fit into this changing world depends upon your personal goals, interests, abilities, and physical traits. Finding the right kind of work will require serious study. You will have many fields from which to choose. Some jobs will require skills of managing people; others will demand creative abilities. Still others will require skills in making parts and handling tools. Whatever career is best for you, you should consider the opportunities for advancing to more responsible work. This is known as "moving up the career ladder."

TechnoFact: In industrial societies, few households are self-sufficient. For example, most people cannot grow, harvest, and cook their own food; design and build their means of transportation; build their own homes; and make their clothes. Instead, they depend on people working in a variety of occupations to design and produce the things they need.

Curricular Connections

Social Studies

Investigate the types of jobs available in your school. Determine the levels of authority and responsibility each has.

Science

Research science careers in private industries, not-for-profit agencies, universities, and governmental agencies. One source for information is the U.S. Department of Labor Internet site. (Use a search engine and enter "Department of Labor.")

Mathematics

Gather data from the U.S. Department of Labor Internet site. Prepare graphs comparing starting wages or salaries for ten different jobs.

Standards for Technological Knowledge

3 **12**

Reinforce
Discuss the role of money in technological activities.

Tech Terms
Money—Any means of exchange widely accepted as payment for goods and services.

Activities

1. Draw a chart listing your abilities, interests, and values. Include what you are good at doing, as well as what you dislike. Try to be honest and look at yourself as you think others might see you.

2. Find several careers you think you might like. Gather information about the requirements of these careers. Describe the duties in terms of working with people, information and ideas, and machines.

3. Compare your abilities, likes, and strengths with the career information you have collected. Do not expect a perfect match; however, the effort will start you on the way to career planning.

4. With two or three other students, create an imaginary business. Decide what product or service you will offer to the public. Write a one-page summary of the company's role in the free enterprise system. Agree on the image you want your company to have, and develop a marketing plan that complies with this image, paying particular attention to ensuring customer satisfaction.

5. Choose a career area that interests you. Research its academic and professional advancement requirements. Prepare a verbal presentation with PowerPoint slides for the class describing the requirements.

6. Develop a schedule for a major assignment in one of your classes. Set deadlines for achieving project milestones. Schedule time to work on the project.

Test Your Knowledge

Do not write in this book. Place your answers to this test on a separate sheet of paper.

1. Why do the major kinds of work people do change over time?
2. Match the statement on the left with the correct type of worker on the right. You will use some answers more than once.

_____ Has a high degree of skill in a narrow range.   A. Skilled

_____ Requires little or no training.   B. Semiskilled

_____ Does routine work.   C. Unskilled

_____ May serve an apprenticeship.

3. What is thoughtware?
4. A company is a(n) _____ institution using _____ to produce a(n) _____, with the intent of making a(n) _____.
5. Match the statement on the left with the correct position title on the right. You will use some answers more than once.

_____ Serves as a major link between engineers and the factory floor.   A. Manager

_____ Organizes resources to produce products.   B. Engineer

_____ Designs products, systems, and structures.   C. Technologist

_____ Solves operational problems.   D. Technician

_____ Often leads a design team.   E. Worker

_____ Constructs products or structures.

6. Summarize the four factors that should contribute to your career choice.
7. The path leading to better jobs is called a career _____.
8. Time is a valuable resource of technology. True or False.
9. Time is only used to measure human labor. True or False.
10. Money is needed to pay for human labor, materials, machines, and energy. True or False.

Activity 2A

Tools and Materials As Resources, Textbook, pages 186–192, and Student Activity Manual, pages 21–23

This is a prescriptive activity introducing tools as an essential input for technology. The activity presents a specific set of procedures to the students. The system used to make the product should be demonstrated. Then, divide the class into three groups:
- Group 1: Makes the game board.
- Group 2: Makes and stains the pegs.
- Group 3: Makes the storage hole cover.

Have each group complete the following steps:

1. Carefully read the directions for making the product.

2. Assign a step or several steps to each member of the group.

3. Obtain the supplies needed to make the product (listed in the textbook).

4. Produce twenty-four parts, using the procedures listed in the textbook.

5. Analyze the output (quality control). When all parts are produced, give each group eight sets of parts. Have each group assemble the products by completing the following steps:

1. Carefully read the directions for the assembly procedure.

2. Assign a step or several steps to each member of the group.

3. Assemble eight products, using these assembly procedures.

Tools and Materials As Resources

Introduction

You have read about tools and technology. Also, your teacher has told you about tools. Now you are ready to use this knowledge. Your teacher will show you how to put this knowledge to work. You will be shown how to use tools to extend your ability to do a job.

In this activity, you will have common hand tools to help you build a simple game. You will be using common measuring, cutting, drilling, pounding, and polishing tools to make a tic-tac-toe board. **See Figure 2A-1.**

Figure 2A-1. This is tic-tac-toe.

Equipment and Materials

➤ 2 x 4 (1½˝ x 3½˝) construction lumber

➤ ¼˝ x ¾˝ wood strips

➤ ⅜˝ dowels

➤ ¾˝ x No. 6 flat head wood screws

➤ Steel rule

➤ Try or combination square

➤ Crosscut or backsaw

➤ Miter box and handsaw

➤ Hand drill

➤ Brace

➤ ½´´ auger bit

➤ ¹⁄₁₆´´, ⁹⁄₆₄´´, and ¹³⁄₃₂´´ twist drills

➤ Countersink bit

➤ Flat wood file or rasp

➤ Block or smooth plane

➤ Abrasive paper and sanding blocks

➤ Screwdriver

➤ Scratch awl or center punch

➤ Hammer or mallet

Procedure

The procedure for the tic-tac-toe game is numbered from 1 to 43, but you do not have to complete the steps in order. You may make the pegs, peg storage hole cover, and game board in any order. Each number is used only once so you and your teacher can easily refer to a specific step without confusion.

Preparing to Make the Product

1. Study the drawings for the tic-tac-toe game. See **Figure 2A-2.**
2. Read the procedure for making the game.
3. Carefully watch your teacher as he or she demonstrates how to make the game.

Qty	Description	Size	Material
1	Game board	1 1/2 x 3 1/2 x 3 1/2	Spruce or hemlock
1	Peg cover	1/4 x 3/4 x 3 1/2	Pine
8	Pegs	3/8 dia. x 5/8	Birch dowel
1	Screw	3/4 x No. 6 flat head	Plated steel

Figure 2A-2. Here is the bill of materials.

Making the Game Board

Select and lay out the material:

4. Select a length of 2 x 4 construction lumber. (Note: The actual size is 1½″ thick x 3½″ wide.)

5. Lay out a line ⅜″ from one end.

6. Lay out a line 3½″ from the first line.

Cut out the game board:

7. Cut the end off to the outside of the ⅜″ line to square the end of the board.

8. Cut off the game part, barely leaving your line on the part.

See **Figures 2A-3** and **2A-4.** Lay out the game board:

9. Draw lines 1¼″ in from the edges and ends of the block.

10. Draw lines ¼″ in from the edges and ends.

11. Draw lines ¼″ down from the face on the edges and ends.

12. Locate and mark the nine peg holes.

13. Locate and mark the peg storage holes on one edge.

14. Locate and mark the peg cover pivot screw anchor hole.

Game Block

Peg Storage Cover

Figure 2A-3. These are working drawings.

Step 1: Draw lines 1 1/4" apart.

Top View

Step 2: Draw lines 1/4" from the edges and the ends.

Top View

Step 3: Locate and mark the peg holes.

Top View

Step 4: Locate and mark the peg storage and cover pivot holes.

Side View

Figure 2A-4. Here is someone laying out the game board.

Produce the game board:

15. Saw kerfs (shallow slots) about ⅛″ deep on the four 1¼″ lines.

16. Drill the nine ¹³⁄₃₂″ peg holes, ⅜″ deep.

17. Drill the two ½″ peg storage holes, 3″ deep.

18. Drill the ¹⁄₁₆″ pivot screw hole.

19. File and plane the ¼″ x ¼″ chamfers around the top of the block.

20. Sand all surfaces.

Making the Pegs

Select and lay out the materials:

21. Select a length of ⅜″ dowel.

22. Check the end to see that it is square.

Produce the pegs:

23. Set a stop block on the miter saw for a ⅝″ cut.

24. Cut the end of the dowel square, if necessary.

25. Cut eight pieces of dowel, ⅝″ long.

26. Sand and lightly break (round) the ends of the pegs.

Making the Peg Storage Hole Cover

Select and lay out the materials:

27. Select a length of ¼″ x ¾″ pine.

28. Draw a line ¼″ from the end.

29. Draw a line 2½″ from the first line.

30. Locate and mark the pivot screw hole.

31. Lay out the radius on each end.

Produce the peg storage hole cover:

32. Cut the end off to the outside of the ¼″ line to square the end of the board.

33. Cut off the hole cover, barely leaving your line on the part.

34. Drill a ⁹⁄₆₄″ pivot screw hole.

35. Countersink the hole for a No. 6 flat head screw.

36. Sand or file the end radii.

37. Sand all surfaces.

Finishing and Assembling the Game

Apply finish to the parts:

38. Stain four pegs a dark color and let dry.

39. Apply a surface finish to the board, peg storage hole cover, and pegs.

40. Allow all finishes to dry properly.

Assemble the product:

41. Place the dark pegs in one storage hole and the light pegs in the other hole.

42. Make the screw hole for the peg storage cover. It should be over the anchor hole in the game board.

43. Attach the cover with a ¾″ x No. 6 flat head wood screw.

Challenging Your Learning

See **Figure 2A-5.** Make a chart like the one shown. List the steps in the procedure in which you used each type of tool.

Type of Tool	Procedure Step
Measuring tool	
Cutting tool	
Drilling tool	
Gripping tool	
Pounding tool	
Polishing tool	

Figure 2A-5. This is a list of where you used each tool.

Safety Rules

➤ Cutting tools have sharp edges. Never carry pointed tools in your pockets.

➤ Always cut and chisel away from your body or the hand holding the part.

➤ Always carry cutting tools with sharp edges pointing down.

➤ Use each tool for its proper purpose.

➤ Mushroomed heads on chisels or other *struck* tools are dangerous. Bits of material can fly off and strike you.

➤ Never use tools with loose handles.

➤ When using a saw, keep your free hand away from the saw blade.

➤ Never rub your fingers across cutting tools.

➤ Remove slivers immediately and sterilize the wound.

➤ Cuts should be allowed to bleed freely for a short time. Then, bandage properly.

➤ Keep working surfaces free of scrap and unnecessary tools.

➤ Sweep up scrap and debris from floors and discard in a waste bin.

➤ If jigs or fixtures have sharp edges, wear protective gloves. Never wear gloves, however, when operating power equipment.

Activity 2B

Tools As a Resource

The Challenge

Develop a device (tool) that will separate marbles according to their diameters.

Background

Since the earliest times, human beings have developed tools to do work. Early tools included stone implements and weapons. Now we have complex tools that do work for us. Machines cut materials with ease and accuracy. Vehicles whisk us to faraway places. Computers process information and control machines. People developed each of these devices to help other people. In this challenge, you are faced with a problem. You have a container holding three different sized marbles. These marbles need to be sorted by size (their diameters). Measuring each marble individually would take too much time. You have decided to use technology to help you, and you plan to build a device to sort the marbles.

Materials

Develop your technological device using any or all of the following materials:

➤ ½˝ x 1˝ wood strips
➤ ¼˝ plywood
➤ White glue
➤ ¾˝ brads
➤ ⅜˝ dowels
➤ Masking tape
➤ Paper clips
➤ Poster board
➤ String

Activity 2C

Materials As a Resource

Materials As a
Resource—Material
Properties and
Applications,
Textbook, pages
194–197, and
Student Activity
Manual, pages
29–30
This is a prescriptive
activity to introduce
students to material
properties. A set of
material samples
should be provided,
containing a variety
of materials, such as
sheet aluminum, cop-
per, steel, acrylic
plastic, plywood, and
lumber. Divide the
class into groups of
two to four students
and have each group
complete the follow-
ing steps:
1. Perform all the
appropriate tests the
procedures in the
textbook outline.
2. Record test
results on the form
supplied on page 29
of the Student
Activity Manual.
3. Present their test
findings to the class.
4. Discuss the impli-
cations of the data for
material selection
and application.

Introduction

The world is made of materials. Science tells us that materials are found in three states: gases, liquids, and solids. Solid materials are often called *industrial materials.* They can be grouped as metals, ceramics, polymers, and composites.

Each material has a unique set of properties. These properties tell us how a material will act or perform under certain conditions. As discussed in this section, there are seven major material properties:

➤ Physical.

➤ Mechanical.

➤ Chemical.

➤ Thermal.

➤ Electrical and magnetic.

➤ Optical.

➤ Acoustical.

In this activity, you will be able to test your knowledge of materials and their properties.

Equipment and Supplies

➤ Material kit

➤ Data recording sheet (A sample is provided. See **Figure 2C-1.**)

➤ Metalworking vise

➤ Pliers

➤ Ball peen hammer

➤ Mill file

➤ Flashlight

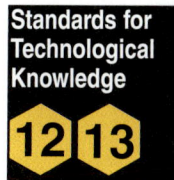

Data Recording Sheet Material Identification and Property Testing						
Group: _____ Date: _____						
Specimen Number	1	2	3	4	5	6
Name of material						
Type of material						
Density						
Hardness						
Ductility						
Light reflectivity						
Torsion strength						

Figure 2C-1. This is a sample data recording sheet. (Do not write in the book.) Fill in the sheet given to you, as materials are tested. On the back of the sheet, list products for which each material is suited. Remember, manufacturers select a material for its properties.

SAFETY: Review with your instructor any safety rules concerning use of tools. If not properly handled, these tools may cause injuries.

Procedure

Your teacher will divide the class into groups of three to five students. Each group should do the following:

1. Obtain a material kit. It will contain a set of six materials. You will need two samples of each material. Samples should be the same size. They should also be numbered.

2. Obtain a data recording sheet.

3. Carefully look at each material. Write down the name of each material (such as copper, wood, or steel) on the chart.

4. List under its name whether the material is a metal, ceramic, polymer, or composite.

5. Select one sample of each material.

6. Arrange the samples by weight. Since they are the same size, this will show which are denser than the others.

7. Put a *1* in the "Density" column for the heaviest sample, a *2* in the column of the next heaviest, and so on.

. Test the hardness of each sample. Take two file strokes on the surface of each sample. See **Figure 2C-2.** The harder the material, the less the file will cut. **SAFETY: Use a file with a handle fitted over the tang. Be careful not to run a hand or finger over filed edges. The edges may be sharp.**

9. Mark a *1* in the column of the hardest material, a *2* in the column of the next hardest, and so on.

10. Test the ductility of each sample. Place one end of the sample in the vise. Grip the other end of the sample with a pair of pliers. Bend the sample back and forth three times. See **Figure 2C-3.** Note how easy it is to bend each sample. Also look for any cracking or breaking along the bend line.

11. Place a *1* in the column for the material that was easiest to bend and did not crack. This is the most ductile sample. Place a *2* in the column of the next most ductile sample and so on.

Figure 2C-2. Here is an illustration of how to test a material for hardness.

Figure 2C-3. This is someone testing the ductility of a material.

12. Take the second set of material samples.

13. Test the light reflecting property of each material. See **Figure 2C-4.** Shine the flashlight on each sample. A material that "shines" is reflecting light.

14. Place a *1* in the column of the material reflecting the most light. Place a *2* in the column of the next best reflector and so on.

15. Test the torsion strength of each material. Place one end of the sample in the vise. See **Figure 2C-5.** Grip the other end of the material with a pair of pliers. Try to twist the material. The material that is the hardest to twist has the highest torsion strength. **SAFETY: Pliers can pinch fingers hard enough to draw blood or raise blisters. Keep your hands away from the jaws. Do not use pliers with damaged or worn jaws. A slip may cause an injury.**

16. Place a *1* in the column of the material with the best torsion strength, a *2* in the column of the next best, and so on.

Figure 2C-4. Use a flashlight to test a material's ability to reflect light.

Figure 2C-5. Here is someone performing a torsion test.

Activity 2D

Energy As a Resource

Energy As a Resource—Motor Schematic, Textbook, pages 198–203, and Student Activity Manual, page 33

This is a prescriptive activity allowing the students to make a common energy converter—an electric motor. The activity presents a specific set of procedures to the students. It may be done individually or by groups of students. If using groups, assign three students to each group. Have the students complete the following activities:
- Student 1—Lay out and produce the base.
- Student 2—Make the armature.
- Student 3—Make two motor bearings.
- All—Assemble and test the motor.

Introduction

People have always looked for ways to do work more easily. They want to do jobs efficiently and quickly. Therefore, humans have invented tools and machines. One of the most important inventions is the electric motor. We can find such motors in use almost everywhere. These motors power machines in factories. They run our refrigerators and freezers. Electric motors move air through furnaces and air conditioners. They move the hands on wall clocks. Think of your home. How many uses of motors can you list?

This activity will let you build a simple motor. With the help of your teacher, you can see how the motor converts energy into power. See **Figure 2D-1.** A motor changes electric energy into mechanical motion.

Device	Energy Input	Energy Output
Electric motor	Electric	Mechanical
Electric generator	Mechanical	Electric
Battery	Chemical	Electric
Electric light	Electric	Light and heat
Oil burner	Chemical	Heat and light
Windmill	Mechanical (linear)	Mechanical (rotary)

Figure 2D-1. These are ways of converting energy from one form to another. Can you add to the list? Try to do this by listing five energy converting devices you see every day.

Equipment and Supplies

➤ Materials listed on the bill of materials. See **Figure 2D-2.**
➤ 72″ of 18 gauge magnet wire
➤ 20″ of 14 gauge solid uninsulated copper wire
➤ Thread to tie the coil together
➤ Coil winding jig. See **Figure 2D-3.**
➤ Bearing bending jig. See **Figure 2D-4.**

➤ 1½ or 6 volt battery
➤ Scratch awl
➤ Hammer or mallet
➤ Wire cutters
➤ Screwdriver
➤ Utility knife
➤ Abrasive paper

Part No.	Qty.	Description	Size	Material
1	1	Base	1/2 x 4 x 6	Plywood
2	2	Bearings	14 ga.	Copper wire
3	1	Coil	2 3/4" I.D.	
4	4	Magnets	1/4 x 1" dia.	Speaker magnets
5	4	Screws	1/2 x 6	Sheet metal
6	2	Connecting wire	18 ga. x 12"	Copper wire

Figure 2D-2. Here is the bill of materials for making an electric motor.

1/2 x 4 x 6 plywood

Woodblock
2 3/4 dia x 3/4 thick

Black mark

Figure 2D-3. This is a coil winding jig. About five are required.

Procedure

Preparing to Make the Motor

1. See **Figure 2D-5.** Gather the materials needed to make the motor:

   A. 1½˝ x 4˝ x 6˝ plywood base

   B. 1 pc. 18 gauge magnet wire, 96˝ long (72˝ for the coil and 24˝ for the two connecting wires)

   C. 1 pc. 14 gauge wire, 20˝ long

   D. 12˝ thread

   E. Four ½˝ x No. 6 pan head sheet metal screws

   F. Four ¼˝ x 1˝ magnets (or a substitute your teacher suggests)

2. Each group of four to six students at a workbench should secure the following:

   A. Coil winding jig

   B. Bearing bending jig

   C. Battery

   D. Scratch awl

   E. Hammer or mallet

Figure 2D-4. Here is a bearing jig.

Figure 2D-5. This is an electric motor. Refer to the bill of materials for part names.

F. Screwdriver

G. Utility knife

H. Wire cutters

Making the Motor

First, you need to make the base. See **Figure 2D-6.**

Magnet location **Screw location**

3/4

4

3/4

1

3

6

Figure 2D-6. Here is a motor base.

1. Lay out the location of the screw holes and magnets.

2. Start the screw holes with the scratch awl.

3. Prepare for attaching the magnet. (Your teacher will show you how. The method will vary with the type of magnet used.)

4. Sand the edges and ends to remove sharp edges and splinters.

5. Start a screw in each of the four holes. Do NOT tighten the screws at this point.

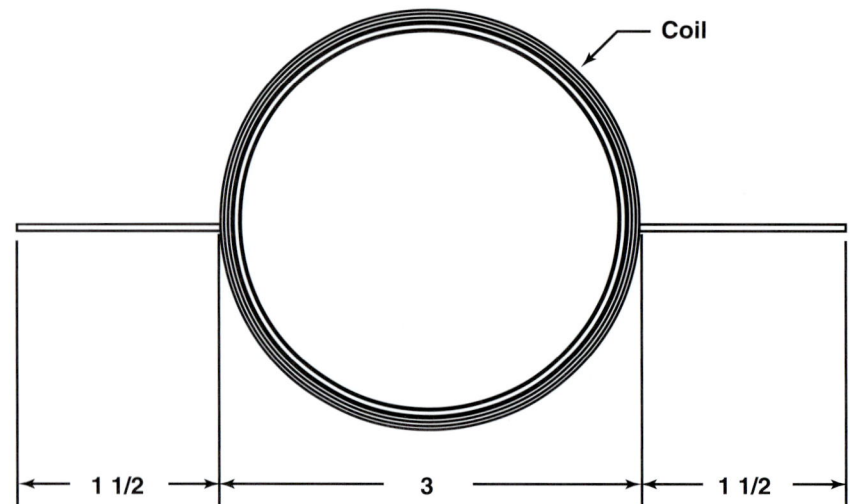

Figure 2D-7. This is the completed coil for the motor.

Then, you must make the coil. See **Figure 2D-7.**

1. Cut about 20˝ off the 18 gauge wire and lay it aside.
2. Mark 1½˝ from one end of the long wire.
3. Using the utility knife, carefully remove the insulation from the end of the wire up to the mark. Be careful to cut away from any part of the body.
4. Place the wire in the coil winding jig. Leave the 1½˝ clean end of the wire extending along one of the black marks.
5. Carefully wind the wire around the center core.
6. Stop winding the coil when you cannot make another turn and leave 1½˝ on the opposite black mark.
7. Extend the wire along the black mark.
8. Cut the excess wire, leaving 1½˝ along the mark.
9. Remove the coil from the jig.
10. Tie thread at four places to hold the coil together.
11. Remove the insulation from the second end of the coil.

Finally, you need to make the bearing. See **Figure 2D-8.**

1. Cut the bearing wire in half.
2. Use the bearing bending jig to form two bearings.

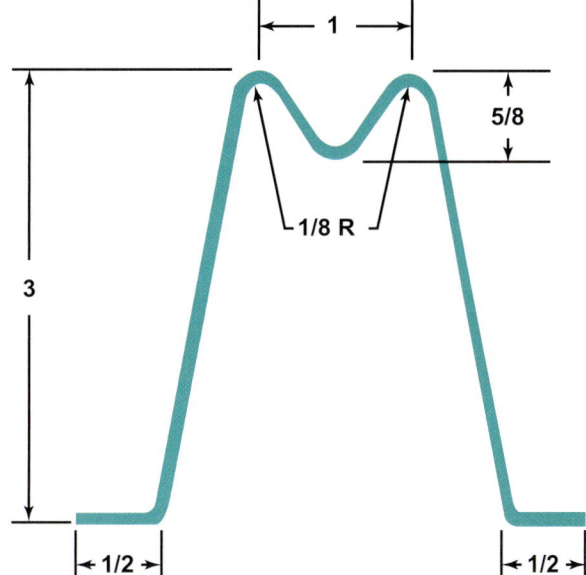

Figure 2D-8. Here is a motor bearing.

3. Cut off extra wire at the ends of the bearings.
4. Carefully remove the insulation from one foot of each bearing.

Assembling the Motor

1. Cut the 20˝ piece of 18 gauge wire in half.
2. Remove the insulation from each end of the wires.
3. Place one bearing under the screws at one end of the base.
4. Tighten the screw on the end that is still insulated.
5. Place the end of one of the connecting wires under the other screw.
6. Tighten the screw on the bearing and connecting wires.
7. Repeat steps 3–6 for the other bearing and connecting wires.
8. Attach the magnets.
9. Place the coil between the bearings.
10. Connect the motor to the battery.
11. Test the motor operation.
12. With the help of your teacher, describe how the motor converts electrical energy into mechanical motion.

Activity 2E

Energy and Machines As Resources

Energy and Machines As Resources, Textbook, page 204

This activity is designed to allow the students to use their creativity to design a system using a simple machine and human strength. There is no correct answer. To use this activity, divide the class into groups of two to four students. Have each group complete the following steps:
1. Define the problem they are attempting to solve.
2. List any restrictions they face, such as materials that can be used and the strength of the person using the machine.
3. Develop a sketch of two to five ways to solve the problem.
4. Select and refine the best solution.
5. Build a prototype of the solution, using only the materials listed in the textbook. (Limiting the materials introduces a common restriction all product designers face.)
6. Demonstrate the solution.
7. List the ways they can improve their designs.

The Challenge

Develop a device (tool) that will use mechanical energy to lift a load.

Background

Average individuals are not particularly strong. They cannot lift loads equal to their own weight. Compare this to the strength of common ants. Ants can carry loads many times their own weight. People have learned to compensate for this weakness. They have developed technological devices to lift loads and move items. These devices are all based on six simple machines: the lever, inclined plane, wheel and axle, wedge, screw, and pulley. See the following page. For this challenge, you have a load (a brick) that must be lifted 18˝ off the floor. Design and build a technological device that will lift the load. It must use less force than the brick weighs.

Materials

Develop your technological device using any or all of the following materials:

➤ 1˝ brads
➤ ½˝ x 1˝ wood strips
➤ ⅜˝ dowels
➤ Heavy string
➤ Pulleys
➤ Scales
➤ Wire
➤ Woodblocks

Activity 2F

Information As a Resource

Information As a Resource—Note Taking, Textbook, pages 205–209, and Student Activity Manual, pages 37–38

This activity is designed to allow students to read, review, and synthesize information. Generally, this activity is done individually, but the students can work in groups of two. To use the activity, have the students complete the four steps outlined on page 205 of the textbook. As an alternative, this activity can be assigned as homework.

Introduction

Information is used to understand technology and solve technological problems. It comes from many sources and can be obtained by talking to people or observing their actions. For most researchers, the printed word has been their greatest resource. This resource may be found in the form of books, magazines, technical reports, and product catalogs. Today, computers and the Internet allow researchers from every field to share information.

This activity will let you gather and apply some technical information. You will read about how a computer modem and home-to-home e-mail on the Internet works. Diagrams and illustrations have been included. You will be asked to do the following:

➤ Outline the information.

➤ Prepare a brief report.

➤ Illustrate the report.

Procedure

1. Study the materials included with this activity.

2. Take notes on the important information.

3. Provide a brief outline for the information, using the following headings:

    A. Historical background

    B. System of operation

    C. Technological importance of the invention

4. Write a brief report with illustrations (sketches).

History

Computers are machines that perform calculations and process information with amazing speed. The first computers were very large and often took up a whole room. They were very expensive to design and maintain, and only the government, big businesses, and universities could afford them. Today, you will find computers almost everywhere. More likely than not, you have one in your own home.

These machines help people find and organize information, create and test models, and solve problems. They even help us to communicate with each other. Charles Babbage designed the true ancestor of the modern computer in the 1830s. This steam-operated machine was called the *Analytical Engine.* It was never completed, but it was designed to perform calculations with a mechanical calculating unit controlled by punched cards. These punched cards were the basis of the card-handling machines Dr. Herman Hollerith developed in the 1880s.

In 1944, Professor Howard Aiken completed his Automatic Sequence Controlled Calculator (ASCC). This machine was over 50′ long and 8′ tall! It took 0.3 seconds to add or subtract, 4 seconds to multiply, and 12 seconds to divide. In 1946, Dr. John Mauchly and J. Presper Eckert completed the first electronic digital computer. This machine was called the *Electronic Numerical Integrator and Calculator (ENIAC).* ENIAC could perform as much work in one hour as ASCC could in a week. Technological advances in the 1950s and 1960s led to the computers you see today. These advances include more reliable transistors, more accurate *analog* systems, and digital computers with miniaturized *integrated circuits.*

How a Computer Modem Works

Computers process information expressed as ones and zeroes, but when the machines exchange information over telephone lines, the data must be converted, or *modulated,* into analog, or wave, form. After the data is sent and received, it is converted back into digital form at the other end. This is the job of the *modem.* The name comes from the jobs a modem performs: *mo*dulation and *dem*odulation.

Modems can be inside your computer's hard drive or added on to your system like a video game is added on to your TV. In order to use a modem, you have to tell your computer to send information to a certain destination (phone number). In this case, let's pretend you are sending your friend some information on your favorite TV show.

After giving your computer the phone number, the computer orders its modem to open a phone line and dial the number. When your friend's modem answers the phone, your modem announces it would like to make contact. It announces this with a *hailing tone.* This is just like when you pick up the phone, hear the dial tone, dial, and then hear the ringing. As the user, you can hear the tones on your computer as the connection is being made.

Figure 2F-1. This diagram shows how a computer modem transmits signals from one computer to another.

Your modem and your friend's modem exchange messages after the connection is made. They agree on how fast the information will be sent and how it will be packaged. The modems also agree on how the information will be checked for errors.

The modem begins transmission. Your computer, the sender, feeds its modem digital data. The modem turns the data into a high-speed series of tones and puts it in the phone line. Your friend's modem, the *receiver,* hears the tones and converts them back into digital form. After the transmission has been completed, your computer instructs its modem to break the connection, or you could say, hang up the phone. See **Figure 2F-1.**

How Home-to-Home E-Mail on the Internet Works

If you live in New York and want to send your cousin in California a letter, you can use the postal system or your computer. The *Internet* allows you to use your computer to send any kind of information over telephone lines. Refer to page 529.

First, you have to type in the information on your computer. Then, you have to type in your cousin's Internet address and use the send command. When the send command is given, the message flows across a telephone wire to a computer directly linked to the Internet. Here, it is expressed as a collection of the ones and zeroes of computer language.

The Internet breaks the message into packets. Each of these packets has your cousin's address on it. The packets make their way across the Internet separately. Computers route them onto the least busy pathways. Each one of the packets can take different routes. This helps the system use the circuits efficiently and provide alternate routes, in case there is a breakdown somewhere. It may take as little as a fraction of a second to send your letter. Everything depends on the traffic!

Your cousin's computer in California puts the packets back together. Then, it inserts the letter into your cousin's electronic *mailbox*. When your cousin turns on her computer and looks into her mailbox, she'll find your letter. She can read the letter on the computer screen or print it out. See **Figure 2F-2.**

Figure 2F-2. This is the path information takes to travel from one telephone line to another through the Internet.

Section 3
Creating Technology

Technology is created to solve problems. The problems may be as simple as creating a device to organize your music collection, or they can be as complex as developing medicines to help fight illnesses.

People can solve problems using several different methods. They may use the invention process to create a new device. People may change an existing invention, creating an innovation. Also, they may use the design process to create a useful solution.

Many people design products and systems. You may have designed a device to solve a problem of your own. If so, you are a designer. Designers use a number of steps to create new solutions. These steps include activities such as identifying the problem, researching, drawing sketches, making models, and testing. In the following ten chapters, you will learn how the design process works.

9 Invention and Innovation

10 The Design Process

11 Identifying Problems

12 Researching Problems

13 Creating Solutions

14 Selecting and Refining Solutions

15 Modeling Solutions

16 Testing Solutions

17 Communicating Solutions

18 Improving Solutions

Technology Headline

Interactive Television

Have you ever wished you could check your e-mail while watching your favorite TV show? You may soon be able to do this and many more exciting things with interactive television. This new technology combines TV with the Internet and allows you to interact with the show you're watching.

Interactive TV will give viewers access to Internet sites and instant messaging on their TV screens. You will be able to use the Internet and watch TV at the same time by activating a picture-in-picture screen. A program guide will be available, which will organize channels into categories, making viewing easier. You will be able to choose a channel by clicking on words and graphics.

With interactive TV, you will be able to play along with TV game shows, see the local weather, and access real-time statistics during sports events. Just by using your remote control, you will be able to print recipes from the cooking show you are watching, shop on-line, and order concert tickets. You will even be able to listen to your favorite radio station and view movie trailers.

In addition to these great features, interactive TV will allow you to watch your favorite shows anytime you want. It will include digital video recording, which will allow you to search for and record shows, using an electronic program guide. Instant replay, fast-forward, and skip ahead features will also be available with this technology.

Many TV and cable networks are already developing applications to be used with interactive TV. This new technology will become more common in the coming years. It may not be very long before interactive TV is in your household!

Discussion Starter
Refer the students to the section title "Creating Technology." Have the students complete the following sentence: "Creating technology means…." Ask volunteers to read their responses to the class. Discuss the responses and record accurate ones on the board. Have the students use the responses to make a concept web of ways to create technology.

Group Activity
Write the word *invention* on the board. Ask the students to work in small groups to brainstorm their own definitions for the term.

Writing Assignment
Ask the students to use the text and any available outside resources to research and write descriptive definitions of the eight steps of the design process. Ask volunteers to read their definitions.

Chapter 9
Invention and Innovation

Did You Know?

➤ A patent is the record explaining an inventor's new invention. The U.S. Patent and Trademark Office grants patents. A patent protects the invention from being stolen from the inventor.

➤ The first patent was granted to Joseph Jenks for creating a better sawmill.

➤ The U.S. Patent Office has issued over 5 million patents since it opened in 1646.

Objectives

The information given in this chapter will help you do the following:

➤ Define and explain *invention*.

➤ Summarize the history of invention.

➤ Explain the difference between invention and discovery.

➤ Identify the three major categories of inventions.

➤ Describe invention as a problem solving process.

➤ List and give examples of the steps of the invention process.

➤ Give examples of some characteristics of inventors.

➤ Name the three main reasons for inventing.

➤ Recognize the advantages of invention teams.

➤ Cite the definition of *innovation*.

➤ Recall the meaning of *adaptation*.

Key Words

These words are used in this chapter. Do you know what they mean?

adaptation

challenge

creative thinking

discovery

financial invention

Industrial Revolution

innovation

invention

invention process

leisure invention

problem

problem solving

scientific discovery

social invention

spin-off

Chapter Outline
Did You Know?
Objectives
Key Words
History of Invention
Discovery
Invention
Problem Solving
The Invention Process
  Identify the Problem
  Collect Information
  Think Creatively
  Experiment
  Review
Inventors
Why Invent?
  Social Inventions
  Financial Inventions
  Leisure Inventions
Invention Teams
Innovations
Adaptation
Summary
Curricular Connections
Activities
Test Your Knowledge

Have you ever used a shoestring for something other than to tie your shoes, created a device to help a friend, or even made up a new dance move? If you have, you are an inventor. Inventors create new and unique products. These products are called *inventions.* Inventions are created through the invention process. This process uses imagination and knowledge to turn ideas into devices, products, and systems. See **Figure 9-1.**

History of Invention

The invention process dates back to the beginning of humanity. When people began to create tools and clothes, they became inventors. Tools and clothes did not exist before people created them. Inventions have followed the materials available at any point in time.

In the Stone Age, axes and tools made of stone and bone were invented. People also invented ways of creating pottery. Next, in the Bronze Age, inventions like the wheel and the plow were created. Irrigation and writing were also invented in the Bronze Age. The inventions of the Stone and Bronze Ages were important to our world. These inventions were the building blocks for later inventions.

The Iron Age came next. New devices were invented in the Iron Age. One important invention was the screw pump, which Archimedes, a Greek inventor, developed. It uses a large

Figure 9-1. The products we use today have been developed using imagination and knowledge. (Harris Corp.)

TechnoFact: Inventors' ideas are often referred to as *intellectual property.* The United States Patent and Trademark Office protects intellectual property in several ways. Utility patents are issued to protect processes, machines, products, and compositions of matter (such as medicines) for up to twenty years. Design patents protect new, original, and ornamental designs for products for up to twenty years. Plant patents guard invented or discovered plant varieties for up to twenty years. Trademarks protect words, names, symbols, sounds, or colors distinguishing goods and services. Trademarks can be renewed forever. Copyrights shield writings, music, and works of art for the life of the author, plus seventy years.

Figure 9-2. Here is a modern example of a screw pump invented thousands of years ago. (Lakeside Equipment Corporation)

screw to move water uphill. The screw pump is still used today, although it may not seem like a major invention anymore. See **Figure 9-2.** At the time in history when it was invented, however, moving water was a major problem. Archimedes solved that problem by using a screw inside of a cylinder. The screw turns and carries the water upward. The cylinder is used to keep the water on the screw. The screw pump is also important because it is one of the first inventions that can be credited to an inventor. Earlier inventions are too old to know who invented them.

There are many early inventions we take for granted. Today, there is paper to write on, compasses to lead our way, and batteries to power our devices. Without these inventions, we would be faced with many of the same problems our ancestors had.

Invention did not stop at the Iron Age. Inventions have been created in every age of time. An important time period for invention is known as the *Industrial Revolution,* which lasted from around 1750 to 1850. The Industrial Revolution was a time when many machines and devices were invented. The invention of the steam engine started the revolution. See **Figure 9-3.** It was used in the invention of factories, on the railroad, and in ships. Also,

Figure 9-3. Steam engines were used in many applications during the Industrial Revolution. (Jerry E. Howell)

TechnoFact: Today, inventions occur quite frequently. New technologies increasing our abilities to communicate, transport, and manufacture are developed every day. However, that has not always been the case. The early men and women of the world created inventions much more slowly. It took humankind thousands of years to harness fire and invent weapons, tools, and the wheel.

TechnoFact: Archimedes, who invented the screw pump, is known for many other things as well. He was born in Sicily around 287 B.C. Archimedes was more of a mathematician than an inventor. His inventions, including the screw and the pulley, were practical uses of the geometry he studied.

Figure 9-4. The space shuttle is just one example of a modern invention. (NASA)

during the Industrial Revolution, the telegraph was invented. Better ways to plant and grow crops were also invented during that time. Many of the inventions of the Industrial Revolution have had lasting effects on the world. Without the Industrial Revolution, there would be no factories to make products. There would be no passenger trains. It would also be much harder to communicate.

Many of the inventions since the Industrial Revolution have been related to information. This recent period of time is known as the Information Age.

In this age, vacuum tubes and transistors have been invented. These inventions led to the television and computer. The telephone, Internet, space shuttle, and compact disc player are all inventions of the Information Age. See **Figure 9-4.**

Over time, inventions have become more complex. The wheel no longer seems like an important invention. Without the invention of the wheel, however, there would be no automobile. Many things we use and create today come from inventions and discoveries thousands of years

TechnoFact: The Information Age relies very heavily on small pieces of silicon. Silicon chips are the "brains" of most modern electronic devices. Invented by Jack Kilby and Robert Noyce, silicon chips are only about half a millimeter thick and can hold hundreds of thousands of electronic circuits.

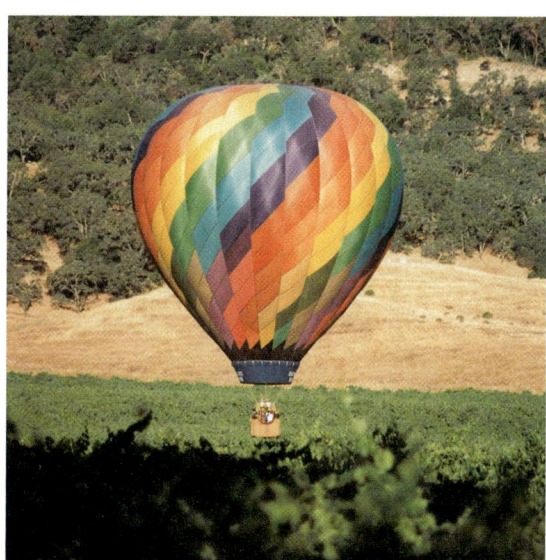

Figure 9-5. Discoveries and inventions have led to the hobby of hot air ballooning. (Napa Valley Balloons, Inc.)

old. Inventors often use logical processes of trials and alterations to create useful products and systems. Many of the inventions and innovations we use today have developed gradually over time.

Discovery

Discovery is a word often confused with *invention.* You may hear the words used together. It may seem they are interchangeable; however, they are not. An invention is a new idea, process, or device. Humans design and create inventions.

Discoveries are different. They are not human-made. Discoveries are naturally occurring. A *discovery* is made when something that naturally occurs is first noticed. Discoveries are often called *scientific discoveries.* Scientists and researchers are the people who

normally make the discoveries. Important discoveries include electricity, vacuums, and Newton's laws of motion. Discoveries are the foundation for scientific knowledge.

Inventions and discoveries are often used together. The Mongolfier brothers, for example, discovered that a balloon filled with hot air would rise. That discovery, in 1783, led to their invention of the hot air balloon. Since that time, more discoveries about propane and other gases have been made. These discoveries have led to more inventions related to the hot air balloon. Because of these discoveries and inventions, hot air ballooning is now a safe and fun sport for many people. See **Figure 9-5.**

Leonardo da Vinci was an inventor who lacked the discoveries he needed. Da Vinci created plans and drawings for bicycles, helicopters, gears, and parachutes. Many of these plans were never built. In his time, the materials he needed to build these inventions did not exist. It would be hundreds of years until the materials would be discovered or invented.

Invention

Inventions, unlike discoveries, are human-made. Humans create inventions. An invention can be as simple as a paper clip or as complex as a satellite. See **Figure 9-6.** Inventions can be broken down into several categories:

➤ Devices and machines. These are inventions with moving parts or electrical circuits used to do work.

TechnoFact: Isaac Newton discovered three laws of motion: 1. All objects will stay at rest, or continue moving, until acted upon by an outside force. 2. The force required to affect a given change on the velocity or vector of an object is directly proportional to the object's mass. 3. For every action, there is an equal and opposite reaction.

Standards for Technological Knowledge

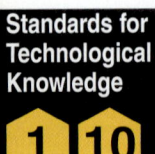

Discussion
Have the students list inventions from each of the three categories: devices and machines, products, and processes.

Discussion
Describe and give examples of the uses of the different problem solving processes.

Tech Terms
Vulcanization—The process of treating crude rubber in order to increase its strength or elasticity.

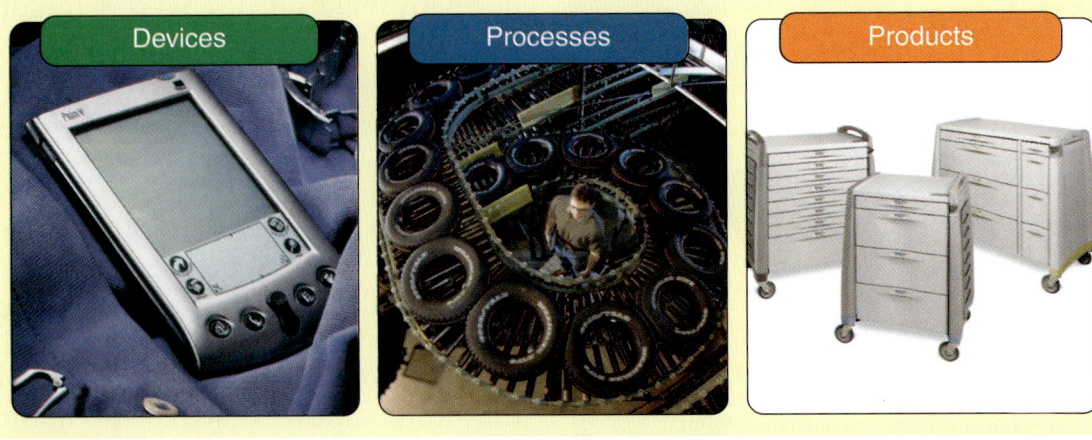

| Devices | Processes | Products |

Figure 9-6. Inventions can be devices, processes, or products. Left—This is an electronic device known as a personal digital assistant. (Photo by Steven Moeder, courtesy IDEO) Center—This process is a tire production line. (Goodyear Tire and Rubber Company) Right—This product is a medication cart. (Design Central, design firm; Artromick International, client)

The lightbulb, compact disc player, and mousetrap are all common devices and machines.

➤ Products. These are manufactured artifacts. Examples of products include ink pens, sporting equipment, and clothing. Even artwork, pieces of music, and food are products.

➤ Processes. These are new techniques and ways of doing things. Processes can be changes in the way products are manufactured. They can also be new techniques for performing surgeries, growing crops, or making medicines.

Problem Solving

Most inventions are created through a planned process. A few, however, are created by accident. Robert Goodyear created, perhaps, the most famous accidental invention. Goodyear worked to create rubber that was not sticky or brittle. In 1839, after years of work, he accidentally dropped a piece of rubber on his oven. When the rubber cooled, Goodyear found that the piece was not sticky or brittle. His invention of vulcanized rubber was accidental. Today, we have shoes, tires, and many other products made of his vulcanized rubber. See **Figure 9-7.**

Accidental inventions are not common. Most inventions are very well planned. Inventing follows a set of steps. It is a process. The process begins with a problem and ends with a solution. It is a *problem solving* process.

The invention process is just one type of problem solving. Troubleshooting, experimentation, and research and development are other problem solving processes. All problem solving processes are useful, but some technological problems can best be solved through a certain method. Troubleshooting is used to find the

TechnoFact: Frank Gilbreth was known for improving many processes. He studied the motion of people while they were working. One of his first motion studies was the motion of bricklayers. He improved the bricklayers' process and nearly tripled their daily output. He and his wife, Lillian, studied many occupations and developed a list of fifteen operations all jobs could be broken down into. This set of operations is known as *therblig*, an approximation of *Gilbreth* spelled backward.

cause of a problem in a technological system. This type of problem often calls for some type of specific knowledge. Once the cause of the problem has been recognized, the next step is to fix it and check if it works. Experimentation is used to carry out tests on technological products and systems. This process strongly resembles the scientific method, in that both are logical and require tinkering, hypothesizing, studying, fine-tuning, testing, and recording. Research and development is the process used to fix problems. This process is used intensively in business and industry to prepare devices and systems for the marketplace. The invention process creates new products to solve problems.

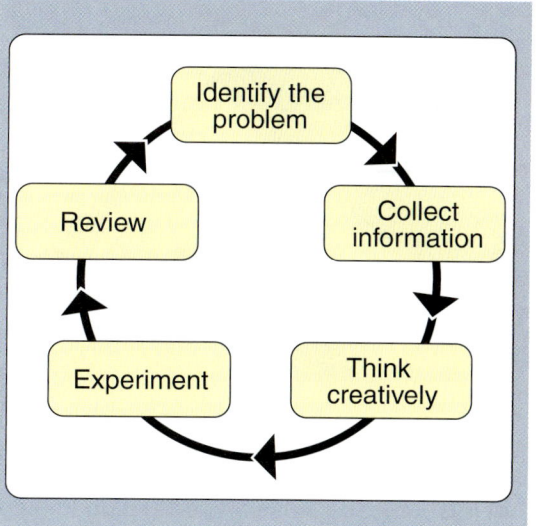

Figure 9-8. The invention process follows this order.

Standards for Technological Knowledge

1 **10**

Extend
Ask the students why it is important that the inventor first identifies the problem.

The Invention Process

Most of the inventions people have produced over the course of history were developed using the *invention process.* See **Figure 9-8.** This process has five steps:

1. Identify the problem.
2. Collect information.
3. Think creatively.
4. Experiment.
5. Review.

Identify the Problem

The invention process starts with a problem or challenge. In this first step, inventors identify the problem. A *problem* is anything that can be made better through change. For example, the time and effort it takes to mow the lawn is a problem. Problems

Figure 9-7. Tires are a product made from vulcanized rubber. (Goodyear Tire and Rubber Company)

TechnoFact: Even accidentally discovered products must be developed using problem solving processes. For example, Dr. Percy Spencer was testing a new magnetron when he noticed that the candy bar in his pocket had melted. He placed other foods next to the magnetron and discovered it heated and cooked the foods. He used this information and an invention process to design and patent the first microwave oven.

Figure 9-9. The danger of using a hair dryer near water was a problem. To solve the problem, manufacturers use a special plug to help prevent shocks.

can be dangerous. For example, using electrical appliances around water is a dangerous problem. See **Figure 9-9.**

Challenges are also problems. A *challenge* is an obstacle or goal that needs to be met. An example of a challenge is to safely travel to the planet Mars.

Collect Information

The next step is to collect information. Inventors research the problem. They learn as much as they can about it. Inventors interview experts. They read books, magazines, and academic journals. Good inventors keep notebooks or personal journals. They write down everything they do and learn.

Think Creatively

The third step is the use of *creative thinking.* Creative thinking requires the human brain. Only humans have the ability to think in creative ways.

Animals are unable to invent devices because they do not have creative minds. Animals act on instinct. Humans use creativity to solve problems. See **Figure 9-10.**

In this step, inventors begin to think of solutions. In the first two steps, they focus only on the problem. The inventors identified and researched the problem. Now they use the results of the research and creativity to design solutions. The inventors come up with as many solutions as possible. They make sketches, drawings, charts, and graphs. At the end of this step, the inventors have a few good designs.

Figure 9-10. Animals build shelters using instinct (top). (Wisconsin Department of Natural Resources) Humans use creativity and technology to build homes (bottom). (Habitat for Humanity International)

Experiment

The next step is to experiment. In this step, inventors build inventions. They use the drawings created in the last step. It is important they know how to use tools and machines. The inventors must build models and test them. See **Figure 9-11.** This is a trial and error process. Inventions often fail at this step, if they do not work or meet the need. When a solution fails, the inventor fixes it and tries again. This step may take weeks or even years.

Review

The final step is review. After the invention works, inventors review their notebooks. They make sure the invention solves the problem. It is possible to invent a device that does not solve the problem. If the invention does not solve the problem, the inventor starts over again with step one. If it does solve the problem, the invention process is complete.

Inventors

The people who use the invention process are known as inventors. Inventors are curious people of all ages. Both children and adults can invent. Children often make good inventors because they are naturally curious. Alexander Graham Bell and Benjamin Franklin are famous inventors known for their curiosity. Benjamin Franklin was curious about lightning and electricity. He flew a kite in a lightning storm to prove the two were related.

Figure 9-11. Inventions are tested to make sure they will function correctly. (IKEA Home Furnishings)

Alexander Graham Bell was curious about how sound traveled underwater. He went to a lake and placed his head in the water. His partner then hit two rocks together. They did this at different distances. He found that sound did travel underwater. Bell's and Franklin's experiments led to major inventions because they were curious people.

Inventors look at things in their lives. They wonder how they can make things better. Some inventors think about things like games and hobbies. See **Figure 9-12.** James Naismith did just that. He was a physical education teacher who wanted a game his students could play inside during the winter. This teacher created the game of basketball. Milton Bradley was also an inventor of games. He invented games like The Game of Life® board game.

TechnoFact: Alexander Graham Bell wanted to be remembered as a teacher of the deaf, not as the inventor of the telephone. He spoke the first words, "Mr. Watson, come here. I want to see you," into his invention in 1876. Much of the money he received from the invention of the telephone went back into research for the deaf. He was one of the founders of the National Geographic Society, in 1888.

Standards for Technological Knowledge

6

Tech Terms
Creativity—Artistic or logical ingenuity.
Technical Knowledge—Knowledge of tools, devices, and machines used to create or operate technology.

Brainstorm
Have the students discuss what inventors like the Wright Brothers and James Watt would have needed to know to invent their devices.

Creative and imaginative people are often inventors. Creativity is an important part of invention. Inventors use their creativity to see situations in new ways. If inventors were not creative, new devices and systems would not be created. The Wright brothers were creative and imaginative. Orville and Wilber Wright imagined flying. They used their creativity to create the first powered airplane.

Inventors need to have a great deal of knowledge. They must know about the science and mathematics used in their inventions. Samuel Morse invented the telegraph in 1840. See **Figure 9-13.** He had to know how to use electricity. Morse also needed technical knowledge. He knew how to use tools to create his invention. Technical knowledge is important for inventors. Many inventors can think of ideas. Some do not have the knowledge of technology to build their inventions.

People who use the invention process must also know about already existing products. James Watt invented the steam engine. He looked at models of steam engines that did not work and

Figure 9-13. Samuel Morse's telegraph sped up the communication. (Jerry E. Howell)

took them apart. Watt then discovered why they did not work. He built the first working steam engine. Inventors, like Watt, enjoy taking things apart and finding how things work.

Why Invent?

People invent for three reasons. Some people invent to help themselves and others. The products and systems they develop can be called *social inventions.* Other people want to make money with their inventions. These inventions are known as *financial inventions.* There are other people who invent because they find it fun. These devices and processes are called *leisure inventions.*

Social Inventions

Social inventions are created to help the inventor or other people. They make our lives better and easier. These inventions impact society. Social inventions can be simple, like a jar opener for the elderly. They can also be complex, like vaccines for children. The first vaccine was invented to stop smallpox. Edward Jenner invented it

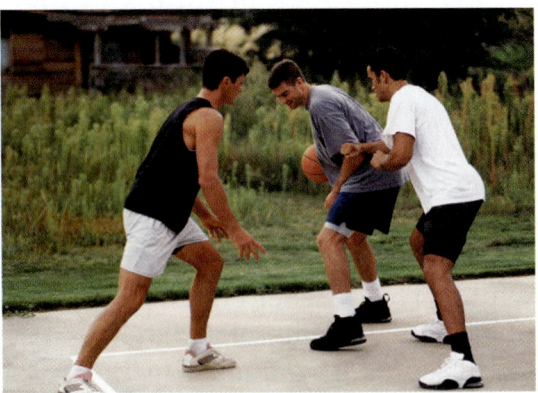

Figure 9-12. Sports, like basketball, are invented for fun and recreation.

TechnoFact: Inventors are often pictured as eccentric, white-bearded men. However, there have been many famous women inventors. Gertrude Belle Elion, for example, was the first woman inducted into the National Inventors Hall of Fame. Her work led to the invention of drugs to fight leukemia and the rejection of transplanted tissue.

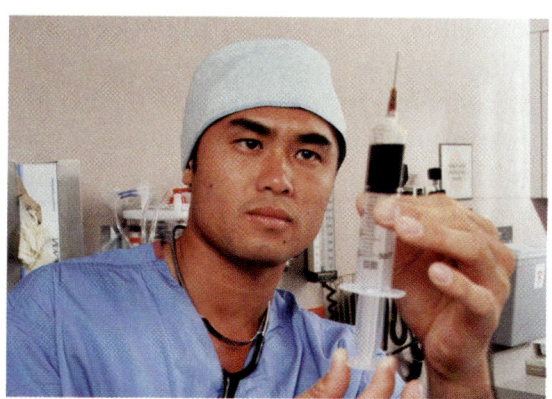

Figure 9-14. Vaccinations are social inventions.

to help society. Vaccines are used to control many different diseases. See **Figure 9-14.** Sometimes social inventions are created to help family and friends. You may have created a social invention for someone you know. A desk for someone in a wheelchair is a social invention.

Financial Inventions

Most inventions are created to make money. These are called *financial inventions.* Financial inventions often make things faster or easier to do. Eli Whitney invented the cotton gin to produce cotton faster. The faster the cotton was produced, the more money the farmers made. The computer and computer software are financial inventions. They help businesses operate faster and more efficiently.

Leisure Inventions

Leisure inventions are created for the pleasure of inventing. Inventors who like to tinker create these. These inventors like to invent as a hobby.

Leisure inventions include small games and toys. Rube Goldberg was a cartoonist who drew funny leisure inventions. His inventions used many steps to solve simple problems. Leisure inventions are often very creative and are even patented. See **Figure 9-15.**

Figure 9-15. This leisure invention of a helium-filled sunshade is patented. (Patent No. 5,076,029, U.S. Patent and Trademark Office)

Invention Teams

Social and financial inventions are usually created in teams. Invention teams are used because each team member has different abilities and knowledge. Some people are better at using tools. Others have more scientific knowledge. Many invention teams work in research labs. See **Figure 9-16.** Thomas Edison created the first research lab in the United States. His

TechnoFact: Some inventions are created for one purpose and end up being used for another. The jigsaw puzzle, invented by John Spilsbury in 1767, began as an educational tool. Spilsbury was a mapmaker who glued a map to a piece of wood and cut out the countries. Educators used such puzzles to teach geography. Today, jigsaw puzzles are fun leisure inventions.

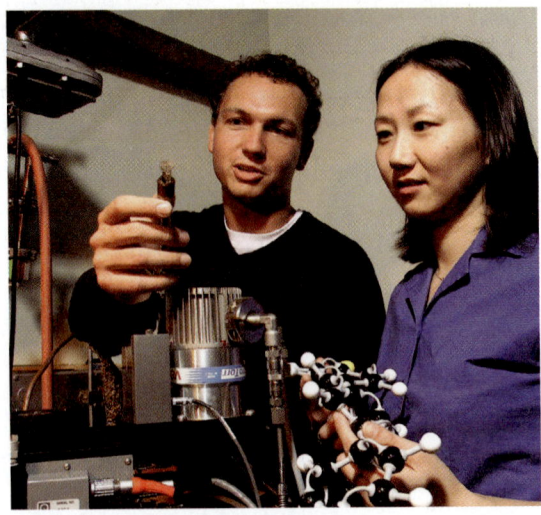

Figure 9-16. The first transistor made from a single material was developed in a research lab. (Lucent Technologies, Inc./Bell Labs)

research lab was built in Menlo Park, New Jersey. Edison's researchers helped invent the phonograph and the electric lightbulb. Research labs can be used to invent many things. For example, they are used to invent new materials, electronic equipment, and medications. Research labs can be found in universities, large companies, and government agencies.

Innovations

Inventors do not always invent new devices. They often create *innovations.* Innovation is the process of altering an existing product or system to improve it. Innovations make inventions better. All technological refinement occurs through the process of innovation. Many of the products created today are innovations. Innovations can do several things for the inventions. They can make inventions work better, be less

expensive, or be built with better materials. Innovations can even be made from several inventions.

Innovations make inventions more useful. They have been made to improve the speed and capacity of computers. These innovations have made the computer more useful. Computers can do more things faster than they could just a few years ago.

Inventions can be made less expensive with innovation. Henry Ford innovated the way automobiles are built. His mass production system made automobiles cheaper to buy. More people could afford automobiles because of Henry Ford. Many innovations have changed the way automobiles are manufactured today. Today, most automobile production lines are very automated. See **Figure 9-17.**

Innovations can be large or small changes to an invention. An inventor may change the material of the invention. The bicycle has gone through

Figure 9-17. Current automobile production lines are innovations. (DaimlerChrysler)

TechnoFact: ENIAC (Electronic Numerical Integrator And Computer) was the predecessor of the modern electronic computer. ENIAC, built between 1943 and 1946, filled an entire room and weighed over sixty thousand pounds. Due to inventions and innovations in electronics, materials, and manufacturing, modern computers are much smaller and more capable.

many innovations. When the bicycle was invented, it was made of wood. New materials helped to innovate the bicycle. Bicycles have been made of steel, aluminum, and titanium. Some bicycles are even made of carbon fiber. These innovations changed the weight, comfort, and ride of the bicycle. They made bicycles easier to ride.

Putting two inventions together can also make innovations. In-line skates are an innovation. See **Figure 9-18.** They are a combination of shoes and skateboards. Innovations are important to our society.

Adaptation

One type of innovation is adaptation. *Adaptations* are developed when inventions are used for something

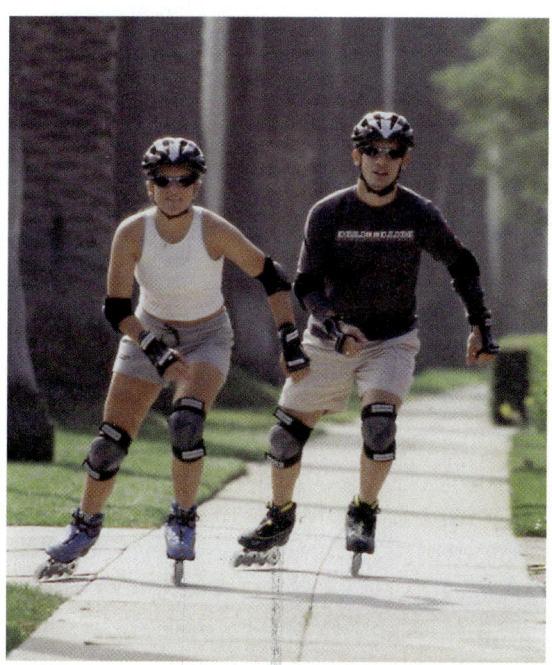

Figure 9-18. In-line skates are innovations used for fitness and recreation. (Rollerblade)

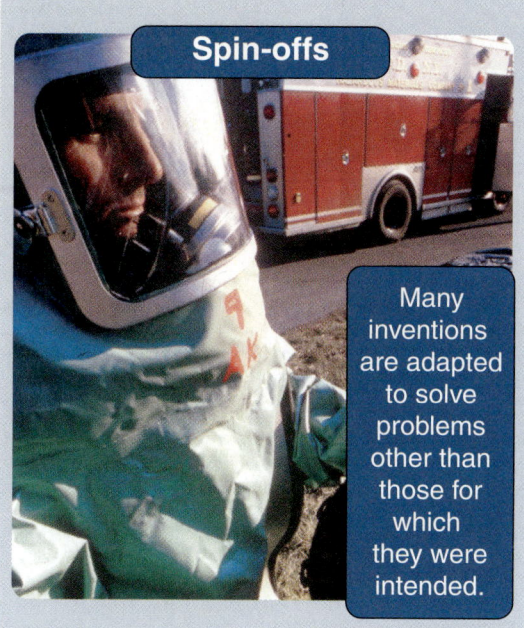

Spin-offs

Many inventions are adapted to solve problems other than those for which they were intended.

Figure 9-19. The protective clothing firefighters wear is a NASA spin-off from the space suit. (NASA)

other than the purposes for which they were intended. Medicines are adapted when they are used to treat ailments besides the one they were meant to cure. The National Aeronautics and Space Administration (NASA) is known for creating adaptations. NASA has created over one thousand inventions that have been adapted to solve other problems. These adaptations are known as *spin-offs*. Racecar drivers now use technology from space suits worn by the Apollo astronauts. Cordless power tools were first used in space. Even the suits firefighters wear are NASA spin-offs. See **Figure 9-19.**

Innovations and adaptations make devices, products, and processes easier to use. They help to make inventions better and more useful. Innovations are an important part of invention.

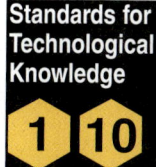

Standards for Technological Knowledge

1 10

Research
Have the students identify NASA spin-offs they have used in their own lives.

TechnoFact: NASA spin-offs are used in many areas of life, including food preparation and medical treatment. Freeze-drying and water purification technologies that were first used on the *Apollo* missions have been adapted for commercial use. Space shuttle fuel pump technology was adapted to create artificial hearts.

Technology Explained

high-rise building: A multi-story residential or commercial building that has a skeleton frame.

The Great Pyramid at Giza in Egypt is a construction marvel, even today. This type of tall structure, however, was not common throughout history. Many years later, the great cathedrals were built in Europe. The cathedrals have high, thick stone walls. This type of wall is called a *load-bearing* wall because it holds all the weight (load) of the roof.

Construction such as this has height limits. Load-bearing walls are not practical in buildings over five stories tall. The walls become too thick and heavy.

In the past, as cities became larger, taller buildings were needed. It took the development of the elevator and the skeleton frame to make tall buildings practical.

The elevator allowed people to move quickly between floors without walking up stairs. For example, the Empire State Building's elevators can move up to 1,400´ (427 m) per minute. At this speed, people can travel from the lobby to the 80th floor in 45 seconds.

A skeleton frame carries the weight of the building, much like the human skeleton carries the body. These developments made high-rise buildings, or skyscrapers, possible. The first high-rise building was built in Chicago in 1885. Built for the Home Insurance Company, it was ten stories tall. The exterior walls provided protection from the weather, but did not carry any load. This type of wall is called a *curtain wall* because it merely hangs from the frame.

Today, high-rise buildings use two types of framework. The first type is reinforced concrete, **Figure A.** This type

of framework is cast on-site. Forms are erected around a network of steel rods. The concrete is poured inside the

Reinforcing rods
Horizontal beam
Vertical column
Wire mesh reinforcing
Foundation pad

Figure A. The photograph above shows a building under construction. The drawing shows how a reinforced concrete structure is put together.

Tech Terms
Invention—A new device, system, or procedure.
Innovation—An improvement on one or more inventions.
Adaptation—An invention used for a different purpose than intended.

Summary

Invention is a problem solving process. Humans have used the process of invention for thousands of years. Inventions are new and unique devices meeting our needs. The people who create inventions are known as inventors. Inventors are creative and imaginative people. They often work together in research laboratories. Inventors follow the steps of the invention process to create new inventions. They also make changes to older inventions. These changes are called *innovations*. Adaptations are inventions

TechnoFact: Several presidents of the United States are known for their inventions, but Abraham Lincoln is the only president with a patent. Although it was never built or tested, the system of adjustable air chambers Lincoln invented was designed to allow steamboats to pass through shallow water or over sandbars without unloading. Thomas Jefferson also invented several devices, such as the swivel chair and the pedometer. However, he never patented his inventions.

Figure B. This is a steel-framed building. Notice the different shapes of the steel members.

forms. After the concrete has cured, the forms are removed.

The weight of reinforced concrete limits its use to buildings of moderate height. Taller buildings use steel skeletons. Steel is fabricated into angles and I beams. Steel columns and beams are bolted or riveted together, **Figure B.**

This allows the beams to expand and contract uniformly with temperature changes.

Once the frame is erected, the floor and roof are installed. The exterior walls are then put in place, **Figure C.** These walls can be made of steel or aluminum panels, bricks, concrete blocks, sheets of glass, or other materials.

The title of the world's tallest building passes from skyscraper to skyscraper as new structures are built. The tallest buildings at the end of the twentieth century were the Petronas Towers in Malaysia, built in 1997. They are each 1,483′ high. The Sears Tower in Chicago had the highest occupied floor (1,431′).

Figure C. The building in this picture is having its exterior walls finished. The brick is in place, and the windows are installed.

used for new purposes. Inventions, innovations, and adaptations all help to make life easier.

Extend
Have the students identify buildings within your community built with a skeleton frame.

Curricular Connections

Language Arts

Create an invention and keep an inventor's log. Use the inventor's log to write a simple patent application.

Science

Choose a scientific discovery (such as electricity, gravity, or friction). Make a list of inventions used to make that discovery. Make a second list of inventions that come from that discovery.

Social Studies

Select an industrialized country (such as the United States, Germany, or England). Research the major inventions developed in that country. Create a display presenting the country's inventions and inventors.

Activities

1. Choose an invention affecting your life (for example, the automobile, telephone, television, or computer). Create a display highlighting the invention. Show the history of the invention and how the invention works.

2. Select a problem in your life or the life of someone in your family. Follow the invention process and create an invention solving the problem.

3. Research an adaptation or spin-off NASA created. Divide a poster board into two sections. On the first half, list and explain an invention NASA created. On the second half, show how the invention has been adapted and how it is used today.

TechnoFact: Most of the world's tallest buildings are relatively new and located in Asian countries. Among the five tallest buildings, the oldest is the Sears Tower in Chicago. It was built in 1974 and is the only one located in the United States. The other four buildings, located in Malaysia, Shanghai, and China, were built in the 1990s.

Test Your Knowledge

Do not write in this book. Place your answers to this test on a separate sheet of paper.

1. Paraphrase the definition of *invention.*
2. Describe why the Industrial Revolution was important to the history of invention.
3. Why are discoveries different from inventions?
4. List the three types of inventions.
5. The invention process is the only problem solving process. True or False.
6. Accidental inventions are the most common type. True or False.
7. The invention process has a set of steps. True or False.
8. Rewrite, in your own words, a definition for the word *problem,* as it relates to the invention process.
9. Summarize three characteristics of inventors.
10. An invention created to help people is a(n) _____ invention.
11. What type of invention is created to increase profits?
12. An invention created for fun is a(n) _____ invention.
13. A team of inventors today creates most inventions. True or False.
14. A(n) _____ is an improvement on an invention.
15. Select the phrase best describing an adaptation.
    A. A new and unique product or system created with imagination and knowledge.
    B. An improvement on an already existing invention.
    C. The initial realization of a natural occurrence.
    D. An invention used for something other than the purpose for which it was intended.

![TSA] Modular Activity

This activity develops the skills used in TSA's Inventions and Innovations event.

Inventions and Innovations

Activity Overview

In this activity, you will research a technological invention or innovation, then design and construct a tabletop display to illustrate your finding.

Materials

➤ Materials appropriate for a tabletop display (will vary greatly)

Background Information

Selection. Before selecting the theme for your project, consider inventions and innovations from each of the major divisions of technology:

➤ Medical

➤ Agricultural and biotechnology

➤ Energy and power

➤ Information and communication

➤ Transportation

➤ Manufacturing

➤ Construction

Use brainstorming techniques to identify several possible inventions and innovations from each area. Select a topic that has had some historic, cultural, social, economic, or political impact.

Research. Use a variety of sources to research your theme. Do not rely solely on information you find on the Internet. Use books and periodicals available at your local library. Research the inventor. Was there an individual inventor, or was the invention developed by a corporation? What were some previous inventions that allowed this invention to become a reality? How was the invention received by the public, and was the response expected? Has the invention changed since it was introduced? If so, how?

Guidelines

➤ The display must address the historical significance and any cultural, social, economic, or political impacts of the invention or innovation.

➤ The display can be no larger than 2´ deep × 4´ wide × 4´ high.

➤ If a source of electricity is desired, only dry cells or photovoltaic cells can be used.

Evaluation Criteria

Your project will be evaluated using the following criteria:

➤ Use of design principles

➤ Page layout

➤ Selection and placement of clip art

➤ Font selection and usage

Chapter 10
The Design Process

Did You Know?

➤ All designers use a design process. Not all processes, however, are the same. Some designers may only use four steps, while others may use twelve. The processes are still very similar. They all start with a problem and end with a solution. All design processes also have research and idea development sections. The design processes all have steps in which the designer creates solutions. They also have steps designers use to share their solutions with others. The steps in this book are one example of a process that can be broken down into many steps.

Objectives

The information given in this chapter will help you do the following:

➤ Define *design.*

➤ Explain the difference between artistic and engineering design.

➤ Describe product and system design.

➤ List and summarize the eight steps of the design process.

Key Words

These words are used in this chapter. Do you know what they mean?

artistic design

content

design

design brief

design process

detailed sketch

engineering design

form

function

mechanical drawing

mock-up

model

product design

prototype

rendering

system design

Chapter Outline
Did You Know?
Objectives
Key Words
What Is Design?
Types of Design
  Artistic Design
  Engineering Design
Steps of the Design Process
  Identifying the Problem
  Researching the Problem
  Creating Solutions
  Selecting and Refining Solutions
  Modeling the Solution
  Testing the Solution
  Communicating Designs
  Improving Solutions
Summary
Curricular Connections
Activities
Test Your Knowledge

Resource
Transparency 10-1: Technological Design

Discussion
Ask the students what types of items are not designed. Ask them if there are examples of technology that is not designed.

Brainstorm
Have the students list occupations that perform design tasks.

Figure 10-1. Advertisements are designed.

Design affects us every day. Look around the room in which you are sitting. Every item in the room has been designed. The desk you are sitting in, the carpet on the floor, and the lights in the ceiling have all been designed. Even the clothes you are wearing were designed. If you look through the window, you may see a vehicle, building, or billboard. All of these have also been designed.

What Is Design?

We live in a designed world. This means the things we use have been designed. *Design* is the act of creative planning used to solve a problem. It results in useful products and systems. A designer generally has a specific problem to solve. Often, the design process takes place in design teams. The members of such a team contribute various types of skills and ideas

toward solving the problem. The problem may be to design a physical object, such as a three-bedroom house. It may be to create a less physical thing, such as an advertisement to sell a product. See **Figure 10-1.** The problem may even be to grow crops that cannot be killed by insects. The solution created is usually unique and new.

Design is an action with a purpose. Many different people use the act of design. Engineers design roads and bridges. Architects design buildings. Fashion designers design clothing. See **Figure 10-2.** Woodworkers design furniture. Artists design paintings and sculptures. Chemists design medicines. Even teachers design lessons.

People use design to help organize their thoughts and actions. The

Figure 10-2. Fashion designers use the design process.

TechnoFact: The design process may include a number of people throughout a company, or it may be limited to a single person. An inventor of a single idea may complete all the steps of the process without assistance. Companies that design products may use a team of people to complete the design. They may also pass the design through a number of different departments in the company.

TechnoFact: The term *design* can be used in several ways. It can be used to describe a set of steps used to create a new product. It can also be used to describe the appearance of an object. Lastly, it can be used to describe the actual drawings and sketches of a product.

thoughts of designers are the ideas they use to create solutions. Designers get ideas from many places. The actions of a designer are known as the design process. A designer uses the design process to solve problems. This chapter will focus on types of design and the design process. The steps of the design process will be introduced in this chapter. They will be explored in more depth in the chapters following.

Types of Design

Design is used in many areas of life. Products, devices, systems, and pieces of art are all designed. All of

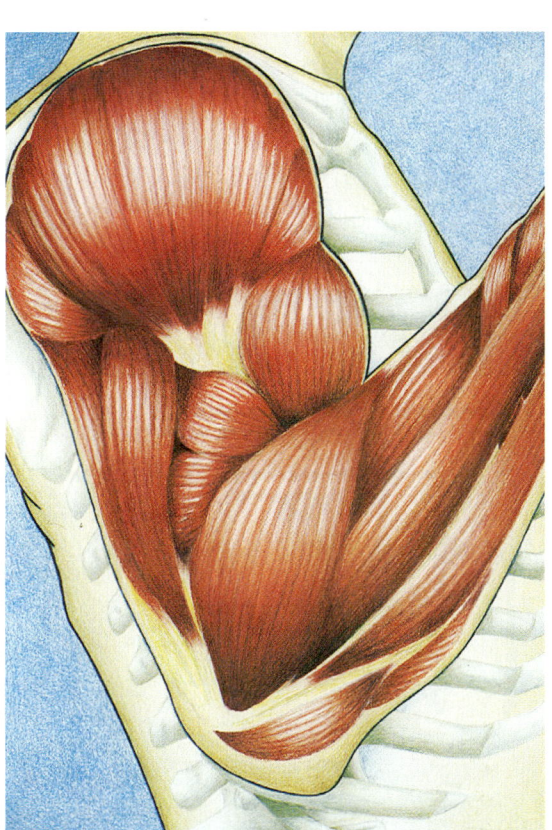

Figure 10-3. Artistic design is different from engineering design. (Mark Chrapla).

these items can be divided into two types of design:

➤ Artistic design.
➤ Engineering design.

Artistic Design

Artistic design is the type of design artists normally do. It may take many forms. Paintings, sculptures, music, and songs are all forms of artistic design. See **Figure 10-3.** Artistic design is used to solve problems. The artists usually choose their own problems to solve. They also choose how to solve the problems. The solution may be a piece of pottery, a drawing, a symphony, or a pop song.

This type of design focuses on two aspects of design:

➤ *Content.* This is the topic, information, or emotion the artist is trying to communicate. The content is often the problem being solved.
➤ *Form.* This is how the solution looks or sounds. Color, shape, and rhythm are all aspects of form.

Artistic designers have only themselves to please. They decide if their solution has answered their problem.

Engineering Design

Like artistic design, engineering design is used to create solutions for problems. Engineering designers, however, are normally given problems to solve. *Engineering design* is used to create plans. These plans can be for buildings, roads, tools, or machines. See **Figure 10-4.** Many types of people

TechnoFact: Some products of artistic designers can be protected by the United States Patent and Trademark Office. Musicians, for example, can copyright music they have written. A copyright lasts for the life of the artist, plus seventy years.

Figure 10-4. Engineers use plans to create products. (Product Development Technologies, PDT)

are involved with engineering design. Architects, fashion designers, industrial engineers, and chemists are examples of engineering designers. Their jobs involve two aspects of design:

➤ *Form.* Like in artistic design, this is how the solution looks.

➤ *Function.* This is how the solution works.

Form and function are very important in engineering design. They are important because people other than the designer decide if the solution solves the problem. These people decide by buying and using the engineer's product. If the product looks bad or does not work, they will not buy or use the product again. Engineering designers are faced with many types of design. The types can be split into two main categories:

➤ System design.

➤ Product design.

System Design

System design involves organizing different parts to solve a problem. By combining different parts, system designers create systems. For example, an electrician combines wires, switches, and outlets to create electrical systems. Audio engineers design entertainment systems. See **Figure 10-5.** They organize speakers, wires, and audio components (including receivers, DVD players, and televisions). Some designers combine smaller systems to create a larger system. Architects combine electrical, entertainment, heating, cooling, and structural systems to create a house.

Product Design

Product design is the creation of products meeting human needs and wants. Automobiles and clothing are both designed products. Television shows and crops are also designed products. There are many types of

Figure 10-5. This diagram shows the components of an audio system. (Monster Cable Products, Inc.)

TechnoFact: The redesign of one part of a system can lead to the redesign of the other parts. For example, if the DVD player in a home theater system has been redesigned to use fiber optic outputs, the stereo receiver used with the system must be redesigned to accept fiber optic inputs.

product designers. Fashion designers, graphic designers, chemists, and engineers are product designers. Their products are clothing, advertisements, medicines, and machines.

Product designers must consider form and function. The product must look nice and work well. A chef can be a product designer. See **Figure 10-6.** Chefs design food to look pleasant and taste good. Industrial designers are product designers that work in many areas of design. They design furniture, computers, and medical and entertainment products. These designers make sure products are easy to use, look nice, and work well.

Designers also make sure products can be manufactured and sold. Toy designers are product designers. They research toys children enjoy so they know if children would want to play with the toy they design.

Steps of the Design Process

Both system and product designers use a design process. The *design process* is a series of steps to design a new product or system. These steps lead a designer from the problem to the

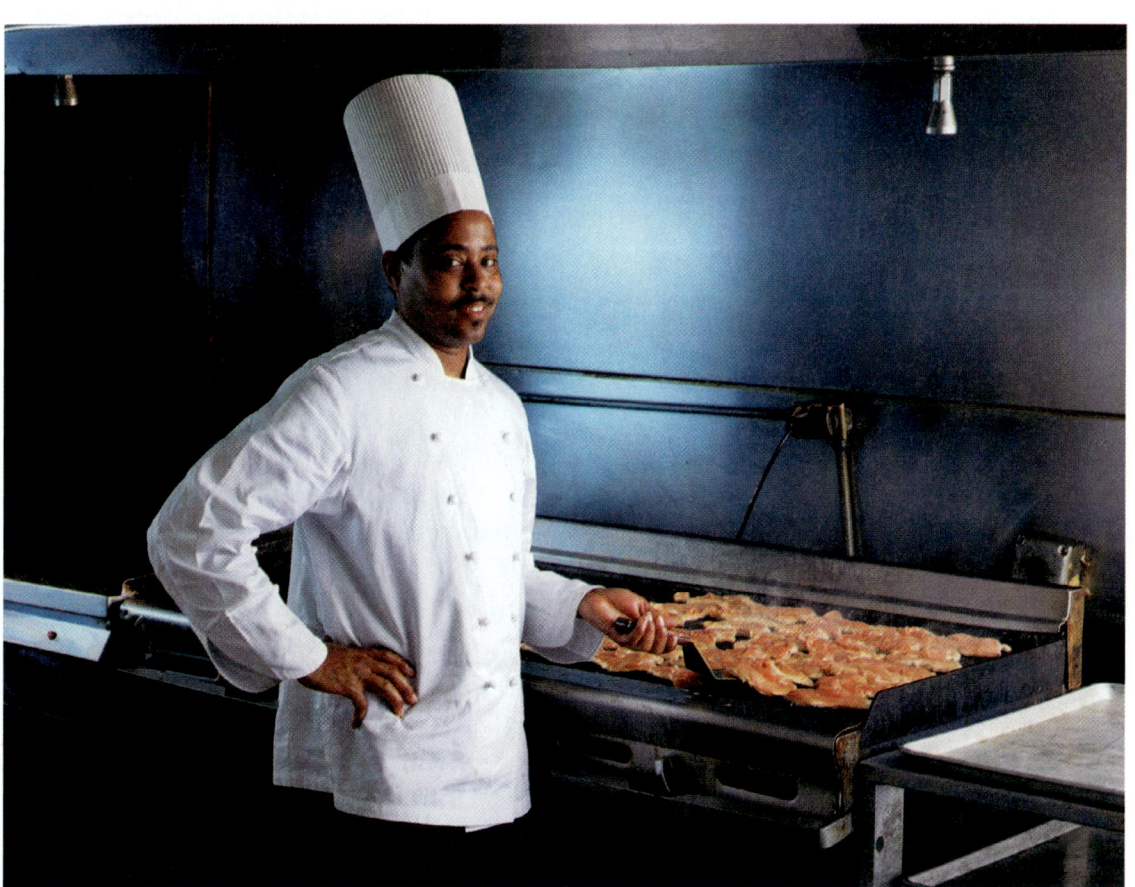

Figure 10-6. Chefs design food that looks and tastes good.

TechnoFact: As our technology advances, so does our need for new types of designers. Fifty years ago, there was no need for software and computer designers. A hundred years ago, there were no aerospace or airplane designers. Today, these designers are very important to our society.

solution. See **Figure 10-7.** A common design process includes eight steps:

1. Identifying the problem.
2. Researching the problem.
3. Creating solutions to the problem.
4. Selecting and refining solutions.
5. Modeling the solution.
6. Testing the solution.
7. Communicating the solution.
8. Improving the solution.

The design process can be used to solve all kinds of problems. After you recognize and choose a need, want, or problem to solve, this process can result in a solution that could lead to an invention or innovation. The goals of the problem to be solved need to be identified. These goals should specify what the desired result is. To apply the process, you need to do research. Then you have to synthesize and study the resulting information you collected.

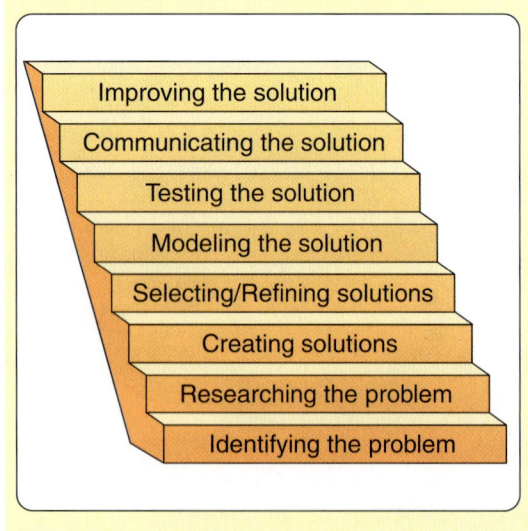

Figure 10-7. The design process uses eight steps.

Improving the solution
Communicating the solution
Testing the solution
Modeling the solution
Selecting/Refining solutions
Creating solutions
Researching the problem
Identifying the problem

The eight steps of the design process are usually completed in this sequence, but they can be carried out in different orders and repeated as required. Each problem is unique and may call for different procedures or for the steps to be completed in a different sequence. For the most part, designers cannot jump forward and skip a step. They can, however, move backward to a previous step. The design process is like a steep set of stairs. You must step on each one to move forward. To go backward, you can turn around and jump down to a lower step. The steps in this process may also vary because designers and engineers have their own inclinations and problem solving approaches, and they may want to handle the design process in different ways.

For example, designers may have to go back to a previous step if they get new information throughout the process that changes the problem. They may have to skip back to the first step and identify the problem again. Another example is a designer creating a solution that fails at the testing stage. The designer cannot move forward because the test was not successful, but must move backward in the process. He or she may choose to move two steps back and select a different solution. With the new solution, the designer moves to the next step and continues the process.

Identifying the Problem

All products and systems are designed to meet a human need or want. In the first step of the design process, the need or want must be identified.

TechnoFact: The design process discussed in this chapter is a common example of the process. However, it is not the only design process. Many industries and designers have their own processes. Most processes generally follow the steps and concepts presented in Figure 10-7. For example, the Royal Institute of British Architects divides the design process into four phases: assimilation, general study, development, and communication. While the titles of the phases are quite different from the process discussed in this chapter, the work completed in each stage is very similar.

Figure 10-8. Some designers specialize in creating landscape plans. (Keith Nelson)

The need or want will become the problem the designer will try to solve. Problems can be found in all areas of life. Designers find problems at work, in the home, in vehicles, and even during leisure time. Most designers focus on only one type of problem. For example, some designers focus on problems with landscapes or floor plans, while others specialize in organizational problems. See **Figure 10-8.**

When designers identify a problem, they write it as a problem statement. The problem statement is very specific and states the exact problem. It is important that the designer focuses on the problem statement and not the solution. For example, a solution is to design a chair. A problem statement is to design a device that would hold a person in a seated position. The solution and problem statement may not seem very different. They are, however, very different. The problem statement is very open and allows the designer the freedom to design many different solutions. See **Figure 10-9.** The solution statement is very narrow and leaves only one solution.

The problem statement is one part of a document called the *design brief.* The designer creates the ***design brief***

Figure 10-9. Open-ended design briefs allow creative solutions. (IKEA Home Furnishings).

TechnoFact: Furniture has been designed for thousands of years. The Egyptians were one of the first designers of furniture. They began by designing chests and trunks to hold their possessions. The chest allowed them to protect their belongings as they traveled. Eventually, as they became less nomadic, they designed and built many of the furniture pieces we have today.

Figure 10-10. Researchers gather information from sources including books, magazines, and videos.

to guide the entire design process. The design brief includes the requirements, or the criteria and constraints, of the design. Criteria identify the desired elements and features of a product or system and usually correlate to the purpose or function of these elements. Constraints, such as size and cost, determine the restrictions on a design. The process of identifying the problem and the design brief are discussed further in Chapter 11.

Researching the Problem

Designers create solutions based on research. Since designs are based on problems, the designer must use research to learn about the problem. Designers learn as much as they can about the problem they are solving. Imagine you are a designer creating a better school locker. You need more information before you begin designing. You need to know the purpose of the lockers. Are the lockers used for books

or for physical education supplies? What size do they need to be? What colors do the people who will use the lockers like? What materials are best for lockers? These are just a few of the questions you have to research.

Designers find the answers to many of their questions by using research. They research in books, magazines, and catalogs. See **Figure 10-10.** Designers also conduct surveys, tests, and Internet searches. They study products that solve similar problems. Research has a process designers follow. Designers begin by listing the problem and finding as much data as possible. In the locker example, designers would gather data such as the average size of lockers. They may also find the percentage of red lockers in all of the schools in the state. The designers would then analyze the data and make conclusions. The conclusions are the final results of the research. The designers will use the conclusions to design the solution. Chapter 12 focuses on researching problems.

Creating Solutions

In this step, discussed further in Chapter 13, designers begin to create solutions. This begins with developing ideas. The process of developing ideas is called ideation. One technique of ideation is brainstorming. It is a group problem solving process in which everyone in the group presents their ideas in an open forum. No one is allowed to criticize anyone else's ideas for any reason. After all the ideas are recorded, the group selects the

TechnoFact: Market research is a type of research conducted to get information about what consumers want in products. Market research is often done in malls, shopping centers, and over the phone.

best ones and further develops them. Questioning and graphic organizers are some other techniques of ideation. The purpose of ideation is to develop as many ideas as possible. The importance is quantity, not quality. Ideas are not judged this early in the process.

As the designers develop ideas, they begin to start sketching. See **Figure 10-11.** Sketching allows designers to show their ideas to others. It is easier to show someone a design than to try to explain it with words. Sketching at this stage is called *rough sketching*. The purpose of rough sketches is to get ideas onto paper. It is not to create exact drawings. These sketches are as simple as shapes and outlines. Designers use

the rough sketches to select several ideas they will further develop.

When the designers have completed all the rough sketches they can, they review what they have done. They select the sketches they feel have the most promise. The designers then create refined sketches. These sketches are more detailed and complete than the rough sketches. There are three types of sketches used in this step of the design process, and each type has certain advantages and disadvantages. All three are examples of pictorial sketches. Pictorial sketches show the object as it would look to the eye. See **Figure 10-12.** The pictorial sketches are shaded to make them look more realistic. At the

Standards for Technological Knowledge

9 **11**

Discussion
Ask the students why it is important that designers generate many ideas near the beginning of the design process.

Tech Terms
Rough Sketch—A basic sketch used to record ideas. Quantity is more important than the quality of rough sketches.
Refined Sketch—A more detailed sketch than a rough sketch. Size, shape, and basic function can be seen in refined sketches.

Figure 10-11. Sketches are the first method of putting ideas onto paper. (Ford Motor Company)

TechnoFact: Grid paper is often used for sketching because it helps designers keep their lines straight. There are three main types of grid paper: square (the most familiar), isometric, and perspective.

Tech Terms
Appearance—The look of the solution.
Function—How the solution works.
Production—How the solution is manufactured.

Discussion
Ask the students to describe instances when they have used graphic models. Have them discuss how graphic models are most often used.

end of this step, the designers should have several good ideas and sketches of possible solutions. The better the ideas, the better the solution.

Selecting and Refining Solutions

The most promising design is selected in this step of the design process. The designer, management, or client selects the design that will be developed into the solution. The selection is based on the aspects of the design. The aspects of design include items like appearance, function, and production. The individual aspects are listed on a chart. The advantages and disadvantages of each design are then added to the chart. Then the design that seems to be the best solution is selected.

The design is also refined in this stage. The first refinement is the appearance of the design. Designers use elements and principles of design to create a design with a good appearance. The designers then create two different types of sketches. The first is called a *rendering*. A **rendering** is a sketch shaded and colored in full color. See **Figure 10-13.** It shows how the design will look in different lights. The second sketch is a detailed sketch. The ***detailed sketch*** is drawn using the sizes of the proposed solution. It also includes dimensions. The dimensions, text, and arrowheads are used to show the size of the object. Selecting and refining solutions is the topic of Chapter 14.

Figure 10-12. Pictorial sketches are easy to understand and are drawn to show objects as the eye sees them. (Design Central, design firm; Artromick International, client)

Modeling the Solution

The fifth step in the design process is the creation of a model. ***Models*** can be used to communicate ideas, experiment with concepts, and organize data. Designers will make several different types of models. Graphic models, like tables, charts, and graphs, are used to show information about the design. Mathematical models are equations and formulas. Charts, formulas, drawings, and sketches are some two-dimensional representations of solutions. The models many designers use the most are physical and computer models. Both of these models are used to create three-dimensional objects. Computer models are three-dimensional computer

TechnoFact: Renderings are created many different ways. In one method, the designer sketches and shades with pencils and markers. In another method, the designer sketches the object with pencils and scans it into a computer. Then, the designer colors the sketch using graphics software.

Figure 10-13. A rendering shows the object in full color. (DaimlerChrysler).

objects. See **Figure 10-14.** Physical models are actual three-dimensional products.

Figure 10-14. The fishing reel is created as a computer model. (Model developed using VX CAD/CAM software, VX Corporation)

Physical models are used very often to help change ideas into useful solutions. Creating and testing physical models allow people to actually feel and touch the object. These models can also be looked at in real light and from any angle. Models are particularly vital for the design of large items, such as cars, spacecraft, and airplanes, because it is cheaper to examine a model before the final products and systems are actually made. The two main types of physical models are mock-ups and prototypes. *Mock-ups* are appearance models. They show how the object will look in real life. They do not, however, function like the solution will. *Prototypes* are physical

TechnoFact: Physical models are often used as displays. The lobbies of many universities and large companies contain models of their campuses. Such models serve as three-dimensional maps, familiarizing visitors with the locations and sizes of the buildings and other features of the campus.

Technology Explained

propeller: A rotating, multi-bladed device that drives a vehicle by moving air or water.

All vehicles have five systems: structure, suspension, propulsion, guidance, and control. The propulsion system causes the vehicle to move. It converts energy into motion.

The propulsion system often uses an engine to create a rotating motion. This motion is transmitted to a device that moves the vehicle. Most land vehicles use wheels to create movement. A different system is required, however, for vehicles traveling on water and in the air. A common propulsion device is a propeller, **Figure A.**

Figure A. Propellers are used on airplanes, both old and new. (Piaggio Aviation).

Propellers are primarily used on airplanes and ships.

A propeller works using two principles of physics. The first principle states the following law of physics: For every action, there is an equal and opposite reaction. The propeller forces air or water backwards as it spins. This, in turn, causes the propeller to be pushed forward. Actually, the propeller moves like a screw. The tips of the blades follow a path like the threads of a screw. In theory, one revolution of the propeller causes the vehicle to move ahead like one turn of a nut would cause it to move on a bolt. A boat or plane does not, however, move that far. Both water and air are yielding fluids. They compress slightly

---

models that actually function. See **Figure 10-15.** They are working models. These models are used to test how well the design will work.

Physical models can be built using many different materials. The materials used depend on the size, shape, and use of the model. Designers and model makers use a process to build models. The models are created by first blocking out the basic shape. The designers and model makers then finely shape the pieces. The pieces are assembled and finished. The models can then be used for testing or a variety of different uses. Models and modeling are discussed further in Chapter 15.

Testing the Solution

The testing stage is the step determining whether or not the solution works. The only good solutions are solutions meeting the design problem. In this step, discussed further in Chapter 16, the designers will test the solution to make sure it solves the design problem. The testing stage begins with the

TechnoFact: Many children grow up building models, whether they are model cars, airplanes, or houses. Many companies specialize in designing building blocks and model sets for children. LEGO, for example, is a leading manufacturer of building blocks. As of 1996, approximately 180 billion LEGO® blocks had been produced since their introduction in 1949. Every year, children spend nearly five billion hours playing with LEGO blocks.

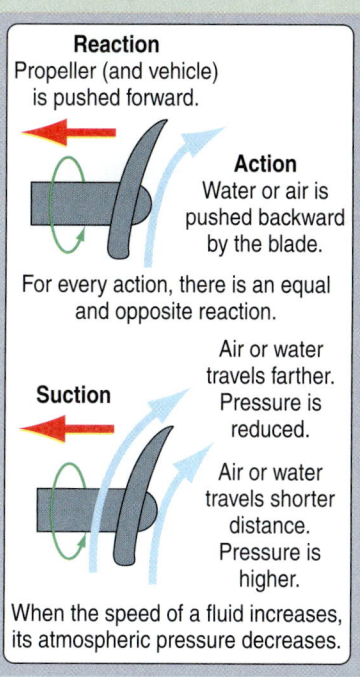

Reaction
Propeller (and vehicle) is pushed forward.

Action
Water or air is pushed backward by the blade.

For every action, there is an equal and opposite reaction.

Suction

Air or water travels farther. Pressure is reduced.

Air or water travels shorter distance. Pressure is higher.

When the speed of a fluid increases, its atmospheric pressure decreases.

Figure B. This diagram explains how a propeller works.

from the pressure of the spinning propeller. Ships generally move about 60 to 70 percent as much as they should in theory.

The second principle states another law: As a fluid travels faster, its pressure is reduced, **Figure B.** When a propeller spins, the water or air travels faster across the front of the blade than it does on the back. This causes the pressure to decrease on the front. The decrease in pressure draws the propeller forward, adding to its efficiency.

designer conducting tests on the model. There are many different tests that can be performed. See **Figure 10-16.** The designer chooses the test fitting best with the design problem and solution. For example, a solution to a bridge design would be tested for strength. It would not, however, be necessary to test a telephone design for strength. A telephone design would be tested to make sure it fits the users' hands.

Designers test the solution using a procedure. They begin by identifying the purpose and type of test they will

use. The designers then gather the equipment and supplies and conduct the test. During the test, they gather data. After the test is complete, the designers analyze the data and evaluate the results. The test and evaluation help to convert the designer's idea into a realistic solution by determining if the proposed solution is appropriate for the problem. Evaluation is used to establish how well the design meets the established criteria and constraints and to give guidance for improvement, if needed. Procedures for evaluation

Standards for Technological Knowledge

9 11 13

Research
Have the students research and identify common tests conducted on products.

Discussion
Explain to the students why product testing is important to the design process.

Figure 10-15. A prototype is a working model that may be used to test product performance. (Ford Motor Company)

vary from visually studying to actually operating and trying the products and systems. Some other strategies include using previous knowledge

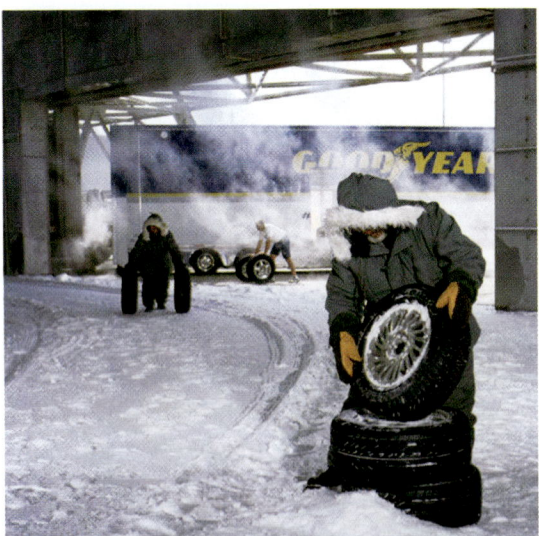

Figure 10-16. Products are tested at extremes to make sure they work properly. (Goodyear Tire and Rubber Company)

and asking questions. The designers use the results of the evaluation to make an assessment of the solution and improve the design. The *assessment* of a design determines whether the solution will be discarded, revised, or sent to the next step. Designs sent to the next step are drawn in various ways to better communicate the solution.

Communicating Designs

If the solution passes the test, it is ready to be manufactured. Designers, however, cannot just give the manufacturer the model of the solution. They must communicate the solution using a set of documents. The documents include a bill of materials, specification sheets, and drawings. The bill of materials lists all of the items

TechnoFact: Testing often forces the solution to extremes to measure the durability and strength of the product. For example, the test in Figure 10-16 places tires in extreme cold to test their function. It is very likely that a similar test is conducted at a very high temperature.

Figure 10-17. Mechanical drawings are accurate technical drawings. (Design Central, design firm; Artromick International, client).

Brainstorm
Have the students think of ways designs could be communicated.

Tech Terms
Orthographic Drawing—A type of drawing used to show three-dimensional objects in two dimensions.
Pictorial Drawing—A drawing showing objects as they would be seen by the eye.

needed to make the product. The specification sheets identify the details of the materials and quality of work. The drawings show the sizes of the solution and how the parts fit together. A designer's ideas, processes, and results can be documented in design portfolios or journals or with drawings, sketches, and schematics. The results of the design process can also be communicated in many other ways, such as on a web page or with a model.

The drawings used are called *mechanical drawings.* See **Figure 10-17.** *Mechanical drawings* are very accurate and precise. They are created using either traditional or computer-aided design (CAD) tools. There are two main classes of drawings: orthographic and pictorial. Orthographic drawings are also known as multiview drawings. They use several views to describe the solutions. These drawings are also dimensioned to show the size of the pieces. Pictorial drawings are created to look three-dimensional. They are used to show how the solution will look when it is complete. These drawings are easier to understand for most people because they look the most realistic. When all of the drawings and documents are

TechnoFact: Mechanical drawing is a process that is thousands of years old. The evidence of the use of drawing tools can be traced back four thousand years to Babylon.

Discussion
Ask the students why design is a never ending process.

Brainstorm
Ask the students what products they have seen improved in their lifetimes.

complete, they are presented to management or the client for final approval. Communicating solutions is the topic of Chapter 17.

Improving Solutions

Improving solutions, highlighted in Chapter 18, may seem to be the last step of the process. It is, however, just the beginning of a new process. Design is never ending because there is no perfect design. All solutions can be improved. The best designs optimize the desired qualities—safety, economy, effectiveness, and dependability—within the specified constraints. A design that was the best solution at one time will eventually become outdated. All designs build on the imaginative ideas of others.

Designers will begin to improve their designs the moment the designs are manufactured. See **Figure 10-18.** They have several reasons to improve their solutions. These reasons range from profit and competition to function and safety. To improve their designs, designers rely on information. Designers receive information from various sources. Customers and competitors are two of the sources of information. Other sources are government agencies and the company itself.

When designers improve solutions, they go back and redo some steps of the design process. Some improvements require the designer to start again at the very beginning and redefine the problem. Only going back one or two steps can solve other problems. Improving designs is the only

way our technology and technological literacy will advance.

Summary

Design is the action of creating a solution for a problem. It always has a purpose. There are two types of design: artistic and engineering. Artists produce artistic designs for themselves. Engineering designs are produced to solve the problems of others. Engineering designs must have a nice appearance and function well.

Products, systems, and structures are designed using a process. The process for designing identified in this chapter and throughout the book has eight steps, but it is not the only design process. It is simply one design process used. All designers use a design process. This process helps reduce the number of product and system failures. It also helps ensure that the products and systems meet human needs and solve problems.

Figure 10-18. Designs must be reviewed as they are developed.

Curricular Connections

Language Arts

Write a letter to a local designer. Enquire about the design process he uses in his own design. Share any responses with your class.

Science

Select a type of designer (such as a chemist, automobile engineer, or architect). Research and list types of job duties they have that include science.

Social Studies

Study how design has changed throughout history. Choose a civilization, study its designs, and show how and when other civilizations influenced it.

Activities

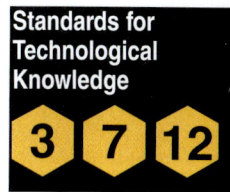

Standards for Technological Knowledge

3 **7** **12**

1. Create a display showing examples of artistic and engineering design. Use the display to explain the differences between the two types of design.

2. Choose an example of a system. Research how it was designed and explain how the parts work together.

3. Use the eight steps of the design process to create a simple product design.

TechnoFact: In 1985, the Coca-Cola Company tried to improve the formula for their Coca-Cola® soft drink. The new formula was preferred to the old drink in blind taste tests. However, consumers did not want the drink to change, and they protested, wrote letters, and called Coca-Cola. After two months, the Coca-Cola Company reintroduced the old formula as Coca-Cola Classic® soft drink.

Test Your Knowledge

1. Name five items in your classroom that have been designed.
2. Design is action with a(n) _____.
3. Architects are the only designers who use the design process. True or False.
4. Paraphrase the definition of *artistic design*.
5. _____ design is used to create plans of buildings and roads.
6. Product designers must consider the _____ and _____ of the products they are designing.
7. Give examples of three types of product designers.
8. List and describe the eight design steps.
9. A mock-up is a working model. True or False.
10. The design process ends with which step?

   A. Testing the solution.

   B. Researching the problem.

   C. Communicating the solution.

   D. None. It never ends.

Answers to *Test Your Knowledge* Questions

1. Answers will vary. (examples: a desk, the carpet, the lights, clothes, the blackboard, decorations)

2. purpose

3. False

4. The type of design used to solve problems the artist chooses.

5. Engineering

6. form, function

7. Answers will vary. (examples: fashion designers, graphic designers, chemists)

8. Identifying the problem (writing the need or want as a problem statement), researching the problem (learning about the need or want), creating solutions (developing ideas), selecting and refining solutions (listing the aspects of design and creating a design), modeling the solution (creating tables, formulas, charts, and mock-ups), testing the solution (determining whether or not the design works), communicating the solution (developing a bill of materials, specification sheets, and drawings), and improving the solution (modifying the design).

9. False

10. D. None. It never ends.

Modular Activity

This activity develops the skills used in TSA's Digital Photography event.

Digital Photography

Activity Overview

In this activity, you will create a digital photo album composed of five sections:

➤ Cover sheet listing name(s), course, and date

➤ Discussion of each image, including camera settings, lighting considerations, editing techniques, and relation to theme

➤ Five prints of the enhanced/edited digital images, one for each of the following themes: portrait (animal or human), landscape, still life, action, and contrast

➤ Resource list, which lists the type of camera and software used

➤ Five prints of unedited images

Materials

➤ Three-ring binder

➤ Digital camera

➤ Computer with photo editing software and color printer

Background Information

General. Check your camera's settings before you take pictures. Most often, using the camera's automatic setting will allow you to "point and shoot." This means you can concentrate on your subject, instead of worrying about camera controls. To avoid blurry photos, hold the camera firmly with both hands, and press the shutter button with a slow, steady motion. Don't "jab" the button with your finger—that will cause the camera to shake.

Portrait. Get close to your subject. Fill most of the viewfinder with your subject, instead of the background. Press the shutter button halfway to bring the subject into focus, then watch for a good expression and take the photo. Try both people and pet portraits, and try them with and without using the camera's flash. If your subject is against a bright background (such as a window), you'll probably need to use flash unless you want to do a silhouette.

Landscape. If the camera has a zoom lens, use the wide-angle setting to show a broad area. Try to get an interesting object (for example, a rock, bush, fence) in the foreground to add depth to your picture. The wide-angle lens will make it easier to keep both near and distant subjects in focus.

Still life. Try for a pleasing arrangement of objects that fit a theme (for example, a vase of flowers, pruning shears, and gardening gloves, or a baseball and fielder's glove surrounded with related material). Don't use your flash, since that would cause harsh shadows. Instead, use room lighting, possibly with a table or desk lamp placed just outside the picture to provide additional light. You can move the lamp around to place shadows where you want them.

Action. Try to shoot action pictures on a brightly lit day, so the camera's automatic mode will choose a fast shutter speed. This will prevent blurring. Try to arrange the situation so your subject is moving toward you, or at an angle, rather than "crossing" in front of you. This will help to stop motion. Practice with the camera so you can adjust for *shutter lag*—the time between pressing the button and the actual shutter operation.

Photo editing and enhancement. Use your computer's photo-editing software to adjust the brightness of your picture (darken or lighten as necessary). If it has a *color cast,* such as a very blue or very yellow appearance, your software should have the ability to remove the cast. Don't be afraid to crop away part of the picture. Many pictures can be greatly improved by eliminating extra background and focusing attention on the subject.

Guidelines

➤ The photo album must include one color and one black and white image.
➤ All images must be 4˝ x 6˝.
➤ Photographic paper or high quality glossy paper may be used for photographic prints.
➤ Images should be enhanced and edited with photo editing software.

Evaluation Criteria

Your project will be evaluated using the following criteria:
➤ Composition
➤ Photo editing
➤ Relation of image to theme
➤ Discussion of images

Regulating Technology

The importance of safety in the workplace cannot be overemphasized. Some aspects of safety include working safely with hazardous materials and wastes, working safely with machinery and tools, working safely with electricity, preventing fires, and understanding how to extinguish any fires that may occur.

There are many codes, laws, standards, and regulations related to safety and technology systems. Often, such standards are adopted into laws. These laws are designed to protect workers and the general public from injuries and other health risks.

In addition to safety standards specified by law, most companies have safety standards developed for their specific systems and processes. Many companies have safety training sessions to ensure that employees are familiar with the hazards and safety regulations in their workplace.

Codes and standards address issues in addition to safety. In many cases, codes and standards establish specifications to ensure that manufacturers produce items consistently and at an acceptable level of quality.

In manufacturing, the International Organization for Standardization (ISO) supports and manages international standards for production. The American National Standards Institute (ANSI) and American Society for Testing and Materials (ASTM) create additional standards.

The construction industry is also subject to many regulations. Zoning laws control the types of structures that can be constructed in each section of a community. Building codes identify the processes and resources that can be used for each aspect of construction. Typical, laws dictate that construction is performed in compliance with the "Americans with Disabilities Act" (ADA), the National Electrical Code (NEC), and Occupational Safety and Health Administration (OSHA) regulations. Periodic inspections are made by government officials.

In transportation systems, state organizations control the use of highway systems. The Federal Aviation Administration (FAA) controls airspace and air safety and distributes piloting licenses. Other agencies regulating the transportation industry set highway construction standards, control the safety of interstate trucking, and endorse shipping on the oceans.

Organizations such as the U.S. Food and Drug Administration (FDA) regulate technology in the agricultural and bio-related industries. Food products and medicines must be tested extensively before they can be sold to consumers to ensure public health.

Technology is powerful, and it has the potential to cause great harm if it is not controlled or used properly. These regulations exist to protect society and the environment. Thus, it is imperative that they be followed. In many cases, not adhering to such regulations not only puts you and others in danger, but it also may result in legal consequences. Before engaging in any technological activity, be sure to research the regulations governing it and follow them precisely.

Chapter 11
Identifying Problems

Did You Know?

➤ Design solutions often create new problems and opportunities.

➤ Before the gasoline automobile, there was no need for gas stations. Once the gasoline automobile was being used, people had a new problem. Their problem was that they needed to refill their cars with gas.

➤ The sliding keyboard tray is a solution to a problem the use of computers created. Before computers became popular, desk manufacturers did not need to design sliding trays into their desks.

Objectives

The information given in this chapter will help you do the following:

➤ Define *need* and *want*.

➤ Describe opportunities and problems.

➤ Explain the three different types of problems.

➤ Identify common places to find problems.

➤ Give examples of problem statements.

➤ Cite the definition of *criterion*.

➤ Recognize the meaning of *constraint*.

➤ Create a design brief.

Key Words

These words are used in this chapter. Do you know what they mean?

constraint

criterion

design brief

external constraint

need

opportunity

personal problem

problem

problem statement

social problem

technological problem

want

Chapter Outline
Did You Know?
Objectives
Key Words
Needs and Wants
Problems and Opportunities
Types of Problems
Recognizing Problems
Problem Statements
Criteria
Establishing Constraints
Design Briefs
Summary
Curricular Connections
Activities
Test Your Knowledge

The design process begins by identifying a problem. This is a very important step in the design process. In this step, designers determine what they are designing. This chapter will explain how problems are identified.

Needs and Wants

All people have needs and wants. Basic human *needs* are the requirements it takes to live. Shelter, clothing, and food are considered the basic human needs. Many problems and opportunities are discovered in human needs. One problem is the need for enough food for the world's population. The need for clean drinking water is another problem. Shelter is a problem found in many communities. See **Figure 11-1.** Many cities have people who have no place to use for shelter. The needs for both shelter and clothing are major problems. Shelter and clothing help to protect people from the elements, like wind, rain, snow, and heat.

Medication can also be a human need. Many diseases and illnesses require certain medications to treat them. If the basic needs are not met, people will not survive. Once basic needs are met, people have wants. *Wants* are things people desire. They are not needed to survive, but they help make life better and easier. Wants can be large or small. A small want could be to talk with someone on the phone. A large want may be to travel to the Moon. Wants are often mistaken as needs. You may say you need a new skateboard. That is not a need. You will survive without a new skateboard. Wants, however, are still important. They can be many types of things. A want could be a change in comfort. A softer chair is a want. A more comfortable bicycle is also a want. Wants can be entertainment, like a DVD player or a game system. They help to make life more enjoyable.

Problems and Opportunities

Wants and needs are problems that new products or systems can make better. The designs for products, devices, structures, and systems all

Figure 11-1. Homes are built to meet the need for shelter. (Habitat for Humanity International)

TechnoFact: John Pemberton was an inventor who wanted to make a fortune by inventing a popular drink that could also serve as a medicine. He set out with his eyes on the coca plant, which had been used in South America for thousands of years for medicinal purposes. After much work and the addition of kola nut, sugar, and citric acid, he had a good-tasting beverage. He redesigned the formulas and tested the product at local soda fountains until he felt it was good enough to sell. He named the product by combining the names of the two major ingredients, Coca-Cola soft drink. Today, it seems that he accomplished part of his goals; Coca-Cola soft drink is a popular drink that has indeed made a fortune.

Figure 11-2. Some problems are solved with the design of medications.

start with a problem or an opportunity. A *problem* is a situation that can be made better through an improvement. One example may be that there is no medication to treat a certain disease. See **Figure 11-2.** A problem could also be that you need to get across town in five minutes. People come across problems every day. Problems often cause difficulty if they are not solved. Arriving at school after the bell rings is a problem. You can see there would be difficulties if that problem were not solved. If you were not at school in time for the bell, you would be tardy.

Opportunities are different than problems. Problems cause people to design new and improved products and systems. *Opportunities* use new or existing products and systems in new ways. Crop farming has presented many opportunities over the years. See **Figure 11-3.** Many farmers found an opportunity in no-till farming. Farmers designed the no-till method to increase the yield of crops. In typical farming, the ground is tilled to break it up and remove any excess crop. After it is tilled, the farmers plant new seeds. No-till farming uses the same seed typical farming does. The farmers, however, do not till the ground before planting. They use special planters that cut through the remnants

Figure 11-3. The innovation of older farming methods presents new opportunities.

TechnoFact: In 1976, Millard and Linda Fuller founded Habitat for Humanity. Their goal was to eliminate poverty-related housing problems worldwide. Since 1976, Habitat for Humanity International has built more than 125,000 houses in eighty countries.

TechnoFact: Farming is a technological process that has been used to solve the problem of food supply. Civilizations began farming nearly ten thousand years ago. Farming remained nearly the same for thousands of years. Today, many of the processes are much like those used thousands of years ago. However, we are much more efficient and can harvest faster and with a greater yield due to our machines and technology.

Standards for Technological Knowledge

9

Resource
Reproducible Master 11-1: The Design Process: Defining the Problem

Research
Explain that a country's standard of living is based, in part, on its social, personal, and technological levels. Have the students find the standard of living for several countries around the world.

Brainstorm
Have the students develop a list of technical problems a person with a physical disability may encounter on a daily basis.

of the last year's crop. This method leaves nutrients in the ground for the new seeds. No-till farming is an opportunity to grow better crops.

Types of Problems

Problems can be found in all areas of life. There are many different kinds of problems. All problems can be placed in one of three categories:

➤ Social problems.

➤ Personal problems.

➤ Technological problems.

Social problems are problems dealing with society. Social problems affect large groups of people. Illiteracy is one example of a social problem. It is the inability of people to read or write. Education is used to solve low literacy. Violence and crime are also social problems. The police try to solve these types of social problems. Some social problems are the results of natural disasters like hurricanes, floods, and tornados. Governments and organizations like the Red Cross try to solve many social problems. See **Figure 11-4.**

Personal problems are problems affecting individuals. Emotional problems are personal problems. Low self-esteem is an example of a personal problem. Academic problems are personal problems with schoolwork. Teachers and counselors work to solve academic problems.

Technological problems are problems that can affect individuals and groups of people. The need for a new building is a technological problem.

The building may be a large structure for many people. It may also be a small house for one person. The need for electricity to power a machine is another technological problem. See **Figure 11-5.** These problems can be solved with devices or systems. Designers, engineers, and architects are some of the people that solve these problems. This book will focus only on how people solve technological problems.

Recognizing Problems

All technological problems must be recognized before they can be solved. A designer cannot create a solution if there is no problem. Problems are often easy to find. Some designers look at their own lives to find problems. People with disabilities often find problems completing daily tasks. Designers create solutions for these types of problems. See **Figure 11-6.**

Figure 11-4. Disaster relief workers help solve social problems after natural disasters. (American Red Cross)

TechnoFact: The International Federation of Red Cross and Red Crescent Societies is the world's largest humanitarian organization. Their mission is to improve the lives of vulnerable people by mobilizing the power of humanity. Their work focuses on four areas: promoting humanitarian values, disaster response, disaster preparedness, and health and community care.

Figure 11-5. Electricity distribution plants help spread electricity to homes and businesses.

Some people find problems at work. Bette Nesmith found a problem while doing her job. She was a secretary who

Figure 11-6. Designers create solutions to solve problems for people with handicaps. (NASA)

had to use an electric typewriter for the first time. Nesmith found she had too many errors and did not have a good way to correct them. Her problem led to her creation of Liquid Paper® correction fluid, which is often called *white out*.

Other people find problems during leisure time. You may have found a problem while playing a sport. Branch Ricky, a former baseball player, was worried about pitches hitting batters in the head. His problem dealt with safety. Ricky designed the first batting helmets for baseball players. See **Figure 11-7.** Soon after he designed the helmets, all players were wearing them.

Bette Nesmith and Branch Ricky both had problems they could solve themselves. Many people have problems they cannot solve themselves. Some of these problems are too complex. Sometimes people do not have

TechnoFact: Christopher Latham Sholes was a mechanical engineer who invented one of the most important business machines prior to the computer. In 1868, he patented the first practical modern typewriter. The typewriter was manufactured and marketed by the Remington Arms Company.

Figure Discussion
Ask the students what problem they think the device in Figure 11-8 is solving.

Figure 11-7. Batting helmets were designed to protect baseball players from injury. (American Baseball Company)

the time to solve their own problems. People often take their problems to designers.

People who hire designers are called *clients.* The designers are paid to solve their clients' problems. Designers must first identify their client's problem and begin by writing what the client says about the problem. See **Figure 11-8.** These people

Figure 11-8. Designers work with clients to determine the clients' problems. (Design Central, design firm; Artromick International, client)

TechnoFact: Some design companies, or firms, specialize in certain products. For example, a design firm may choose to design only toys or software. However, other firms design products for many different industries. These firms hire designers with various backgrounds so they can handle many different problems.

review the statements from the clients and create a problem statement.

Problem Statements

A *problem statement* is a clear definition of the problem. For example, a problem may be to "design a device that will evenly spread paint on a wall." This is a clear definition. It is an open-ended statement stating a problem, not a solution. A statement like "design a paint roller" states the solution. This statement is too narrow. It would be hard to develop a new design with a statement to design a paint roller. The first statement allows the freedom for the designer to go in many directions. See **Figure 11-9.**

Criteria

After the problem statement is written, the designer will list the criteria for the design. *Criteria* are elements of the problem needing to be solved. They are some of the requirements, or parameters, placed on the development of the product or system. Criteria are listed before the designer starts to develop ideas. They are identified

Standards for Technological Knowledge

8 9 11

Resource
Transparency 11-3: Define the Problem

Brainstorm
Have the students select several products and write the possible problem statements used to design the product.

Tech Terms
Requirement—
Something demanded or needed.

Figure 11-9. This paint sprayer is one solution to the problem of applying paint. (Eastman Chemical Company)

TechnoFact: Paint is made up of a dye, a binder to hold it together, and appropriate thinners to make it easy to use. As technology has advanced, so have the dyes, binders, and thinners in paint. Primitive paints were made from animal fat, indigenous dyes, clay, and charcoal. For much of the nineteenth and part of the twentieth centuries, lead was used in the making of paint. Nowadays, there are a number of substances (latex, for instance) added to paint to allow it to be used in a variety of situations.

Standards for Technological Knowledge

2 **8** **11**

Resource
Reproducible Master 11-2: Technological Design Constraints

Brainstorm
Have the students select a product and explain what the design criteria for it would have been.

Extend
Explain the difference between a criteria and a constraint.

during this first step of the design process. Criteria answer the following questions:

➤ Who will use the device or system?

➤ Where will it be used?

➤ What will it do?

➤ How well will it work?

➤ How reliable will it be?

The list of criteria would look something like this: "Homeowners will use the device. The device will be used on the inside walls of their home. It will spread the paint evenly and thick enough that only one coat is needed." The criteria are a very important part of identifying the problem. Changing the criteria can change the design of the solution. Imagine if the criteria above were changed to "A professional will use the device on the outside of the house." The solution would be very different.

It is important the criteria are accurate. When a solution is complete, the designer will evaluate the design against all of the criteria. If the design meets the criteria, it is successful.

Establishing Constraints

After preparing a problem statement and criteria, the designer must begin to list the design constraints. The *constraints* are the limitations for the design. A limitation is a boundary. Like criteria, constraints are parameters placed on the development of the design. Constraints tell designers how

far they can go with a design. See **Figure 11-10.** There are several categories of constraints providing guidelines for design:

➤ Technical constraints.

➤ Financial constraints.

➤ Time constraints.

➤ Production constraints.

Technical constraints are the simplest to understand. Technical constraints deal with the size and shape of the object and the materials that will be used. Whether or not a device uses batteries or is hand powered is a technical constraint. An example of a technical constraint is "The product must be less than six inches in height and three pounds in weight. It must also be round and made of plastic."

Financial constraints involve the costs of the design. The first cost is the initial cost. The initial cost is the money it takes to design the solution. This is often called the *design budget.* The next cost is how much money it will take to make the product. The last

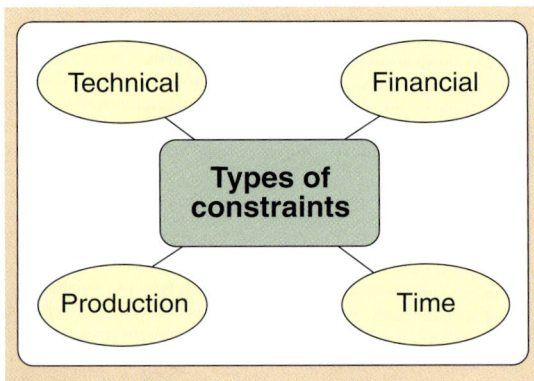

Figure 11-10. Problems can have four types of constraints.

TechnoFact: Constraints may change, depending on who uses the product. For example, when designing content for the Internet, the designer must take into consideration the computer technology the user will use to view it. It would be ineffective to design a highly graphic page if the users viewing the material are using very slow modems.

financial constraint is how much the product can make if it is sold.

Time constraints are limits to the amount of time the design process can take. A time constraint may be that the design process can only take three months. For many products, this is a short time. Other time constraints are how long the product will last. Designers create products to last only a certain amount of time. A lightbulb is a product designed to last a certain amount of hours. See **Figure 11-11.**

Production constraints are the restrictions on how the device can be built. A designer must look at the available tools and materials. A production constraint might be that the

Figure 11-12. Electrical wires, outlets, and switches are placed in buildings to meet the local building codes.

solution must be able to be produced using hand woodworking tools.

The designer places some constraints on the design. Lawmakers and government agencies establish other constraints. Most designers have these types of legal constraints to follow. These are called *external constraints.* External constraints are limits people other than the designer and client set. External constraints can be technical, financial, time, or production constraints.

If the problem is to construct a shelter, local building codes are some relevant external constraints. The building codes limit the types of materials that can be used. They also place constraints on the location of objects like electrical outlets and switches. See **Figure 11-12.** Other external constraints are the location and shape of the site where the shelter will be built.

Figure 11-11. Lightbulbs are designed to last a certain amount of time.

TechnoFact: External constraints in the form of building codes go back thousands of years. The Code of Hammurabi, c. 1700 B.C., placed responsibility for the quality of construction on the builder. If the building collapsed and lives were lost, the builder would also lose his own life.

Standards for Technological Knowledge

2 11

Figure Discussion
Have the students discuss the possible constraints for the design of the fish tank in Figure 11-13.

Brainstorm
Have the students write a design brief for the example (fish feeder) in Figure 11-13.

Design Briefs

In this step of the design process, the designer creates a problem statement, criteria, and constraints. The designer puts them together in one document called the *design brief.* The design brief helps the designer completely understand the problem. It lets the designer see the problem, its requirements, and its limits together. The design brief is created after a long process. The designer must work through the problem to completely understand it.

Put yourself in the place of a designer. You have a problem:

Your brother is going off to college. He has asked you to feed his fish for him. He likes to feed his fish once a day at exactly 10:00 a.m. You have to be at school at 8:00 a.m., and you don't get home until 3:00 p.m. He leaves in two weeks and will be back to pick up his fish in two months. He will pay you a total of five dollars a week to feed the fish.

After thinking about the problem, you must create a design brief. The design brief may look similar to the one shown in **Figure 11-13.**

After writing the design brief, you must review it. See **Figure 11-14.** You

Example Design Brief: Fish Feeder

Problem Statement
Create a device or system that will automatically distribute food to fish at an exact time every day.

Criteria
The device or system
has the following criteria:
- You will set it.
- It will work automatically.
- It will distribute food.
- Every day, it will work properly.
- It will work at the set time.

Constraints
The device or system
has the following constraints:
- It must cost less than forty dollars to design and build.
- It must be able to be produced with materials on hand.
- It must fit on top of the fish tank.
- It must be designed and built in less than two weeks.

Figure 11-13. Solutions are often designed to fit existing problems. Here is an example design brief for a fish feeder. The problem is clearly stated, and then the known criteria and constraints are listed. (All-Glass Aquarium)

TechnoFact: The term *design brief* can be used in several ways. The Technology Student Association uses design briefs as an introduction and basis of information for many of its activities. Its design briefs lay out the situation and problem to be solved. They often include a design statement, criteria, and constraints.

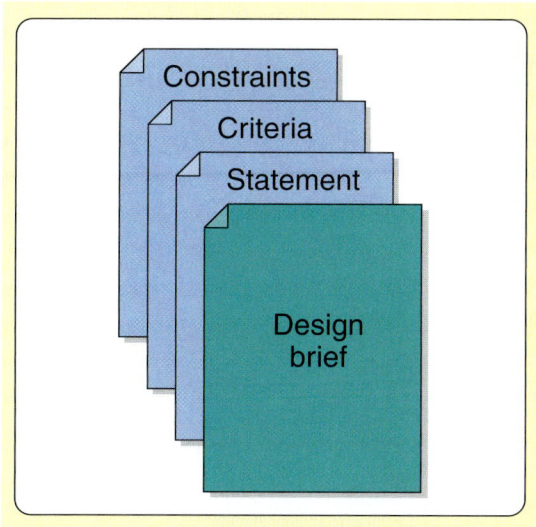

Figure 11-14. Design briefs include the problem, criteria, and constraints of a design.

need to ask yourself some of the following questions: Does anything need to be added to the problem statement? Is this statement clear? Does it state the problem? Is it open-ended? Are the criteria clear? Was anything left out? Are there more requirements? Are there any other constraints? Once the design brief is complete, you move to step two of the design process. You have to start gathering information to help you solve the problem.

Summary

Identifying the problem is the first step of the design process. This step is very important because it starts the process. Problems can be found in various places. Many problems are found in human needs and wants. Once a problem is recognized, it must be written clearly. Problems are rarely found in clear form. The designer must work to find the true problem.

The true problem is written as a problem statement. A problem statement is a clearly defined problem. Once the problem statement is written, the designer must identify the requirements of the design. The requirements are called *criteria*. The criteria will be used to judge the solution at the end of the process. The designer also lists constraints. Constraints are limitations of the problem. Money, time, and materials are examples of constraints.

After the statement, criteria, and constraints are identified, they are written as a design brief. A design brief is a document clearly describing the problem, requirements, and limitations. When the design brief is complete, step one is finished. The designer, however, may need to come back to step one and make changes later in the process.

Curricular Connections

Language Arts

Most companies and organizations work to solve problems. Write a letter to managers of a local company or organization. Ask them about the types of problems they work on solving and how they identify problems.

Science

Research a personal problem. Create a poster board showing how the problem is identified and treated. Include whether or not the problem is treated with an artifact of technology.

Social Studies

Examine the technological problems of various countries. Compare the problems in other countries to the problems in your own.

Standards for
Technological
Knowledge

3 **12**

Activities

1. Identify and list possible needs and wants of the following types of people:

   A. A baby girl.

   B. A teenage boy.

   C. A middle-aged woman.

   D. An elderly man.

2. Examine the problems existing around you. Create a list of problems you find in the following places:

   A. Your school.

   B. Your home.

   C. Your community or neighborhood.

3. Write a design brief including a problem statement, criteria, and constraints. The problem should be related to your own life.

TechnoFact: It is estimated that pharmaceutical companies spend over 20 percent of their sales income just in the research and development of drugs that can solve the problems of life-threatening illnesses and diseases.

Test Your Knowledge

Do not write in this book. Place your answers to this test on a separate sheet of paper.

1. A need is a requirement to live. True or false.
2. List four examples of a want.
3. A(n) _____ is a situation that can be made better through an improvement.
4. The use of a new or existing product or system in a new way is a(n):
   A. Problem.
   B. Opportunity.
   C. Need.
   D. Want.
5. Name and describe the three types of problems.
6. Problems can only be found at home. True or false.
7. It is important to write a problem statement clearly. True or false.
8. What is the function of criteria?
9. List and explain the four types of constraints.
10. A design brief includes a problem statement, criteria, and constraints. True or false.

Answers to *Test Your Knowledge* Questions

1. True

2. Answers will vary. (examples: to talk to someone on the phone, travel to the Moon, receive a new skateboard, sit in a softer chair)

3. problem

4. B. Opportunity.

5. Social problems (those dealing with society), personal problems (those affecting individuals), and technological problems (those that can affect individuals and groups of people).

6. False

7. True

8. To list the requirements of the solution.

9. Technical constraints (the size and shape of the object and the materials that will be used), financial constraints (the costs of the design), time constraints (the amount of time the design process can take), and production constraints (how the device can be built).

10. True

Chapter 12
Researching Problems

Did You Know?

➤ The glue on Post-it® notes was created while Dr. Spence Silver was researching another product. In 1968, Silver was a scientist researching acrylate adhesives. Through his experimental research, he found an adhesive that was not very sticky. It was not until several years later, in 1980, that a use was found for it. Art Fry put Silver's research to use. Fry added the adhesive to small pieces of paper and created the Post-it note.

Objectives

The information given in this chapter will help you do the following:

➤ Define *research*.

➤ List and explain the three types of research.

➤ Describe the research process.

➤ Recognize that the problem being researched may be the problem statement from the design brief.

➤ Give examples of how data is gathered.

➤ Summarize the use of graphs to analyze data.

➤ Explain how conclusions are developed.

Key Words

These words are used in this chapter. Do you know what they mean?

aesthetic

bullet point

conclusion

data

experimental research

graph

information

library research

market

market research

research

research process

scientific method

survey research

Chapter Outline
Did You Know?
Objectives
Key Words
Types of Research
  Library Research
  Survey Research
  Experimental Research
The Research Process
  List the Problem
  Gather Data
  Analyze the Data
  Make a Conclusion
Summary
Curricular Connections
Activities
Test Your Knowledge

Have you ever tried to solve a problem, write an essay about a famous person, or design a new device? If you have done any of these things you have probably done some research. Research is an important part of solving problems, writing biographies, and designing products. See **Figure 12-1.**

Researching a problem is the second step of the design process. In the first step, the designer identified a problem. In the second step, the designer finds as much information about the problem as possible. Take the essay about a famous person, for example. If you are writing a paper about Thomas Jefferson or Eli Whitney, you need to research that person's life. You will be prepared to write about him if you gather a lot of information. Being well informed will help you develop a good essay.

Similarly, being well informed will help designers. The more information they can collect, the better their designs will be. If architects are designing an apartment complex, they need to research other apartments. They may talk with people who live in apartments. It is important to remember that, at this step, the designers are only researching the problem. They are not creating a solution. The designers are simply gathering information about the problem. They use the design briefs created in step one to guide their research. At the end of this step, the designers will develop conclusions to help with creating solutions, which is the next step of the design process.

In order to design new products and systems, new knowledge must be developed. Without new knowledge, there would be no improvements in the world. New knowledge is gathered using *research.* Researching is scientifically seeking and discovering facts. It is used in many different areas. Doctors use research when they are treating patients. When doctors have patients with new symptoms, they read journals and reference books to find a solution. See **Figure 12-2.** Companies use research to sell products. A company selling children's toys uses research to find the things children like. Designers use research to find data and facts about a problem.

Types of Research

Finding facts, or researching, can be done in a few different ways. Each way is a different type of research. Researchers choose the type of research best fitting each of their problems. Sometimes, facts about a problem can

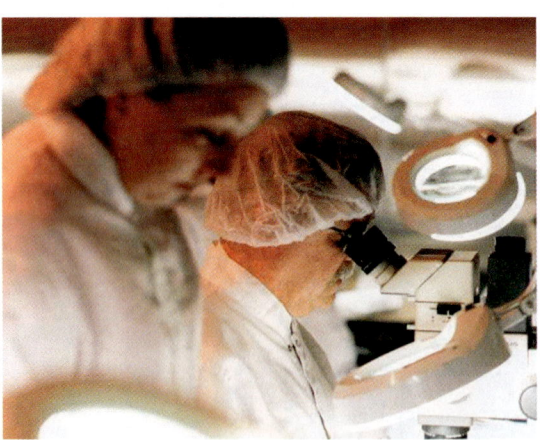

Figure 12-1. Products must be researched before they are manufactured. (NASA)

TechnoFact: The field of research contains many potential jobs, including research scientists, research analysts, market researchers, and research study specialists. Research scientists often work on the development of new scientific products. They plan and carry out scientific experiments. Research analysts review the data gathered through research. They create reports to show the outcomes. Market researchers work one-on-one with the subjects of the survey. They interview people in person and by phone to gather data. Research study specialists develop the framework of the research. They create surveys and research instruments.

only be found using a certain type of research. At other times, information can be found using several types of research. See **Figure 12-3.** There are three major types of research:

➤ Library research.

➤ Survey research.

➤ Experimental research.

Library Research

The first type of research is called *library research.* Library research is used to get background information on the problem. This type of research uses different forms of printed materials. See **Figure 12-4.** Books are one example of printed materials. They are a good place for designers and researchers to begin their research. Designers use books to read about the

Standards for Technological Knowledge

12

Resource
Transparency 12-1: Gather Information

Brainstorm
Ask the students how they have used library research.

Demonstrate
Show the students how to find a book or resource using the library's card catalog or online catalog.

Figure 12-3. Three types of research are used.

subjects of their designs. Suppose you are asked to design a child's toy. You may read books about children and how they play. The more you can learn about children, the better your design

Figure 12-2. Reference materials are important research tools.

TechnoFact: Research is a major portion of many companies' budgets. General Motors spent $7.4 billion on research and development (R&D) in 2001. In the same year, Eli Lilly and Company spent $2.2 billion on R&D.

TechnoFact: Libraries date back more than four thousand years to a collection of clay tablets in Babylon. Years later, the Alexandria Library was built and, until its destruction in 415 A.D., was one of the largest libraries of early civilization. Today, the largest U.S. library is the Library of Congress. Thomas Jefferson established the Library of Congress in 1802 as a library for the president and members of Congress. It is viewed today as the national library and is visited by millions of people every year.

Figure 12-4. Libraries have large collections of research materials.

will be. Designers also use books to see how others have solved similar problems in the past. A person designing a chair may use books to see how chairs were made over the last two thousand years. They will make notes and record information about the design of chairs.

Catalogs and magazines are other types of printed material. They are more current than many books. These materials can help a designer learn how other people are currently solving similar problems. Current ideas and trends can be found in catalogs and magazines. Fashion designers can use magazines to make sure their designs are in style. Designers often look through the catalogs of their competitors. This helps them stay up-to-date.

The Internet is another form of printed material and library research. See **Figure 12-5.** It can be used to find news and magazine articles, books, and catalogs. There is a large amount of information on the Internet. Designers and researchers, however, must be careful when using the Internet. It is easy for people to put false information on the Internet, and it is important all of the information designers use is accurate. Bad information can lead to bad designs.

Some other types of related research materials are manuals and

Figure 12-5. Internet research is a common form of library research.

TechnoFact: Over 95 percent of all U.S. public and school libraries use the Dewey Decimal Classification system to organize their collections. The system was developed by Melvil Dewey and first published in 1876. It is used in more than 135 countries, in over thirty languages.

Figure 12-6. Market researchers gather reactions of potential customers to new products.

Standards for Technological Knowledge

Brainstorm
Ask the students if they have ever completed a survey. Ask them to think of some of the advantages and disadvantages of survey research.

Research
Have the students select two or three products and determine the market those products are designed for.

experienced people. The information these resources provide can be used to see and understand how things work. This information is helpful in learning how to use a product and determining if it works well. In addition, a lot of manuals give guidelines on how to troubleshoot a product or system.

Survey Research

The second type of research is *survey research.* It is used to find people's reactions to designs and events. Have you ever been stopped in a shopping mall and asked several questions? If you have, you have been part of survey research. The questions probably started with your age and income. Then the researcher may have asked if you have ever used a certain product. The researcher then might have asked you questions about a product. Those questions were asked to learn your reaction to that product. See **Figure 12-6.**

This type of survey research is called *market research.* The groups of people using certain products are called *markets.* Market research helps designers determine what types of

TechnoFact: It is very possible that a person may be taking part in market research without even knowing it. Numerous Internet sites track the users' behavior. Researchers examine what site the person came from and what the person clicks once they are inside the site. This helps them to identify the changes needed to the site and the kind of advertising to place on the site.

Standards for Technological Knowledge

10

Research
Have the students write a survey to find out what their classmates think of a product.

Tech Terms
Experiment—An action or process used to determine if something is effective or valid.
Hypothesis—An unproved theory, or hunch, on which further investigation is based.

people (markets) like their product. Market research is done in person, like in the shopping mall. It is also done on the telephone and through the mail. Have you ever bought a product that had a registration card? Did the card ask questions about why you bought the product? If so, your registration card was also used for market research. See **Figure 12-7.**

There are companies dedicated to conducting market research. They do not make any products themselves. These companies just survey people about the products of other companies. The survey companies work through the mail and Internet. They send surveys asking about a number of products. These companies ask if you like certain products. They also ask why you do or do not like the products. The survey companies may even ask about colors or sizes of the

Figure 12-8. Researchers use laboratories to learn new information. (Eli Lilly and Company)

products. Once they have surveyed many people, they sell the results to other companies. This information helps designers make good decisions and better products.

Experimental Research

Experimental research uses tests to learn new information. You may think of a scientist doing this type of research. Scientists run tests and experiments to learn about problems like diseases and illnesses. Designers, however, also use experimental research. See **Figure 12-8.** Designers conduct tests to learn about materials and procedures.

The tests, or experiments, start with a hypothesis. A hypothesis is a hunch about what a test will prove. Designers may have a hunch (hypothesis) that blue telephones sell better than white ones. They might set up an experiment to prove their thoughts. The designers can only test one variable at a time. A

Product Registration Card

Name_____

Address_____

City _____ State _____ Zip_____

Name of product_____

Serial number_____

Date of purchase_____ Store_____

How did you hear about the product?_____
_____

Is the product for home or work?_____

Have you purchased other products from this

company?_____

If so, which products?_____

Figure 12-7. Registration cards are used to gather information.

TechnoFact: Telemarketing laws have negatively affected the use of the telephone as a market research tool. Many states have developed "no-call" lists. Residents can place their name on the list, which makes it illegal for telemarketers to call them. A national "no-call" registry is currently in the process of being created.

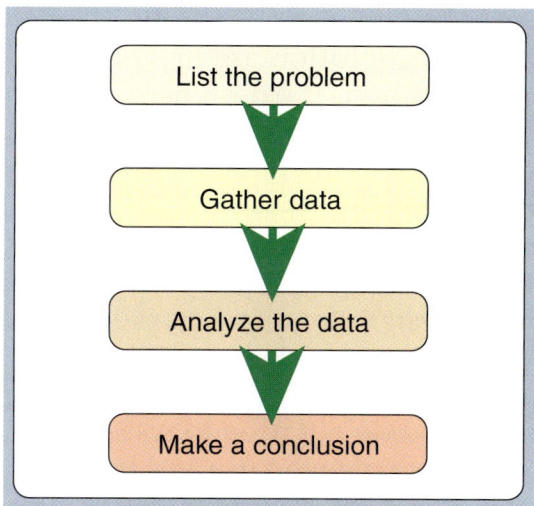

List the problem

Gather data

Analyze the data

Make a conclusion

Figure 12-9. The scientific method is a research process.

Figure 12-9. There are four steps in the scientific method:

1. List the problem.
2. Gather data.
3. Analyze the data.
4. Make a conclusion.

List the Problem

A research project starts by identifying the problem. The problem being researched may be the problem statement from the design brief. The designers may need to review the design brief. They need to make sure they understand the problem.

Gather Data

Once the problem is identified, *data* is collected. Data is raw facts and figures. It can be a group of numbers or a listing of facts about the problem. Data collected at this point will be analyzed later. It will then be used to make conclusions about the problem.

Data is gathered by designing and using tools, such as data-collection devices for interviews, surveys to be mailed, and computer-based documents on the World Wide Web. Assessment tools also can include devices designed to perform tests on such things as water quality, air purity, and ground pollution. Once the appropriate instruments are designed, they are used to gather data.

The designers can gather data in several different ways. They may choose library, survey, or experimental research. The designers may use more

variable is a difference between the items being tested. The color, size, and shape of the telephone are all variables. The designers will make sure only the color of the telephone is different. They will probably even place the different colored telephones in the same place in the store. The designers will then run the experiment. If more blue telephones are sold, the designers' hypothesis is correct. If the designers' hypothesis is wrong, the designers will do more research. They will try to determine why they were wrong.

The Research Process

Research is a process. When people research, they follow certain steps. The *research process* used is often called the *scientific method*. The *scientific method* is a set of steps used to find all of the facts about a problem. See

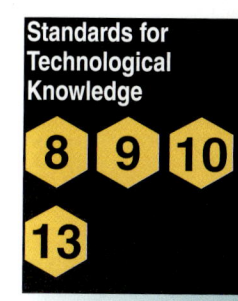

TechnoFact: The use of the research process has been around for many years. Wilbur and Orville Wright used a scientific research process in the development of the airplane. They used wind tunnels and flight-testing to collect data about their airplane.

Standards for
Technological
Knowledge

8 9 13

Brainstorm
Ask the students how the needs of a married couple with small children would differ from the needs of an elderly person with no children.

Research
Select an age group and find as much information as you can about its interests, needs, and wants.

than one of the types of research. Teams of people are often involved with gathering data. One team may survey people, while another team reads books and catalogs. The more data gathered, the more information the designers will have. The data gathered helps the designers make good decisions. See **Figure 12-10.**

Designers use design briefs to guide their research. The criteria from the design brief help direct the designers. They state who will use the solution and how well it needs to work. The criteria also tell what the solution will do and where it will be used. The designers must research these statements.

Who Will Use the Solution?

Designers research different aspects about the people who will use the solution. The designers gather data showing the lifestyles of the users. They may find that the average users of a solution are married with small children. The designers will also research the people's interests. They may identify the hobbies and sports the users enjoy.

Some companies design products for only one market. Other companies create designs for many markets. Baby food is a product designed for one market. The companies designing baby food do not need to research the type of people who use and buy baby food every time they create a new type of food. A company designing shoes, however, may need to research different users of its product. If the company had always made tennis shoes and decided to design work boots, it would need to research a new market. The people buying tennis shoes may have different lifestyles and interests than people who buy work boots.

Figure 12-10. Teams of designers often work together. (General Motors)

TechnoFact: The use of social sciences can be a benefit to designers as well. Physical anthropology, which studies the size of humans, can be used to ensure the product fits a high percentage of the market. Cultural anthropology can be used to understand why people buy one product over another.

Designers must study the properties of the materials in their solution. They must also make sure their product is safe for people to use.

Figure 12-11. Designers gather information on new materials for use in their products. The materials used in this leg brace are high strength and lightweight. (NASA)

How Well Does the Solution Need to Work?

When designers research how well the solution will work, they study the materials used to build the solution. Materials are important to the design. Some materials are better for certain things than others. See **Figure 12-11.**

Concrete would not be the best material for a spoon. Steel would not make the best backpack. Designers will research and gather data on different types of material. The data will be related to the materials' properties. Properties are characteristics such as density, smoothness, hardness, and reactions to chemicals and sound.

Designers will also research safety. See **Figure 12-12.** Designs must be safe for people to use. Designers will study the safety of other products like theirs. They will also interview safety experts. Safety includes the effects on people and the environment. The design must be safe for the environment when it is used and discarded. Designers will research the safest designs for their solutions.

What Will the Design Do?

What the design will do is the function of the design. When researching the function of the design, the designers will interview and survey people. They will collect data showing how people want the product to function and showing the function of other

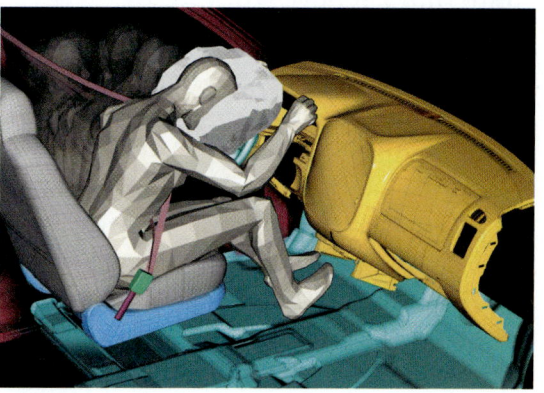

Figure 12-12. The safety of products must be researched. (Ford Motor Company)

Standards for Technological Knowledge

8 9 13

Research
Assign a product to the students and then have them determine whether the materials currently used in the product are the same materials that were used in the past.

Brainstorm
Have the students list the ways they may be damaging the environment by discarding certain products (for example, batteries and plastics).

TechnoFact: As new materials are being developed, products are being innovated to use the new materials. Guitars have been made of wood for hundreds of years. However, they are susceptible to the effects of heat, water, and humidity. New material advances have led to the design of guitars made from engineered materials that are resistant to heat and moisture.

Research
Have the students take measurements of the different desks around the school. Have them describe the ways in which a student's desk, teacher's desk, and principal's desk differ.

Tech Terms
Analyze—To study in detail.
Data—Raw numbers and figures.
Information—Organized data.

products that solve similar problems. The data collected on the function will be in the form of a list of features. In the design of a bulletin board, a feature may be that the design must hold a piece of paper vertically. Another feature might be that the piece of paper must be read from twenty feet away. Features can also be artistic or aesthetic. *Aesthetic* features relate to how the design looks. Texture and color are aesthetic features.

Where Will the Design Be Used?

Research on where the design will be used will determine the size and shape of the design. If designers are designing a desk chair, it is important to research desks. The designers need to measure or find common sizes and heights of desks. They may also research the average space the users have around their desks. This data would be useful when determining the size of the desk chair. The environment is also important. A chair for an executive of a large corporation would be designed differently than a chair for an elementary school student.

Analyze the Data

Data alone is hard to understand. Imagine looking at a list of twenty cities, each with twenty-four numbers next to them. This list may not make much sense. Now imagine seeing a map of the United States showing the average temperature of twenty cities. See **Figure 12-13.** You would probably have no problem understanding the

map. The list of numbers and the map both show the temperatures of twenty cities. The map, however, is easier to understand because the data (temperatures) has been analyzed. The data has now become *information.* Information is organized data. Once all of the data has been gathered, the designer's job is to analyze it. The designer must convert the data to information. Information has more value to consumers because it is easier to understand.

Reviewing data and making graphs are some ways designers analyze data. Graphs are used to turn numerical data into information. They make data more

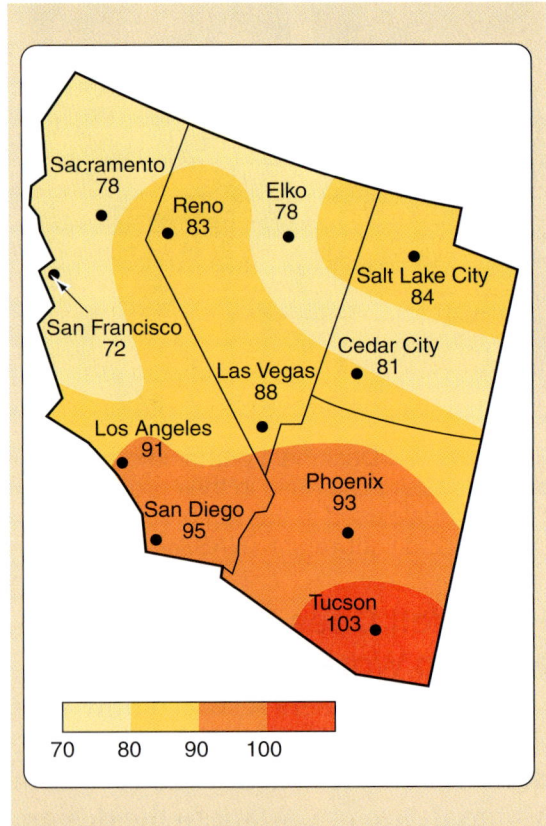

Figure 12-13. Weather maps make complex data easy to understand.

TechnoFact: Guides are available to help designers determine the size and shapes of products. Human factors guides include information on the physical capabilities and limitations of humans. Designers can use these guides instead of conducting their own research.

understandable. A *graph* is an illustration showing how two or more items compare to each other. Line, bar, and pie graphs are the most common types.

A line graph is often used to show trends. See **Figure 12-14.** A trend is how something has changed over time. Line graphs use dots and lines to show trends. Dots are placed at intersecting points. The lines connect the dots to make them easier to see. A line graph has two axes (x and y). The x-axis is horizontal and usually represents a period of time. The y-axis is vertical and represents the item being studied.

Bar graphs are used to compare different amounts. See **Figure 12-15.** Bar

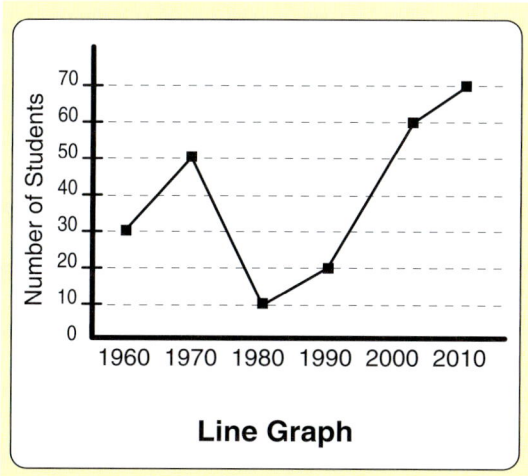

Line Graph

Figure 12-14. Line graphs are used to show data over a period of time.

graphs have an x-axis and y-axis like line graphs. They can be laid out horizontally or vertically. The most common bar graphs use thick bars of color to show the amounts. Some bar graphs, however, use pictures instead of bars. A bar graph could be used to show the density of different materials.

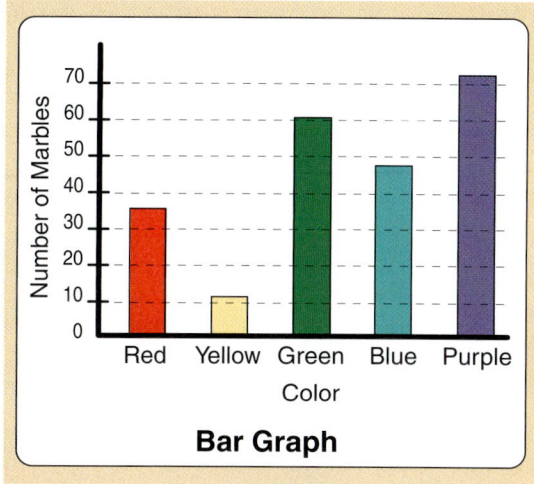

Bar Graph

Figure 12-15. Bar graphs use different colored bars to show data.

Pie graphs are used to show how an object has been divided. See **Figure 12-16.** A pie graph is drawn in the form of a circle. Sections are then divided inside the circle, much like cutting a pie. Pie graphs often deal with percentages. The types of materials

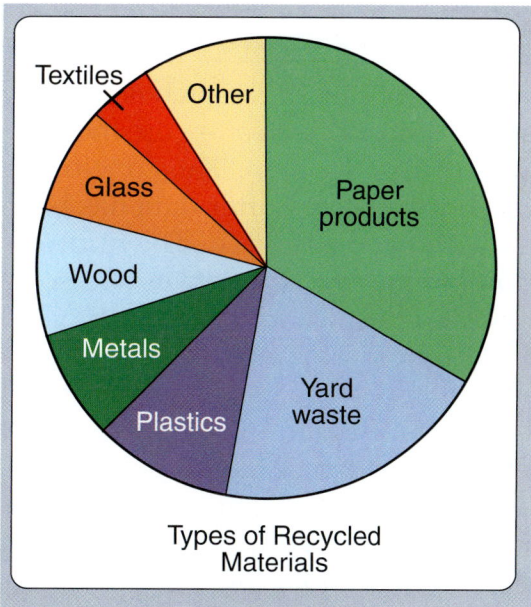

Types of Recycled Materials

Figure 12-16. Pie graphs are useful for breaking data into pieces.

TechnoFact: A pictograph is a graph that uses cartoons or drawings of objects to show quantities of objects. It is often used in the place of bar graphs. It is good for young audiences because it visually represents the data.

Example
Show the students some examples of physical measuring devices, such as thermometers, scales, rulers, and barometers.

Extend
Make conclusions regarding the two graphs you produced. Encourage class participation.

recycled can be represented in a pie graph.

Graphs work well for numerical data from surveys and experiments. They do not work as well, however, for some data gathered from library research. Much of the data from library research is in the form of notes. Designers take notes while reading printed materials. This data is organized and analyzed differently. One method of analysis is to organize all notes according to the topic of research. The designers rewrite the data as *bullet points.* A bullet point is a single piece of information, like one item from a list. The designers then circle the data they feel is important to their designs. This process has created information from data. The notes were data. Once they were organized and the important parts were circled, they became information.

Make a Conclusion

The last step of research is to make a conclusion. *Conclusions* are the findings of the research. The conclusion can only be made after data has been gathered and analyzed. The information coming from the data must also be read and reviewed. The accuracy of the information obtained must be interpreted and evaluated. It is important to decide if the information is useful before drawing any conclusions.

Developing specific standards for what is useful is important in making decisions about the information's usefulness. Sometimes determining accuracy is simple—taking information from physical measuring devices, for example. At other times, accuracy is harder to determine, as when assessments are based on public opinion, which can vary significantly from group to group and from time to time.

The designers make the conclusions after they have all of the information they need. They make a list of their conclusions. If designers are researching a new design to staple paper together, their findings may be something like this:

➤ The average manual stapler costs between two and fifteen dollars.

➤ The average size of a stapler is eight inches long and two inches wide.

➤ Staplers can be either manual or automatic.

➤ Most people like black staplers.

➤ Most stores carry three to five different brands of staplers.

➤ Staplers, on average, can staple fifteen sheets of paper.

➤ The typical stapler punches the staple through the paper and folds the staple over.

These findings are the results of research. Library, survey, and experimental research would have been used to develop these conclusions. The size and cost of the stapler could be generated through library research. Survey research could discover the facts that people like black staplers and that most stores carry three to five brands. The number of pages a stapler can staple could be gathered through experimental research. See **Figure 12-17.**

TechnoFact: Any time data is collected, there is a chance it may contain a certain amount of error. This is known as the survey's margin of error. The margin of error is always displayed with the findings as a percentage. For example, if the margin of error is ±0.1, it means that if the entire population was surveyed, the findings would be correct 99 percent of the time.

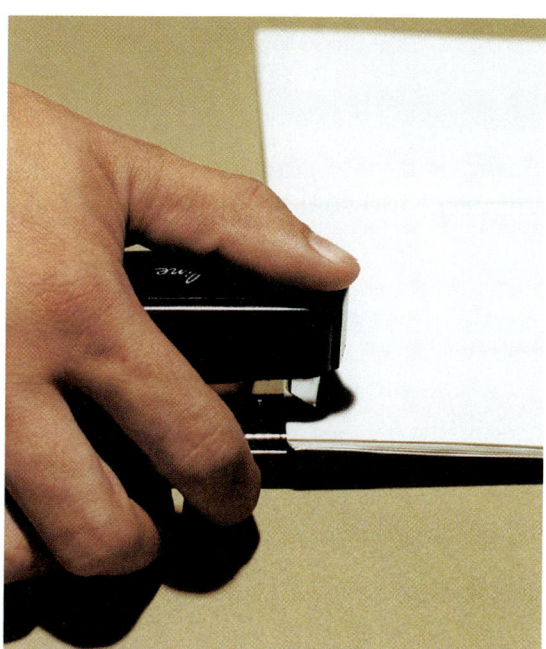

Figure 12-17. Testing available products is part of experimental research.

Summary

Step three of the design process is researching the problem. The rest of the design is based on the conclusions made in this step. There are three types of research designers can use. Library research is used to develop background information. It is also useful in finding how others have solved similar problems. Survey research uses questionnaires to find the reactions of consumers. Experimental research is used to conduct tests and find data through experiments.

The research done in this step is a process. This process resembles the scientific method. It begins by listing the problem identified in the previous step of the design process. The designers then gather raw numbers and facts, or data. They research the who, what, how, and why of the problem. After the data is collected, the designers analyze it. They make graphs and lists of the data. These new lists and graphs are information. The information is then reviewed. The designers report the findings, or conclusions. All of the research done in this step will be used to develop ideas and sketches for a new product or system. At the end of this step, the designers may not be satisfied with the research or the design brief. They can always go back and redo either of them.

Extend
Explain the three different types of research.

Research
Have the students interview a researcher and discuss the process he or she uses to reach conclusions.

Curricular Connections

Language Arts

Learn how to use the school or public library to conduct research. Learn how to use card catalogs and search engines.

Science

Choose a scientific advancement (such as the polio vaccine or sheep cloning). Create a poster board describing the experimental research done to create that advancement.

Mathematics

Choose an article from a newspaper with numerical data. Create a graph related to the information.

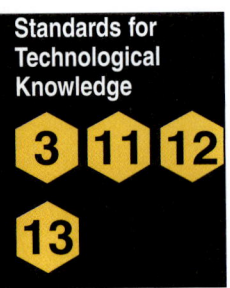

Standards for Technological Knowledge

3 11 12

13

Activities

1. Give examples of how the following companies would use the three types of research to gather data for the products they design.
   A. A national soft drink company.
   B. A small newspaper company.
   C. An automobile manufacturer.
   D. A fast-food company.

2. Gather data related to a problem in your community and create a graph informing the public about the problem or potential solutions.

3. Using the fish feeder design brief in Chapter 11, follow the steps of the research process and report your conclusions to the class.

TechnoFact: Survey researching is a career in which a person designs and conducts surveys. These researchers collect data and information used to make decisions. They may work in a number of different industries, including product design, market research, and economic policy.

Test Your Knowledge

Do not write in this book. Place your answers to this test on a separate sheet of paper.

1. Designers are the only people who conduct research. True or false.
2. _____ research uses printed materials to gather data.
3. _____ research is used to find people's reactions.
4. A hypothesis is a result of research. True or false.
5. Explain the four steps of the research process.
6. Where can you find the problem to be researched?
7. Raw facts and figures are examples of _____.
8. Organized data is called _____.
9. Graphs are used to turn data into information. True or false.
10. The findings of research are known as _____.

Chapter 13
Creating Solutions

Did You Know?

➤ Mechanical drawings use symbols that help explain features of the drawings. These symbols are standard so everyone who looks at a drawing can understand it. Different types of drawings use different sets of symbols.

➤ The symbols used in welding drawings explain how the parts should be welded together. Some common welding symbols include the following:

⊖ Seam

⋎ Weld "V" groove

⌐ Weld all around

Symbols used in engineering drawings include the following:

⌀ Diameter

⊤ Drill through

⌴ Countersink

Some common architectural symbols include the following:

⊓ Door

═ Window

Objectives

The information given in this chapter will help you do the following:

➤ Define *ideation*.

➤ Explain brainstorming, graphic organizers, and questioning.

➤ Describe a rough sketch.

➤ Describe a refined sketch.

➤ Summarize the procedure for creating sketches.

➤ Identify the difference between shading and shadowing.

➤ Create an isometric sketch.

➤ Produce an oblique sketch.

➤ Explain and create a perspective sketch.

Key Words

These words are used in this chapter. Do you know what they mean?

box
brainstorming
cone
cylinder
elements of design
graphic organizer
ideation
isometric sketch
oblique sketch
orthographic drawing
pictorial drawing
pyramid
questioning
refined sketch
rough sketch
shading
shadowing
sketch
sphere
thumbnail sketch

Chapter Outline
Did You Know?
Objectives
Key Words
Exploring Ideas
Ideation
  Brainstorming
  Graphic Organizers
  Questioning
Sketching
  Rough Sketches
  Refined Sketches
  The Sketching Process
  Refining Techniques
    Shading
    Shadowing
  Types of Sketches
    Isometric Sketches
    Oblique Sketches
    Perspective Sketches
Summary
Curricular Connections
Activities
Test Your Knowledge

The design process can be divided into two sections: problem seeking and problem solving. The first two steps of the process are problem seeking. In step one, the problem is identified, and a design brief is created. The problem is then researched in step two. Both of these steps deal with seeking information about the problem. Step three is the first step to involve solving the problem. In this step, the designers begin to think of ways the problem can be solved. They will develop many ideas and draw sketches for each of the ideas. See **Figure 13-1.** The sketches will help to explain their ideas to others. The ideas and sketches developed will then be refined, modeled, and tested in the next steps of the design process.

Exploring Ideas

Some people may come to this step and think they already know what their final solution will be. They might

Figure 13-1. Sketches are ideas on paper.

Figure 13-2. Broad thinking leads to new solutions to problems. (Motorola, Inc.)

feel that developing a number of solutions is a waste of time. This is the wrong attitude to have. Designers must explore a number of ideas. The more ideas explored, the better the design will be. Designers should generate original and creative ideas. They have to think broadly and develop a wide variety of ideas. If designers use narrow thinking, their designs will not be unique. For example, imagine you are designing a telephone. You should explore many different ideas (broad thinking). One idea may be a phone fitting in your ear like a hearing aid. Another idea could be a digital phone with a computer video display. See **Figure 13-2.** These are two unique solutions to the problem of a new telephone. You would not want to create ideas like a red desk phone and a black desk phone. That would be narrow thinking. These ideas are essentially

Figure 13-3. Unique products are the results of creativity and imagination. (Patent No. 5,050,855, U.S. Patent and Trademark Office)

the same thing, with only the color differing. The small changes, like color, can be made later. It is important that designers explore many ideas. Exploring ideas is called *ideation.*

Ideation

Creating a number of new ideas to solve a problem is ideation. The process of ideation is creative and imaginative. See **Figure 13-3.** Creativity leads to new ideas and solutions. The solutions created at this step should be simple and rough. Later in the process, the solutions will be revised and refined. Ideation begins by reviewing the problem and design brief. Remember, the solutions developed must solve the problem. The solution must also fit the criteria and

limitations. After reviewing the design brief, the research conclusions must be reviewed. This review will help the designer understand what others have done and what people want from the solution.

Ideation is a free flowing activity. Ideas must be able to flow without others shooting them down. In ideation, all ideas are good ideas. There are no right or wrong answers. It is important to record (sketch or write down) all ideas. Ideas seeming silly or wild at first may make more sense later. If you do not record them, you may forget some good ideas.

The environment you use to create solutions is very important. You must be able to concentrate on the problem. In front of the television at home may not be the best place to focus on creating ideas. Many designers find quiet places to work. They may leave their offices and work outside or in a library. See **Figure 13-4.** The only

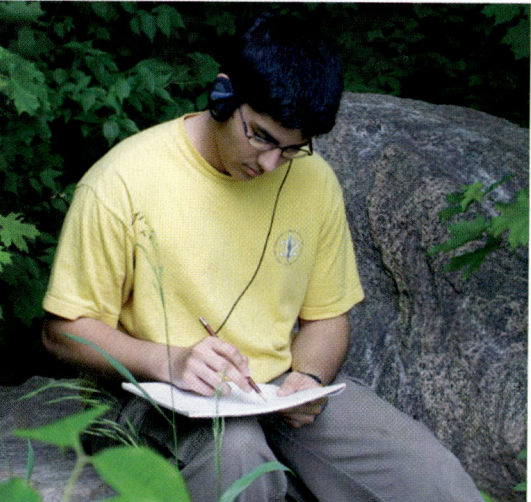

Figure 13-4. Designers often work best in quiet areas away from their offices and desks. (Tony Gothard)

Reinforce
Discuss the importance of reviewing the design brief before creating ideas.

Brainstorm
Ask the students to identify the environments they work best in.

TechnoFact: Creative thinking leads to many new ideas, designs, and inventions. In 1960, Alfred Fielding and Marc Chavannes created a textured wall covering that was not very successful. They went back to the ideation stage and generated ideas for other potential uses. They decided that their textured wall covering would work better as a packing material; we now know their invention as Bubble Wrap® material.

things needed to create ideas are a pencil and paper. Ideation can be done almost anywhere. When creating ideas, designers often flip through magazines or books. They may even listen to inspiring music. Designers often see pictures or hear words, which spark ideas for solutions to their own problems.

There are several different methods of ideation. Some people find one works better for them than others. Three of the main methods are brainstorming, graphic organizers, and questioning.

Brainstorming

Brainstorming is a method used to develop many ideas. It is useful in small groups for coming up with unusual solutions. Brainstorming is a process beginning with the design problem. The process has several steps.

In the first step, the small group must choose a leader and a recorder. The leader will start and end the brainstorming and make sure the group stays on task. The recorder will be in charge of recording all the ideas shared. See **Figure 13-5.**

Next, the leader will set a time limit. Brainstorming is done best when a time limit is set. The group members will stay more focused if they only have a certain amount of time. At this point, the leader will share the design brief with the group. The leader will make sure everyone understands the problem. The ideas generated must solve the problem. The group leader will also share the research conclusions.

It is important that the research is used to come up with ideas.

Next, the members of the small group all focus on the problem. The group leader then asks for ideas. The group members try to list as many creative ideas as possible. Solutions can become very creative when people get ideas from each other. One person may present a new idea. The new idea may spark an idea in your mind. You may have never thought of it without the other person's idea. Brainstorming works well when the group members use each other's ideas to add to their own. In brainstorming, like other forms of ideation, it is important that there is no criticism. If people are criticizing or making fun of ideas, people will not want to share their thoughts. The best thoughts may not be shared if people are afraid to give their ideas.

Throughout the whole session, the recorder makes notes of all the ideas developed. When the time is up, the leader must stop the session. The

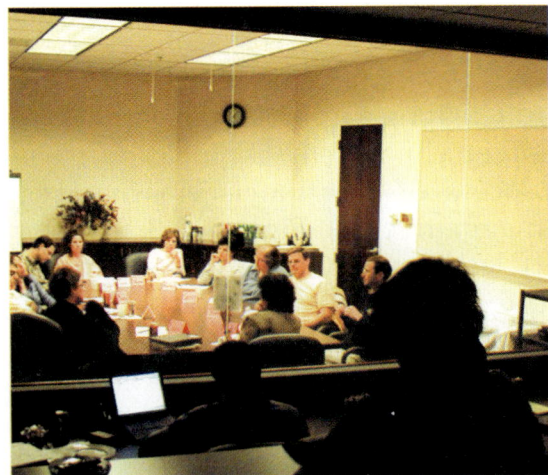

Figure 13-5. A typical brainstorming session can generate many ideas. (Product Development Technologies, PDT)

TechnoFact: In 1941, Alex Osborn created the technique of brainstorming. He was an advertising executive who felt the typical meetings were not allowing the creation of ideas. He developed four rules for his meetings: No criticism of ideas, go for large quantities of ideas, build on each other's ideas, and encourage wild and exaggerated ideas.

Figure 13-6. Graphic organizers help generate ideas.

Standards for Technological Knowledge

Discussion
Ask the students why brainstorming is only used to generate ideas and not to choose a final solution.

Demonstrate
Select a product and show the students how to create a graphic organizer of its function or parts.

leader may ask all of the group members for one final idea. Once the group is done, the brainstorming is over. The notes will be reviewed later in the design process. The main purpose of brainstorming is to develop a list of ideas. It is not the purpose to choose the final solution.

Graphic Organizers

A *graphic organizer* is another method of developing ideas. It is a diagram that helps to organize thoughts. One person or a group of people can create a graphic organizer.

To create a graphic organizer, designers begin with a blank sheet of paper. They write the problem in the middle of the sheet and circle it. Then they create branches from the circle. The branches are the major features or functions of the design. If the design problem is to design a new camera, the word camera goes in the middle circle. See **Figure 13-6.** The branches are features, such as type of camera, loading the film or media, focusing the picture, taking the picture, and changing the settings. Below the branches, the designer lists the possible ways to design the features or functions. Under the *taking the picture* branch, the solutions might be a button on the top of the camera, a voice activated control, an automatic timer, and a handheld switch.

The graphic organizer helps to make sure all the features of the problem are considered. It is also helpful because all the solutions are listed on the same sheet of paper. It is easy to see all the ideas you have developed.

Questioning

Questioning is the third method of ideation. It is done differently than brainstorming and graphic organizing.

TechnoFact: Graphic organizers are used for a number of purposes. They are useful in clarifying the relationships among concepts. They are often used as aids for comprehension and vocabulary acquisition.

Standards for Technological Knowledge

9

Demonstrate
Complete a simple process, such as opening a door or tying a shoe, and show the students how to use the questioning process to record the steps.

Discussion
Ask the students which method of ideation they would most likely use to generate ideas. Have them explain their answers.

Questioning is a process in which the designer asks the question "Why?" It is done while working with existing products. See **Figure 13-7.** The designer asks why things are done the way they are in the existing product. Imagine you are designing a better way to toast bread. If you were using the questioning method, you would locate a toaster and a loaf of bread. You would use the toaster and ask yourself questions like the following:

➤ Why is the bread loaded from the top?

➤ Why do I push a lever to start the toaster?

➤ Why does a knob control the darkness scale?

The key to questioning is that the questions are written. Once the questions are asked, they need to be solved. Designers often find creative solutions to problems because they asked "Why?"

Brainstorming, graphic organizers, and questioning are all useful methods of creating ideas. These approaches to ideation all result in lists of solutions. Lists, however, are not always the best way to describe solutions. Have you ever tried to describe an idea you had using a list? It can be very difficult to do. Imagine if someone asked you to describe how your bicycle works or what the White House looks like. Would these things be easy to describe? They would probably be

Figure 13-7. Questioning relies on the written answers to the steps involved in the process.

TechnoFact: Several people have developed computer-aided brainstorming applications. These applications are designed for people who lack creativity, but are in need of a creative solution to a problem.

Figure 13-8. Sketches are used to communicate ideas between one person and another. (Product Development Technologies, PDT)

hard to describe without drawing the gears of your bike or the front of the White House.

Sketching

Using a *sketch* can make describing ideas much easier. The methods of ideation can be very helpful in coming up with ideas, but sketches are useful for describing the ideas to other people. See **Figure 13-8**. Sketches are tools that help designers communicate their ideas. They are a way to record the designers' thoughts on paper. As thoughts flash into the heads of designers, the designers sketch them so others can see them. Sketching is usually done in two steps:

➤ The first step is often called *design, preliminary,* or *rough sketching.* These sketches are quick and simple sketches.

➤ The second step is often called *working, final,* or *refined sketching.* These sketches are more detailed and require more time to complete.

Rough Sketches

During the design process, designers always have ideas that "pop" into their heads. This is especially true during the ideation stage. Designers may have finished a brainstorming session and have an idea on the way home or during dinner. For this reason, many designers carry sketchpads and notebooks to record such ideas. When designers forget their sketchbooks, you may see them make sketches on napkins or small pieces of paper. At this point, it is not important what type of paper or writing utensil the designer uses. All that matters is that the designer gets the ideas onto paper. The designers are not concerned with the quality of the sketches. Good designers try to capture as many ideas as they can.

These sketches are called *rough sketches.* See **Figure 13-9.** The word *rough* may seem to suggest that the sketches are hard to understand. This, however, is not the case. Rough just means the sketches are in the early stage of development. They are not completely thought out. Rough sketches are basic ideas showing shapes and outlines. They do not show a great amount of detail.

Rough sketches can be ideas of a complete solution, like an entire bicycle. The bicycle designer would create rough sketches of bicycles of different

Standards for Technological Knowledge

8 **9** **11**

Resource
Transparency 13-1: Sketching

Resource
Transparency 13-2: Develop Solutions

Resource
Transparency 13-3: Rough Sketches

Discussion
Ask the students why sketches make it easier to understand an idea.

Extend
Have the students create sketches of ideas they have for new products.

TechnoFact: Sketching can be traced back to the prehistoric use of natural materials and surfaces, like mud, chalk, and clay.

Standards for Technological Knowledge

8 9 11

Resource
Transparency 13-4: Thumbnails

Tech Terms
Elements of Design—The tools controlling the appearance, or aesthetics, of a design.

Research
Have the students use the Internet to find various sketches developed by designers.

Figure 13-9. Rough sketches are early sketches showing the basic shapes and outlines. (Ford Motor Company)

sizes, shapes, and designs. Size and shape are two *elements of design* developed using rough sketches. The sketches can also be ideas of smaller parts of the whole solution. See **Figure 13-10.** In the bicycle example, the rough sketches could be different ideas for the brakes and gears of the bicycle.

These sketches are also known as *thumbnail sketches* because of their size. Most rough sketches are fairly small, with several rough sketches fitting on a sheet of paper. Good designers are able to create rough sketches fairly quickly. These sketches may take beginners a while to create. The time spent on creating rough sketches

is not what is most important. The importance of rough sketches is that many ideas are generated.

Figure 13-10. Some sketches show solutions of only part of a design. (Design Central, design firm; Artromick International, client)

TechnoFact: The use of sketches can even be seen in artistic design. Many painters use the technique of underdrawing. Underdrawing is the creation of a sketch on the canvas before the artist begins to paint.

Refined Sketches

After ideation, the designer should have a large number of rough sketches. These sketches may contain many good ideas. There may be ideas on how to solve the entire problem or how to solve small pieces of the larger problem. The sketches are simply ideas on paper. They make up a library of solutions. The rough sketches need more work done to them before they become proposed solutions to the problem. From the rough sketches, the designer must select the most promising solutions. This is the first time in the entire design process the ideas may be reviewed. In the brainstorming and rough sketching stages, all ideas were valid. No ideas were criticized or discarded.

The designer now creates new sketches based on the best ideas from the rough sketches. The new sketches are called *refined sketches.* See **Figure 13-11.** A refined sketch may focus on one rough sketch. It may, however, combine parts of several sketches to make one new idea. The sketches combine the ideas ideation created. The purpose of the refined sketch is to narrow down the design ideas. The designer will create several different refined sketches. The best solutions will be chosen in the next step of the design process.

Standards for Technological Knowledge

9 11

Demonstrate
Show the students some examples of refined sketches.

Figure 13-11. Refined sketches are more detailed than rough sketches. (DaimlerChrysler)

TechnoFact: In graphic design, a rough layout is the next step after the thumbnail drawing. A rough layout is a sketch that begins to include the details and sizes that will be found in the final product.

The Sketching Process

Sketches are not always easy to create. See **Figure 13-12.** Sketching is a technique. There is a certain way to sketch. Sketching has several steps:

1. Visualizing the object.
2. Blocking out shapes.
3. Adding an outline.
4. Drawing design features.

Visualizing the Object

The technique of sketching begins with visualizing the object. Visualizing or seeing the object may seem like a simple thing. You may be able to look at an object and sketch it. In rough sketching, however, you may not be able to look at something real. If you are creating a new and innovative product, there is no image at which to look. You may only have the idea in your mind and have to "see" the object in your imagination. Sketching is often thought of as seeing and thinking with a pencil. Good designers can easily see their ideas and use pencils to recreate them. Seeing your

Figure 13-13. Orthographic sketches show each side as a separate view. (Keith Nelson)

ideas takes practice. The more designers sketch, the better they become.

There are two ways designers see objects. They can either see objects in two or three dimensions. When designers see in two dimensions, they see the object in six different views. They see the front, top, left side, right side, back, and bottom. Sketches made in two-dimensional views are called *orthographic sketches.* See **Figure 13-13.** *Orthographic drawings,* which are more precise than sketches, will be discussed more in Chapter 17. People

Figure 13-12. Sketches help develop ideas. (General Motors)

TechnoFact: A three-view orthographic drawing uses two planes of projection. The front view uses the frontal plane of projection and shows the height and width of the object. The top view shows the width and depth using the horizontal plane. The side view shows the height and depth on the profile plane.

who are not trained to read them may find orthographic sketches and drawings hard to understand. At this stage in the design process, it is easier for designers to see ideas in three dimensions. Three dimensions are the easiest because that is how the eye sees things. It is also best because almost all people can understand a drawing in three dimensions. These types of drawings are called *pictorial drawings.* The three major types of pictorial drawings are isometric, oblique, and perspective drawings, all of which will be discussed later in this chapter.

Once the designers can see the idea, they begin by breaking down the object. Breaking down the idea requires the designers to divide their ideas into basic shapes. Some designers are able to see the basic shapes easily. Other designers must work harder to see the shapes. See **Figure 13-14.** All objects, however, can be broken into one or more of these basic shapes:

➤ *Box.* A box can be either a cube or a prism. A cube is a three-dimensional square. All sides of the cube are the same length. A six-sided die is an example of a cube. A prism is very similar to the cube. The difference is that a prism has at least one side that is a different length. A cereal box is a prism.

➤ *Cylinder.* A cylinder is a round shaft. The cylinder has one circle on each end, like a tube. A soda pop can is an example of a cylinder.

➤ *Sphere.* A sphere is a perfectly round object. Baseballs and basketballs are spheres.

➤ *Cone* or *pyramid.* A cone is a shape that is round at one end and comes to a point at the other. An ice cream cone is an example of a cone. A pyramid is like a cone because it comes to a point at one end. The other end of the pyramid, however, is a square. The Egyptian pyramids are examples of this shape.

Blocking out Shapes

After the object or idea has been broken down into pieces, the sketching begins. The sketching starts by blocking out shapes. Blocking out means drawing light lines where the basic shapes will be. These lines will serve as guidelines used to make the sketch. Blocking out is very important for beginners. The more advanced designers are able to block out shapes very quickly. Beginners must take their time to make sure the shapes are correct.

To begin blocking out, the designer draws a vertical line. Next, two lines are drawn at roughly 30°, starting from the bottom endpoint of the vertical line. These three lines are called the *isometric axis.* The isometric axis will form the front corner of the sketches. The shapes can now be blocked into the sketch. Each shape is blocked in a little differently.

Boxes

The box uses the three lines of the isometric axis as its front corner. See **Figure 13-15.** The vertical line of the axis should be drawn to the same height as the object. The line drawn to the left should equal the length, and the

Standards for Technological Knowledge

9 11

Resource
Transparency 13-5: Pictorial Drawings

Brainstorm
Have the students think of objects representing each of the basic shapes listed on this page.

Demonstrate
Show the students how to create the isometric axes.

TechnoFact: Sketching and geometry are closely related. The shapes used in sketching are commonly encountered in the study of geometry.

Shapes in Designs

A Box

B Cylinder

C Sphere

D Cone

Figure 13-14. The basic shapes can be seen in many common objects. A—A file cabinet is a box shape. (Sauder Woodworking) B—The shape of a cylinder can be seen in a can. (The Coca-Cola Company) C—Baseballs are spheres (American Baseball Company) D—Ice cream cones are a cone shape.

line to the right should equal the width of the box. Once the axis is drawn, the designer will draw a line the same length as the vertical line at the other two endpoints. Then the designer will connect the three vertical lines. This forms the front and right side of the object. A line is drawn the same length as the two angled lines from the top and right side corners. This creates the top of the box. In a cube, all the lines drawn would be the exact same size.

TechnoFact: The first aluminum can entered the beverage market in 1958. Today, all beverage cans in the United States are aluminum. Steel is still used in other countries. Over the years, the weight of the cans has decreased; they are 30 percent lighter than they were twenty years ago.

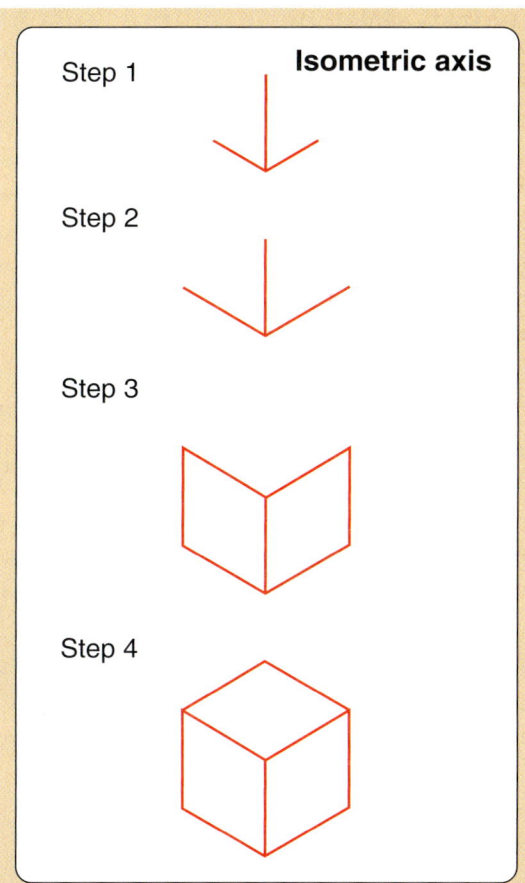

Figure 13-15. Boxes are sketched using these four steps.

Cylinders

The cylinder is drawn using a square at the axis. See **Figure 13-16.** To begin the cylinder, designers use the angled lines of the axis and draw a square. They then draw an ellipse. An ellipse is an isometric circle. The ellipse is drawn through the midpoint of each line of the square. This forms the bottom of the cylinder. To make the top of the cylinder, designers draw vertical lines from the corners of the square. They then create a square positioned at the top of the cylinder. The designers draw an ellipse in the top square.

To finish the cylinder, they draw a horizontal line from one corner of the bottom square to the other corner. Two vertical lines are then drawn where the line crosses the ellipse. These create the sides of the cylinder.

Spheres

Spheres are drawn by first creating a cube. See **Figure 13-17.** The cube is created as described above. Once the cube is drawn, a circle is created inside of it. The circle is drawn so it touches the midpoint of the outside lines of the

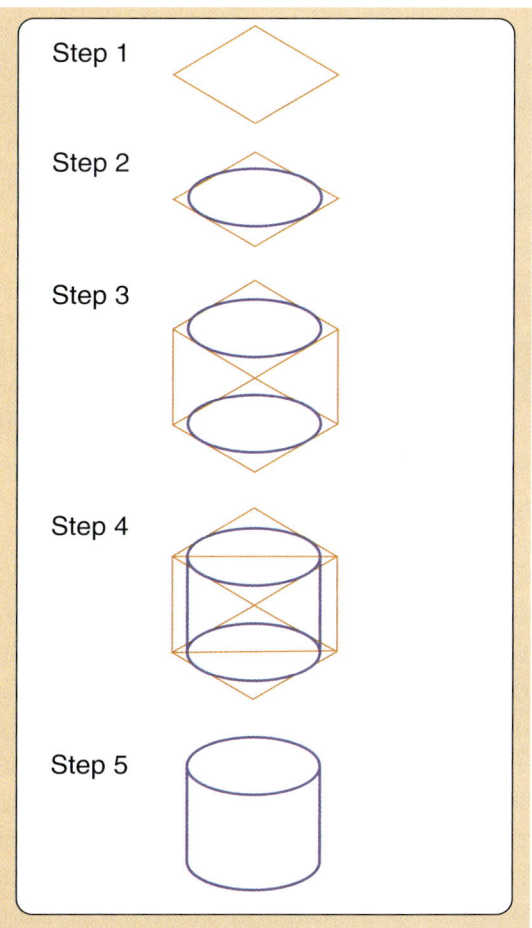

Figure 13-16. Cylinders are sketched using these five steps.

TechnoFact: Cylinders have been used in many inventions. In 1811, Friedrich König and Andreas Bauer revolutionized the printing industry by creating a printing press that used a cylinder. The paper was wrapped around a rotating cylinder as the type moved underneath it.

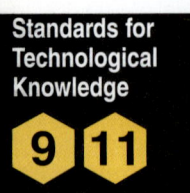

Extend
Explain the difference between a cone and pyramid.

Demonstrate
Show the students how to draw a pyramid and a cone.

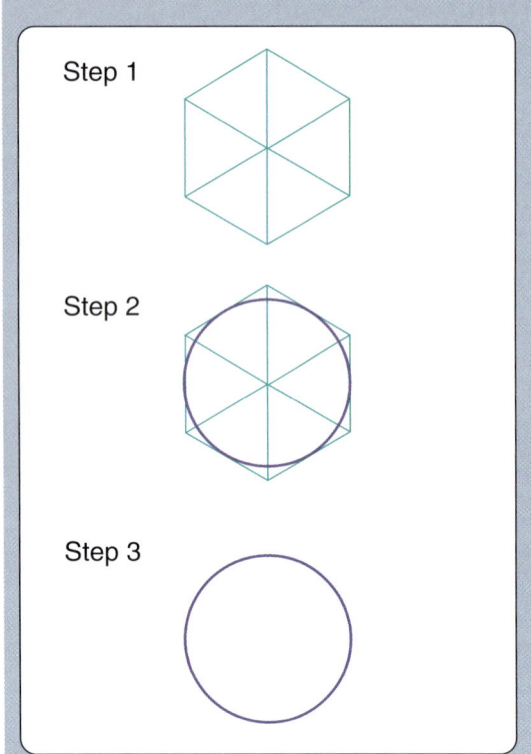

Figure 13-17. Spheres are sketched using these three steps.

cube. When the sphere is drawn it looks like a plain circle. In order for the sphere to look three-dimensional it must be shaded. Shading will be discussed later in this chapter.

Cones and Pyramids

The cone and pyramid are drawn using the rectangular box. The box is drawn as shown in **Figure 13-15.** Next, an *X* is drawn on the top of the box. The center of the *X* is the center of the top of the box. The center point is the point where the top of the cone and pyramid will come to. To draw the pyramid, the designer draws lines from the corners of the bottom square to the center point. See **Figure 13-18.**

To draw the cone, an ellipse must be drawn on the bottom square, just like creating a cylinder. See **Figure 13-19.** Lines are then drawn from the sides of the ellipse to the top center point.

Combinations

Unless the designs are simple products like ice cream cones or basketballs, the sketches created will have more than one basic shape in them. Each of these shapes must be blocked out. You have learned how to block

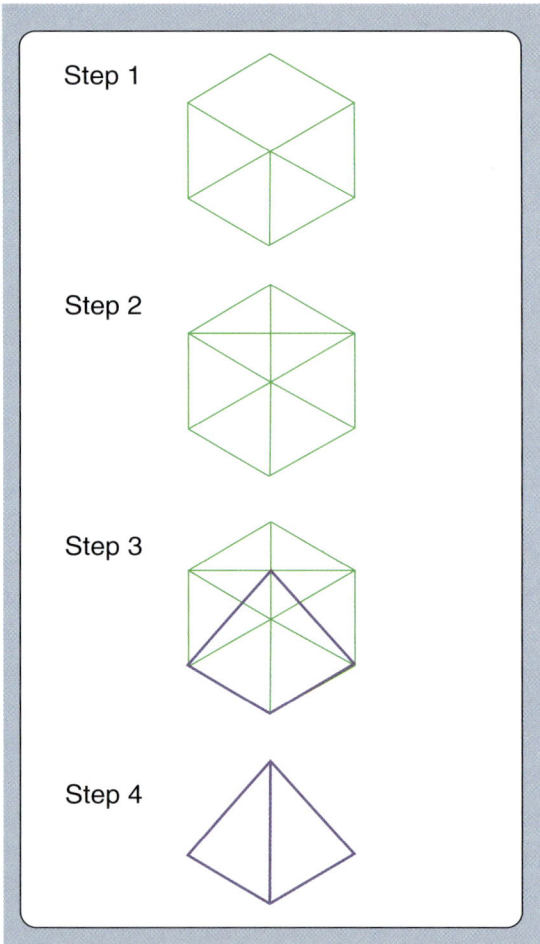

Figure 13-18. Pyramids are sketched using these four steps.

TechnoFact: The most famous examples of the pyramid shape are the pyramids of Egypt. The Great Pyramid of Giza is the largest and most popular. The base covers thirteen acres, and it originally stood 481' high. The pyramid served as a tomb for Khufu.

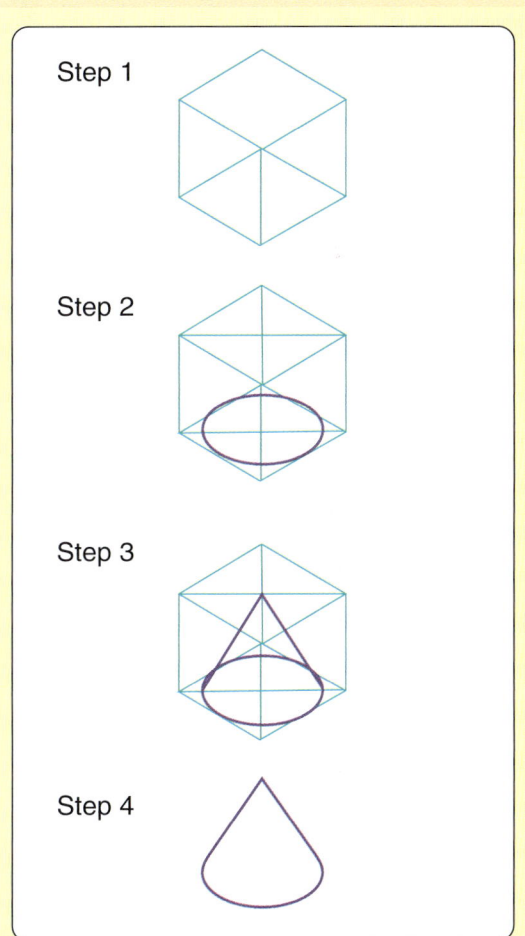

Figure 13-19. Cones are sketched using these four steps.

out shapes using the isometric axis. Sketches with more than one shape will have more than one axis. Imagine designing a guitar. See **Figure 13-20.** You would have several boxes making the shapes of cylinders and one long box representing the neck of the guitar, and you would create the boxes lightly so they could be erased. It may not look much like a guitar at this stage. The look of the guitar will come together in the next stage, adding the outline.

Adding an Outline

The blocked out sketch may not look very much like the designer's idea. It should be, however, the basic shape and proportion of the idea. The designer must add details to the shapes to make the shapes look the way they want them to look. The first detail added is an outline. Creating the outline may include subtracting parts of the shapes. The guitar in the previous example is several cylinders and boxes. To create the outline, a designer would subtract parts of cylinders to make the guitar rounded. See **Figure 13-21.** It is important that the blocked out shapes are drawn lightly because they are guidelines. When the designers add the

Standards for Technological Knowledge

Extend
Have the students block out a product that uses more than one shape.

Extend
Have the students add an outline to the sketch.

Figure 13-20. A guitar would be blocked out using boxes and cylinders.

TechnoFact: Historians speculate the guitar was invented in Spain during the 1300s. The original guitars were smaller and had four groupings of two strings. It was not until around 1800 that the current arrangement of six strings was developed.

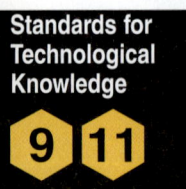

Figure Discussion
Point out to the students that the added external features in Figure 13-22 make the sketch look more realistic.

Extend
Have the students add design features to their sketches.

Figure 13-21. This is an example of the outline of a guitar.

outline, they will draw the lines darker because these lines will not be removed.

Drawing Design Features

The next details to be added are the external features. The external features are objects not included in the outline, but ones that can be seen when looking at the design. The number pad on a telephone is an external feature. These features help in understanding the sketch. The tuning knobs, frets, and sound hole are all external features of a guitar. See **Figure 13-22.** External features also help to set designs apart from each other. A designer may have a number of sketches for a product that have the same outline. The only differences are the external features. For rough sketches, after the external

features are added, the sketches are finished. For refined sketches, additional techniques can be used to add more detail and "life" to the sketches.

Refining Techniques

There are some techniques that can be used on refined sketches to make them appear more realistic. Shading and shadowing are the two most common techniques. Both of these techniques illustrate how the objects would look when placed in light. They add depth and dimension to flat sketches.

Shading

Shading relies on a light source. Most designed objects come in contact with either sunlight or interior lighting. Shading helps to show how the

Figure 13-22. Here is a guitar with the external features.

TechnoFact: In many types of design, the design features are the aspects that set different sketches apart. For example, interior designers may have many sketches of the same room. The design features, like furniture, wall coverings, moldings, and colors may be the only differences in the sketches.

object will look in either of these two light sources. Shading also allows the designer to see the shape and form of the object. To begin shading, the designer must determine from which direction the light source is coming. The direction of the light source will change how the object is shaded. The standard is to imagine the light source is located over the designer's left shoulder.

There are several techniques used in shading. The most common is using different tones to create the shading effects. A tone is a shade of color. Gray is a tone between black and white. In this method of shading, the area of the object closest to the light source will be the lightest tone. The area furthest from the light will be the darkest. This effect can be created with the edge of a pencil or markers. When using a pencil, designers create the darker areas in one of two ways. They may either press harder in the darker areas or go over them several times. When using markers to

Figure 13-24. Drawings can also be shaded dots. (Keith Nelson)

shade, the designer selects markers of different tones to create light and dark areas. See **Figure 13-23.**

Other shading techniques use either dots or lines instead of different tones of color. The dot method uses different amounts of dots to show light and dark areas. See **Figure 13-24.** The more dots an area has, the darker it is. Pens are normally used for dot shading. Line shading is done with different width lines. The darker areas are shaded with thicker lines. Both of these methods work well for shading a rounded object.

Shadowing

Once the object is shaded, the designer adds the shadow. *Shadowing* is normally done using a pencil. The shadow is placed on the opposite side of the light source. If the light is coming from the front left side of the object, the shadow would be placed on the

Figure 13-23. Shading and shadowing help designers see the shape and form of the object. (Design Central, design firm; Ingenico, client)

TechnoFact: Many designers have their own methods of shading and shadowing. There are also a variety of tools designers can use. The tools range from very fine-tipped to very wide-tipped markers and from airbrushes to watercolor brushes. Experienced designers use a combination of tools and techniques to create the look they want.

Standards for Technological Knowledge

9 **11**

Demonstrate
Show the students some examples of oblique sketches.

back right side. The shadow should follow the rough shape of the object. Shading and shadowing take a lot of practice to perfect. The best way to practice shading and shadowing is to look at objects around the room and sketch them. Examine the objects to see how the light hits them. Then try to add shading and shadows to the sketches.

Types of Sketches

There are several ways sketches can be drawn. The three most common types of sketches are isometric, oblique, and perspective sketches. Any of these types of sketches can be used to create rough or refined sketches. All of these types are drawn using the same basic process.

Isometric Sketches

The most popular type of pictorial sketch is an *isometric sketch.* Isometric drawings show the front, top, and sides of an object, just as the eye sees them. Isometric sketches can be drawn with great detail and accuracy, or they can be drawn in rough form. See **Figure 13-25.** Isometric sketches use the isometric axis described earlier in this chapter. The sketches are created using different shapes. Experienced designers are able to create isometric sketches quickly and easily.

Oblique Sketches

Oblique sketches are similar to isometric sketches. See **Figure 13-26.** In isometric sketches, the front and side are both at 30° angles. In oblique

Figure 13-25. Isometric sketches are angled to show three sides of the object. (Design Central, design firm; Artromick International, client).

sketches, only the side view is at an angle. The angle is usually 45°. It can, however, be any angle.

These sketches show the front view in its true shape. Isometric sketches tend to distort shapes like circles and arcs. In an isometric sketch, all circles are shown as ellipses. In oblique sketches, circles in the front view are actual circles. For this reason, circular objects are better drawn as oblique sketches. Also, products having one surface that is the most important are drawn in an oblique sketch. For example, the front views of stoves, radios, and televisions are what most people see. Therefore, oblique

TechnoFact: Isometric sketches are the most common form of axonometric projections. The other types of axonometric drawings are dimetric and trimetric. The differences between the three types are the angle at which they are projected.

Figure 13-26. A—Oblique drawings show the front view directly and the side at a 45° angle. B—Isometric drawings are drawn with both the front and side views at 30° angles.

sketches show these designs better than isometric sketches do.

Oblique sketches are produced using the same steps as isometric sketches. See **Figure 13-27.** The designer begins by drawing the axis. In oblique drawings, the designer creates the oblique axis. The oblique axis has one vertical line, one horizontal line, and one 45° line. There are two types of oblique axes, right and left oblique. In the right oblique, the 45° line is drawn back and to the right. In the left oblique, the line is drawn back and to the left.

The designer uses the oblique axis to block out the shapes. The same basic shapes are used in oblique sketches. The box, cone, pyramid, and sphere are all created the same way as in the isometric sketches. The cylinder is the only shape created differently. The cylinder is created by first drawing a box. A circle is then drawn in the front square, instead of an ellipse.

Once the shapes are blocked out, the designer adds the details. The designer again adds the outline and external features. The last details to be added to refined sketches are shading and shadowing. These details help to make the sketch appear more real.

Perspective Sketches

The third type of pictorial sketch is a *perspective sketch.* Perspective sketches are often used in making refined sketches. They are used to show how objects look to the eye. Imagine yourself standing in the middle of a road. As you look down the road, the buildings get smaller. It looks as though all the buildings and even the road narrow down to one point. This is what perspective drawings look like.

Figure Discussion
Point out the differences between the isometric and oblique sketches in Figure 13-26.

Extend
Have the students create an oblique sketch.

TechnoFact: Oblique sketches are typically created at an angle of 30° or 45°, showing the top and right side of the object. However, they can be created at any angle. If the bottom and left side have more detail, the drawing can be created to show those sides.

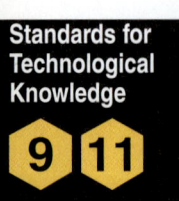
Extend
Explain the difference between the three types of pictorial sketches.

Brainstorm
Have the students list objects that would best be drawn with each of the one-, two-, and three-point perspectives.

Step 1
Oblique axis

Step 2
Block out shape

Step 3
Add outline shape

Step 4
Add external features

Figure 13-27. Oblique drawings are drawn with these four steps.

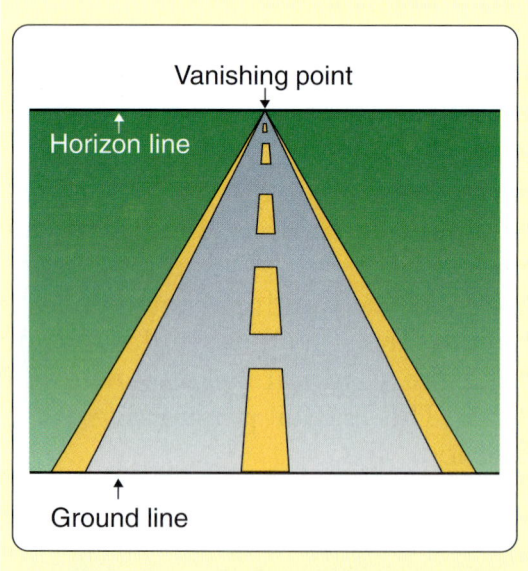

Vanishing point

Horizon line

Ground line

Figure 13-28. Perspective drawings have three basic elements.

When you look down the road, the road and the buildings seem to come from a single point. This point is called the *vanishing point*. See **Figure 13-28.** The line where the sky meets the road is the horizon line. Vanishing points are always located on the horizon line. The ground line is the line formed at the front of the sketch along the bottom of the objects.

There are several types of perspective sketches. See **Figure 13-29.** The most common are one-, two-, and three-point perspective sketches. The two-point perspective sketch is one that has many applications and is used regularly. See **Figure 13-30.** To create a two-point perspective sketch, the designer follows these steps:

TechnoFact: Leon Battista Alberti, an Italian artist, was the first to write about using lines to create perspective drawings. He was also the first to create a grid that could be used in perspective drawing.

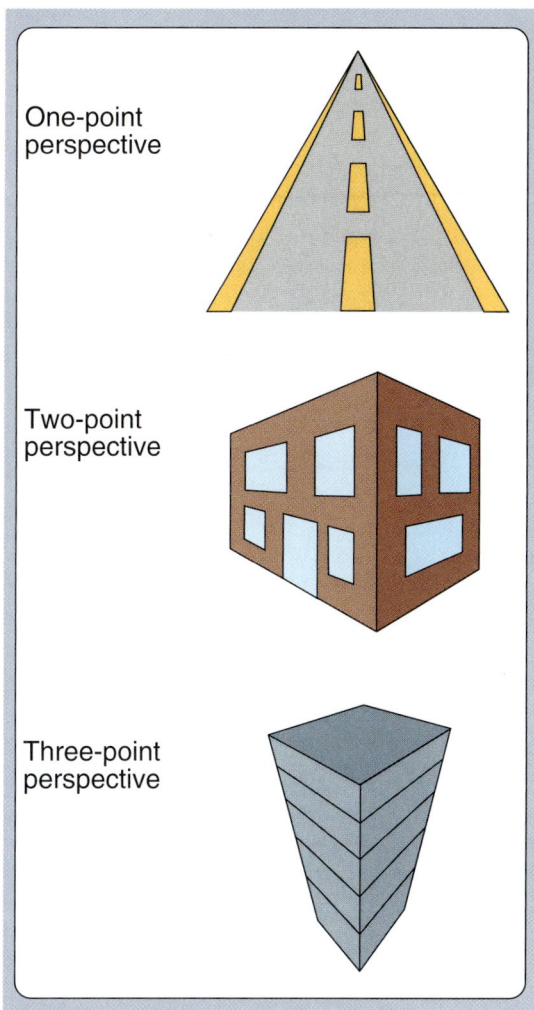

Figure 13-29. There are three main types of perspective drawings.

1. Draw the horizon line and the ground line.
2. Add the vanishing points.
3. Sketch a vertical line from the ground line the height of the object.
4. Sketch perspective lines from the vanishing points to the top and bottom of the vertical line. These perspective lines will form the left and right sides of the sketch.

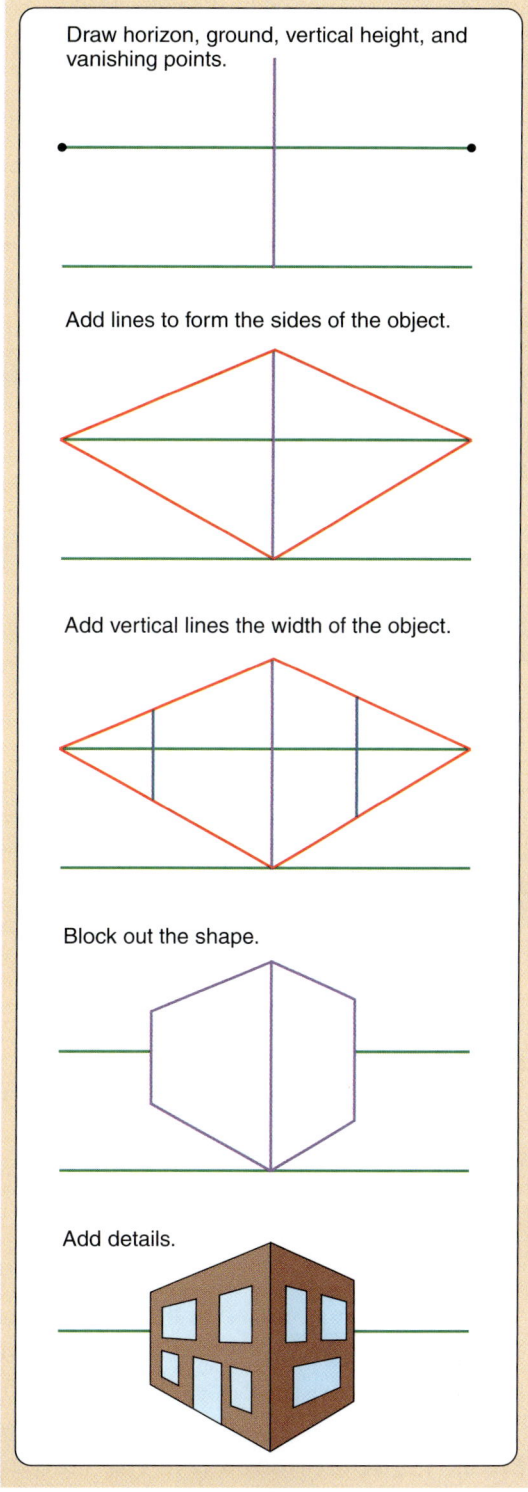

Figure 13-30. Perspective drawings are easy and fun to draw.

TechnoFact: Tall buildings and skyscrapers are often drawn with three-point perspectives. The first skyscraper was built in 1885 on the corner of LaSalle and Adams Streets in Chicago. The building, called the Home Insurance Building, was originally nine stories with a basement. Several years later, two additional stories were added.

Technology Explained

> **elegant solution:** A product meeting a human need in the simplest, most direct way.

Complex products and devices surround us. Many of these have all kinds of frills and little add-ons designers thought were necessary. Many of these add-ons, however, complicate the designs and make the products hard to use.

The opposite of this is the product meeting a need in the simplest and most effective way. Engineers call these *elegant solutions.* These solutions should be the goal of every designer of technological artifacts.

Figure A. The paper clip is an example of an elegant solution.

Think of an adhesive bandage, commonly known as the Band-Aid® bandage. It is designed to hold gauze over a small cut or scratch. Can you think of a better solution? Over the years, no one has thought of one, and therefore, the Band-Aid bandage is an elegant solution.

Another commonplace product is the paper clip, **Figure A.** It temporarily holds sheets of papers together. Everyone uses paper clips, and no one has improved on them. Again, this is an elegant solution.

Think of a number of other devices that work so well that improving them is a challenge. These devices include the zipper, safety pin, stapler, ballpoint pen, pipe cleaner, and thong

Standards for Technological Knowledge

9 **11**

Demonstrate
Show the students how to create one-point and three-point perspective drawings.

Reinforce
Review the different types of ideation processes.

5. Sketch vertical lines the width of the object.

6. Block out the shape.

7. Add details.

8. Shade the drawing.

One-point perspective sketches are created almost the same way as the two-point perspectives are created. The difference is that all lines are either vertical or angled to the vanishing point. In the three-point perspective, all the lines are drawn from one of the three vanishing points. There are no vertical lines in a three-point perspective.

Summary

The process of creating solutions begins with ideation. Designers use brainstorming, graphic organizers, and questioning to develop as many ideas as possible. It is important that all ideas are taken seriously during

TechnoFact: The invention of the ballpoint pen is credited to two brothers. Ladislao Biro developed the ball tip. His brother Georg, a chemist, designed the smudge-proof ink. This elegant solution was designed and produced in 1938.

sandal. Small details may change, such as the types of materials used, but the basic design remains the same. Elegant solutions do not, however, have to be simple products. There are many examples of complex technological products that elegantly solve problems.

Consider the *Apollo* spacecraft, **Figure B**. It carried three men into space and back with relative ease. Its complex systems propelled and navigated the capsule. The spacecraft provided heat, light, and fresh air for the three astronauts inside. The capsule also protected the astronauts from the outside environment.

Elegant solutions are everywhere. Have you ever tried to read and record large volumes of numerical data? It is a time-consuming process and one prone to error. A bar code reader, such as the one shown in **Figure C,** makes this task

Figure B. The *Apollo* spacecraft is an example of an elegant solution. Complex devices can also be elegant solutions.

Figure C. The bar code reader is an elegant solution to a modern problem.

quick and accurate. The unit shown is the size of a hand-held calculator. Despite its small size, the device efficiently reads and processes the information coded in the bars. Bar code readers are widely used to take inventory in retail stores.

ideation. No idea is wrong when the designers are developing ideas.

While the designers are creating ideas, they begin to sketch. Sketching is the process of putting ideas onto paper. The ideas begin as rough sketches. The rough sketches are simple quick sketches that help to communicate the designer's ideas. Designers generate as many rough sketches as possible.

The designers then review all their rough sketches. The ideas coming from the review are then drawn as refined sketches. These sketches are more detailed and better than the rough sketches. The refined sketches are the best ideas developed through ideation and rough sketches.

Isometric, oblique, and perspective sketches are all popular pictorial sketches. They are helpful and allow others to understand the thoughts of the designer. These pictorial drawings are all used in creating solutions. The refined sketches will be taken to the next step of the design process.

Extend
Have the students interview a designer about the ways designers create solutions.

Curricular Connections

Language Arts

Examine the methods of ideation used in creative writing.

Science

Study the difference in mass and volume of different shapes.

Mathematics

Measure several simple objects. Use isometric grid paper to draw an isometric sketch of the different objects.

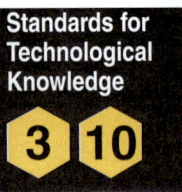

Standards for Technological Knowledge

3 **10**

Activities

1. Use a method of ideation to develop ideas about how the layout of your school could be made better.

2. Find and cut out images from magazines and newspapers showing isometric, oblique, and perspective sketches and drawings. Create a poster board showing the different types of pictorial sketches.

3. Use the design brief and the research gathered in the last two chapters to develop ideas and create sketches of possible solutions.

TechnoFact: The Mercury Project put the first Americans into space. The *Apollo* program placed the first men on the Moon and led to many technological advancements on Earth.

Test Your Knowledge

Do not write in this book. Place your answers to this test on a separate sheet of paper.

1. The process of exploring ideas is called _____.
2. A diagram that helps to organize thoughts is known as a(n) _____.
3. A thumbnail sketch is larger and more detailed than a rough sketch. True or false.
4. A refined sketch uses the ideas from ideation, rough sketches, and the elements and principles of design to create a possible solution. True or false.
5. Give two examples of types of pictorial drawings.
6. _____ is the technique used to show how light hits an object.
7. The process used to show how light hits the space around an object is called _____.
8. Draw an isometric sketch of each of the following:
   A. A box.
   B. A cone.
9. Create an oblique sketch of each of the following:
   A. A rectangular box.
   B. A pyramid.
10. Paraphrase an explanation of a perspective sketch.

Answers to *Test Your Knowledge* Questions
1. ideation
2. graphic organizer
3. False
4. True
5. any two of the following: isometric, oblique, and perspective
6. Shading
7. shadowing
8. Answers will vary.
9. Answers will vary.
10. Perspective sketches show how objects look to the eye.

Chapter 14
Selecting and Refining Solutions

Did You Know?

➤ The study of the measurement of the human body is called *anthropometry*. Human factors engineers use anthropometric information. The information is displayed in charts. The engineers use the information to design products that will fit the most people.

➤ Have you ever wondered how companies know their shirts will fit "all"? The size "one size fits all" was developed using anthropometric information. The shirts are designed to fit the middle 90 percent of people. There are five people in every one hundred who are too small. There are another five out of one hundred who are too big. So the shirts do not really fit all people.

Objectives

The information given in this chapter will help you do the following:

➤ Describe the difference between internal and external selection.

➤ List and explain the five aspects of design.

➤ Define *design sheet*.

➤ Identify the meaning of *evaluation grid*.

➤ Give examples of the six elements of design.

➤ Paraphrase the five principles of design.

➤ Cite the definition of *rendering*.

➤ Summarize the three types of information provided in a detailed sketch.

Key Words

These words are used in this chapter. Do you know what they mean?

aesthetic
appearance
aspect of design
balance
client
color
contrast
design sheet
design style
detailed sketch
dimension
ergonomics
evaluation grid
external selection
finance
form

function
human factors engineering
in-house design team
internal selection
line
outsourcing
production
proportion
rendering
rhythm
shape
texture
unity
value

Chapter Outline
Did You Know?
Objectives
Key Words
Selecting Solutions
  External Selection
  Internal Selection
Aspects of Design
  Appearance
  Function
  Human Factors
  Production
  Finances
Design Sheets
Evaluation Grids
Refining Solutions
  Elements of Design
  Principles of Design
Renderings
Detailed Sketches
Summary
Curricular Connections
Activities
Test Your Knowledge

The fourth step of the design process is selecting and refining solutions. Before designers can select solutions, they must have completed the previous steps. In those steps, the designers identified a problem. They also created a design brief and researched the problem. The designers then made conclusions. They wrote problem statements and created possible solutions. These designers developed and refined a number of ideas and sketches. See **Figure 14-1.**

The designers now have several sketches. The sketches are a collection of the most promising solutions. These are good solutions to the problem. It is important to remember that the designers have gone through each of the first three steps to reach these solutions. Beginning designers often want to skip steps. Completing each task, however, helps to make sure the design is well planned. Expert designers take the first steps very seriously. They know if they spend time creating many solutions at first, the final one will be very good. If designers simply create their first ideas, the solutions might not be successful.

Selecting Solutions

Now it is time for the designers to choose the one solution they feel is the best. This step begins with designers selecting their best idea. This is the second time in the design process a decision must be made. The first decision was made when the designer picked the ideas to refine. This decision is made to select the best solution to the problem. There are two ways the best design can be selected. External or internal processes can be used to select the solution.

External Selection

The first way a solution can be selected is by someone outside of the design process. This is known as *external selection.* Designers often work for *clients,* or customers, who will buy the designs. In external selection, the clients select the best design. An industrial designer may design a lightbulb for an electric company. In this case, the electric company is the client. It may select the design it feels best meets its needs.

Other times, management chooses the best design. Many large companies have *"in-house"* design teams. An *in-house design team* is a group of

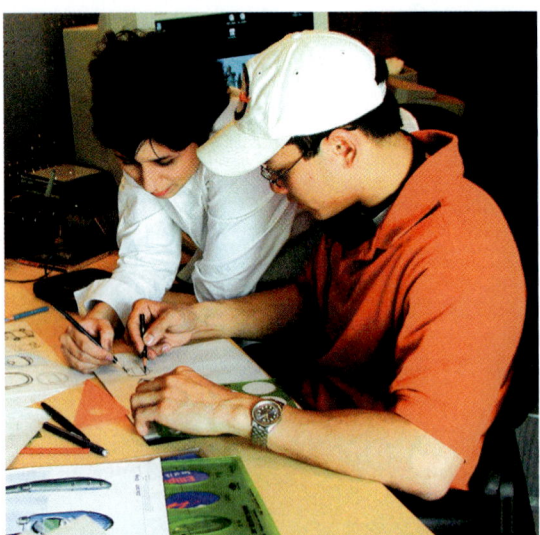

Figure 14-1. In the previous step of the design process, designers created sketches. (Product Development Technologies, PDT)

TechnoFact: Many companies employ human factors engineers not only to design products, but also to review the company's own environment. Human factors engineers may conduct audits of the employees' use of computers. They would ensure that all employees have their computer equipment in the proper places and that their seating is correct. They may also review the layout, lighting, and noise levels in the workplace.

Figure 14-2. Designers must present their ideas to clients and management. (Design Central, design firm; Artromick International, client)

Standards for Technological Knowledge

Extend
Form a review board and have the students examine the ideas of their classmates.

Extend
Explain the difference between external and internal selection.

people working for the company. These designers specialize in designing products for only their company. Many automobile manufacturers have in-house design teams. These design teams often have a review board that approves their designs.

Designers must present their ideas to the client or review board. See **Figure 14-2.** The presentation shows all the work they have done. It shows the design brief, research, rough sketches, and refined sketches. The designers will use the presentation to inform the clients or managers about the features of the refined designs. After the presentation, the clients or managers make the decision.

The clients or managers choose the most promising design. They choose designs based on what they feel is important for their company. The importance of a solution can be placed on several criteria. The amount of money a product may generate, known as profit, is one of these criteria. Another is how well a product satisfies the customers' needs. Lastly, clients and managers want to select products that will help their company become leaders in their field. The customers and managers make decisions as representatives of their companies.

Internal Selection

Internal selection is the second way to select the best solution. The design team does internal selection. A designer, or a team of designers, makes the choice. The people who must choose review all the designs

TechnoFact: The selection of the final solution is a very important design decision. The final solution should be the solution that best fits the criteria. The solution may not be the best looking, fastest, or cheapest to manufacture. However, it may be the best in every other category, making it the best choice in that situation.

TechnoFact: The review board, or planning board in some corporations, not only approves the design, but often also approves funding for the design work. They will choose which design to further develop and how much money you will have in your budget.

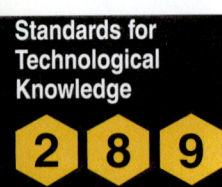

Discussion
Ask the students why it is important that all solutions are evaluated the same way.

Research
Have the students choose a design style and research its characteristics and history.

using a set of criteria. They evaluate the designs and select the solution they feel is the best.

Internal selection has a process that helps to ensure the selection is fair. All the solutions must be evaluated the same exact way. The first step of the evaluation process is to identify the aspects of design.

Aspects of Design

There are many *aspects of design.* The aspects of design are the design criteria. They may be different for every design. A designer creating a soda can will examine different aspects than a designer designing a space shuttle. Aerodynamics, for example, are not important for a soda can. See **Figure 14-3.** Some of the common aspects designers look at are

appearance, function, human factors, production, and *finances.* The designers choose the aspects they will use to evaluate the design.

Appearance

Appearance is the look of the solution. A *design style* often determines the look. Design styles are periods of design in which many of the same appearance features are evident. They can be found in many societies. Queen Anne, Elizabethan, and Victorian are all English design styles. Louis XV and French Regency are design styles from France. American styles include Colonial and Federal styles. Products are often built to match a certain style.

The Art Deco style is an easy design style to recognize. It was popular during the 1920s and 1930s. The Art Deco style uses shapes and straight lines to

Aspects of Design

- Appearance
- Finances
- Production
- Function
- Human factors

Figure 14-3. Aerodynamics is an important aspect of design for a space shuttle. (NASA)

TechnoFact: Furniture is a product that is almost always designed with a specific style in mind. The Shaker style is a popular style in early American furniture. It is influenced by the concepts of order, utility, and durability. Every feature in Shaker furniture serves a purpose.

Figure 14-4. The Art Deco style can be found in many places, including inside buildings. (Photo: Timothy Hursley, Courtesy: Frist Center for the Visual Arts)

Standards for Technological Knowledge

8

Discussion
Have the students discuss the advantages and disadvantages of mechanical, hydraulic, and pneumatic solutions.

Example
Give the students some examples of hydraulic systems, such as car brakes, backhoes, and log splitters. Then, give them some examples of pneumatic systems, such as nail guns and air tools.

produce a certain look. Every kind of product, from toasters to buildings, has been designed in the Art Deco style at some time. See **Figure 14-4.** Popular design styles today are the Modern and Contemporary styles. These styles use materials like metal, wood, and glass to create a clean and up-to-date appearance.

Function

The *function* of the solution is how it works. Solutions can have mechanical, electrical, hydraulic, or pneumatic features. Mechanical solutions use moving parts to function. Gears, pulleys, springs, and other moving objects are used in mechanical solutions. Electrical solutions use the movement of electrons, called *electricity*. Designs using electricity often

have batteries, transistors, lights, and speakers.

Hydraulic and pneumatic solutions are very similar to each other. They both use the power of fluids to operate. Both solutions have pistons, valves, and gauges. The main difference is the type of fluid used. Pneumatic solutions function using compressed air. Hydraulic solutions use liquids like water and oil. See **Figure 14-5.**

Human Factors

Humans use many design solutions. These products must be designed with the human users in mind. A designer would not create a fifty-pound bowling ball. It would be too heavy to use. This is called human factors engineering. *Human factors*

TechnoFact: In order to use pneumatic tools, you must have a source of compressed air. In most instances, that source is an air compressor. An air compressor has a gas or electric engine that operates a piston to pull in air and compress it. The air is then stored in a tank until it is used by the power tool.

Standards for Technological Knowledge

8

Extend
Discuss the importance of ergonomics in design. Encourage class participation.

Research
Have the students research products that have been redesigned to improve ergonomics.

Figure 14-5. This device uses a hydraulic piston to bend exhaust pipes. (WB Automotive)

engineering is used to make sure people can use the solutions.

Human factors engineering is also known as *ergonomics.* There are several types of ergonomic issues. The first issue is the size and shape of the solution. The solution must be the right size for people to use. For example, a telephone must fit in a person's hand. It must also be able to be held up to the user's ear. So, the telephone cannot be too big or heavy.

Comfort is another ergonomic issue. Chairs and seats are designed to be comfortable. See **Figure 14-6.** Many car seats have different settings. These settings help make the seat comfortable. There are adjustments that change the angle and location of the seat. There are also settings that support the user's back.

Body movements are also an ergonomic issue. Humans can only move in certain ways. This is why shirts do not have pockets on the back. People would not be able reach into their pockets very easily if the pockets were on the backs of their shirts. Buildings are also designed with this in mind. It would be

Figure 14-6. Comfort and ergonomics are important design aspects. (Mante chair for ADI [Art Design International] of St. Hubert, Quebec. Designed by Leon Goldik and Jeff Sokalski 2000–2001, Computer Model by Jeff Sokalski, Software: Ashlar Vellum 2000)

TechnoFact: Henry Dreyfuss was a pioneer in industrial design; he did a great deal of human factors research. He wrote *The Measure of Man: Human Factors in Design,* which explored the size, shape, and abilities of the human body.

very hard to get to the second floor if stairs were four feet tall.

Production

Production involves the building of the solution. Designers must think about how the solution will be made. See **Figure 14-7.** There are several considerations going into the production of a design. First, the designer must select the material. Different designs call for different materials. Is stone the best material for a mechanical pencil? No, plastic would probably be a better choice.

Designers must think about tools and machines. They must determine what is needed to build the solution. Designers must also decide if their

Figure 14-7. Designers must consider the tools and machines that will be used to produce the solution. (DaimlerChrysler)

company and clients have the abilities to build the product. Some designers do not have the tools, machines, and abilities to build a certain solution. Design companies, or firms, often specialize only in the design of the products. They pay companies specializing in production to produce the products. This is called *outsourcing.* Outsourcing means design companies send the designs outside of their company to be built.

Finances

Finances involve money, including the money both spent and collected. The money spent is known as expenses. The main expenses are found in design and production. Design costs are the money spent to pay designers and buy design materials. The production cost is the amount of money used to manufacture the solution. This includes paying workers, buying materials, and using tools and machines.

The money collected is the income. Income is gathered through product sales. See **Figure 14-8.** The most important figure in a business is profit. Profit is income left after the expenses are paid. It is the money the company makes.

Design Sheets

Designers will use the aspects of design to select the best solution. The aspects must first be identified in each of the refined sketches. Designers may

TechnoFact: Even companies manufacturing products that have been designed by design firms may outsource some parts. For example, the manufacturer may not have the capabilities to produce all of the parts, so it will have other companies make them. Once the manufacturer receives the parts, it adds them to its production line and assembles the final product.

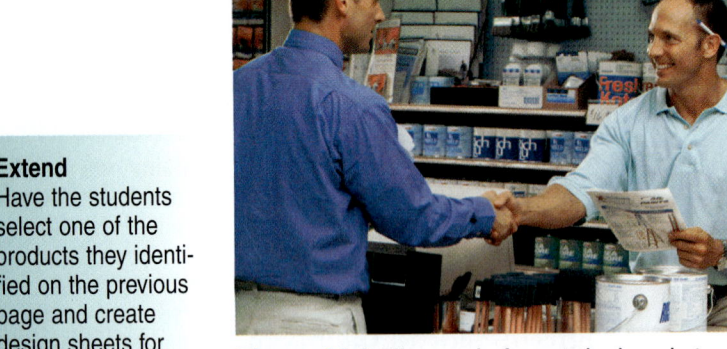

Figure 14-8. The goal of most designs is to make a profit when consumers buy the product.

create a *design sheet* for each of the promising solutions. See **Figure 14-9.** The design sheet shows the refined sketch of the design. It also lists the aspects of design. The designers then write descriptions for each aspect. They create a design sheet for each of the possible solutions. The design sheet acts as a summary of the different solutions. It helps the designers choose the best solution.

Evaluation Grids

Once the design sheets are complete, the designs are evaluated. The designers read the design sheets and begin selecting the best design. It is often hard, however, to pick the best one. Designers can use an *evaluation grid* to help them determine the best solution. See **Figure 14-10.**

An evaluation grid is a chart used to record data. The possible solutions are listed across the top of the grid. The aspects of the design are listed along the side. This forms a grid. The grid is used to make the choice objective. An

objective evaluation means the designs are evaluated fairly. To make sure the evaluation is objective, the designs must be compared to a standard. The standard could be an old product. For example, if you are creating a new ladder, the old ladder could be the standard. Each new ladder design is compared to the existing ladder design. Many designs, however, are brand new. For these design problems, the designers use the design brief as the standard. They compare the new design to the criteria of the design brief.

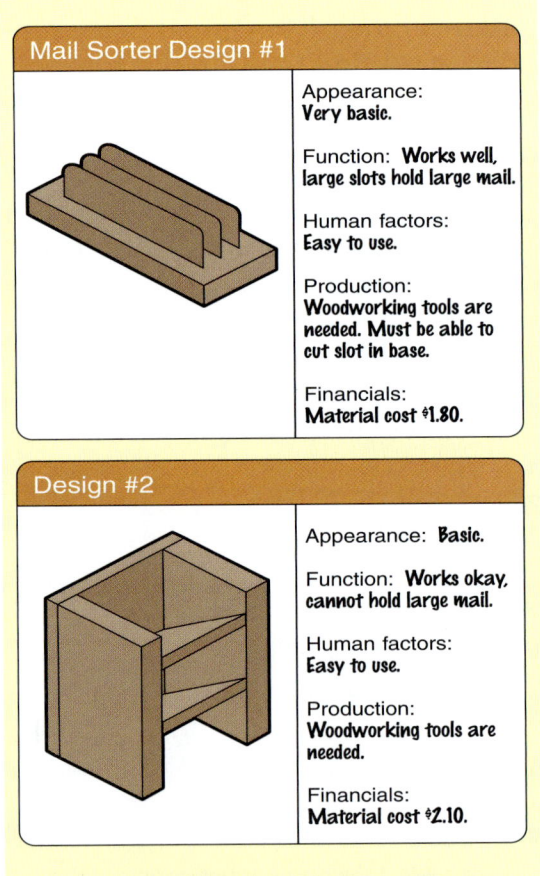

Figure 14-9. Design sheets help designers review the aspects of each design.

TechnoFact: Design sheets are a useful form of reflection. They allow the designer to step back from the design and understand how it performs.

	Design #1	Design #2	Design #3
✓ Above average — Average ✗ Below average			
Appearance	✗	—	✓
Function	✓	✗	✓
Human factors	—	✗	✗
Production	—	✓	—
Financials	✓	—	✗
Design score	1	-1	0

Design score = Number of ✓ − Number of ✗

Figure 14-10. Evaluation grids are used to help select the best solution to the design problem.

The designer fills in the grid by placing symbols in the boxes. The symbols they use could be numbers from one to ten or pluses and minuses. Another set of symbols is the check mark, *X*, and dash. A check mark means the solution is better than the standard. An *X* means the design is below the standard. The design is worse than or does not meet the standard. A dash means the design is even with the standard.

When rating the solutions, the designers may ask themselves questions. The questions could be as follows: Does the design style match the type of product? Does the design look nice? Will it function well? Will it require a lot of maintenance? See **Figure 14-11.** Does the design fit the user? Will it be comfortable? Does the manufacturer have the right tools to build the solution? Would it be too expensive to buy the right machines? Does the manufacturer have the knowledge to work with

these materials? Would this solution cost too much to produce? Would the company make a profit? When the designers have answered these questions, they fill in the grid.

The designers then calculate a score. For example, they might list the number of check marks, dashes, and Xs. Then they subtract the number of Xs from the number of check marks. The result is the design score. The design with the highest score is the best design. If several designs are tied for first place, they can be combined into one design.

Refining Solutions

The solution chosen is never perfect. There are always changes that can be made. Many of the changes are found when the designs are evaluated. The designers may see features in

Figure 14-11. Computers are designed and built to require very little maintenance. (Dell)

Standards for Technological Knowledge

Resource
Reproducible Master 14-2: The Design Process: Refining Selected Ideas or Solutions

Tech Terms
Objective Measurement—A type of measurement that removes bias and prejudice. The judgment is made using a norm.
Subjective Measurement—A type of measurement subject to bias and prejudice. There is not a definitive correct answer.

Extend
Have the students use an evaluation grid to select their best idea from the two designs they have developed.

TechnoFact: Evaluation grids are used in many forms and for many purposes. Ben Franklin suggested an evaluation grid that can be used in many situations. To use Franklin's evaluation grid, a person would divide a sheet of paper in half and write "pro" on one side and "con" on the other. He would then list all the positive and negative aspects about the decision in the appropriate columns. This type of evaluation grid is often called the *Franklin Decision Maker.*

other designs that they like. This is the time when the designers can make those improvements. One area often refined is appearance. Appearance is very important to the sale of the solution. Products that look nice sell better than products that look bad. The appearance of a product can be refined using the elements and principles of design.

Elements of Design

When designers are refining solutions, they begin to examine the look of the design. The design must be functional. It must work properly and solve the problem. The design should also be pleasant to the senses. The appearance of a product is called the *aesthetics.* Aesthetically pleasing products are nice to see. See **Figure 14-12.** Designers try to create aesthetically pleasing designs. To make sure they are on the right track, they use the elements of design. The elements of design are tools controlling how an object looks. The elements are *line, shape, form, value, color,* and *texture.*

Line

Line is the most basic of all the elements. A line in mathematics is the shortest distance between two points. In design, a line is described as a stretched dot. Lines can be used to create motion. A smooth line on a sports car is meant to show the car is fast. See **Figure 14-13.** Lines can also be structural. On a building, lines are often beams and columns holding up the building. Lines can make objects seem bigger or smaller. It is said that vertical stripes on clothing make people look slimmer.

Shape

Shape is the space made by enclosing a line. Shapes have lengths and widths. A shape is a flat, or two-dimensional, element. Shapes can be regular. Street signs are regular shapes

Figure 14-12. Designers use the elements of design to create aesthetically pleasing products. (Design Edge)

TechnoFact: The elements of design can be found in all types of design. Interior designers, product designers, fashion designers, and architectural designers all use the elements of design to create pleasing designs.

Figure 14-13. Sleek lines can give the appearance of a fast vehicle. (General Motors)

that all have different meanings. Shapes can also be irregular. Leaves are irregular shapes.

Form

Forms, like shapes, have lengths and widths, but they also have heights. A form is three-dimensional. Form is also referred to as volume. It is the space an object takes up. The form of some objects is standardized. Standardized means the object can only take up a certain amount of space. A car stereo has a standardized form. All car stereos must fit in the same space in all cars. Designers of car stereos cannot change the form of their product.

Value

Value is the degree of light and dark in the design. It is the amount of white, black, and grays in between. Value can be seen on an achromatic scale. An achromatic scale shows white at one end, black at the other, and the shades of gray in between. See **Figure 14-14.** Value can make things stand out and is helpful when creating a pattern.

Figure 14-14. The achromatic scale shows the shades of gray between white and black.

TechnoFact: Compact discs are an object that must be designed within a standard size. The discs can be no larger than 12 cm in diameter and 1.2 mm thick. If they were any larger, they would not fit into CD players or cases. They can, however, be smaller.

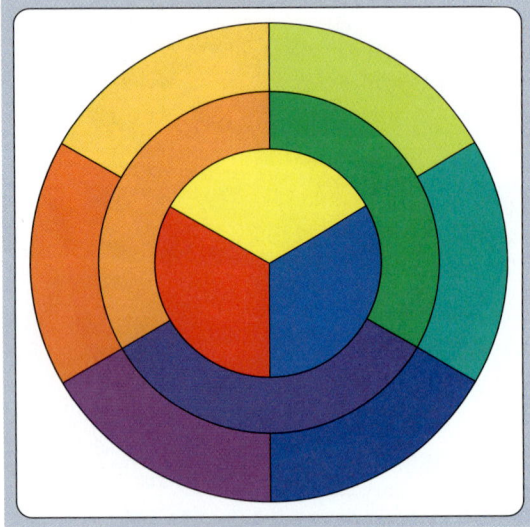

Figure 14-15. The color wheel is based on the three primary colors. All other colors are mixtures of these three.

Color

Color is a property of light. The light reflected off an object determines the color. There are several mixtures of color. When a color is added to white, the color is called a *tint*. When black is added to a color, the result is a *shade*. Black and white themselves are not colors. There are different types of color:

➤ Primary colors. These are the three colors that can be mixed to produce any other color. Red, yellow, and blue are the primary colors.

➤ Secondary colors. These are the colors produced when two primary colors are mixed. The secondary colors are orange, green, and violet.

➤ Intermediate colors. These are mixtures of a primary and a secondary color. Yellow-green, blue-green, blue-violet, red-violet,

red-orange, and yellow-orange are the six intermediate colors.

A color wheel is used to show the primary, secondary, and intermediate colors. See **Figure 14-15.** Colors are used for many different reasons. They can affect our moods. Warm colors, like red, yellow, and orange, can make us excited and happy. Cool colors, like blue, violet, and green, can make us feel comfortable and relaxed. Can you guess why most sports cars are warm colors?

Texture

Texture is the last element of design. It is the surface on an object. See **Figure 14-16.** Texture can be both the appearance and feel of the product. For example, the grit of a piece of sandpaper can both be seen and felt. A certain texture is often needed on

Figure 14-16. All objects have textures. Some objects can be recognized by seeing or touching the textures.

TechnoFact: Complementary colors can appear very vibrant next to each other. Designers often use complementary colors in logos to make the company seem exciting. The Chicago Bears, Los Angeles Lakers, and New York Mets all use complementary colors in their logos.

products. Plates and bowls should have a smooth texture so food can be wiped off easily. Hand tools should be ribbed or have an easy texture to grip.

Principles of Design

The design elements are easier to use when the designer also examines the principles of design. The principles of design are the ways the elements are arranged. The principles control how the elements work together. The principles are *balance, contrast, unity, rhythm,* and *proportion.*

Balance

Balance is the weight of the elements on each side of the design. There are two types of balance: symmetrical

Figure 14-17. This lamp is symmetrically balanced; each side is a mirror image of the other. The red line shows the center of the object. (Fine Art Lamps)

and asymmetrical. Symmetrical, or formal, balance is produced when both sides of the design are the same. See **Figure 14-17.** If you drew an imaginary line through the exact center of a symmetrical object, the two sides would be mirror images. Asymmetrical, or informal, balance is created when the two sides are not the exact same, but yet the two sides have an equal amount of elements.

Contrast

Contrast is used to add variety to a design. It helps bring attention to a certain part of the design. Value and color are often used in contrast. The use of black and white to make something stand out is the use of contrast. It is the contrast of black text on white paper that makes this book easy to read. Two unlike colors can also be used to bring attention to something.

Unity

Unity is the opposite of contrast. It helps the design blend together. The use of either warm or cool colors creates unity. Cool colors on the color wheel range from yellow-green to violet. The warm colors are those on the other half of the wheel. In graphic design, using the same typeface throughout a design can create unity.

Rhythm

Rhythm is the effect of motion. Rhythm is created with repetition of elements. The repetition leads the eye through the design. This movement gives the feeling of motion.

TechnoFact: Formal balance is easy to find in architecture. Most cathedrals and Greek temples were built with formal balance. The Taj Mahal is also an example of formal balance in architecture.

Standards for Technological Knowledge

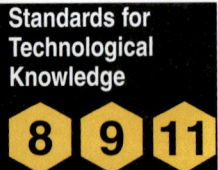

8 9 11

Resource
Transparency 14-2: Rendering

Brainstorm
Select several products and have the students describe how the elements and principles of design are used in those products.

Extend
Have the students create a rendering of the product idea they have developed through the previous activities.

Proportion

Proportion relates to the size and shape of the object. See **Figure 14-18.** If an object is too tall or too wide, it is said to be out of proportion. The other principles of balance, contrast, and unity help to proportion objects.

Renderings

The overall appearance of the solution is best shown in a *rendering*. A rendering is a type of sketch. It is the best representation, on paper, of how the final solution will look. See **Figure 14-19.** Renderings are often shaded with color. A series of renderings showing the solution in different lights is often prepared. This will show how the object will look when it is being used.

Figure 14-18. The size of the buttons and the shape of the card reader are in proportion. (Design Central, design firm; Ingenico, client)

Figure 14-19. Renderings are created in color to show how the final design will look. (Keith Nelson)

Renderings are created by first drawing the solution. The rendering can be drawn as an isometric, oblique, or perspective sketch. The color is added in a similar way as in shading. A light source is selected. The lightest area is closest to the light. As the object gets darker, the color used becomes darker. In a rendering, the area closest to the light may be a soft yellow color. The color furthest from the light could then be a dark red. Renderings are very helpful for visualizing the actual object. They are used in many types of design, especially packaging, fashion, and architecture.

Detailed Sketches

Renderings are an excellent way to show the appearance of a design. They do not, however, show the size of the solution. It is important for the

TechnoFact: The term *rendering* is also used in the field of computer graphics. In 3D computer graphics terms, a rendering is a computer-generated, colored image of the product. The designer is able to place lights on the image, assign materials to the product, and change the angle at which the object is viewed. The software determines how the design appears with those materials, under those lighting conditions. The designer can easily change the elements and re-render the image.

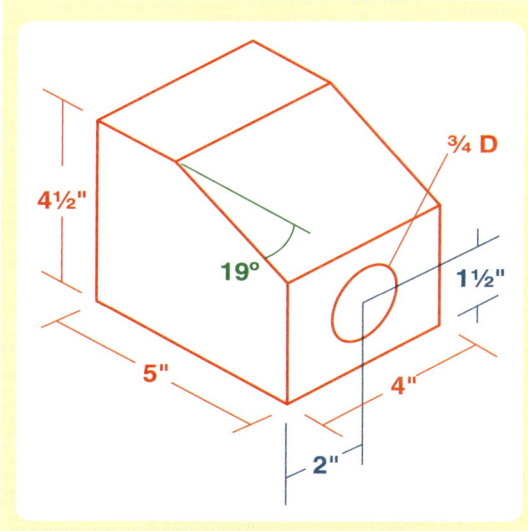

Figure 14-20. Detailed sketches use dimensions to show the size (in red), location (in blue), and shape (in green) of the object.

designers and clients to know how big the object will be. *Detailed sketches* are used to show the size of the solution. Detailed sketches use text to describe the object. The text is called *dimensions*. See **Figure 14-20.** The dimensions show three different types of information:

➤ Size information. Dimensions are used to describe the size of the object. Size dimensions show the length, width, and depth of rectangular objects. They also list the diameter of circles and the radius of arcs.

➤ Location information. This describes the placement of objects. Location dimensions describe where the parts of the object are located. In a floor plan, the location dimensions show the placement of doors and windows.

➤ Shape information. Some dimensions show the shape of the object.

Angle measurements show shape information.

The detailed sketches are often pictorial sketches. The designers first create the sketch. They then add the dimensions. The information in detailed sketches is very helpful. The designers will use the detailed sketches in several of the next steps. These sketches will be used to construct models and prototypes. They will also be used to create drawings of the solution.

Summary

The detailed sketches and renderings are the final outputs of this step. This step involves both selecting and refining a solution. Either the designer or a person outside of the design process makes the selection. Designers use a process to select the best solution. They review the solutions from the last step. These designers identify the aspects of design. They then use a grid to determine the best solution.

Designers improve the appearance of the solution. They use the elements and principles of design. The elements of design are the tools designers use. They help to create products with good aesthetics. When the appearance is finished, the designers create two types of sketches. The first is a rendering showing the solution in full color. The second is a detailed sketch. It has dimensions showing the size, location, and shape of the solution.

The designers will take the work they have done into the next step. The

Standards for Technological Knowledge

Resource
Reproducible Master 14-3: The Design Process: Selecting the Best Solution

Resource
Transparency 14-3: Detailed Sketches

Reinforce
Review shading and shadowing from Chapter 13.

Extend
Have the students create a detailed sketch of the product idea they have developed.

TechnoFact: Detailed sketches are known as rough comprehensives. The comprehensives show the exact dimensions and placement of objects. The typefaces and images in the comprehensive should be as finalized as possible.

Technology Explained

bar code reader: A system used to read numerical codes on packages and tags.

Business and industry strive to be as efficient as possible. This allows companies to produce and sell products at competitive prices. One laborsaving device is the bar code reader, **Figure A.** Most people have seen a bar code reader at the supermarket or discount store checkout. The reader, **Figure B (left),** is connected to a cash register (called a *point of sale terminal*) and one or more computers, **Figure B (right).**

The bar code system uses binary numbers. Binary

Figure A. This bar code reader is reading bar codes on boxes before they are shipped to customers.

numbers represent numbers and letters with a series of ones and zeros. Bar codes use white spaces to represent zeros and black bars to represent ones. To see an example of a bar code, look on the back cover of this book.

A laser beam reads the bars. The beam hits a spinning mirror, which spreads the beam out so it can read the bar

renderings and detailed sketches will help to create models. If the designers feel they do not have a good enough design, they may choose to go back to an earlier step. If they change things at an early stage of design, it may help create a better solution later in the process.

codes. The laser beam strikes the package as it is moved past the reader. The black bars absorb the laser light. White spaces reflect back into the reader and onto a detector.

The detector receives the reflected light and changes the light into pulses of electricity, which are amplified and decoded.

The bar code information is fed to the in-store computer. This computer has the description and price for each product in the store in its memory. Price information is sent to the point of sale terminal, where it is printed on the receipt and shown to the customer on a digital display. The in-store computer keeps track of the sales information it receives from the scanner. A record of this information can be sent to the store's central office. The central office can use this information to track sales and order new inventory.

Figure B. The diagram on the left shows how a bar code reader works. A bar code reader sends a laser beam toward the bar codes and then receives the reflections. The detector turns the reflections into electrical signals. The diagram on the right shows how the signals from the scanner are fed to the in-store computer. The in-store computer can send the information from the scanner to the store's central office.

TechnoFact: There have been several types of barcode systems invented. During the 1940s–1960s, three main types of barcodes were developed: the bull's-eye code, the numerical code, and the Uniform Product Code (UPC). The UPC code was the code adopted by the National Association of Food Chains in April of 1973. This is the standard for barcodes today.

Standards for Technological Knowledge

Curricular Connections

Mathematics

Calculate human factors. Measure the height and arm span of all the students in the class. Calculate the averages. Create graphs showing the data.

Science

Find examples of the elements and principles of design in nature. Create a display of your findings.

Activities

1. Research design styles in different areas of design (such as fashion, architecture, and furniture). Create a poster board showing the different styles.

2. Gather various examples of a product (for example, flashlights, mechanical pencils, or drinking cups). Develop an evaluation grid for the product. Use the grid and determine the best design.

3. Create a rendering and a detailed sketch of a product idea.

Test Your Knowledge

Do not write in this book. Place your answers to this test on a separate sheet of paper.

1. The designers on the design team do external selection. True or false.
2. Name the five aspects of design.
3. Another name for human factors engineering is _____.
4. What is the purpose of a design sheet?
5. A(n) _____ is used to record data when selecting the best solution.
6. The element of design made by enclosing a line is called a(n) _____.
7. Value is the amount of white, black, and gray in a design. True or false.
8. Match the principles of design below with the right descriptions:

   The use of two opposite colors to bring attention to something. _____
   The effect of motion. _____
   The weight of elements on each side of a design. _____
   The blending of a design. _____

   A. Balance.
   B. Contrast.
   C. Unity.
   D. Rhythm.

9. A rendering is a black-and-white drawing. True or false.
10. List the three types of information dimensions can show.

Answers to *Test Your Knowledge* Questions
1. False
2. Appearance, function, human factors, production, and finances.
3. ergonomics
4. To show the refined sketch of the design and list the aspects of design.
5. evaluation grid
6. shape
7. True
8. B. Contrast. D. Rhythm. A. Balance. C. Unity.
9. False
10. Size, location, and shape information.

Chapter 15
Modeling Solutions

Did You Know?

➤ Styrofoam® material is a type of polystyrene foam. Dow Chemicals produces it. This foam was first created in the 1940s and has never been used to create cups or plates. The objects typically called "Styrofoam cups" are another type and brand of foam.

➤ Many animation movies use computer modeling. The models are created as wireframe models, and the frames are colored. It can take hours or even days for a computer to turn the wireframe models into surface models.

➤ Small figurines and dolls have been found dating back to 13,000 B.C. The dolls even had clothing.

Objectives

The information given in this chapter will help you do the following:

➤ Define model.

➤ List and describe the three purposes of a model.

➤ Explain the differences between graphic, mathematical, computer, and physical models.

➤ Give examples of the uses of different modeling materials.

➤ Identify the types of tools used in modeling.

➤ Describe the process of model making.

Chapter Outline
Did You Know?
Objectives
Key Words
What Is a Model?
Why Are Models Used?
  Communication
  Experimentation
  Organization
Types of Models
  Graphic Models
  Mathematical Models
  Computer Models
  Physical Models
Modeling Materials
  Sheet Materials
  Found Objects
  Sculpting and Molding Materials
  Adhesives
Modeling Tools
Making a Model
Summary
Curricular Connections
Activities
Test Your Knowledge

Key Words

These words are used in this chapter. Do you know what they mean?

acetate	mathematical model
acrylic	mock-up
bristol board	model
chart	mounting tape
chipboard	paraffin wax
clay	physical model
computer model	plastic cement
contact paper	plywood
corrugated cardboard	polystyrene
double sided tape	polyurethane
foam	prototype
foam core board	schematic
formula	solid model
graph	spray adhesive
graphic model	surface model
hardboard	table
hot-melt glue	veneer
illustration board	wax
machineable wax	white glue
matboard	wireframe model
	wood glue

Standards for Technological Knowledge

9

Resource
Transparency 15-1: Models

Resource
Transparency 15-3: Functions of Models

Discussion
Ask the students to discuss the ways in which they have used models.

Brainstorm
Have the students think of as many careers using models as possible.

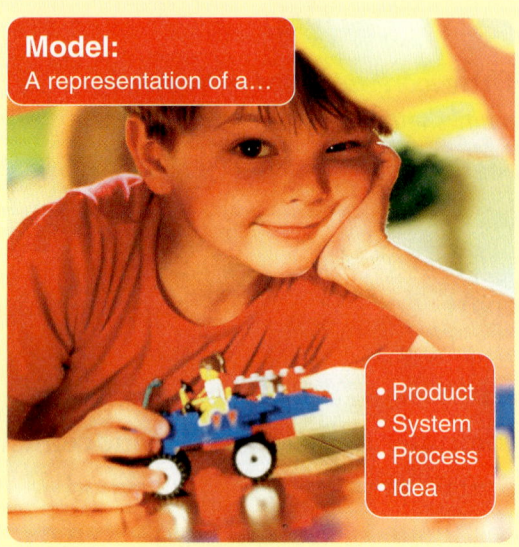

Model:
A representation of a...

• Product
• System
• Process
• Idea

Figure 15-1. People of all ages can use building blocks to build models. (LEGO and the LEGO brick configuration are trademarks of the LEGO Group ©2002 The LEGO Group. *The LEGO® trademarks and products are used with permission. The LEGO Group does not sponsor or endorse the publication.*)

Have you ever built anything with building blocks? See **Figure 15-1.** Have you ever played with an action figure or doll, seen your family tree, or looked at a weather map? If you have done any of these things, you have used a model. Models are an important part of the design process. The fifth step of the process is designing and building a model. The designers use the sketches and renderings from the previous step to construct a model of the solution.

What Is a Model?

A *model* is a representation of a product, system, process, or idea. When you think of a model, you probably think of a small object resembling a bigger, real-life object. A model car or dollhouse may come to mind. These are both good examples of models; however, they are not the only examples of models. There are many types of models used in many different professions. Engineers use models to show products and systems, like roads and bridges. Architects use models to show buildings, monuments, and subdivisions they have designed. See **Figure 15-2.** Amusement parks use models of people made of wax to entertain us. Weather forecasters use models to show weather patterns. Scientists use models to show small objects, such as atoms and DNA.

Why Are Models Used?

There are three main reasons models are used. Models are used to communicate, experiment, and organize. Some models may even be used for more than one of these reasons.

Communication

Models can be used to communicate. They allow people to see solutions in ways that are easy to understand. Just like sketches were used to visualize what the designer was thinking, models help people to see the designer's thoughts. Models used to communicate are often used when the idea or product is hard to understand. Engineers may create a

TechnoFact: Many colleges, universities, and technical schools have programs and classes in modeling and model making. Many of the schools have student clubs affiliated with professional organizations in the field. For example, the Association of Professional Model Makers (APMM) has student chapters in many post-secondary schools. Other organizations, like the International Game Developers Association (IGDA), offer students special pricing for membership and products.

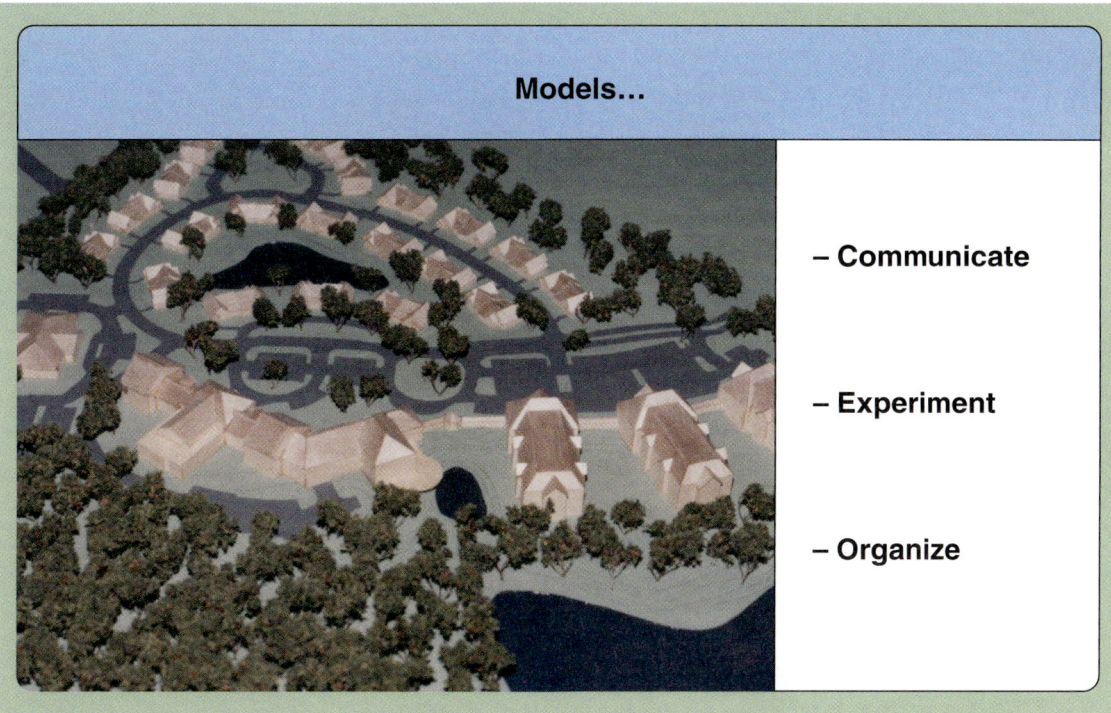

Models...

– Communicate

– Experiment

– Organize

Figure 15-2. City planners and architects use models to show what their designs will actually look like. (Custom Modeling & Graphics Studio, Inc.)

Standards for Technological Knowledge

9

Figure Discussion
Ask the students to explain the purpose of the model in Figure 15-3.

Discussion
Ask the students how models are used in experimentation.

model showing how a product is built. See **Figure 15-3.** The model would communicate better than the raw data they collected. These models may also

Figure 15-3. Models help communicate the design and construction of solutions. (DaimlerChrysler)

be used to get feedback from other people. A designer may make a model of a product to make sure the client approves of the design. Any changes the client wants would be less expensive to make at this stage of the design process.

Experimentation

Models demonstrating how the product will work can be built. These models are used to experiment with the product's operation and use. Models could show that the designers' ideas actually work. They are good ways for designers to test their ideas. For example, a model of a new hinge may be created to determine if the hinge would work correctly. Models may be created

TechnoFact: John Lloyd Wright, the son of Frank Lloyd Wright, invented Lincoln Logs® toys, the building blocks used to build model log cabins. John developed the idea while visiting the construction site of his father's Imperial Hotel. His log set was first sold in 1916.

TechnoFact: Matchbox® toys were first created in 1953. Jack Odell created a prototype of an Aveling Barford Road Roller for his daughter. He put the vehicle in a matchbox so she could take it to school. Matchbox toys today are still small models of actual vehicles.

to test the size and shape of solutions. They also might be created to experiment with different methods of production. See **Figure 15-4.**

Organization

The last use of models is organization. Organizational models are often used when designers are creating new systems. If manufacturing engineers are designing a new manufacturing line, they might create a model. Their model would contain all the tools, machines, and conveyors used in the manufacturing line. They could use the

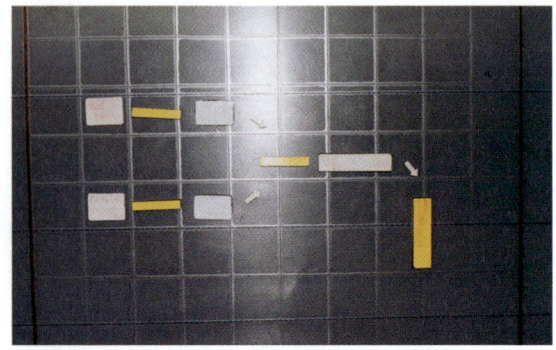

Figure 15-5. This magnetic modeling board helps manufacturing engineers organize production lines.

model to move the objects around to design the plant layout. See **Figure 15-5.** Companies also use models to show the

Below 180°F   **235°F**   **260°F**

Below 180°F   **235°F**   **260°F**

Figure 15-4. These models were produced at different temperatures, as an experiment to determine the best way to manufacture them. (Photo: Rick English, Courtesy: IDEO)

TechnoFact: The model of DNA developed by Francis Crick and James Watson is one of the most recognizable models of a molecule. Physical models are often used to show the structure of chemicals and molecules.

organization of their business structure. A corporate organizational chart is this type of model.

Types of Models

All models are representations of other objects. Some models are visual representations. Models can also be mathematical or even three-dimensional representations. All models fit into one of the four major types: graphic, mathematical, physical, or computer models.

Graphic Models

Graphic models are visual representations on paper. They are used to organize and communicate information. One example of a graphic model, discussed in Chapter 12, is a graph. *Graphs* are visual representations of data. Data is gathered in several steps of the design process. Designers collect data during the research step and gather a great amount of data in the testing step. The data can be modeled in a graph.

A chart is the second type of graphic model. *Charts* can be used to model how a set of information is arranged or to highlight a series of events. They can also show sets of data. An organizational chart is used to show how objects are arranged. A family tree is an example of an organizational chart. It shows how a family is organized back through history. Organizational charts are created like trees. See **Figure 15-6.** The chart begins at the top of the tree, and there are branches and leaves below it. Organizational charts branch out (get larger) as they go down.

A process or flow chart is used to show a series of events. Many restaurants have process charts informing the cooks of the order in which to prepare

Resource
Transparency 15-2:
Model the Solution

Resource
Transparency 15-4:
Types of Models

Resource
Transparency 15-5:
Graphic Models

Reinforce
Review the information about graphs from Chapter 12.

Research
Have the students find a company's organizational or flow chart.

Figure 15-6. Organizational charts are graphic models showing structure and organization.

TechnoFact: Organizational charts are used in many companies and government organizations. The charts are helpful in determining the chain of command.

the food. Manufacturing plants use flow charts to plan the operations in a production line. Flow charts are discussed further in Chapter 23. Charts showing sets of data are known as tables. A *table* is a graphic model made of rows and columns. Tables are often used to compare sets of data. The evaluation grid discussed in Chapter 14 is an example of a table.

The last type of graphic model is a schematic. A *schematic* is used to show a process, like a flow chart. Schematics, however, use pictures and symbols to show the process, rather than words. They are also known as diagrams. One common use of a schematic is to show electrical circuits. See **Figure 15-7**. In an electrical schematic, the electrical components are described with symbols. The electrical wires are shown as solid lines. Electrical schematics show how the electricity flows through a circuit.

Figure 15-8 box contents:

Mathematical Formulas

$$\lambda = cf$$
$$p = mv$$

Chemical Formulas

$$C_8H_{10}N_4O_2$$
$$AgNO_3 + KCl \rightarrow AgCl + KNO_3$$

Figure 15-8. Formulas and equations are common mathematical models.

Mathematical Models

Graphic models are not the only type of model using symbols. *Mathematical models* also use symbols, along with numbers and letters. See **Figure 15-8**. Mathematical models use the symbols to represent values in a formula. *Formulas* are the most common type of mathematical model. Formulas help us to understand complex math, science, and technology concepts. The concept of momentum can be written as a mathematical model: $p = mv$.

In this formula, p stands for momentum. Mass is represented by m, and v is equal to velocity. The formula for momentum uses letters of the alphabet to represent the concepts. Other formulas use symbols to stand for items in the formula. The formula for the length of a wave (such as a light or sound wave) uses the Greek letter lambda, λ. The

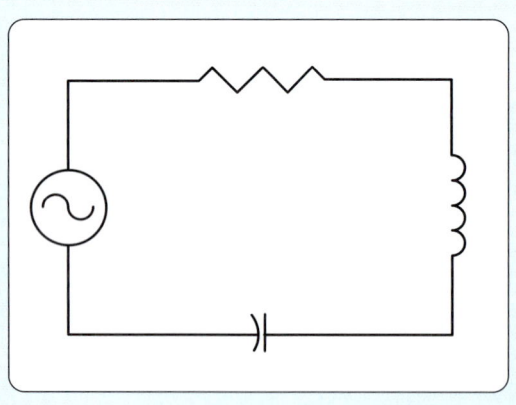

Figure 15-7. Electrical schematics are diagrams showing the layout of electrical systems. (Forrest M. Mimms III)

TechnoFact: The periodic table is one popular graphic model. The periodic table used today closely resembles Dmitry Mendeleyev's period table, published in 1869. He listed the elements by atomic weight and left gaps for elements that had not yet been discovered.

formula is $\lambda = cf$. In the formula, λ is the wavelength, c is the speed of light, and f is the frequency. The formulas for momentum and wavelength are used to make it easier to calculate the concepts.

Concepts are not the only use for mathematical models. Mathematical models can represent materials and compounds. These models are often called *chemical formulas.* Chemical formulas use the symbols from the periodic table to represent compounds.

Figure 15-9. Mathematical models are used in space travel. (NASA)

For example, the chemical formula for caffeine is $C_8H_{10}N_4O_2$.

Two or more chemical formulas can be used together to show a chemical reaction. Chemical reactions can be found in all aspects of life. One chemical reaction in the area of technology occurs in photography. The first step in making film is creating a light-sensitive emulsion. This can be done by combining a soluble silver salt, such as silver nitrate ($AgNO_3$), with a soluble halide, such as potassium chloride (KCl). This results in a light-sensitive halide, silver chloride ($AgCl$), and potassium nitrate (KNO_3). The chemical reaction is modeled as $AgNO_3 + KCl \rightarrow AgCl + KNO_3$. The silver chloride is light-sensitive, and now the film can be used to take photographs.

Chemical and mathematical formulas are often used, along with many other formulas. They can be used to develop large complex mathematical models. These models can be used to determine the aerodynamics of a vehicle, the predictions of an election, and even weather patterns. Weather forecasters rely on mathematical models to determine the weather forecast. NASA researchers use mathematical models to plan the routes rockets and space shuttles take. See **Figure 15-9.**

Computer Models

Computer models are models created using a computer to represent a design, situation, or event. See **Figure 15-10.** The main type of computer model is the three-dimensional

Research
Have the students find a formula they think they could use in a technology class.

Research
Have the students find the chemical formulas for several common chemical reactions.

TechnoFact: Physical models of algebraic and geometric surfaces were first produced in the late eighteenth century. These models were used to show the shape and form of mathematical concepts, like oblique cones and ellipsoids.

Figure 15-10. This computer model shows the function of a push-button latch without building an actual model. (Design Central, design firm; Artromick International, client)

computer model. It is created with the use of a computer-aided design (CAD) system. CAD helps designers create

models that look how the real design will look. There are three different types of three-dimensional computer models. Each one has advantages and disadvantages.

The first is a *wireframe model.* Wireframe models use lines to represent the edges of objects. See **Figure 15-11.** Every edge of the design is shown as a line. There is no covering over the model, so the model is see-through. For this reason, wireframe models are not good for showing the appearance of complex objects.

The next type of three-dimensional computer model is a surface model. *Surface models* are much like wireframe models, except they have coverings. The covering is known as the surface. The color of the surfaces can be used to show how a certain design will look. Surface models can

Figure 15-11. Wireframe models show objects as if the objects were made of thin wire.

TechnoFact: Wireframe technology was first developed in the 1960s. The early wireframe models had to be drawn point by point. Today, there are many functions that automate the process of creating wireframe models.

be shown in different light and angles. Surface models are good for showing the appearance of designs. They cannot, however, be used to show interior details. The insides of surface models are hollow.

The most complete three-dimensional computer model is a solid model. *Solid models* have edges and surfaces like wireframes and surface models. They also have interior features. Solid models are used to represent the insides and outsides of designs. See **Figure 15-12.** Solid models are created using objects like boxes, cones, and spheres. They can be given the properties of different materials and tested to determine the properties of the design. Other computer models are used to predict situations and events, such as the weather and economic activity.

Physical Models

The last type of model is the physical model. *Physical models* are actual three-dimensional replicas of the design. See **Figure 15-13.** Physical

Figure 15-12. Both surface and solid models can be designed with the same colors and textures the actual solution will contain. (Custom Modeling & Graphics Studio, Inc.)

Figure 15-13. Physical models are real, three-dimensional replicas of the design. (Jerry E. Howell)

Resource
Transparency 15-7: Physical Models

Research
Have the students research the most popular 3-D modeling programs.

Discussion
Have the students discuss the advantages and disadvantages of physical models.

models are built so the designers and clients are able to see what the design will look like. They are often the most complex type of model and take the most time to complete.

Different types of designers have many different names for physical models. Architects use the terms *study* and *presentation models.* Engineers may call them *experimental* or *operational models.* There are also *conceptual, plot,* and *scale models.* The two terms best describing all models are *mock-up* and *prototype models.*

Mock-ups

A *mock-up* is a model concerned with the look of the solution. A mock-up model is an appearance model; it looks like the final solution. Mock-ups

TechnoFact: One advantage of a computer solid model is the ability of the computer to analyze the model. Most solid modeling programs can analyze physical properties, like area, density, mass, and volume.

are helpful in assessing the elements and principles of design found in Chapter 14. Elements like color, form, and shape can be examined with a mock-up. See **Figure 15-14.** If designers are creating a new design for a computer monitor, they will create a mock-up of the design. The mock-up will help the designers and design team determine whether or not the design has the right look.

In many cases a mock-up is created at the same size as the final product. This, however, is not always possible. If a product is too big to be created full-size, the model is built on a smaller scale. Architects often build models on a scale of one-quarter inch to one foot. In a model built to this scale, everything measuring one-quarter inch will be one foot when the real building is built.

Architects often create mock-ups of buildings to see how they will look. See **Figure 15-15.** These architects are able to move the mock-ups to different lights to see how the buildings will look at different times of the day. If the mock-up includes other buildings around their design, they can determine if their building will affect any others. They may find that when a light hits their building, it reflects back on the building next to it. The designers use this information to make changes to their design.

Mock-ups only look like the solution. They do not function like the final product. Mock-ups may not even be built out of the same material as the final product. It would be a waste to build a mock-up of an automobile out

Figure 15-14. Several mock-ups are developed to explore different solutions to the same problem. (Photo: Rick English, Courtesy: IDEO)

of steel. Clay, cardboard, and foam are often used for mock-ups because they are inexpensive and can be formed easily.

Prototypes

Physical models that do function like the final product are called *prototypes.* A prototype is the most realistic representation of a design. Prototypes are working models. They are usually full-size and made from the same materials as the final product. Prototypes look, feel, and work just like the final solution will.

Prototypes are used to test the function, operation, and safety of the design. See **Figure 15-16.** To test aerodynamics, prototypes of automobiles and aircraft are placed in wind tunnels. Prototypes of crops are planted in test fields. To make sure flashlights fit comfortably in a user's hand, prototypes are used. Some prototypes are also called *betas.* Beta versions of software are given to computer users to test their operations.

TechnoFact: Model railroads are created in several different scales, denoted by different letters. HO, N, and S scales are among the most popular. In the HO scale, 3.5 mm equals 1′. The S scale is slightly bigger: 4.76 mm equals 1′. The N scale is smaller, with 1.9 mm equaling 1′.

Figure 15-15. This architectural mock-up was developed at a one-quarter inch to one-foot scale. (Laura Trimble)

Figure Discussion
Ask the students why a prototype, rather than a mock-up, is needed in Figure 15-16.

Discussion
Ask the students whether or not it is always practical to build a prototype. Have the students explain their answers.

Figure 15-16. This engine prototype is being tested to make sure it functions correctly. (Boeing)

TechnoFact: The prototype of the X-36 Tailless Fighter was used in a total of thirty-one test flights by NASA. The X-36 prototype is a 28 percent scale model of the designed aircraft. The prototype is 19′ from front to back, has an 11′ wingspan, and weighs about 1,250 pounds.

Discussion
Have the class discuss the difference between a comprehensive and a focus prototype. Have them think of instances when each type might be used.

Brainstorm
Ask the students why time and budget help determine the materials used in a model.

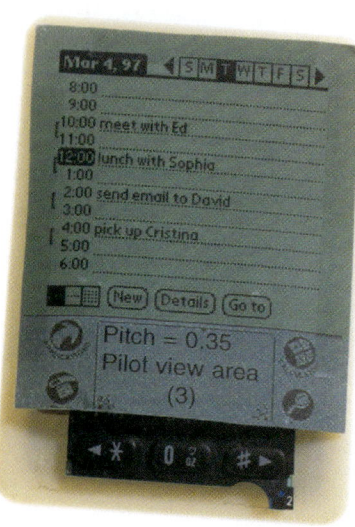

Figure 15-17. These focus prototypes were created to examine different types of buttons. (Photo: Rick English, Courtesy: IDEO)

Prototypes can either be comprehensive or focus prototypes. Comprehensive prototypes examine the entire design. A comprehensive prototype of a computer has all the components in the design (such as the processor, memory, disk drives, and video card). A focus prototype only examines one part of the design. See **Figure 15-17.** A common focus prototype on an automobile is the door latch. A focus prototype of the door latch might be created to test how it functions. If the designers are only concerned about one object, like the door latch, it is unnecessary to build a prototype of the entire design.

It is common for designers to discover unexpected results when they are building prototypes. They may find things they did not think about in the sketches and drawings. The design may be too big or small. It may not work the way they expected it to work. The designers may even find that it is impossible to build the design they wanted to or that they need to create different tools or machines to build their design. This is one of the reasons modeling is so important.

Modeling Materials

Physical models can be made from many different types of materials. The materials used depend on a number of factors. Some of these factors include the use and shape of the

TechnoFact: Museums employ model makers to research and build exhibits. The Smithsonian Institution has designers and model makers at all of their individual museums. They also have a staff at their central exhibits office. The function of their designers and model makers is to create high-quality, compelling exhibits.

model, the budget and time the designer has to work, and the abilities of the designer.

The use of the model will help determine the best materials to use. The model's use includes who will be viewing the model. See **Figure 15-18.** Different materials may be used if the client will view the model from a distance, rather than handle it. The shape of the object also determines the materials to use. A model of a skyscraper would best be constructed with a flat material, like foam board. It would be much more difficult to use a material like clay, which is not flat, because the designer would have to sculpt the sides until they were flat.

The budget of the modeling process also controls the materials used. It is possible that the best material may be too expensive to use in the modeling stage. For example, jewelers often make models of rings out of inexpensive materials. Making models out of gold and platinum would cost too much. The time the designer has to make the model is another important issue in determining the materials to be used. Some materials take more time to work with than others. For example, some types of foam must be coated with a sealer before they can be painted. See **Figure 15-19.** Other materials can be painted on without the sealer. If there is a shortage of time, certain materials may be used to save time.

The ability of the designer is also an important consideration. Some designers are better at working with certain materials. They may not be able to create high quality models with other

Figure 15-18. This model was built for a science fiction movie. The materials used to build it were chosen for their appearance on screen, rather than for their function. (AMS Phoenix/Jim Dore)

materials. These designers may not even have the right tools and machines to work with some materials.

All of these factors go into the decision of the right materials to use. There are hundreds of materials designers use to create models. In this chapter, a few of the most common materials will be discussed. Modeling materials can be broken down into several categories: sheet materials, found objects, sculpting and molding materials, and adhesives.

Sheet Materials

Sheet materials are commonly found as two-dimensional objects. These materials are used because of their flat surfaces. They can be found in most building material and hardware stores in standard sizes. Sheet

TechnoFact: Models used in film and television are often much smaller than they appear. The *Star Wars* films used models of spacecraft ranging from several inches in length to full size. Even the materials used on the models may not be as they appear. For example, shiny tape is often used to give the appearance of metal strips.

Figure 15-19. This foam needs to be sealed before it is painted.

materials each have their own characteristics and are very useful for different types of models. They can be divided into four categories: wood, paper, plastic, and metal.

Wood Sheet Materials

Wood sheet materials are used for larger models or for the bases of smaller models. See **Figure 15-20.** There are three main types of wood sheets:

Figure 15-20. There are many types of wood sheet materials. (Weyerhaeuser Co.)

TechnoFact: Plywood was first used in North America in 1830. The first use was in the piano industry.

Figure 15-21. Plywood is made of thin layers of wood. (Weyerhaeuser Co.)

plywood, hardboard, and veneer. Plywood and hardboard are very similar. They can be purchased from a hardware store in 4′ by 8′ sheets. Hardboard comes in thicknesses of either ⅛″ or ¼″. Plywood ranges from ⁵⁄₁₆″ to ¾″. Hardboard and plywood are both attached to other materials using glue, nails, or screws. Both plywood and hardboard can be cut using any type of hand- or power saw. The difference between the two is in how they are manufactured.

Plywood is a material made by gluing thin sheets of wood together. See **Figure 15-21**. Plywood is not very attractive and is often used as a base for models. It is normally used in places that will not be seen. Plywood does, however, come in different grades. It can be purchased to look nice if it is needed.

Hardboard is made by compressing and rolling wood fibers together. Hardboard does not have a grain like pieces of wood do. It has a solid dark brown color. It is very hard and stiff and can be used in a number of ways in models.

Veneer is a thin sheet of wood. It is less than ⅛″ thick and comes in many different widths. It is available in various types of wood, including oak, cherry, and maple. Veneer is used to give the appearance of a high quality wood. When cost is an issue, inexpensive plywood or hardboard may be used for the model. The veneer is then glued on top. The veneer gives the model a very nice appearance.

Softwoods like pine, balsa, and basswood are often used in modeling. Each of these woods is easy to cut and shape. Balsa and basswood are the easiest with which to work. They are found at hobby stores in many small sizes. These woods are very lightweight and have a nice appearance. Pine is heavier and used for larger models. It can be found in different grades. The best grade is known as appearance grade and can be used in places where it will be seen. Often several types of wood sheet materials are used together in the same model. See **Figure 15-22**.

Paper Sheet Materials

Paper sheet materials are used for many applications in models. They are often used for walls in architectural

TechnoFact: Balsa wood is extremely strong for its weight. In one test, a piece of balsa, which weighs ten pounds per cubic foot, was given a stiffness strength value of 100. When tested in the same manner, oak received a strength rating of 295, and pine received a strength rating of 230. However, the oak and pine were much heavier. When the strength to weight ratio is calculated, the pine is 6.1, the oak is 8.2, and the balsa is 10.

Standards for Technological Knowledge

Discussion
Discuss the different uses of wood versus paper sheet materials.

Demonstrate
Show the students how to properly use a utility knife. Make sure you cover the safety considerations.

Figure 15-22. This model was built from various wood sheet materials. (Custom Modeling & Graphics Studio, Inc.)

models. These materials can also be used as flat panels on industrial models. All of the paper sheet materials can be cut with hobby or utility knives. See **Figure 15-23.** They can be attached to each other with tape or almost any type of glue. There are several different types of paper sheet materials. See **Figure 15-24.** They are different in size, thickness, color, and flexibility. The paper materials used in modeling are thicker than normal notebook paper. Two of the thinnest paper products used are bristol board and chipboard.

Figure 15-23. Paper sheet materials can easily be cut with a utility knife.

Safety
When using a utility knife, always adhere to the following precautions:

➤ Use sharp blades.
➤ Wear eye protection.
➤ Keeps your hands out of the path of the blade.
➤ Cut in a direction away from your body.
➤ Cap the blade of the knife when it is not in use.

TechnoFact: Around 105 A.D., the Chinese invented paper as we know it today, as an alternative to silk. This paper was made from hemp and other plant fibers.

Figure 15-24. Paper sheet materials come in various sizes and thicknesses. Each one has different applications.

Bristol board is similar to common poster board. It is usually found in sheets ¹⁄₁₆˝ thick. This product can easily be curved, but is not very rigid. It is smooth and shiny on both sides. Bristol board can be found in different colors, but is most common in bright white. *Chipboard* is a little different than bristol board. It has a gray color and is produced from two sheets of gray paper pulp. It can be found in thicknesses ranging from ¹⁄₃₂˝ to ¹⁄₈˝. Chipboard is more rigid than bristol board, but is not as attractive. Chipboard is very common; it is often used for backers in calendars, notepads, and packages of paper.

The next two types of paperboard materials are illustration board and matboard. Both of these materials are around ¹⁄₁₆˝. They are both sturdy and come in various colors. *Illustration board* is colored the same throughout the entire board. It comes in handy because designers do not need to color the edges they cut. This material can easily be drawn on to add features to the model. *Matboard,* or museum board, comes not only in many colors, but also in various textures. The color on matboard is only placed on the top of the board. The inside of matboard is always white.

The two thickest paper sheets are corrugated cardboard and foam core board. Each of these materials has two outer sheets attached to an inner core of material. *Corrugated cardboard* has outer sheets of heavy paper. The inner core is made of heavy paper bent into ridges. *Foam core board* has an inner layer of foam covered on each side with paper. Both materials are very sturdy and come in sizes from ¹⁄₈˝ to ¹⁄₂˝. They can be bent if they are scored with a knife first. Corrugated board and chipboard are less expensive than other sheet materials and are used for models in early stages. See **Figure 15-25.** Foam core board is used for final mock-ups and can be used for some prototypes.

Figure 15-25. Early development models are created with chipboard and corrugated board. (Keith Nelson)

Standards for Technological Knowledge

12

Figure Discussion
Have the students identify the paper sheet materials in Figure 15-24.

Extend
Have the students build a model using paper sheet materials.

TechnoFact: Corrugated cardboard is most popular in the packaging industry. There are several different types of corrugated cardboard, including single-faced, single wall, twin wall, and triple wall. Many designers and engineers work in the corrugated cardboard industry and design the corrugated cardboard and boxes for each specific use.

Standards for Technological Knowledge

12

Demonstrate
Show the students how to properly use a band saw. Make sure you discuss the safety considerations.

Example
Collect samples of contact paper to show students the broad range of patterns.

Plastic Sheet Materials

Plastic materials are often used to represent glass. They can also be used to make a model see-through. *Acrylic* is a plastic material that comes in sheets and can be purchased at either a hardware or hobby store. The most common acrylic sheets are clear. They can, however, be found in various colors. Acrylic is usually too thick—⅛″ to ½″—to cut

Figure 15-26. Acrylic is best cut with a band saw.

with a knife. Band saws work well for cutting acrylic. See **Figure 15-26.**

Thinner plastics, like *acetate,* can be used for windows in architectural models. Acetate can also be used for clear screens in product models. Transparency sheets are a common example of sheets of acetate. Acetate and acrylic can both be secured with plastic cement or tape. *Contact paper* is also a plastic sheet material. It is a vinyl material that is sticky on one side. The non-sticky side is printed with a color, pattern, or texture. It is applied onto another material to give the model a certain look. For example, contact paper with a brick pattern may be used to simulate a brick walkway.

Metal Sheet Materials

The last type of sheet material is metal. Metal sheets come in a variety of types. Common types of metal used in models are aluminum, copper, and tin. These metals can be found in

Safety

When you are working with a band saw, always adhere to the following precautions:

➤ **Before cutting your material, check to make sure the blade is tight and tracked between both guides. Set the upper guide to ¼″ above the material. Allow the blade to get up to speed before you begin cutting the material. Plan the cuts you are going to make. If you are cutting a curve, plan to make relief cuts.**

➤ **While you are cutting your material, wear eye protection. Keep your hands at least 2″ from the blade. Place your hands out of the path of the blade. Keep the material flat on the table. Push straight through the blade. Never back the material out while the machine is running.**

➤ **After cutting your material, turn the machine off immediately. Use a scrap piece of material to clear away any scraps from near the blade.**

TechnoFact: The first clear plastic material was created by John Hyatt in 1870. The invention was based on cellulose and called *celluloid.* Celluloid was used in photography and in products ranging from toys to clothing.

Figure 15-27. Soldering is a technique often used in electronics.

Standards for Technological Knowledge

Brainstorm
Give the students some examples of found objects, such as toothpicks and buttons, and have them list the possible modeling uses for each.

Demonstrate
Show the students the process of soldering. Be sure to thoroughly explain the safety considerations.

sheets and are easy to use. They are strong, but also bendable. Thin metals used in models cannot be glued together. Instead, they are soldered together using solder and a soldering iron. See **Figure 15-27.**

Safety
When you are soldering, always adhere to the following precautions:

➤ **Work in a well-ventilated area.**
➤ **Wear eye protection.**
➤ **Avoid touching the material shaft or tip of the soldering iron.**
➤ **Never leave the soldering iron unattended.**
➤ **Use pliers or a vise to hold the workpiece and avoid burns.**
➤ **Wash your hands after you are finished.**

Found Objects

Objects found or manufactured for other purposes can also be very useful in models. Model makers keep a variety of objects around their shops that can be used to represent things on their models. For example, dowel rods can represent columns on an architectural model. Cork or rubber stoppers may be used as round buttons on a model of a telephone. Other useful objects are film canisters, hobby sticks, matches, straws, milk cartons, wire of different sizes, and fishing line. These objects help the designers add details to their models.

Sculpting and Molding Materials

Sculpting and molding materials are much different than sheet materials. They are used to create shapes that are not flat. Sculpting and molding

TechnoFact: Play-Doh® modeling compound was first introduced in 1956. Originally it was only available in an off-white color. Today, there are many colors and activity sets available. Since 1956, over 700 million pounds of Play-Doh modeling compound have been sold.

Figure 15-28. This designer is sculpting a clay model. (DaimlerChrysler)

materials are used when the solution has curves or rounded areas. They are easier to work into irregular shapes. You may have worked with these types of materials as a child, if you made sand castles with mud or played with Play-Doh® modeling compound.

One of the most common types of sculpting materials is clay. See **Figure 15-28.** *Clay* is a natural material found in the world around us. Today, there are several types of clay used to create models. There is polymer clay, which is human-made; water clay; and oil-based clay. Different clays have different uses. Some types are workable at room temperature, while others must be heated. Some clay dries out when it is left in the air. Other clay dries when it is placed in an oven. There is even some clay that never hardens.

Clay can be used for two different reasons. It can be used to create a complete model. This is called *modeling clay.* Models built with modeling clay are solid clay. These models are usually small. Larger models use styling clay. Styling clay is placed on top of a blank. The blank (wood or foam) is used as filler. Large models would be

too heavy if they were solid clay. The styling clay gives the appearance of a solid model, but is only used on the top layer of the model.

Foam is another type of sculpting material. See **Figure 15-29.** It is used in many types of models for several reasons. This material is easy to shape. It is sturdy enough that it can be shipped and stored. Foam is also available in lumberyard and craft stores. There are two main types of foam: polystyrene and polyurethane.

Polystyrene is the less expensive of the two foams. It can be purchased at hardware stores because it is used as a construction material. This foam is used in construction as sheet insulation. Common names for this foam are *bead board* and *blue board* (because of its light blue color). Polystyrene can be cut and shaped with everything from saws to sandpaper. Even a hot wire cutter can be used to cut through this foam by melting it. Hot glue and many types of model glue will, however, melt the foam. Wood glue and

Figure 15-29. Foam is an excellent material for these early study models. (Photo: Rick English, Courtesy: IDEO)

TechnoFact: Eduard Simon was the first to develop polystyrene, in 1839. However, it was too brittle and served no function. In 1937, Robert Dreisbach refined the invention, and it was soon developed into many useful products.

Figure 15-30. These models were made from urethane foam. They were then sanded and painted. (Design Central, design firm; Ingenico, client)

white glue can be used to attach polystyrene pieces together. The pieces can then be painted with latex paint. Spray paint melts the foam.

Polyurethane, commonly known as urethane foam, is used more often for larger or thicker models because it breaks easier if it is thinly shaped. See **Figure 15-30.** It is expensive and not sold in common places, like hardware stores. Urethane foam does have its advantages. It is more rigid. This foam can be cut more easily with larger tools, like table and band saws. Any type of glue or paint can be used to finish the urethane. Urethane can also be purchased in different densities. This is important for prototypes that must function a certain way.

Another widely used sculpting and molding material is wax. *Wax* is a material that can come from several different sources. Waxes can come from animals, vegetables, or minerals. Mineral waxes, like paraffin wax, are used most often in modeling. *Paraffin wax,* made with petroleum, can be carved for small models. It can also be heated and cast into a mold. Molds are often made of plaster or silicone rubber. The *mold* is a reverse of the product. It allows material, like wax, clay, or plaster of paris to be poured into it and dried. When the material is dry, the mold is opened, and the material is in the shape of the product.

Another type of wax, called *machineable wax*, is often used in modeling. See **Figure 15-31.** Machineable wax is a very dense material cut and shaped in milling machines and lathes. It is used for small models or parts of prototypes. This wax is fairly inexpensive and works well in modeling.

TechnoFact: Polyurethane foam is used as insulation in homes. The foam is sprayed out as a liquid and then expands and hardens. In a typical 4″ thick wall, the foam would have an R-value of around 24. Fiberglass insulation in the same wall would have an R-value range of 11–15.

Figure 15-31. Machineable wax comes in different standard sizes.

Adhesives

There are many different types of adhesives designers use when they are making models. Some of these adhesives have already been discussed. Each one has advantages and disadvantages. All adhesives are best used with certain materials and may not work with other materials. Some may even damage certain materials.

One of the safest adhesives is *white glue.* You have probably used white glue for many projects at school and home. White glue is water-based and very easy to use. It is also easy to clean up. White glue sets fast and dries clear. It can be used on paper products, foam, and even some wood. *Wood glue* is normally used, however,

to glue pieces of wood. It is very similar to white glue. This adhesive also dries fast (in twenty to thirty minutes) and is easy to use. One difference is that it is yellow when it dries.

Hot-melt glue is also used in modeling. See **Figure 15-32.** Hot-melt glue dries very fast and is clear. It can be used on most materials, except certain foams. Hot-melt glue comes in the form of glue sticks. It requires a glue gun to heat and apply the glue. A disadvantage of hot-melt glue is that it is not very strong. Even after the glue is dry, the joint that has been glued is flexible. It can also be messy to use. This glue comes out of the glue gun in an uneven flow and leaves thin strings of glue coming off of the model.

Other adhesives commonly used include *spray adhesive* and plastic cement. Both of these adhesives are very sticky and messy. Spray adhesive is a type of glue applied by spraying it out of an aerosol can. The area around the model must be prepared before it is sprayed. The over-spray from the can

Figure 15-32. Hot-melt glue is one type of adhesive used in modeling.

TechnoFact: Epoxy resins, first marketed in 1946, are very strong and typically clear adhesives. Epoxy is sold as two separate parts. One part is the resin and one part is the hardener. Once the two parts are mixed, they form a very strong chemical bond.

will get on the area surrounding the model. It is very hard to clean up. The adhesive should be sprayed onto both pieces before they are connected. *Plastic cement* is also applied to both surfaces being glued together. It is used on plastic pieces. This adhesive is very strong and dries clear and fast. Both spray adhesive and plastic cement can be dangerous. They have very strong odors and should only be used outside or in well-ventilated areas.

Tape is another adhesive used in the construction of models. It should only be used in places hidden from view. Masking, transparent, and duct tape are used on paper and cardboard models. They can be used to tape corners together. *Double sided tape,* or carpet tape, is sticky on both sides. It is used to stick two materials on top of each other. Other types of tapes have special uses. *Mounting tape* is made of a thin piece of foam that is sticky on both sides. It can be used like carpet tape when a gap between the materials is needed. Aluminum tape is an adhesive that is tacky on one side and shiny on the other. It can be used to give the appearance of metal on a model.

Modeling Tools

Designers build models in special areas designed for model making. They have all the supplies, materials, and tools they need in one place. Designers often specialize in one type of modeling (such as cardboard or foam), so they design their modeling shops to fit their needs for the materials with which they work. For example, a designer specializing in building cardboard models does not need metalworking tools and machines.

The tools used most often in paper and cardboard models are basic tools. Model makers use scissors, utility knives, and hobby knives to cut the material. Hobby knives are very useful because the blades can be changed to make different types of cuts. See **Figure 15-33**. The designers use rulers and scales to measure and draw straight lines. Power tools, like a band saw, may be used for making long cuts in cardboard. The band saw may also be used to cut pieces of acrylic for use in the models. Designers using plastics have strip heaters used to bend and shape plastic.

Designers working with foam have utility and hobby knives. They also use hot wire cutters and small power tools to cut the foam. See **Figure 15-34**. To shape the foam, they use small hand tools, like files and rasps. Designers modeling with clay have specialized tools just for working with the clay. Their tools allow them to shape and carve the clay. They even have fine finishing tools to make a smooth surface.

Model makers building wood models often have the most power and machine tools. Some of their machines may be bench-top size (small enough to fit on the top of a table), while other machines are full industrial size. Many modeling shops have table saws, band saws, and thickness planers as standard woodworking machines. They also have hand power tools, like routers, drills, and scroll saws.

TechnoFact: 3M has long been a manufacturer of adhesives. In 1930, a young engineer named Richard Drew solved the problem of a waterproof seal for the food industry. His invention was the first transparent tape. Today, the tape is known as Scotch® Magic™ tape.

Figure 15-33. Hobby knives are very useful and come with many types of blades. (Hunt Corportation)

The size of the model affects the size of the area the designer needs to work. Some model making areas are in the corner of the designer's studio. Other designers have whole buildings dedicated to making models. They may even have teams of people working on different models.

Figure 15-34. Most designers have work areas including all the tools they need to build models. (Product Development Technologies, PDT)

Making a Model

Modeling is a process using tools, materials, and human knowledge to build a model. Models can range from very large, full-size models of automobiles to small models of jewelry. It does not matter how big or small the models are; they all follow the same basic steps. See **Figure 15-35.** The model building process has five basic steps to complete:

1. Determine the type of model and material.
2. Rough out the shape.
3. Smooth the shape.
4. Assemble the model.
5. Finish the model.

The first step of the modeling process is to determine the type of model you want to build. To do this, the designer examines the purpose of the model. If the model is to show what the solution will look like, the designer creates a mock-up. The designer builds a prototype if the model is to test the function of the solution. The type of model will help to determine the materials used to construct it. Paper sheet materials are normally only used for mock-ups. Prototypes, however, are often constructed with plastic.

Once the type and materials are selected, the shape is roughed out. In this step, the basic shape of the model is created. Designers rough out shapes in many different ways. An architect may cut out the walls and floor of the structure. Fashion designers cut out the fabric to be used in the model. Clay

TechnoFact: Hot wire cutters are tools used in making models. The cutters use electricity to heat a thin nichrome (nickel and chrome) wire. The hot wire melts and separates the foam. Most hot wire cutters have a variable transformer that can be turned up or down to change the temperature of the wire.

Standards for Technological Knowledge

8 **9** **11**

Discussion
Discuss the ways CNC and rapid pro-totyping machines have changed the model making process. Encourage class participation.

Research
Have the students find different types of rapid prototyping processes and list the advantages of each.

Figure 15-35. Models go through a number of steps before they are finished products. (Photo: Rick English, Courtesy: IDEO)

modelers might cut a foam blank and lay the clay on all sides. Industrial designers cut foam into the rough shapes using the hot wire cutter. Other designers may use a computer numerical control (CNC) milling machine to cut pieces of machineable wax. See **Figure 15-36.** Laser cutting machines are even becoming popular in model making. They use lasers to cut through material and create nice, clean cuts.

The next step is to smooth and assemble the model. In some models, this step can have two separate actions. The first action is to smooth the model. Designers use finishing tools and sandpaper to create a nice,

smooth model. In some modeling procedures, this step is done in the same action as shaping the material. Rapid prototyping is a process that does the shaping and smoothing in one step. See **Figure 15-37.** It creates a plastic model from a three-dimensional computer model. Fuse deposition modeling (FDM) is one type of rapid prototyping. FDM is a process that builds the model by creating many thin layers of plastic.

Most models have more than one piece. The designer must assemble the pieces to make the whole model. The model is assembled by using adhesives discussed earlier in the chapter.

TechnoFact: Laser cutting machines can be used to separate or engrave a number of different materials. They can cut through materials like fabric, leather, plastic, rubber, cardboard, paper, and wood. The machines can etch text and designs in many materials, including glass, metal, stone, and the materials mentioned previously.

Figure 15-36. Computer numerical control (CNC) machines are used to produce models and parts that were first designed on computers.

Figure 15-37. This rapid prototyping machine allows the designer to create a plastic or wax model very quickly. (Stratasys, Inc.)

The designer uses the adhesives carefully to ensure the model looks nice and fits together well. The assembly of the model could include putting the walls of a house model together. It could also be done by sewing pieces of fabric together for a clothing model.

Once the model is assembled, the finishing touches are placed on the model. Small details, like buttons or knobs, may be added to the model. Designers may draw on many of the details. They may also add model furniture or scale-sized people. Color or decals may be applied to the model. The model could be painted or stained. After the details have been added, the model is complete.

Summary

Models can be found in many places. We have all used models at different times in our lives. We have seen models in stores, on television, and at school. A model can be any representation of a product, system, process, or idea. Models can be used to communicate ideas, experiment with concepts, or organize information. There are four types of models we come in contact with every day.

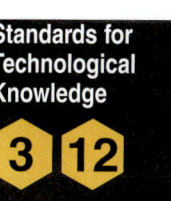

Graphic models are used to make visual representations. Mathematical models help us understand mathematical and scientific concepts, like formulas and chemical reactions. Computer models allow us to construct three-dimensional objects on computers. Physical models are the actual three-dimensional objects representing the final solution.

Physical models can either be mock-ups (appearance models) or prototypes (working models). They can be built from a variety of materials. The use of certain materials depends on the size, shape, and use of the models. Once the materials are determined, the modeling process begins. Models are created by first blocking out the shape and then finely shaping the pieces. The pieces are assembled and finished. The models are then used in a variety of applications.

TechnoFact: The 2003 North American International Auto Show in Detroit featured thirty-eight prototypes of new concept cars. U.S. manufacturers produced seven of the prototypes. The remaining thirty-one prototypes were produced by manufacturers outside the United States.

Curricular Connections

Mathematics

Create a log of the different mathematical models you see and use in one day. Share your log with the rest of the class.

Science

Select a number of different types of adhesives. Conduct an experiment on the properties of the adhesives. Prepare a display comparing their qualities.

Social Studies

Create a model of an ancient civilization. Model the buildings and landscape of the time.

Activities

1. Choose a technological device (such as a battery, radio, or computer). Prepare a display highlighting the uses of graphic, mathematical, computer, and physical models to explain your device.

2. Select a company known for innovative design. Write to the company, requesting information about its modeling techniques and asking for images showing the models it constructs.

3. Use the sketches and drawings you have produced in the previous chapters to construct a model.

Test Your Knowledge

Do not write in this book. Place your answers to this test on a separate sheet of paper.

1. A model is a representation of a _____, _____, _____, or _____.
2. The three uses of a model are to _____, _____, and _____.
3. A _____ _____ is a visual representation on paper.
4. Formulas and chemical reactions are which type of model?
5. Summarize each type of computer model below.
   A. Wireframe model.
   B. Surface model.
   C. Solid model.
6. Two types of _____ _____ are mock-ups and prototypes.
7. A mock-up is a full-size, fully functioning model. True or false.
8. Match each type of material below to its description. Give an example of how each could be used in model making.
   _____ Material made by gluing thin sheets of wood together.
   _____ A single thin sheet of wood.
   _____ A material made with paper on the outside and foam on the inside.
   _____ Also known as museum board.
   _____ A plastic material too thick to cut with a knife.
   _____ Two different types include modeling and styling.
   A. Veneer.
   B. Clay.
   C. Acrylic.
   D. Plywood.
   E. Matboard.
   F. Foam core board.
9. All model makers use the same tools. True or false.
10. Paraphrase the five steps for making a model.

Chapter 16
Testing Solutions

Did You Know?

➤ Sometimes it is too expensive to test the entire solution at once. When large solutions, like aircraft, space vehicles, and buildings, are designed, they are tested in sections. For example, the wing of an aircraft is first tested without the rest of the plane. After the wing and other sections have all passed their tests, they are combined and tested all at once.

Objectives

The information given in this chapter will help you do the following:

➤ Paraphrase the need for testing solutions.

➤ Define *test* and *assessment.*

➤ Describe how testing solutions relates to the design criteria.

➤ Explain the five main characteristics tested.

➤ List and summarize the three testing environments.

➤ Identify the six steps of the testing procedure.

➤ Conduct a test on a solution and assess it.

Key Words

These words are used in this chapter. Do you know what they mean?

appearance

assessment

computer test

consumer safety

cost

durability

field test

function

human engineering

laboratory

market

material property

performance

safety

test

trade-off

variable

Chapter Outline
Did You Know?
Objectives
Key Words
Testing and Assessing
Criteria
Characteristics
  Function
  Human Engineering
  Economics
  Appearance
  Safety
Testing Environments
The Testing Procedure
  Determine the Purpose
  Choose the Test
  Collect Materials and Equipment
  Conduct the Test and Gather Data
  Analyze the Data
  Evaluate the Results
Assessing the Solution
Summary
Curricular Connections
Activities
Test Your Knowledge

At some point in school, you have probably been given a test. See **Figure 16-1.** The test may have been an aptitude test used to show your abilities. It may have been a test of your skills in a game or subject area. The test may have even been a test to assess how well you have learned information. Whichever kind of test it was, it was given for one reason: to measure something about you.

Designers use tests for the very same reason. They want to measure something about their solution. Testing is necessary in design because the solution must solve a problem. Remember, design is not just creating an object. It is creating a solution to a problem. Designers use tests to determine if their solution solves the problem.

Testing and Assessing

In this step, designers perform two different tasks. The first task is to test the solution, and the second is to assess the solution. The two terms, *test* and *assess*, are often used together.

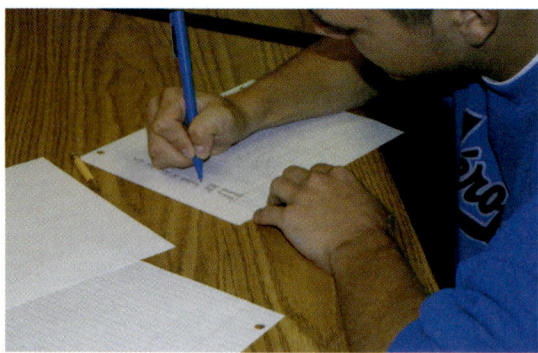

Figure 16-1. A classroom exam is a common example of a test. (Tony Gothard)

Figure 16-2. Scientists conduct tests on chemicals and medicines. (Eli Lilly and Company)

Sometimes they are even used in place of each other. They are, however, two different concepts. A *test* is an experiment or examination. It can be a very scientific experiment on the properties of a solution. See **Figure 16-2.** A test can also be a written document describing how a person feels about a product. Tests are used to generate results.

These results are then used to assess the solution. An *assessment* is a judgment made after reviewing the results of a test. For example, determining that a material is not strong enough for the solution is an assessment. The final assessment of the solution will determine its outcome. It may be found that the solution needs to be redesigned. Testing can lead the designers back to an earlier step in the design process. The designers can always go backward in the design

TechnoFact: The job of testing solutions is conducted by people in several positions. Testing designers or engineers create and determine the tests used to evaluate their product. For example, a software testing engineer would determine the test that best evaluates the effectiveness of the software. Testing administrators and analysts conduct the tests and study the results.

Figure 16-3. Food is evaluated in taste tests.

Standards for Technological Knowledge

9 11

Figure Discussion
Have the students list possible criteria used to design and assess the cookies in Figure 16-3.

Discussion
Discuss the importance of using criteria to assess designs.

process. It is necessary to go back when a solution fails the tests.

Criteria

An automobile design may require aerodynamic testing. A solution for a stereo needs to be tested for sound quality. A new recipe requires taste testing. See **Figure 16-3.** Every design solution requires testing. All solutions, however, do not need the same types of tests. It would be silly to taste test an automobile or test the sound quality of a piece of food. The design criteria determine the tests needed.

The design criteria were set in the early stages of the design process. See **Figure 16-4**. The criteria state how

well the solution should work and who will use it. They also determine what the solution will do and where it will be used. The tests used in this

Design Criteria

The solution will do the following:

✓ Function as a television antenna.

✓ Weigh less than 0.5 lbs.

✓ Be placed on top of a TV.

✓ Look attractive.

✓ Cost less than two dollars to build.

Figure 16-4. The criteria from the first step of the design process should be reviewed to choose the correct tests.

TechnoFact: Tests evaluated by examining the results of all test takers are known as *norm-referenced tests.* IQ tests are an example of a norm-referenced test. Norm-referenced tests are used to compare the results of an individual to the results of a sample group. The scores are typically reported as a percentile, ranging from first to ninety-ninth.

TechnoFact: Tests based on criteria are known as *criterion-referenced tests.* A criterion-referenced test measures the subject against a standard. In criterion-referenced tests, all products being tested could receive the same score.

Standards for Technological Knowledge

8 **11**

Reinforce
Review the aspects of design from Chapter 14.

Research
Have the students develop a list of both destructive and non-destructive tests.

Figure 16-5. NASA astronauts use a prototype of the space station to test maneuvers they will use in space. The model is placed underwater to simulate space. (NASA)

step of the design process must reflect the criteria. They will show whether or not the solution answers the problem and fits the criteria.

If the criteria are written well, they help the designers decide which tests to conduct. If the criteria are not written well, the designers may choose to rewrite the criteria to make sure they are very clear. Since the criteria will help evaluate the solution, any rewriting of the criteria must be done before the testing begins.

Characteristics

The models created during the design process are used for testing. These models give the designers actual objects to test. See **Figure 16-5.** The designers can test the models for a variety of characteristics. Some of the characteristics tested are similar to the aspects of design. These aspects were the principles used to select the best design (discussed in Chapter 14).

Testing is much different at this step. When the best solution was

selected, the designers had only drawings and sketches. The decision was based on how the designers felt the design would function. In this step, the assessment of the solution will be based on data and information showing how the solution actually performs. The main characteristics tested are function, human engineering, economics, appearance, and safety.

Function

Function is a characteristic describing how the solution works. It is tested with two different types of tests. The first type is called a *non-destructive test.* It tests the function of a product without destroying it. The second type is a *destructive test.* Destructive tests are often used to test the limits of a product. A fire test is a common destructive test. See **Figure 16-6.** When testing

Figure 16-6. Researchers are using a destructive fire test on this glass wall. (Underwriter's Laboratory)

TechnoFact: The early Vikings are known to have used destructive tests on various materials. Unfortunately, one material they commonly tested was emeralds. They would strike the emeralds with a hammer to determine their hardness. Today we know a blow from a hammer will destroy any emerald.

Figure 16-7. Sophisticated tools perform tests to ensure the highest quality processor. (Intel)

the function, researchers conduct tests in three categories: performance, durability, and material properties.

Performance is how well the solution works. It is how fast the solution goes, or how much weight it will hold. Performance can also be the quality of a picture or sound. Civil engineers test performance when they build bridges. They make sure the bridges will hold different amounts of weight before they open the bridges to the public. Computer companies test the performance of computer processors before they rate the processors' speed. See **Figure 16-7.**

Durability is how long the solution will work. All solutions with moving parts will eventually wear out. The designers test durability of the solution to determine how long it will last. The durability (life expectancy) of a product can be planned. Sometimes designers plan for the products to wear out after a certain amount of time. For example, a paper plate is designed to be less durable than china. A short life expectancy usually makes products less

TechnoFact: Automobiles undergo a large number of performance tests. Some of the results, like fuel efficiency, are listed on the sales sticker. Other performance tests are conducted and published by automobile periodicals. You can pick up a copy of a car magazine and read the results of performance tests, like braking distance, top speed, and zero-to-sixty times.

expensive. Durability is tested using machines that repeat the same motion over and over.

Material properties are characteristics determining how materials function. Materials have many different properties, as discussed in Chapter 5. Designers test the material properties to make sure the materials they selected will work correctly. There are many different tests used to test material properties. Designers may apply a sudden load to the design solution to test the tensile strength. See **Figure 16-8.** Designers might also test how well the product conducts electricity. Sometimes, designers even test to determine how much sound the solution makes.

Testing must be carefully designed to adequately determine the reliability of the product. Required reliability is defined in the design criteria and describes the product's performance and durability.

Human Engineering

Human engineering examines the way people interact with the solution. The two main areas in human engineering are physical and environmental.

Figure 16-8. Designers can apply a load to a solution to test its strength.

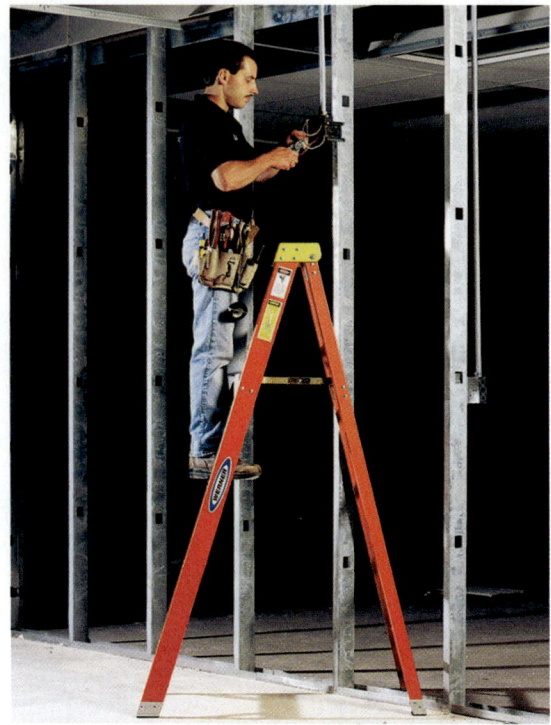

Figure 16-9. Ladders must be designed to fit the human user. (Werner Co.)

When designers test their solutions, they are concerned with both areas.

The physical component deals with how the user handles the solution. The size of both the user and the solution are very important. For example, ladder designers must test their solution to make sure it fits the user. The length of the user's arms and legs are important sizes. The steps of the ladder must be the right distance apart to fit most people. See **Figure 16-9.** The ladder must also be able to be carried to and from the work area.

Human engineering also tests repeated movements. This is done often in the design of systems. Think of a manufacturing line. The engineer designing the line must test all the jobs along the line. It is important that the jobs are safe for the workers.

TechnoFact: Human engineering testing often results in standards and codes to be followed. The codes help designers know the limits and typical abilities of humans, without having to conduct the same tests for every product. For example, designers do not need to test the human ability and range of motion when designing stairs. They can use a building code, like the BOCA National Building Code, which states the maximum height of a step is 8-1/4", and the minimum depth is 9".

Figure 16-10. This chart shows the levels of sound humans can be exposed to for different durations of time. (Data from Occupational Safety and Health Administration Standard No. 1910)

The second concern in human engineering is the environment. The environment is the heat, light, and noise occurring along with the solution. Designers creating a new computer monitor must be concerned with the environment in which it will be used. They test the monitor in different lights. The glare created in some lights may be harmful to the eyes of the user.

There are many charts and books that help designers test for human engineering problems. The books show tables and charts of the types of things human can and cannot do. They also show the levels of light and noise to which people should be exposed. See **Figure 16-10.** These books help designers test their solutions without placing humans in dangerous situations.

Economics

Most designs are created to solve a need and make a profit. Designers test the economics of a solution to make sure they will make money. They test two variables in economics: cost and market.

The *cost* of the solution is divided into two areas: preproduction and production. The preproduction costs include money spent to design the product. By the testing stage, most of the money has been spent on research and the development of sketches, drawings, and models. More money will be spent to create drawings in the next step.

The production costs include money that will be spent manufacturing the solution. They include everything from the price of materials to the wages of the workers. The costs can be calculated by examining the model. Designers will use the model to figure the amount of materials that will be used. They can also figure the amount of time it will take to manufacture the solution.

The second economic factor is the *market.* The market includes those people buying and selling the product. See **Figure 16-11.** Consumers are a major part of the market. They are the ones who will choose whether or not to buy the product. The designers test the solution by surveying consumers. They will again use the model to help the consumers visualize the solution. The second part of the market is the competitors. Competitors are those who make products competing with the solution. Designers may test the products of their competitors to see how their solution compares. They will then advertise the features making their solution better than the competition. The designers will also study the prices of the competitors' products.

TechnoFact: The costs of a product can also be divided into direct and indirect costs. The direct cost is the money required to build each product. Direct costs include materials and wages. The indirect costs are the costs spread out among all the products being manufactured. They include design costs, utility costs, and equipment expenses.

Standards for Technological Knowledge

8

Extend
Have the students rate the appearance of several products using a scale from 1 to 5. Compare the results as a class.

Research
Have the students locate information on the types and uses of crash test dummies.

Figure 16-11. Consumers will choose to buy the products they like the best.

Appearance

The *appearance* of a product is one of the most important characteristics. It is also one of the hardest to test. There is no lab test that can give the designers data on how nice a product looks. Appearance is a characteristic people see and have feelings toward. You may have heard the phrase "Beauty is in the eye of the beholder." The same is true for the appearance of a product. Many people may look at the same solution and have different reactions. Some may like the shape, but not the color. Others may think the texture is unpleasant, while others like it.

The best way to test appearance is to use the reactions of people. Designers often ask many people about their reactions to a product. The questions may be very general, such as "Do you like the product?" They may also be very specific, like "On a scale of 1 to 5, how would you rate the size of the display screen?" The answers from the survey are then gathered. The results of the questions will be used as data to evaluate the appearance.

Safety

Safety is a major characteristic of a design. It involves more than you may think. The solution must be safe for consumers to use. It must be safe for workers to produce. Lastly, the solution must be safe for the environment.

Consumer safety is the most obvious need for safety. Solutions must be tested to ensure they are safe to use. Safety testing can be done by using objects to simulate humans. Automobiles are safety tested with crash test dummies. See **Figure 16-12.** The dummies have sensors placed around their bodies to record data. The data tells where the dummy was injured during the test. This allows accurate tests without putting humans at risk of being hurt. Some products are safety tested using computer simulations. The computer acts as if a human is using the solution and records safety data.

All solutions must be safe for those using them. They should also be safe for those around them. After testing, age ratings are given to some solutions. The age rating identifies the lowest age a person should be to use the product safely. See **Figure 16-13.** Games and movies with adult content

TechnoFact: The Likert technique, developed by Rensis Likert, is often used to gather the attitudes and feelings of a set of people. The respondents complete a survey by selecting from statements like agree, strongly agree, disagree, and strongly disagree. Each response has a numeric value assigned to it. This allows the researcher to quantify the respondent's attitudes. Once the surveys are complete, the data is compiled.

are not rated for children. Designs with small parts are not rated for very young ages because children can swallow the parts.

There are several government organizations supervising product safety. The U.S. Consumer Product Safety Commission (USCPSC) watches over everything from household appliances to children's toys. The National Highway Traffic Safety Administration (NHTSA) tests the safety of motor vehicles. The Food and Drug Administration (FDA) oversees the testing of food, medicines, and cosmetics, like hair spray and makeup.

The organization concerned with how the solution affects the safety of the environment is the Environmental Protection Agency (EPA). Solutions

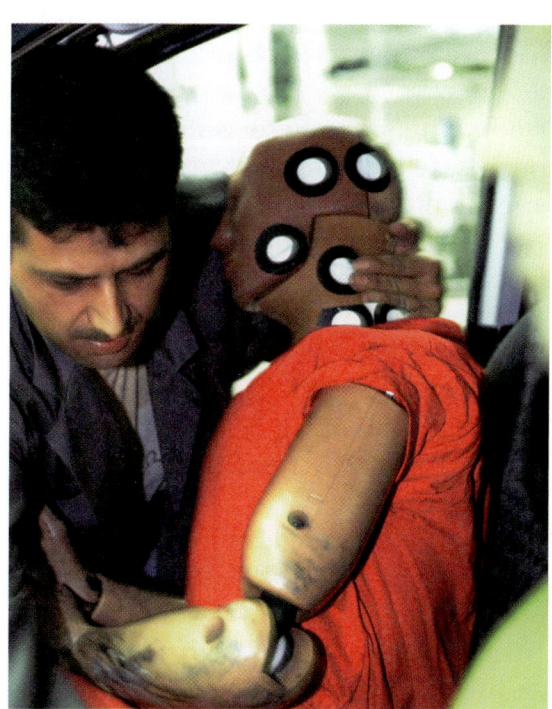

Figure 16-12. Crash test dummies allow the safety of automobiles to be tested without hurting humans. (DaimlerChrysler)

Figure 16-13. This product has an age rating, which tells consumers that it is intended for children 18 months and older.

must not harm the environment when used. Testing is done on products to determine their impacts on the environment. Aerosol cans, for example, are products that were found to hurt the environment. The cans released chlorofluorocarbons (CFCs). After testing was done, the CFCs were found to damage the environment. Now, most aerosol cans have no CFCs, or they have been converted to spray pumps.

Products can also be unsafe for the environment when they are thrown away. Solutions are tested to find how they should be disposed. If some products, like chemicals and batteries, are poured down the drain or thrown away, they can be very unsafe. These products can contaminate the air and water around them.

TechnoFact: The Entertainment Software Ratings Board (ESRB) is a consumer safety organization that tests and rates electronic games. They have developed a set of six rating symbols ranging from Early Childhood to Mature. The ratings are placed on the packaging of games and allow consumers to easily see the age group for which the game is intended.

Standards for Technological Knowledge

10

Research
Have the students find a company specializing in one type of testing.

Discussion
Ask the students why three different testing environments are needed.

Testing Environments

Tests are conducted in three different environments. Tests can either be completed in the laboratory, the field, or virtual space. Each environment has advantages and disadvantages.

The *laboratory* is used when a controlled area is needed. In laboratories, variables, like temperature, moisture, and amount of light, can be controlled. A *variable* is any object that can be changed. Many variables can affect the outcome of a test. For example, if designers are testing a sound, they do it in a lab where they can control the amount of sound. See **Figure 16-14.** They make sure the only sounds in the room are coming from the device they are testing. If there are other sounds, they will distort the results of the test.

The labs used in testing are often very specialized. It is common for a lab to be used for only one type of testing. Because the labs are specialized, it is too expensive for all designers to have their own labs. Some designers have companies that specialize in testing their solutions. This also helps to make sure the testing is accurate.

The second testing environment is field-testing. *Field tests* are conducted in the environments in which the solutions will be used. The solutions are used the same way the consumers will use them. Airplanes are field-tested by flying. See **Figure 16-15.** Field tests can be conducted in areas that are off-limits to the public. Automobile tests are done on closed-circuit tracks. The car being tested may be the only car on the

Figure 16-14. This researcher uses a soundproof laboratory to test the noise the tire produces on the two different surfaces. (Goodyear Tire and Rubber Company)

track. This is done for safety. It would be unsafe to test an automobile at high speeds on roads with other vehicles.

Actual consumers do some field-testing. Prototypes of small appliances, for example, may be given to a group of people to test. These are tested in the users' homes or offices. When the testing is complete, the users report the results of the test to the designers.

The last testing environment is the virtual setting. Virtual testing, or *computer testing,* is a very popular form of testing. See **Figure 16-16.** It requires a computer model and testing or simulation software. Virtual testing can be less expensive than lab testing and field-testing. It can also

TechnoFact: Most consumer electrical products have a mark printed on them by Underwriter's Laboratories (UL). UL is an independent, not-for-profit, safety-testing organization. UL inspection centers can be found in sixty-five countries. In 2002, more than 17 million UL marks were applied to products across the world.

Figure 16-15. This experimental aircraft is tested in test flights. (NASA)

Standards for Technological Knowledge

9 **10**

Resource
Reproducible Master 16-1: The Design Process: Testing Design Solutions

Figure Discussion
Ask the students what environments the product in Figure 16-15 could be tested in. Have them explain their answers.

Tech Terms
Accurate—Free from mistakes and errors.
Valid—Well-grounded in principle and able to withstand criticism.

generate very accurate data. Computer testing is very useful for large and complex solutions that are hard to test in reality. The space shuttle is a solution that has been tested in a computer environment.

The Testing Procedure

Whether the test is conducted in a lab, in a field, or on a computer, it must follow a certain procedure. Following the procedure helps make the test valid. A valid test is one that is accurate and can be trusted. The testing procedure includes six steps:

1. Determine the purpose.
2. Choose the test.
3. Collect materials and equipment.
4. Conduct the test and gather data.
5. Analyze the data.
6. Evaluate the results.

DROP IMPACT, 36" DROP HEIGHT ON REAR CORNER.
PRINCIPAL STRESS ON OUTER SURFACE OF REAR HOUSING.
RED AREAS HAVE EXCEEDED THE YIELD STRESS.

Figure 16-16. The use of computers in testing is very helpful and accurate. (Whirlpool Corporation)

TechnoFact: The aircraft picture in Figure 16-15 is the *Helios* prototype, developed by AeroVironment, Inc. as a part of NASA's Environmental Research Aircraft and Sensor Technology (ERAST). The test flights were conducted at the NASA Dryden Flight Research Center and U.S. Navy's Pacific Missile Range Facility in Hawaii. The aircraft is remotely piloted and has achieved the goal of sustaining flight for more than forty minutes at over 96,000'.

Technology Explained

> **canal lock:** A device allowing ships to change elevation as they travel through a canal.

Canals are part of some inland waterway transportation systems. They must have three features. First, there must be an adequate source of water to keep them full. The canals can be fed from rivers or reservoirs. Second, there must be a way to cross ravines and streams. A "water bridge" called an *aqueduct* is used. Third, there must be a method for raising or lowering a ship to different canal levels, as the elevation of the land changes. A device called a *lock* fills this need, **Figure A.**

A canal lock is essentially a trough with a gate at each end.

The operation of the lock is simple. **Figure B** shows how a boat moves from a higher elevation to a lower elevation in the lock.

First, the lock is filled to the upstream water level. The upstream gate is opened, and the boat enters the lock. Second, the upstream gate is closed to seal the lock. Third, sluice gates are opened in or

Figure A. These large gates are part of a lock on the Illinois River. (Jack Klasey)

around the lower gates. The sluice gates allow water to flow out of the lock until the level of water inside the lock is equal to the downstream level. Fourth, the downstream gates are opened, letting the boat sail out of the lock.

To move a boat upstream, the procedure is reversed. The boat enters the lock when the water level equals the downstream level. Then, the lock is closed, and the sluice gates are shut. The lock fills, lifting the boat to the upstream level. The upstream gate is opened, and the boat sails out of the lock.

The Saint Lawrence Seaway is an inland waterway that has several canal sections. The seaway rises more than 180' (54 m) from Montreal to Lake Ontario, through two lakes,

Determine the Purpose

The purpose of the test is the reason it is being conducted. The purpose should relate back to the criteria and characteristics of the solution. For example, the purpose of the test may be to evaluate the appearance of the solution. Each test should only have one purpose. It would be very hard to conduct a test of both the function and economics of a solution. Conducting more than one test on a solution may be necessary. It may not always be necessary, however, to test all the characteristics of a design.

Choose the Test

Once the purpose is determined, the test must be chosen. There are many different tests that can be conducted

TechnoFact: Many canals, including the Panama Canal, continue to operate by collecting tolls from vehicles passing through the locks. The highest toll paid to pass through the Panama Canal was $141,349.97 by the *Crown Princess*. The least expensive toll was $0.36, paid by Richard Halliburton to swim the canal.

three separate canals, and seven locks. Once the seaway reaches the Great Lakes, ships pass through Lake Ontario to the Welland Ship Canal. This canal has eight locks to raise ships to the Lake Erie level. The difference between the two lakes is about 330' (100 m). The channel from Lake Erie to Lake Huron and on to Lake Michigan is the same elevation. No locks are needed. Five more locks are required, however, to raise ships 22' (7 m) up to the level of Lake Superior.

Lock Area

A

B

C

D

Figure B. This drawing shows how a canal lock works. A—The boat enters the lock. B—The upper gates are shut. C—Water is let out of the lock until the boat is even with the downstream water level. D—The lower gates open, and the boat leaves the lock.

on products. Crash tests and vibration tests are common tests done on automobiles. See **Figure 16-17.** Appearance and clinical tests are typical medication tests. Many products are tested using customer satisfaction tests.

Tests can be conducted using existing tests, or the researcher can create a new test. Testing organizations have collections of tests that have already been developed. The

Figure 16-17. This vibration test is conducted to test the durability of the automobile. (DaimlerChrysler)

Standards for Technological Knowledge

9

Research
Have the students research the typical tests conducted on automobiles.

Extend
Have the students build a model of a canal lock.

TechnoFact: The standard test dummy for a frontal crash test is the fiftieth percentile male Hybrid III. The dummy is 5'10" tall and weighs 170 pounds. The dummy is built to represent the average adult male. The Hybrid III family also includes other dummies representing different ages and genders.

Standards for Technological Knowledge

9

Research
Have the students select a specific product test (for example, testing resistance to chipping in lacquer paint). Next, have them visit the ASTM website (www.astm.org) and find standards for that test.

Extend
Discuss how variables are controlled in tests.

Figure 16-18. A wind tunnel is a piece of testing equipment commonly used in many applications. (Colorado State University)

American Society for Testing and Materials (ASTM) is one of the largest organizations that write testing standards. ASTM has over ten thousand standards for materials and tests. Many of the tests ASTM identifies are very specific. One example of an ASTM test is the *Standard Test Method for Measuring Maximum Functional Wet Volume of Utility Vacuum Cleaners.* In other words, the test is used to check how much fluid a utility vacuum cleaner can handle.

ASTM and other testing organizations are excellent sources for finding existing tests. The second option is to create a new test. If the designers create a new test, it must be evaluated to make sure it is valid. It has to test the right features. The designers must also make sure they control the variables.

For example, designers testing the "bounce" in different basketballs must have the same air pressure in each ball.

Collect Materials and Equipment

Whichever test is chosen must be documented. The written procedure for the test should include any necessary information about the test. The test will have sections for materials, procedures, and observations. The materials include the equipment, or items used during testing. Product samples are common testing materials. Testing equipment consists of the devices needed. Bunsen burners, clamps, and wind tunnels are pieces of testing equipment. See **Figure 16-18.**

TechnoFact: Frank H. Wenham is credited for designing the first wind tunnel in 1871. His device consisted of a fan, steam engine (used to turn the fan), and a tube 12′ long. The device propelled wind down the tube and around a model at the end of the tube.

Conduct the Test and Gather Data

After the materials and equipment have been collected, the next step is to conduct the test. The test should be conducted by following the written procedure. The procedure is often called the *testing method.* The method must be followed step by step, so the test can be repeated with the same results. If the researchers change their procedure, they must make note of it.

It is important that the data is recorded during the test. The data can come in many different forms. Data can be the amount of times the solution was used before it broke. It can be the reactions to the appearance of a solution. The data can even be the force with which a crash test dummy hit the steering wheel.

Analyze the Data

The designers must analyze the data, which can be very overwhelming. The data from a single test may cover tens or even hundreds of pages. The designers must break down all the data. They begin by doing any necessary calculations. Sometimes, averages are calculated. Once the calculations are complete, the designers display the data with graphic models (charts, graphs, or tables).

Evaluate the Results

The graphic models are used to understand the results of the tests. The researchers can evaluate the results by reviewing the tables and graphs. They may find that the solution passed the test. The researchers may also find that the solution failed the test. The results of the test will help the designers determine if the solution is acceptable.

Assessing the Solution

The final stage in the testing of a solution is assessment. The assessment is the designers' final conclusion on the solution. It can only be made after the results from all the tests are completed. The designers review the results from tests done on the function, human engineering, economics, appearance, and safety of the solution and create an assessment.

The assessment usually includes a trade-off. *Trade-offs* are decision processes the designers must face. They recognize the need for sensible compromises among opposing factors. A trade-off can occur when a solution does not pass all the tests conducted. For example, the test results may show that a solution is safe and functions well, but has some appearance problems. The trade-off comes when the designers have to decide if safety and function are more important than appearance. When trade-offs are made, there is a preference or an exchange for one feature or idea in favor of another.

Once the designers take into account the test results and any trade-offs, they determine a final assessment. The designers have three basic choices. See **Figure 16-19**. They can choose to

TechnoFact: Repeatability is an important factor in testing. It is the ability of a test to be repeated with the same results. To ensure repeatability, test designers make accurate notes and write down every step of the test. The person conducting the test must then follow the directions exactly and use the proper equipment.

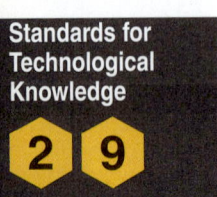

Standards for Technological Knowledge

2 9

Extend
Have the students conduct a product test of your specification.

completely abandon the solution. This assessment leads the designers back to the first step of the design process. There they start over from the beginning. The second choice is to rework or revise the solution. This is selected when the solution did not do well in one or two tests. This may lead designers back to any of the previous steps of the design process. The last assessment that can be made is that the solution meets the expectations. This assessment moves the solution forward to the next step of the design process.

Summary

The testing stage of the design process has two different functions. The first is to test the solution. There are hundreds of different tests that can be conducted on solutions. Five main areas (function, human engineering, economics, appearance, and safety) are tested. Each of these areas can be tested in three different environments: in the laboratory, in the field, and on the computer.

Testing a solution has a procedure. It begins by determining a purpose. Then a test is chosen, materials and equipment are selected, and the test is conducted. The data is gathered during the test. It is analyzed, and the results are evaluated. Once the tests have been evaluated, the designers can make an assessment. The assessment of a design determines whether the solution will be discarded, revised, or sent to the next step.

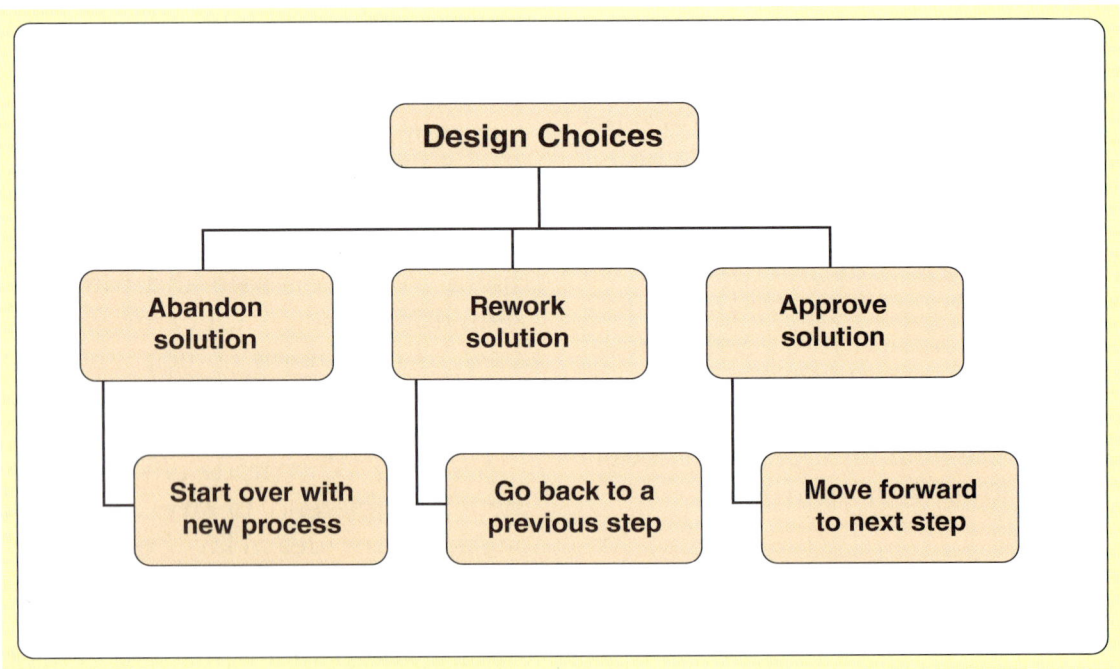

Figure 16-19. Once the testing is complete, the designers must make an assessment. They can abandon, rework, or approve the design.

Curricular Connections

Standards for Technological Knowledge

3

Language Arts

Write a test that can be used to test the function of a solution. Include the purpose, materials, and procedures in the test.

Science

Conduct a test on a material. Record the data and report the results.

Social Studies

Research how tests are used to study humans and human behavior. Create a display showing how testing products and testing human behavior are both different and similar.

Activities

1. Create a display highlighting the main characteristics tested in product testing. List possible examples of tests that can be used to test each characteristic.

2. Research a major testing or safety organization. Create a presentation explaining the organization and give examples of the testing it conducts.

3. Use the testing procedure to test a solution you have developed. You may research a test or develop your own. Once the testing is completed, make an assessment of your solution.

4. Develop a budget for a product you could make in your classroom lab. Some possible projects include bookends, coatracks, wooden pens, or computer monitor supports. First, determine the amount of materials needed to make the product. Include a consideration for waste and determine the cost of materials. Then, determine the time and equipment needed to make the product. Estimate a reasonable hourly wage for labor and a reasonable hourly cost for tools, equipment, and overhead. Develop a marketing plan and estimate the advertising costs. Calculate the cost per item to produce the product, and then suggest an estimated price for the product.

TechnoFact: The Pepsi Challenge® game material was a marketing strategy set up like a field test. The design of the Pepsi Challenge material was to conduct a blind taste test between Pepsi® soft drinks and Coca-Cola® beverages.

Answers to *Test Your Knowledge* Questions
1. It is necessary because the solution must solve a problem.
2. False
3. After reviewing the results of a test.
4. True
5. Function describes how the solution works. Human engineering examines the way people interact with the solution. Economics check to make sure the designers will make money. Appearance is how the product looks. Safety ensures that the solution is safe for consumers, workers, and the environment.
6. B. It is how well the solution works.
7. Laboratory, field, and computer.
8. True
9. Determine the purpose, choose the test, collect materials and equipment, conduct the test and gather data, analyze the data, and evaluate the results.
10. True

Test Your Knowledge

Do not write in this book. Place your answers to this test on a separate sheet of paper.

1. Why is it necessary to test solutions?
2. A test is a judgment made toward a solution. True or false.
3. When is an assessment made?
4. Design criteria are used in the testing stage. True or false.
5. List and summarize the five main characteristics of solutions tested.
6. Select the statement best describing performance:
   A. It is how the solution works.
   B. It is how well the solution works.
   C. It is how long the solution will work.
7. Name the three testing environments.
8. A valid test is one that can be trusted. True or false.
9. Write the six steps of the testing procedure.
10. A trade-off is a compromise with which designers are often faced. True or false.

Modular Activity

This activity develops the skills used in TSA's System Control Technology event.

System Control Technology

Activity Overview

In this activity, you will build a computer-controlled mechanical model representing controls for an elevator serving two levels. You will also prepare a report explaining your design and listing directions for operation.

Materials

➤ Pencil and paper
➤ Two (2) touch sensors
➤ Two (2) lights
➤ Two (2) motors
➤ Computer hardware and software control system

Background Information

General. You will use the sensors, lights, and motors to model an elevator serving two floors. The sensors represent the elevator call button. The motors represent the elevator doors—one representing the first-floor elevator doors, the other representing the second-floor elevator doors. A running motor represents an open elevator door. The lights represent the location of the elevator car. When the first-floor light is on, the elevator car is at that floor. When the second-floor light in on, the elevator is at that floor.

Control conditions. Your control system must adhere to the following guidelines:

➤ The elevator car can be at only one floor at any given time.
➤ The elevator car must be at the floor before the doors can open.
➤ The elevator car cannot leave the floor while the doors are open.
➤ When a button is pushed, the elevator car is called to the floor, the doors open and close, the car moves to the other floor, and the doors open and close.
➤ The car remains at the current floor until called.

➤ There must be a two-second delay between when the car arrives at the floor and when the doors open.

➤ The doors remain open for five seconds and then close automatically.

➤ If the elevator call button is pressed while the doors are open, the doors remain open for five seconds from the time when the button is pushed.

➤ The car must wait two seconds after the door closes before going to the other floor.

Report. Your report must include the following:

➤ Description of solution

➤ Instructions for operation

➤ Printout of control program

Guidelines

➤ Create rough schematics of design solution and control logic.

➤ Sketch the final design before constructing it.

➤ After constructing the model and writing the control program, test the model. Before you begin testing, create a list of the various conditions and situations that need to be tested.

Evaluation Criteria

Your project will be evaluated using the following criteria:

➤ Report

➤ Model functionality, dependability, and ingenuity

➤ Computer program logic and functionality

The Ergonomic Environment

Workers who spend long periods of time doing detailed work with a computer may be susceptible to eyestrain, back discomfort, and hand and wrist problems. *Ergonomics* is the science of adapting the workstation to fit the needs of the drafter. Applying ergonomic principles results in a comfortable and efficient environment.

There are many types of ergonomic accessories that may improve a computer workstation. In addition, a few things can be done to create a comfortable environment and help prevent injury or strain to the operator's body. The following sections list specific guidelines to help prevent injuries for various body parts.

Eyes

- Position the monitor to minimize glare from overhead lights, windows, and other light sources. Reduce light intensity by turning off some lights or closing blinds and shades. You should be able to see images clearly without glare.
- Position the monitor so that it is 18″ to 30″ from your eyes. This is about an arm's length. To help reduce eyestrain, look away from the monitor every 15–20 minutes and focus on an object at least 20′ away for 1–2 minutes.

Wrists and Arms

- Forearms should be parallel to the floor.
- Periodically stretch your arms, wrists, and shoulders.
- Try using an ergonomic keyboard and mouse. The keyboard keeps the wrists in a normal body position, and a mouse will fit the hands more comfortably.

Neck

- Adjust the monitor so that your head is level, not leaning forward or back. The top of the screen should be near your line of sight.

Back

- Use a chair that is comfortable and provides good back support. The chair should be adjustable and provide armrests.
- Sit up straight. This maintains good posture and reduces strain. Think about good posture until it becomes common practice.
- Try standing up, stretching, and walking every hour. This will also reduce strain.

Legs

- Keep your thighs parallel to the ground.
- Rest your feet flat on the floor or use a footrest.
- When taking a break, walk around. This will stretch the muscles and promote circulation through your body.

Chapter 17
Communicating Solutions

Did You Know?

➤ Copies of mechanical and architectural drawings are often referred to as *blueprints*. Sir John Herschel introduced the blueprinting process in the 1840s. In this process, the original drawing is created on semi-transparent paper. The original is placed on top of a sheet of paper covered with a light-sensitive iron salt and exposed to ultraviolet light (such as sunlight). In areas on the original that do not have lines, the light travels through the original drawing and causes the iron salt coating to turn blue. The lines on the original drawing prevent light from reaching the blueprint, so the lines on the blueprint remain white. Thus, the completed blueprint is a blue sheet with white lines. The blueprinting process is rarely used today, but the term *blueprint* is still common.

Objectives

The information given in this chapter will help you do the following:

➤ Identify the three types of documents used to communicate solutions.

➤ Explain the difference between engineering and architectural drawings.

➤ List and describe the three types of working drawings.

➤ Define *orthographic drawing* and *pictorial drawing*.

➤ Describe traditional and computer-aided design (CAD) drawing tools.

➤ Create a mechanical drawing.

➤ Summarize the process of receiving approval for a solution.

Key Words

These words are used in this chapter. Do you know what they mean?

alphabet of lines
architectural drawing
assembly drawing
bill of materials
compass
computer-aided design (CAD)
detailed drawing
dimensioning
drawing board
engineering drawing
45° triangle
isometric drawing
mechanical drawing
multiview drawing
oblique drawing
orthographic drawing
perspective drawing
pictorial drawing
plotter
scale
schematic drawing
specification sheet
tablet
template
30°-60° triangle
T square
working drawing

Chapter Outline
Did You Know?
Objectives
Key Words
Communication Documents
  Bills of Materials
  Drawings
  Specification Sheets
Drawings
Working Drawings
  Detailed Drawings
  Assembly Drawings
  Schematic Drawings
Drawing Classifications
  Orthographic Drawings
  Pictorial Drawings
Mechanical Drawing Tools
  Traditional Tools
  Computer-Aided Design (CAD) Tools
Making Mechanical Drawings
Presenting the Solution
Summary
Curricular Connections
Activities
Test Your Knowledge

Standards for Technological Knowledge

8

Resource
Reproducible Master 17-1: The Design Process: Interpreting the Solution

Resource
Transparency 17-1: Document the Solution

Demonstrate
Show the students examples of blueprints and technical drawings.

Brainstorm
Have the students list examples of design documents they have seen or used.

Figure 17-1. Builders use blueprints to communicate solutions. (Habitat for Humanity International)

Have you ever seen or used blueprints? See **Figure 17-1.** Have you ever used a schematic drawing or seen a drawing showing how a product is put together? If you have, then a design has been communicated to you. The objects used to communicate solutions are created in this step of the design process.

The goal of almost every design process is to make a profit. The solution has been designed with this goal in mind. To make a profit, however, most solutions must be mass-produced. The designers do not manufacture the product. They give the solution to others skilled in manufacturing processes. The people who will produce the product may not have been part of the design process, so they may not know very much about the solution. The designers must communicate all aspects of the solution to those who are making the product. See **Figure 17-2.**

Communication Documents

The best way for designers to communicate a solution is through a set of documents. These documents describe everything the manufacturer must know to produce the solution. Consumers also use design documents when they assemble and service products. There are three major types of design documents:

➤ Bills of materials.

➤ Drawings.

➤ Specification sheets.

TechnoFact: Builders and contractors must be able to read and understand construction documents. They learn on the job or take classes on reading blueprints.

Figure 17-2. Drawings are used in manufacturing of products. (Product Development Technologies, PDT)

Each document has a specific purpose and is needed to produce the solution.

Bills of Materials

A *bill of materials* is a document listing and describing the parts needed to build the solution. It is created in the form of a table. See **Figure 17-3.** This document lists information about the various parts needed to produce the solution. The columns of the table divide the different types of information. The information given is a part number, the number of items needed, and the name of the part. The size of each part and the material to be used are also listed on the bill of materials. Each row of the table provides information for a different part. Every part used to make the solution is listed on the bill of materials. The main pieces are listed first. The hardware and fasteners, like hinges, screws, and bolts, are listed last.

Drawings

Drawings are used to communicate the appearance of the solution. They can be used to show the size and shape of the solution or how the pieces of the solution fit together. The drawings are created as mechanical drawings. *Mechanical drawing* is a type of drawing also known as drafting. These drawings are much different than drawings artists create or those designers sketch. Mechanical

Part Number	Quantity	Part Name	Size	Material
1	1	Base	3/4" x 4" x 4"	Pine
2	1	Slider	3/4" x 3/4" x 3"	Pine
3	2	Rails	3/4" x 3/4" x 3"	Pine
4	1	Top	1/4" x 2 1/2" x 3"	Hardboard
5	1	Jar	Standard size	Mason jar

Figure 17-3. A bill of materials lists all the parts needed to manufacture a gumball dispenser.

TechnoFact: Tables (such as bills of materials) included in drawings are often called *schedules*. For example, a door and window schedule describes the sizes of the doors and windows used in a plan.

Standards for Technological Knowledge

Research
Have the students find an MSDS for a household product or chemical.

Discussion
Ask the students why quality specification sheets are used instead of verbal agreements.

drawings are very neat and accurate. See **Figure 17-4.** Mechanical drawing requires specific sets of tools and skills.

Specification Sheets

The last document used to communicate is a specification sheet. *Specification sheets* are also called *spec sheets.* They are used to describe items that cannot be shown in drawings. There are two main types of spec sheets:

➤ Material sheets.

➤ Quality of work sheets.

Material specification sheets describe the materials that will be used in the product. The bill of materials was used to list the types and sizes of the materials. The spec sheets go into greater detail. They list the exact materials, as well as the quality of materials to be used. For example, lumber can be

purchased in several different grades. A spec sheet lists the grade that should be used. Spec sheets are also used to describe the properties of the materials. One type of material spec sheet is a Material Safety Data Sheet (MSDS). See **Figure 17-5.** An MSDS describes properties like color, odor, density, and melting point. It also lists how to handle and store the materials. These spec sheets describe any health risks in using the materials.

Quality of work specifications explain how well the product will be produced. They list the type of finish on the materials. This can include how well the materials are sanded, painted, or stained. Quality spec sheets also list who is responsible for different stages of the product. This is important in construction. The spec sheet may state that the owner is responsible for laying the carpet and painting the interior.

Figure 17-4. This mechanical drawing shows three views and a rendering of the gumball dispenser.

TechnoFact: Specifications are included in drawings in the form of notes. Notes can communicate the material used or general information about the object (for example, MATERIAL: 1/4″ BRASS and ALL FILLETS AND ROUNDS 0.125″).

Section II - Emergency and First Aid

Primary Route of Entry:
Inhalation
Eyes:
Flush with water for several minutes.
Skin:
Wash with soap and water.
Inhalation:
Remove from exposure.
Ingestion:
Dilute stomach contents with several glasses of milk or water.

Symptoms of Overexposure:
Minimal respiratory tract irritation may occur as with exposure to large amounts of any nontoxic dust.

Medical Conditions Generally Aggravated by Exposure:
None when used as described by product literature.

Additional Information:
None

Section IV - Physical Data

Appearance/Odor: Black powder / faint odor
Boiling Point: N.A.
Solubility in Water: Negligible
Evaporation Rate: N.A.
Vapor Density (Air=1): N.A.
Volatile: N.A. % (Wt.) N.A. % (Vol.)

Softening Range: 43.3-60°C (110-140°F)
Melting Point: N.A.
Specific Gravity (H_2O=1): ~1
Vapor Pressure (mm Hg): N.A.
pH: N.A.

Section V - Fire and Explosion Data

Flash Point (Method Used): N.A.
Flammable Limits: LEL: N.A., UEL: N.A.
NFPA 704: Health - 0, Fire -1, Reactivity - 0

Figure 17-5. MSDS sheets explain all the safety hazards associated with materials. Shown are sections from a MSDS for printer toner. (Xerox)

The bill of materials, drawings, and specification sheets must be exact. These documents are often used when companies make a bid to produce the solution. A bid is a price a company charges to make a product. If a document is wrong, it may cost the designer a lot of money to have it changed once a company has made a bid. The documents are used as part of the contract between designers and manufacturing and construction companies.

Drawings

Mechanical drawings are created to communicate solutions. They can be divided into two major categories:

➤ Engineering drawings.

➤ Architectural drawings.

The two categories are used for different types of objects. Tools, machines, toys, and other manufactured products are drawn with engineering

TechnoFact: Specification sheets are included with all medicines. The sheets inform the consumer of the ingredients, potential side effects, and correct usage of the medicine.

Standards for Technological Knowledge

17

Figure Discussion
Figure 17-6 is the interior arrangement of a *Mercury* capsule. Discuss the Mercury Project with the students.

Demonstrate
Show the students examples of engineering drawings.

Figure 17-6. Engineering drawings are used in many different industries, including aerospace. (NASA)

drawings. Buildings and structures are drawn with architectural drawings.

Engineering drawings are used to communicate products that will be manufactured. The products can be anything from toasters to space shuttles. See **Figure 17-6.** Many types of designers and engineers create and use engineering drawings. Aerospace, automobile, electrical, and mechanical engineers all use engineering drawings to create their products.

Industrial and product designers also use engineering drawings.

Architectural drawings communicate buildings and structures that will be constructed. Floor plans and building details are common architectural drawings. See **Figure 17-7.** Houses, buildings, bridges, and towers are all drawn with architectural drawings. Land surveyors, architects, and city planners use architectural drawings.

TechnoFact: Geometry is used quite frequently in mechanical drawing. Geometric concepts like quadrants, tangents, bisectors, and inscribed circles are applied in mechanical drawing. For example, the midpoint of a line can be found by striking an arc from each of the end points with a compass. Both arcs must have the same radius and be large enough to intersect. A line drawn between the intersections of the two arcs bisects the original line at the midpoint.

Figure 17-7. Floor plans are examples of architectural drawings.

Working Drawings

Engineering and architectural drawings are drawn as working drawings. *Working drawings* are the most complete drawings produced. These drawings display all the information needed to build the solution. They also show how the products are assembled.

Working drawings are drawn to a certain standard. Most designers follow the standards the American National Standards Institute (ANSI) sets. The ANSI has developed a list of standards describing how objects should be drawn and sizes should be shown. Following the ANSI standards ensures that all working drawings can be read and understood. Designers create three types of working drawings:

➤ Detailed drawings.

➤ Assembly drawings.

➤ Schematic drawings.

Detailed Drawings

Detailed drawings are produced from each piece of the solution. See **Figure 17-8.** The detailed drawings show the exact size and shape of the pieces. The drawings are complete enough that the pieces can be built using the drawings. Detailed drawings use different views to show the object. The designer must choose the views best describing the part.

These drawings describe the size and shape of the piece using dimensions. Dimensions use arrows and numbers to describe size and location features. A floor plan is a detailed drawing including dimensions. A detailed drawing also gives information on the material and finish, using notes. Notes are lines of text describing an object. They give the viewer necessary information about the piece.

TechnoFact: The American National Standards Institute (ANSI) is a not-for-profit organization that develops voluntary standards. The ANSI was first founded in 1918, as the American Engineering Standards Committee (AESC). Its creation was a joint venture by several organizations, including the American Institute of Electrical Engineers (now IEEE), American Society of Mechanical Engineers (ASME), American Society of Civil Engineers (ASCE), American Institute of Mining and Metallurgical Engineers (AIMME) and American Society for Testing Materials (ASTM).

Discussion
Ask the students how they have used assembly drawings.

Example
Show the students some examples of assembly drawings used by consumers.

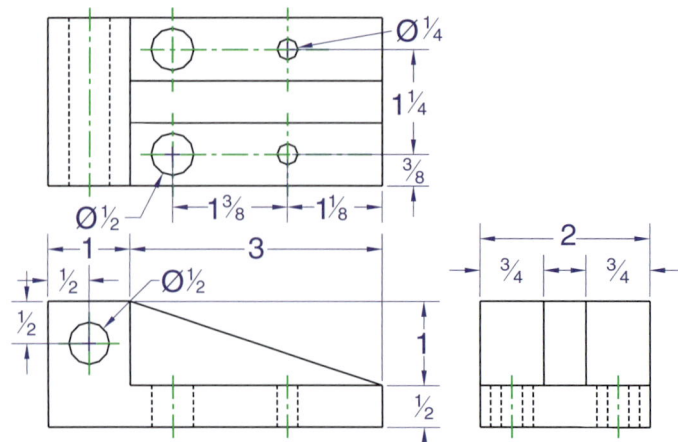

Figure 17-8. Detailed drawings are used to explain the sizes and shapes of individual pieces of the solution.

Assembly Drawings

Assembly drawings show how parts fit together. See **Figure 17-9.** These drawings can be used to help put the pieces together. Products requiring consumers to assemble them often include assembly drawings. The drawings help visualize how all the parts fit with one another. They are drawn to look three-dimensional with all the pieces pulled apart. Dotted lines are used to show the location of many of the pieces.

These drawings can also be drawn to show an assembled product. They can help show how the device or product functions. The parts are labeled with numbers or letters and arrows,

Figure 17-9. Assembly drawings show how products fit together. (Design Central, design firm; Artromick International, client)

TechnoFact: There are several types of assembly drawings. An outline assembly drawing shows the outline of the completely assembled product. An exploded-view assembly drawing shows each piece of the product pulled apart from the center, as in Figure 17-9. A sectional assembly drawing shows the assembled product as if it were sliced in half to reveal the inside.

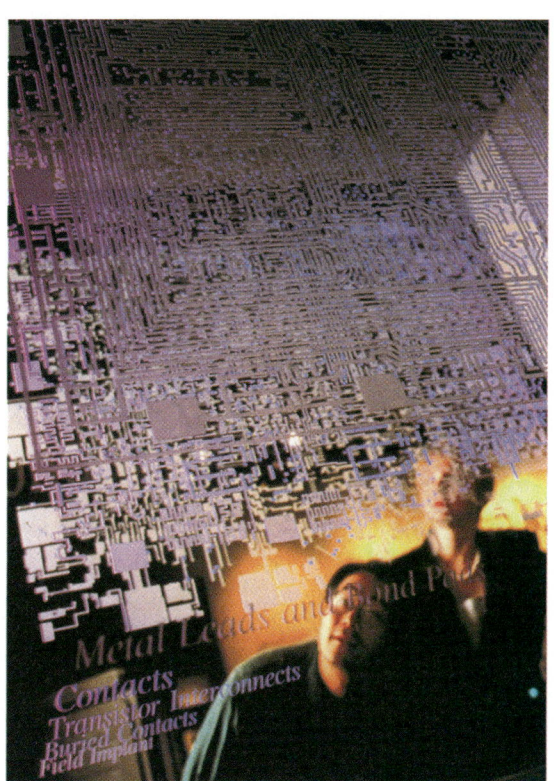

Figure 17-10. This is a large-scale schematic of the layers of circuits in a computer processor. The final processor is only a few inches in length. (Intel).

called *leaders.* The numbers or letters match the bill of materials included with the drawing. This helps people identify each part. Assembly drawings do not include dimensions. If the exact sizes of the parts are needed, the detailed drawings are used.

Schematic Drawings

Schematic drawings are used to show systems. A system is a group of objects working together for a common goal. Schematic drawings are used to show how the parts are connected to form a system. See **Figure 17-10.** These drawings are used for electrical circuits,

water pipes, and production lines. Schematic drawings are not drawn to actual scale. They do not communicate size and shape like other working drawings.

These drawings are used to show the relation of the parts to one another and how the product flows. Electrical drawings show the flow of electricity between parts of the circuit. Piping drawings show how water moves from one location to another. Drawings of production lines communicate how the product moves from one end of the production line to the other.

Schematic drawings use symbols for the components of the system. See **Figure 17-11.** ANSI sets the symbols so all designers can read the schematic drawings. Lines representing paths

Standards for Technological Knowledge

Example
Show the students an automobile's wiring diagram as an example of a schematic.

Research
Have the students identify the symbols used in HVAC, hydraulic, electrical, or pipe fitting schematics.

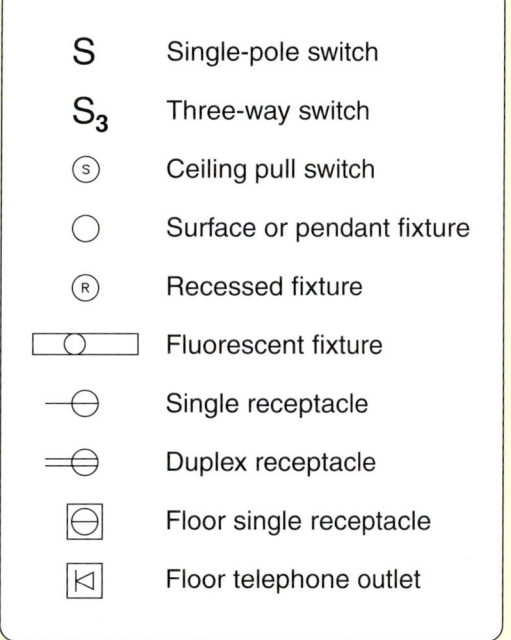

S	Single-pole switch
S_3	Three-way switch
ⓢ	Ceiling pull switch
○	Surface or pendant fixture
Ⓡ	Recessed fixture
⊏○⊐	Fluorescent fixture
─⊖	Single receptacle
═⊖	Duplex receptacle
⊟	Floor single receptacle
◁	Floor telephone outlet

Figure 17-11. Standard symbols are used to make schematics easier to draw and understand.

TechnoFact: Schematic drawings are used by several types of designers and engineers. Manufacturing engineers use schematics to design the flow of products through a manufacturing line. Electrical engineers use schematics to design electrical components and circuits.

are used to connect the symbols. In an electrical drawing, the lines represent wires or copper on a circuit board.

Drawing Classifications

Most design solutions are physical objects having three dimensions. The objects have a width, a height, and a depth. In this step, the designers' job is to decide the best way to display the three dimensions. Designers have two basic ways to communicate solutions.

➤ Orthographic drawings.
➤ Pictorial drawings.

Orthographic Drawings

Orthographic drawings are one way of communicating solutions. They show objects in separate two-dimensional views. They are often called *multiview drawings* because they use several views to describe the object. When creating multiview drawings, designers must select the best views to describe the object. All objects can be shown in up to six different views. Imagine a typical house for a moment. The house has six different sides. You can easily walk around the house to see the front, back, and the right and left sides. If you were able to fly over the house, you would see the top. If the house had a crawl space, you would be able to look at the bottom of the house. In an orthographic drawing, each of the six views of the house can be drawn as individual two-dimensional drawings.

Because there can be six different views, multiview drawings can be very confusing. To help lower the confusion, each view has a specific location. The placement of the views is standard, which means all designers place the views in the same location. To understand the location of the views, imagine placing the solution in a glass box, then rotating the box and drawing exactly what you see on each of the six sides. See **Figure 17-12.** Now imagine the box is hinged on the front, on each side. When the box is opened, each view is in the right place. See **Figure 17-13.**

Using the glass box to see all six sides is an example of orthographic projection. *Orthographic projection* is the term used to create six views of an object. All six views are not, however, always needed. Some objects can be completely described in one or two views. It is actually uncommon to see

Figure 17-12. The views of the object are projected on the sides of the glass box.

TechnoFact: Orthographic drawings can contain views other than the six orthographic views of an object. Special views are used to communicate objects that have odd shapes. Auxiliary views are used to show the true size and shape of inclined surfaces. They are placed adjacent to an orthographic view, and their line of sight is perpendicular to the inclined surface.

orthographic drawings with all six views. Designers do not waste the time to create unnecessary views. Drawings with one, two, or three views are the most common and have specific uses.

➤ One-view drawings. These are used to show flat parts and objects. See **Figure 17-14 (top)**. Sheet metal and hardboard pieces are drawn using one view. The top of the object is the view normally shown.

➤ Two-view drawings. These are used when drawing cylindrical objects. Soda cans and baseball bats are drawn with two views. In two-view drawings, the front and side or end are shown. See **Figure 17-14 (center)**.

➤ Three-view drawings. These are used for rectangular objects. They are usually created using the front, top, and right side of the object. See **Figure 17-14 (bottom)**. Three-view drawings are often used to draw complex parts.

When creating orthographic drawings, the first step for designers is to identify the front of the object.

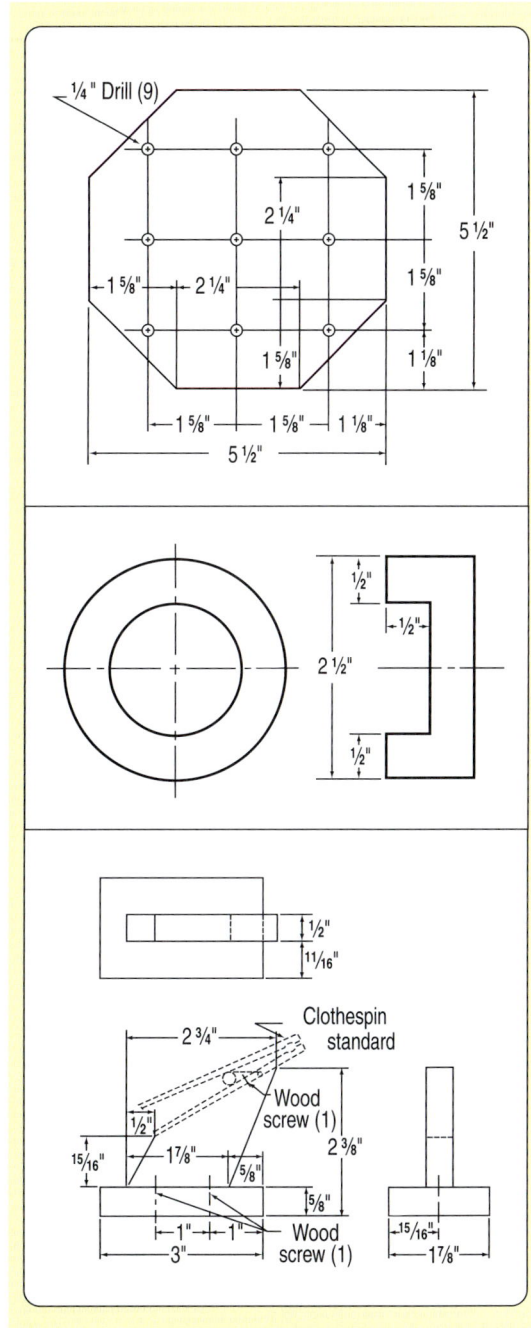

Figure 17-14. Orthographic drawings can be one-, two-, or three-view drawings.

Standards for Technological Knowledge

Extend
Have the students sketch the front and end views of a soda can or bottle.

Discussion
Discuss the best way to choose the front view of an object.

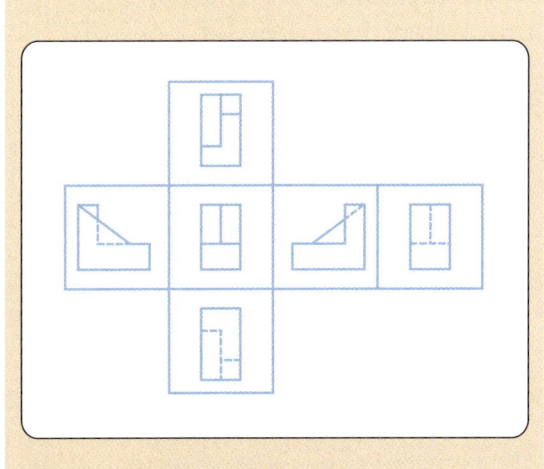

Figure 17-13. When the box is open, the views are in the right place.

TechnoFact: Section views are used to communicate objects that have complex internal features. A section line is drawn through the object and serves as a cutting plane. The inside of the object is then shown in a separate view. Section patterns, often a series of parallel diagonal lines, indicate the surfaces intersecting the cutting plane.

The front is always the side showing the most detail. Sometimes the front of the object is very obvious. For example, the front view of a television set would be the side the television screen is on. Other times, however, it is harder to determine which view should be the front view. An automobile, for example, would be drawn with the side of the vehicle as the front of the object. It is the best front view because it shows the most detail. The side shows the slope and aerodynamics of the car. Once the front of the object is selected, the remaining views are easy. The top view is found by rotating the object down 90°. The right side view can be drawn by turning the object 90° to the left. All the views should be 90° from the view next to them.

Line Types

One challenge with orthographic drawings is that some objects have parts that cannot be seen in all the

Figure 17-15. Hidden lines are used to show the slots in the toaster.

Figure 17-16. The alphabet of lines includes the standard lines used in technical drawing.

views. Looking at the front view of a toaster, for example, you are not able to see the slots for the bread. If an object has holes drilled into it, you do not know how deep they go into the object. For this reason, there is a type of line used to show features that cannot be seen. See **Figure 17-15.** These lines are called *hidden lines.*

A hidden line is just one type of line used in multiview drawings. There are many different types of lines. They are known as the ***alphabet of lines.*** See **Figure 17-16.** The most common lines in the alphabet are the construction line, object line, border line, hidden line, and centerline. Each line is drawn differently and has a specific purpose:

➤ Construction lines. These are solid lines drawn lightly. They are used to lay out drawings.

➤ Object lines. These are the lines used to show the edges of the object. They are heavy and dark lines.

➤ Border lines. These are the thickest and darkest lines. They are used to surround the edges of the paper. Borders are an important part of drawings.

TechnoFact: The lines making up the alphabet of lines not only have specific functions, they also have specific sizes. Visible lines should be solid, with a thickness of 0.030". A typical hidden line should be 0.015" thick, with the 1/8"-long dashes evenly spaced 1/32" apart.

Figure 17-17. Dimensions include extension lines, dimension lines, and dimension text.

➤ Hidden lines. These show the location of objects that cannot be seen. Hidden lines are drawn with short dashes.

➤ Centerlines. These are used to show the centers of any objects with center points, like arcs, ellipses, and circles. Centerlines are drawn with alternating long and short dashes.

Dimensions

There are other types of lines used in drawings. *Dimensioning* is a process using two types of lines: extension lines and dimension lines. See **Figure 17-17.** Extension lines are lines drawn from an object to the outside of a view. They show the edge or center of an object. Dimension lines are drawn between two extension lines. The dimension lines have an arrowhead on each end and a measurement in the middle. Together, the extension and dimension lines are used to describe the dimension of an object. The dimension of an object, as described in Chapter 14, is a measurement giving one of three types of information:

➤ Size dimensions. These describe the length of objects.

➤ Location dimensions. These show the distance between two features (such as circles or arcs) or between a feature and the edge of the object.

➤ Shape dimensions. These show the angles between different parts of an object.

Dimensions must be displayed clearly so they can be understood. See **Figure 17-18.** The designers must include all the dimensions needed to produce the object. They should not, however, include more dimensions than are necessary. There are many rules designers use when dimensioning drawings:

➤ Dimensions should not be placed inside the object.

➤ Dimensions should be placed between the views, not along the outside of the drawing.

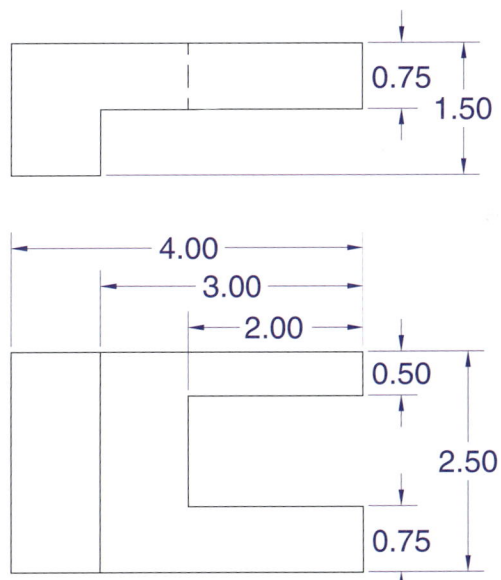

Figure 17-18. Dimensioned drawings should be easy to read and should follow the rules of dimensioning.

Standards for Technological Knowledge

Reinforce
Explain the three types of information that can be provided in dimensions.

Discussion
Ask the students why there are rules for dimensioning.

TechnoFact: Tolerances are also part of dimensioning. Tolerances are the total amount of variance allowed in the manufacturing of the object. For example, a part may call for a thickness of 2.25", with a tolerance of 0.03". The part would be acceptable as long as the thickness was between 2.22" and 2.28".

Figure 17-19. Pictorial drawings appear as the eye sees them. (Keith Nelson)

➤ Dimensions should be placed on the view best showing the measurement.

➤ The location and size of all circles and arcs must be shown.

Circles and arcs are dimensions using a leader. A leader is a line with an arrow on one end and a measurement on the other. The measurement given for a full circle is always the diameter of a circle. The radius measurement is shown for any incomplete arcs or circles. The dimensions used should not distract viewers. They are used to better communicate the sizes and shapes of objects.

Pictorial Drawings

Pictorial drawings, as the name suggests, are like pictures. See **Figure 17-19.** They show how objects look to the eye. Pictorial drawings are easier to understand than orthographic drawings. They are often used when the viewer does not understand orthographic drawings. Orthographic drawings can be confusing because they are separated into different views. In pictorial drawings, the views are drawn together. Instead of a flat drawing of the sides, pictorial drawings are drawn to

look three-dimensional. Because they look real, they are easy to understand.

There are several types of pictorial drawings. See **Figure 17-20.** These are the three main types:

➤ Isometric drawings.

➤ Oblique drawings.

➤ Perspective drawings.

You may remember these types from the pictorial sketches in Chapter 13. Pictorial drawings, however, are different than sketches. The sketches created

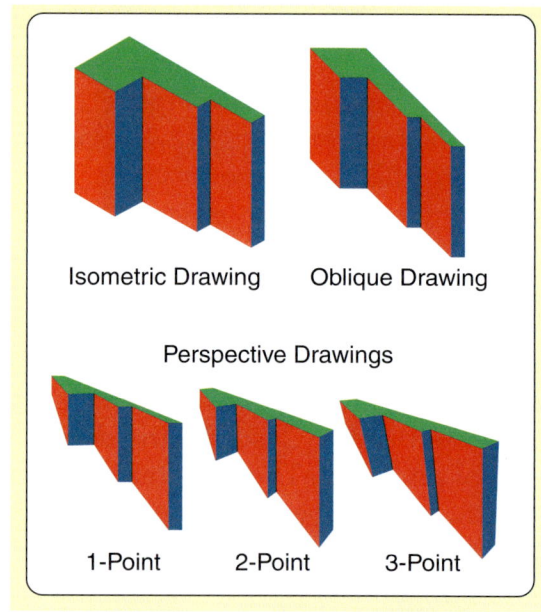

Figure 17-20. Pictorial drawings include isometric, oblique, and perspective drawings.

TechnoFact: Isometric drawings are the most common example of axonometric drawings. All axonometric drawings are rotated on an axis. Other axonometric drawings include trimetric drawings, in which no angles are equal, and diametric drawings, in which two of the angles are equal.

Figure 17-21. Isometric drawings are drawn at 30° on each side.

in the earlier steps were simply ideas, and they were not very detailed. The pictorial drawings in this step are very detailed and accurate. They represent the final solution of the design process.

Isometric Drawings

Isometric drawings are one of the most common types of pictorial drawings. See **Figure 17-21.** The word *isometric* means "equal measurements." Isometric drawings are created with angles equal to each other. The three lines making the front corner are always 120 degrees from each other. The lines along the bottom of the object are each 30 degrees from horizontal.

All lines in an isometric drawing are drawn to scale, which means each line is accurate when it is measured. The length, width, and height of the object are all drawn to scale, or accurate, in isometric drawings. They may be drawn full-scale. Full-scale drawings

are used when the object will fit on the paper using the actual measurements. These drawings can also be drawn to other scales. Scales are labeled using two numbers. The first number is a length on the paper, while the second number is the actual length in real life. For example, a scale of ½"=1" means that every ½" on the paper equals 1" on the object. So, if the object is 3" long, it is drawn at 1½". Scales are used when the object is either too big for the paper or too small to be seen.

Oblique Drawings

Oblique drawings are not used as often as isometric drawings. They are used to show objects in which one view is the most important, such as kitchen cabinets or television sets. See **Figure 17-22.** These drawings are

Figure 17-22. Objects with most of the detail in the front view are often drawn with oblique drawings.

TechnoFact: Mechanical drawing is used in pattern development. A pattern development is a flat drawing that could be transferred to a material and folded to make a product. Packaging designers create patterns of boxes and packages. Pattern development is also used in the sheet metal industry.

Standards for Technological Knowledge

Resource
Transparency 17-4:
One-Point
Perspective

Resource
Transparency 17-5:
Two-Point
Perspective

Reinforce
Review the terms
vanishing point,
horizon line, and
ground line and
discuss how they
are used in perspective drawings.

Example
Show the students
examples of one-
point, two-point, and
three-point perspective drawings.

created using the front view from an orthographic drawing. The front of an oblique drawing is the only true size view. Lengths and angles can be measured only on the front view. To give the front view depth, lines angling backward are drawn from the front of the object. The angle of the lines can be drawn at any angle. The most common, however, are 30° and 45° angles.

There are three main types of oblique drawings. The only difference between the three types is the depth of the object. When the depth is drawn the true length, the drawing is a cavalier oblique. A drawing showing half of the length is called a cabinet oblique drawing. A general oblique drawing is drawn with a depth of anywhere from one-half to full-size. Oblique drawings are mainly used for display drawings. They are not often used for drawings that will be measured because only the front is accurate.

Perspective Drawings

Perspective drawings are the drawings most often used in the presentation of ideas. They are the pictorial drawings most like what the eye sees. Perspective drawings are very similar to perspective sketches, discussed in Chapter 13. These drawings, however, are created with drawing tools and are much more accurate than the sketches. They rely on vanishing points and vertical lines. A vanishing point is a spot toward which the lines are pointed.

The number of vanishing points in a perspective drawing determines its type. A one-point perspective is drawn using only one vanishing point. The one-point perspective drawing resembles the oblique drawing. The front view is flat on the paper, while the top and sides of the drawing are at an angle. The difference is that in the oblique drawing, all the angled lines are parallel, or the same angle. In the one-point perspective, all the lines are either vertical or angled back to a vanishing point.

A drawing with two vanishing points is called a *two-point perspective*. See **Figure 17-23.** It looks similar to an isometric drawing. In a two-point perspective, the vanishing points are on opposite sides of the horizon line. Each

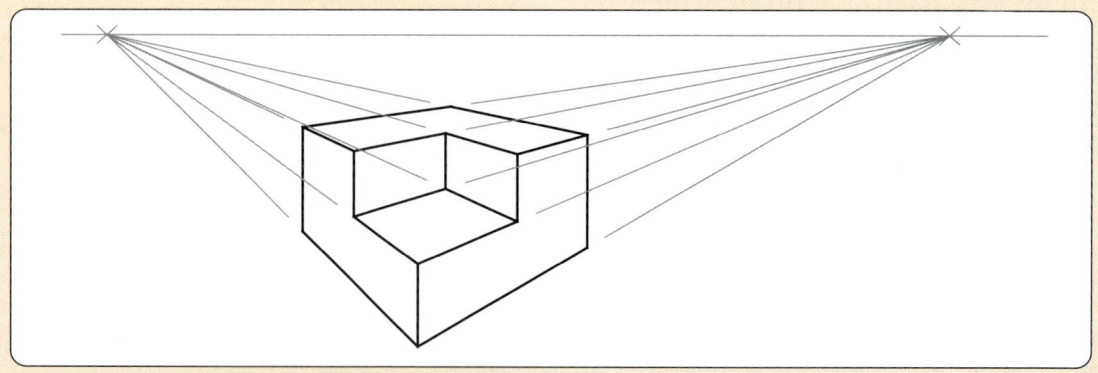

Figure 17-23. This is a two-point perspective drawing.

TechnoFact: Perspective drawings are used in the field of artistic design, as well as in product design. Both fields often use perspective drawing boards. The boards are made specifically for creating perspective drawings and include straight edges that pivot from the vanishing points.

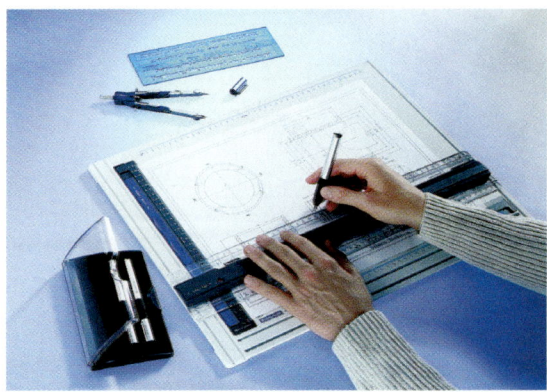

Figure 17-24. Many designers still use traditional drawing tools. (Staedtler Mars)

side of the object in the sketch extends toward one of the vanishing points.

A three-point perspective drawing is a drawing that has three vanishing points. Three-point perspective sketches can be used for tall buildings and structures. Three-point perspective drawings are not used as much as one- and two-point perspectives.

Mechanical Drawing Tools

Tools are involved in producing mechanical drawings, as they are involved in almost all uses of technology. They are used to help designers draw straight and accurate lines. See **Figure 17-24.** These tools are also used to make mechanical drawings easier to create. Without mechanical drawing tools, it is difficult to draw perfect circles or create parallel lines. There are two types of technical drawing tools:

➤ Traditional tools.

➤ Computer-aided design (CAD) tools.

Traditional Tools

You may already be familiar with a few traditional technical drawing tools. When you think of architects or engineers, you may imagine them using T squares, triangles, and compasses. All these tools are traditional technical drawing tools. Traditional tools are used when designers and draftspeople create drawings using pencils and paper.

A basic set of traditional tools includes a drawing board, a T square, a 45° triangle, a 30°-60° triangle, a compass, an eraser, templates, and a pencil. See **Figure 17-25.** The *drawing board* provides a smooth surface on which to draw. It also includes one straight edge for the T square to move along. The *T square* is a tool made up of two pieces, the head and the blade. The two pieces are perpendicular to each other. The head rests along the side of the drawing board and keeps the blade horizontal across the paper. The T square is used to draw all horizontal lines in the drawing. It is also used as a ledge for other tools, like triangles.

A standard set of traditional drawing tools includes two triangles. One

Figure 17-25. These tools make up a traditional set of drawing tools.

TechnoFact: Drafting machines are tools that can take the place of the T square and the triangle. The drafting machine has two straight edges, one perpendicular to and one parallel to the bottom edge of the paper. These straight edges can also be rotated to any angle.

of the triangles, called the *45° triangle,* has one 90° corner and two 45° corners. The other triangle is called a *30°-60° triangle* and has one 90° corner, one 30° corner, and one 60° corner. The triangles are used with the T square to draw vertical lines. They are also used to draw lines and 30°, 45°, and 60° angles. If both triangles are used together, angles of 15° and 75° can also be drawn.

The *compass* is the drawing tool used to create circles and arcs. The compass resembles the letter *A*. One leg of the A has a small, pointed tip, and the other holds a pen or pencil. The compass is held at the top and rotated around to draw circles. See **Figure 17-26.** A tool that is very similar to a compass is a divider. The main difference is that dividers have two pointed tips instead of one. Dividers are used to transfer measurements without using a scale.

Figure 17-27. Scales help engineers and designers draw large objects on paper.

A *scale* is a tool used to make measurements. It is a specialized type of ruler. Scales have different sides that help designers draw objects larger or smaller than they are in real life. Designers use different types of scales, depending on the industry in which they work. Architects and interior designers use architectural scales. Civil engineers and land surveyors use engineering scales. See **Figure 17-27.** Industrial designers use mechanical drafting scales. Each scale has sides that are useful for the type of drawings they will create. Architectural scales have sides used in drawings where one-eighth, one-fourth, one-half, and three-fourths of an inch equal one foot. Engineering scales are used for drawings measuring ten, twenty, thirty, forty, fifty, or sixty feet for every inch. Mechanical drafting scales have a side for full-scale, divided in sixteenths of an inch, and sides for one-eighth, one-fourth, one-half, and three-fourths scales.

Figure 17-26. Compasses are used to create perfectly round circles.

TechnoFact: Designers create drawings on a number of different media. Many thicknesses, grades, and colors of paper are used. Designers also use vellum and Mylar™ films to copy their drawings with blueprint machines. The films are clear and allow the light of the machine to pass through the empty areas.

Figure 17-28. Templates make drawing standard objects easier.

Another tool that is a great help to designers is a template. A *template* is a piece of plastic with shapes and symbols cut into it. Designers place the template on their drawings and trace around the edge. Common templates are circular and architectural templates. See **Figure 17-28.**

All the traditional tools are used together to create mechanical drawings. Many designers use them in many industries. Some designers feel it is important to understand how to use traditional tools. Other designers feel these tools are not as important as computer-aided design (CAD) tools.

Computer-Aided Design (CAD) Tools

Computer-aided design (CAD) utilizes the computer to create technical drawings. See **Figure 17-29.** Many types of designers use CAD tools that are part

of a computer system. A typical CAD system includes a computer, a monitor, a printer or plotter, storage systems, input devices, and CAD software.

The computer used for CAD tools is a bit different than the typical computer used in a household. CAD computers require a large amount of memory and a large processor. Some computers used for CAD even have two processors. The computer processor speeds up the time it takes for the computer to perform commands. The memory in a computer allows the computer to handle more functions at once.

Another difference in CAD systems is the way these systems store files. Many CAD drawings are large and take up a lot of disk space. So, designers use tape backup drives and CD drives on which they can write information. Using these drives allows the designers to save the drawings they create, which is one advantage of CAD drawings. Designers are able to save

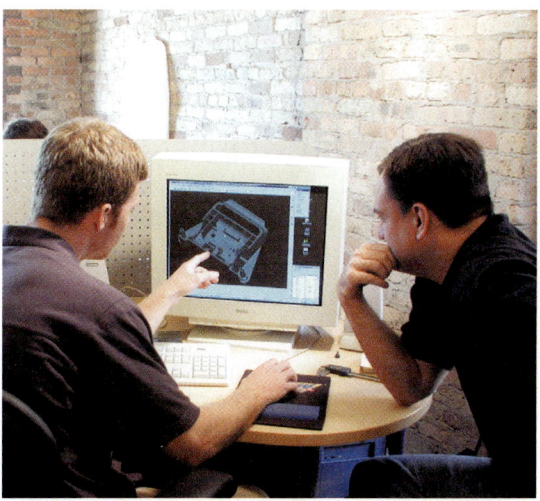

Figure 17-29. Designers often use computer-aided design (CAD) tools. (Product Development Technologies, PDT)

TechnoFact: Computer monitors display images by illuminating collections of individual pixels. Increasing the number of pixels in the monitor increases the quality of its display. The number of pixels in a display is known as the *resolution.* A resolution of 1208 x 1024 indicates there are 1208 columns and 1024 rows of pixels in the display.

Figure 17-30. Large computer monitors with high resolutions help designers see large drawings.

the drawings and easily make changes after the drawings are complete.

Because the designers must be able to see the drawings they are creating, they often use large computer monitors. See **Figure 17-30.** The bigger monitors allow them to see more of their drawings. The designers are also able to zoom into different areas of the drawings to look more closely. CAD computer monitors must also have a high resolution. The resolution controls how much information can be displayed on the screen at one time.

CAD systems use a different type of input device than most systems. Like most other computers, a CAD system uses a keyboard and a mouse. Some also use a *tablet,* however, which is a device that has a pad and a puck. The puck resembles a mouse. It is used to draw objects and select commands from the tablet pad.

The tools used in a CAD system often include a type of printer other

computer users do not use. A *plotter* is a printer able to print on large paper. Many architectural drawings, for example, need to be printed on paper twenty-four inches wide and thirty-six inches long. Plotters allow designers to make prints this big.

The actual computer requirements depend on the CAD software used. There are many types of CAD software. Software ranges in price from a few dollars to several thousands of dollars. The quality and limitations of the software also range from low quality to outstanding quality. Designers purchase the right software for their needs. Some software is created for a typical homeowner to create floor plans and landscape designs. Other software allows designers to create complex parts or even design entire buildings and cities.

Making Mechanical Drawings

Mechanical drawings can be produced with either traditional or CAD tools. Each type has advantages and disadvantages. Traditional drawings are often easier to create because the designer does not have to learn a computer software package. CAD drawings, however, are easier to make accurate and easier to correct or change. Regardless of the type of drawing tools used, the process of creating a drawing is very similar. For example, the following steps are used to create an orthographic drawing of a picture frame using either traditional or CAD tools. See **Figure 17-31.**

TechnoFact: CAD software is based on a coordinate system. Most software uses the same Cartesian coordinate system used in mathematics. The coordinate system allows the location of any point in space to be given as a set of three numbers, the distances from the origin along the X, Y, and Z axes. For example, a point with the coordinates (4,2,3) is 4 units from the origin on the X axis, 2 units from the origin on the Y axis, and 3 units from the origin on the Z axis.

Standards for Technological Knowledge

9

Extend
Have the students create a bill of materials and specification sheets for a product they are designing.

Figure 17-31. These steps are used to create mechanical drawings.

1. Decide how many views are needed to show the features of the object.

2. Sketch the views on a sheet of paper.

3. Draw construction lines outlining the front and top views.

4. Add a 45° line and transfer lines outlining the right side view.

5. Locate the centerlines of any circles or arcs.

6. Add the rest of the lines in all the views.

7. Erase or trim the construction lines.

8. Add any hidden lines and additional centerlines needed.

9. Add dimensions.

Figure 17-32. Designers display their work in presentations to clients or management.

Presenting the Solution

The final stage in communicating solutions is to present the final solution. All products must be approved before they are sent to production. To receive approval the designers must present the final solution to management. See **Figure 17-32.** The presentation includes a written report highlighting the features of the design. The report covers all the expectations and costs of the design. It also includes all the final drawings of the solution. The designers distribute the written report along with an oral presentation. If the presentation and reports are approved, the solution goes into production.

Summary

Solutions must be communicated to several groups of people. Management must approve the solution before it is manufactured. Manufacturers must understand the solution in order to produce it. Consumers must be able to assemble and use the solution. Designers use three different documents to communicate the solution. They create bills of materials, drawings, and specification sheets showing and explaining the solution.

The drawings created are called *mechanical drawings.* They are created using either traditional or CAD tools. The drawings are very accurate and precise. They can be detailed, assembly, or schematic drawings. Each drawing type has specific uses. Detailed drawings are usually created using orthographic drawings. Orthographic drawings have several views used to describe the object. Assembly drawings are normally pictorial drawings. Pictorial drawings ar created to look three-dimensional and are much easier to understand than orthographic drawings. When all the drawings and documents are complete, they are presented to management or the client for final approval.

Curricular Connections

Language Arts

Research a type of material. Write a specification sheet explaining the common uses, material properties, and health risks of the material.

Science

Create a drawing that would be used with a science lab report. The drawing should show the equipment and supplies used in the experiment.

Mathematics

Draw a 1″ grid on a simple cartoon image. Using a scale, recreate the cartoon image at three different scales (for example, 1″ = ½″, 1″ = 2″, and 1″ = ¼″).

Activities

1. Bring a toy or household object with three or more parts from home. Create a bill of materials for the product.

2. Create a detailed drawing of an object in your classroom (for example, a desk, bookshelf, or chair). Use the correct line types and add dimensions.

3. Create a set of documents for the solution you have created in the previous chapters. Once the documents are complete, present your solution to the class.

TechnoFact: The methods of creating mechanical drawings are very similar when using traditional tools and 2-D CAD software. However, 3-D CAD software is a bit different. With 3-D CAD software, the designer draws an object in three dimensions, and the software generates the individual orthographic views.

Test Your Knowledge

Do not write in this book. Place your answers to this test on a separate sheet of paper.

1. A bill of materials contains information about the health risks of a material. True or false.

2. A mechanical drawing is an accurate drawing created using drawing tools. True or false.

3. Give three examples of objects that would be communicated by engineering drawings and three that would be communicated by architectural drawings.

4. Name the three types of working drawings.

5. A schematic drawing can be used to show the flow of a product. True or false.

6. Orthographic drawings are also called _____ _____.

7. List four examples of traditional drawing tools.

8. Computers used for CAD are the same as most computers used as household computers. True or false.

9. The steps used to create drawings are very similar when using traditional tools and when using CAD tools. True or false.

10. Explain the process of presenting a solution.

Modular Activity

This activity develops the skills used in TSA's Video Challenge event.

Video Production

Activity Overview

In this activity, you will develop a storyboard, script, production plan, and finished video for one of the following themes:

➤ Profile of your school
➤ Profile of a student or faculty member
➤ Profile of a local business or organization
➤ Discussion of a community issue

Materials

➤ Paper and pencil
➤ 3´´ x 5´´ note cards
➤ Three-ring binder
➤ Video camera
➤ Computer with video production software

Background Information

Planning the video. Plan your video as a story, making sure it has a beginning, middle, and ending. List the important points you want to cover, then use those points to create a simple outline. Think visually—remember that your story will be told mostly with pictures, instead of words.

Storyboard. To help yourself and the others involved in the project visualize the video, make a *storyboard.* This is a series of simple pictures, almost like a comic book, that will show each change in what the viewer will see. Make each individual storyboard sketch on a separate note card. This will allow you to try different combinations to get the most effective sequence for the video. When you have a final sequence, number your cards.

Scripting. Even though the pictures (video) will tell most of the story, the spoken words (audio) will tie-together the pictures. Your script will follow the sequence of the storyboard, with a brief description of each numbered shot and the actual words that will be recorded to accompany that shot. Sometimes, audio is not fully scripted.

Production planning. Video is almost never shot in the final program sequence, so careful planning is needed to make the most effective use of time and resources. For each numbered shot in your storyboard, your production plan should list the location of the shot, any on-camera people involved, any props or special materials needed, and (where relevant) the time of day the shot must be made. For initial planning, note cards with the information for each shot are useful. They can be sequenced to make efficient use of your resources—for example, grouping together all the shots at one location or involving the same people.

Shooting the video. To provide visual interest in the finished video, be sure to vary your types of shots—don't shoot all close-ups or all wide shots. Suit the shot type to the subject and to the flow of the video as shown on your storyboard. To help identify each shot, make a large card with the scene number written in bold marker. Shoot a few seconds of the card at the beginning of the shot. To provide editing flexibility, shoot a variety of *cutaway* shots. These are used as transitions between other shots. For example, if you are showing a school assembly program, include a few close-ups of members of the audience listening to the speaker. When shooting interviews, you would normally concentrate on the person being interviewed. When the interview is finished, shoot some close-up shots of your reporter asking questions and some of he or she just looking interested and nodding.

Camera and microphones. If you are shooting in a location where you depend on battery power for the camera, be sure to have one or more fully charged "backup" batteries available. When lighting conditions change (such as moving from indoor to outdoor settings), be sure to check the camera's *white balance* and adjust if necessary to avoid adding a color cast to your video. When shooting general scenes, such as a cafeteria or sporting event, record the "live" audio, which can later be used in post-production. For recorded narration or interviews, avoid using the camera's built-in microphone, if possible. Much better audio quality will result from using a microphone (wired or wireless) placed close to the subject being recorded.

Postproduction. When editing your scenes, use various visual effects, such as dissolves, fades, and zooms. Music, sound effects, and additional voice-over narration can be added during postproduction. Use cutaway shots to avoid disturbing *jump cuts* in the video. These occur, for example, when an interview must be edited to eliminate unwanted material. If the camera is focused on the speaker, changes in expression or head position between the adjoining shots will make the cut obvious. By inserting a second or so of video showing the interviewer looking interested, the change will not be noticed.

Guidelines

➤ Final video must be 3 minutes to 5 minutes in length.

The following items must be included in your final printed report:
➤ Cover page
➤ Table of contents
➤ Storyboard
➤ Script
➤ Description of editing techniques

Evaluation Criteria

Your project will be evaluated using the following criteria:
➤ Storyboard, script, and production plan
➤ Correlation of storyboard and video
➤ Technical quality of video

Chapter 18
Improving Solutions

Did You Know?

➤ Companies often update and improve their logos. A logo is a symbol of a corporation's identity. Logos can be traced back to ancient Greece, where craftspeople marked their work with their own symbol. Today, they are used as the main symbols of companies.

➤ Logos are designed so they are easy to recognize and remember. They should also be pleasing to look at and in good taste. It would hurt a company's image if their logo were disrespectful and unsightly. Most importantly, the style of the logo must be up-to-date. Think of several major companies. Have they updated their logos in your lifetime? Are their new logos more up-to-date?

➤ Also, logos must be able to be reproduced easily. For this reason, most logos use only one, two, or three colors. The easier the logos are to reproduce, the less expensive it is to print them on a number of different items.

Objectives

The information given in this chapter will help you do the following:

➤ Explain the importance of improving solutions.

➤ Identify the four main reasons for improvement.

➤ Describe the four major sources of information about products.

➤ Summarize the process of improving solutions.

Key Words

These words are used in this chapter. Do you know what they mean?

brand name

building code

competition

competitor

consumer

function

internal source

patch

profit

recall

regulation

standardization

Chapter Outline
Did You Know?
Objectives
Key Words
Reasons for Improvements
  Profit
  Competition
  Function
  Safety
Sources of Improvement
  Consumers
  Competitors
  Government Organizations
  Internal Sources
Making Improvements
Summary
Curricular Connections
Activities
Test Your Knowledge

Have you ever played a game you wish had more options, had trouble finding information on a website, or taken a toy back to the store because you did not like it? If so, designers want to know about your dissatisfaction. They are concerned about how well people like their products. Designers work hard to improve their designs, even after the designs have been produced.

In this step of the design process, designers search for different ways to improve their products. It may seem to you that once designers reach this stage, they are done with the design. The designers have gone through many steps to get to this stage. They have done everything from identifying a problem all the way through solving it. This included conducting research and tests and creating sketches, models, and drawings. The solution has probably even been manufactured and sold by this point. See **Figure 18-1.** You may think the designers are finished with the product by now.

Designers, however, are never really done with a solution. Design is a process that does not stop. Designers can always go back and make changes to a design. The design process is like a ladder. See **Figure 18-2.** At this step of the ladder, designers may choose to step backward to a previous step. When the designers went through the process the first time, they created a solution. It was the solution that seemed to be the best at the time. The designers used research, surveys, and tests to determine the solution they designed. As time goes on, however, many things change. The best design at one moment may not be the best solution at a later time. For example, imagine an automobile designed in 1945. See **Figure 18-3.** Is this automobile the best solution in today's world?

Figure 18-1. This is a finished image of the Palm V® handheld computer. (Photo: Steven Moeder, Courtesy: IDEO)

TechnoFact: The keyboard currently used on computers and typewriters is the product of innovation. The first keyboard layout caused the typewriter to jam because the type bars for commonly used letters were close together, causing them to collide frequently. In 1872, Christopher Sholes rearranged the keyboard to separate frequently used letters. This layout is commonly referred to as the QWERTY layout and is the standard today.

Figure 18-2. The steps of the design process can be seen as a ladder.

The antique car does not have air conditioning, power steering, a CD player, or an airbag. This car is not desirable and would not sell very well. The design is obsolete, or out-of-date. Some designs take a long time before they become obsolete. Other designs are obsolete in a very short amount of time. Consider a computer. A computer designed several months ago is already outdated. In a matter of a few months, there are new computers that are faster and better.

Reasons for Improvements

Keeping a product from becoming outdated is a concern for designers. If their product is out-of-date, the company will suffer. Designers must improve their solutions to keep them current. The concern to improve products can be grouped into four categories:

➤ Profit.

➤ Competition.

➤ Function.

➤ Safety.

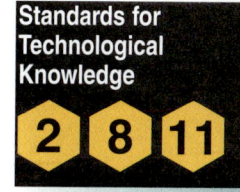

Standards for Technological Knowledge

2 8 11

Research
Have the students select a product and create a timeline of improvements that have been made to the product.

Discussion
Discuss why it is important for companies to improve their designs.

Figure 18-3. This antique auto does not have many of the features found in cars today. (Ford Motor Company)

TechnoFact: The automobile is a product that is constantly being improved. Cadillac has improved its vehicles to include sensing technologies. It has developed infrared technology that detects objects in low visibility conditions and displays them on a heads-up display on the windshield. It also included sensors in the rear of the car to alert the driver of the presence of objects as the vehicle backs up.

Profit

Profit is a major factor in product improvement. It is the amount of money left over after the company's bills are paid. Designers work to increase their profit in two different ways. They can raise the price of the product. This increases the income from the solution and is the easiest way to increase the profit. People, however, may choose not to pay the extra money for the product. If you went to the store to buy a game costing five dollars more than it did the day before, would you pay for it? You would probably find another product to buy with your money. If all people stopped buying the product, the company would lose money instead of making more. See **Figure 18-4.** If the company raises the price, it should also improve the function of the product. Improving the product helps people feel that the extra money is worth spending.

The second option designers have to increase profit is to lower their expenses. If it costs less to produce and sell the product, they will have

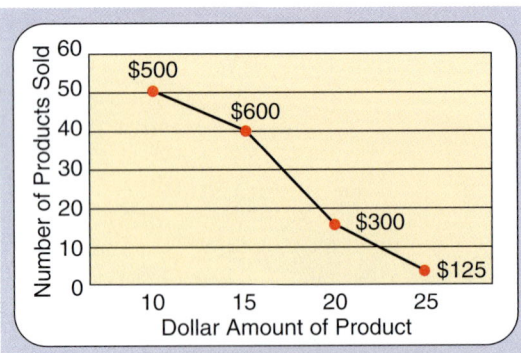

Figure 18-4. This graph shows how many products will sell at each price and how much money they will generate.

Figure 18-5. Manufacturing is a major part of a company's costs.

more profit. There are several places designers may try to lower their costs. One place is in manufacturing. See **Figure 18-5.** Designers may hire a manufacturing engineer to examine how the product is produced. They may find that the manufacturer has waste materials that can be used to make more products or that they have more workers than they need; unfortunately, this can cause companies to lay off workers. The designers may also be able to use less expensive materials to product the products. If they can lower the cost of materials, they can increase profit.

Competition

The second reason for improvement is competition. Designers improve their products to keep ahead of their competitors. *Competition* is an important part of business. It is the

TechnoFact: The role of a manufacturing engineer is to design the most effective way of producing a product. These engineers focus on both time and cost efficiency. They design the layout of the production line and the tooling used in the manufacturing.

term used when two or more companies sell the same types of products. Businesses compete to gain customers. Competition is great for consumers because it helps keep prices down. It also keeps the quality of products at a relatively high level. Designers improve their designs to stay in front of their competitors. They want consumers to choose their products over the products of another company. To ensure this happens, they make their products more appealing. They may begin by improving the design of a product's package. The package is the first thing that catches the consumer's eye. See **Figure 18-6.** Many consumers are willing to buy the first product they see. Some consumers, however, need more than a nice package. Designers also improve the function of the solution. They include more or better features than the competitors' products.

Designers also improve products to build a brand name. A *brand name* is the title of a line of products. If a consumer likes one product with a certain brand name, they are more likely to buy another product with the same name. For example, someone who likes Ruffles® potato chips, which Frito-Lay makes, would be more likely to buy another type of chip Frito-Lay makes than chips another company makes.

Function

Function is the third reason to improve a solution. It was mentioned as a part of profit and competition. Function is also, however, a major reason for improvement on its own. Throughout the design process, sketches and drawings were analyzed. Prototypes and mock-ups were tested and evaluated. It is still possible, however, that the solution does not function as well as it could. There may be certain uses that were not tested. The product may fail when it is used in those manners. All products are subject to failure. Once these failures are discovered, the designers must improve the solution to eliminate the problems.

When a product has problems or issues needing to be addressed, they are handled quickly. This is very common in software design. When software designers discover a problem, they improve the solution and create a patch. A *patch* is a small piece of software that fixes a known problem. The patch can be installed on top of the current software. When larger changes are made to software, designers create new versions. New versions

Figure 18-6. Packages are designed to be attractive and grab your attention. (Design Central, design firm; Medlife, client)

Standards for Technological Knowledge

Research
Have the students select a product and compare the features found in the product to the features in competing products.

Research
Have the students choose a company brand name and create a display showing the products available in that brand.

TechnoFact: Software patches and fixes are often available on the manufacturer's website. Microsoft has developed an update program the user can configure to automatically search for and download the newest patches for Windows® computer programs.

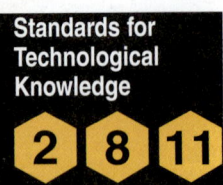

of software are created after all the problems from the original software have been identified. They also have changes in the functions or features of the program. If the changes are major, the version is given a new whole number (like version 6.0). See **Figure 18-7.** If the changes are small, it is given a decimal number (like version 3.2).

Safety

Safety is the last major reason for improving solutions. The safety of the solution was tested during the design. Some safety issues, however, are found after the product has been produced. When a major safety problem is found, a product recall is issued. A *recall* is a request that those who purchased a product bring it back. In return, the consumers receive a new product, or their original product is altered to make it safe. Recalls are common in products like food and children's products. If it is found that food may be contaminated, the company recalls it. Companies want to ensure all their products are safe. Many companies rely on product registration to issue recalls. They will send a notice to the registered owners of the product.

When safety concerns that can lead to death are found, mandatory recalls are issued. A mandatory recall means if you do not return the product, you are using it at your own risk. Not all safety issues require mandatory recalls. Some issues are very minor and are handled using voluntary recalls. A consumer with a product

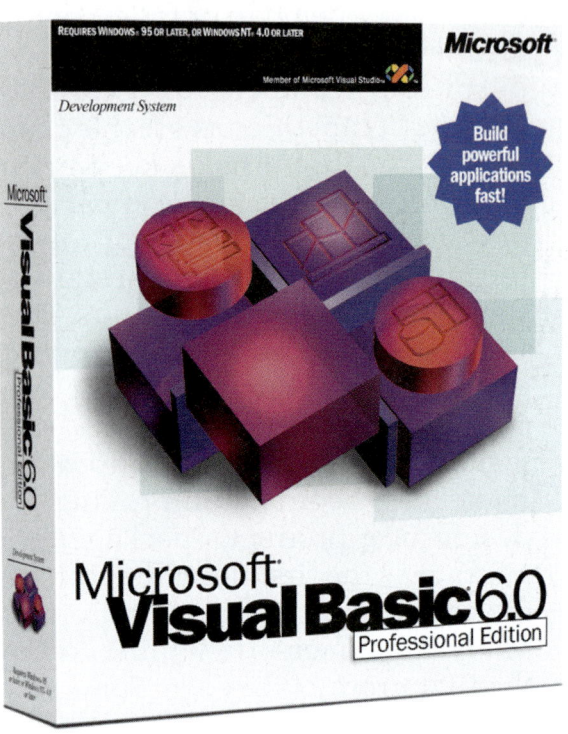

Figure 18-7. New software is often given a new version number. (Box shot reprinted with permission from Microsoft Corp.)

that has been voluntarily recalled has the option of returning it for a new and improved product. Consumers are not risking serious injury or death, however, if they choose not to return the product.

Sources of Improvement

Designers rely on a number of different sources of information about their products. These sources help them determine the types of improvements needed. They include competitors, customers, government organizations, and the companies themselves.

TechnoFact: The U.S. Consumer Product Safety Commission is an excellent source for information about product recalls. Its website (www.cpsc.gov) allows consumers to search by product type or company. The recalls posted on the site give detailed information about the product defect and the best way to contact the company.

Figure 18-8. The users of the product can be great sources of information. (LEGO and the LEGO brick configuration are trademarks of the LEGO Group ©2002 The LEGO Group. The LEGO® trademarks and products are used with permission. The LEGO Group does not sponsor or endorse the publication.)

Extend
Have the students write a letter to a company regarding the improvements they would like the company to make in its product(s).

Research
Have the students find consumer feedback pages on the websites of several companies.

Consumers

Consumers may be the best source of information for improving solutions. See **Figure 18-8.** *Consumers* are the users of products. After using a product, they know what they like and do not like about it. Designers receive feedback from consumers in several ways.

One source of feedback is when customers return items. Designers and companies collect a list of reasons products are returned. They may find that many people are returning the product for the same reason. The designer will use this information to redesign and improve the solution.

Another way they receive information is through complaints. Companies and designers take complaints seriously. Companies want their customers to be satisfied and to enjoy their products. When customers are unhappy with a product, the company tries to improve the product. With the use of the Internet, it is much easier for customers to reach companies and voice their opinions. Most companies have websites including the phone numbers, street address, and e-mail addresses that can be used to contact them.

TechnoFact: The Consumer Electronics Association (CEA) found that, on average, 7 percent of consumer electronics products sold are returned to retailers or manufacturers. In 2002, that amounted to $6.7 billion.

Technology Explained

cornflakes: A breakfast cereal processed from corn.

Cornflakes were one of the first cereals that appeared in the late 1800s. Their popularity remains after more than 100 years, **Figure A.** The story of producing cornflakes starts with selecting the proper grain. The corn should have a strong, yellow color. Its moisture content should be between 10 and 14 percent. Corn that is too dry is difficult to change into flakes. Moist corn can spoil while it is stored.

Figure A. Cornflakes are part of many people's breakfasts.

The first step in the production process is cleaning. Foreign materials and dirt are removed. The clean corn is then screened to remove small kernels.

The remaining corn is steamed to soften it. Then, the germ (inner core) and husk (outer shell) are removed. The material remaining is called *grit*. It is cooled and dried. This part of the grain is the raw material for making cornflakes.

The raw grits are placed in large, rotating pressure cookers, **Figure B.** Water and nutrients, such as niacin, iron, and riboflavin, are added. The

Brainstorm
Ask the students where they have seen or received product samples.

Discussion
Discuss the reasons positive feedback is important to designers.

Consumers should know it is also important to provide positive feedback. See **Figure 18-9.** Through returns and complaints, designers receive negative feedback. Designers also need information, however, on the positive features of the solution. This information helps the designers change only the bad parts of the design.

One way designers identify the good and bad features of a solution is through product samples. Companies often distribute free samples of their products. Comment cards are handed out along with the products. Consumers are asked to try the product and fill out the comment cards. The comment cards are written so the company can get as much feedback as possible. The company can then make improvements without taking away good features.

Figure 18-9. Both positive feedback and negative feedback are helpful for designers.

TechnoFact: Corn Flakes® cereal was created by W.K. Kellogg. He was the business manager of a hospital in Battle Creek, Michigan, and his brother was the superintendent. They developed many different types of food to add to the patients' vegetarian diets. In 1894, W.K. Kellogg invented Corn Flakes cereal. The cereal was an instant success. By 1906, Corn Flakes cereal was available to the general public.

material is cooked for about two hours, while its moisture content is stabilized.

The cooked grits are passed through a drier to separate them into small clumps. This process also reduces the material's moisture content. The dry grits are put through a series of sieves to obtain the correct sized grits to make the cornflakes.

Steam is added to the grits to make them moist and warm. The grits are run between large rollers to produce flakes. The output is screened, and undersized flakes are removed.

The flakes pass on a conveyor belt through an oven. In a few minutes, the new flakes are toasted. They develop a crisp texture and light brown color. The flakes are cooled as they leave the oven, after which they are packaged for delivery. The special plastic bag in the package keeps the flakes fresh for about 12 months.

Figure B. This is the process used to make cornflakes.

Competitors

Competitors are an excellent source of information. They are the companies selling the same types of products. Companies use their competitors' products to improve their own solutions. Designers compare their own products to competing products. They must be careful, however, that they do not create the exact same products.

Most products are either copyrighted or patented. Copyrights and patents are legal rights making it illegal to copy or reproduce a product. The process of receiving a copyright or patent usually takes place after a product has been designed and tested. This is why some design projects are kept secret during their developmental stages. Many solutions are uncovered at a public event. Automobiles are one example of a product revealed at large events. See **Figure 18-10.**

Government Organizations

Government organizations are a source of improvement different than the others. When designers use consumers and competitors as sources of information, they choose whether or not to make the changes. Government

Reinforce
Review the importance of competition from Chapter 17.

Research
Have the students search the U.S. Patent and Trademark Office website (www.uspto.gov) for a patent or trademark of a product or company of interest.

TechnoFact: In 2003, more than 60 vehicles made their international or North American debut at one event. The North American International Auto Show (NAIAS) in Detroit, Michigan was visited by over 810,000 people and included over 750 vehicles.

Figure 18-10. Auto shows are used to uncover new products and vehicles. (General Motors)

organizations, however, do not give the designers an option. Many government organizations make regulations. *Regulations* are rules and laws designers must follow.

Building codes are common sets of government regulations. They are regulations local governments set. These codes inform builders and designers of the rules they must follow in designing and building structures. See **Figure 18-11.** Architects and other designers must follow the building codes as they design buildings. They must also follow the codes when they improve or renovate buildings. When builders renovate old homes, they must update everything to meet the new building codes. For example, if you are renovating an older home, you may need to replace parts of the electrical wiring because the current wiring does not meet today's building codes.

All designers must deal with government regulations of some sort. Some designers, like chemists, who design medicines, have many regulations set by the Food and Drug Administration (FDA). The U.S. Department of Agriculture (USDA) sets the regulations for those who raise animals and crops. The Federal Communications

Figure 18-11. Builders must check local building codes when constructing new buildings. (Habitat for Humanity International)

TechnoFact: The U.S. Food and Drug Administration (FDA) regulates more products than one may think. It regulates products in seven different areas, including food, drugs, medical devices, biologics (such as vaccines), animal feed and drugs, cosmetics, and radiation emitting products (for example, microwaves and cellular phones).

Commission (FCC) regulates television and radio programs.

Internal Sources

Internal sources are improvements coming from inside the design company. They try to improve products to help their company. The need to lower manufacturing costs can come from internal sources. Lowering manufacturing costs can have many benefits for the company.

Standardization is another improvement coming from internal sources. It is a process used to make standard parts. Standard parts can be used for more than one product. See **Figure 18-12.** For example, a company making television sets can design the same remote control to work on all their TVs. It is less expensive to design and manufacture one remote control than many different models.

Making Improvements

Making improvements on a design is a continuous process. It uses the same design process that was used to create the product the first time. Once the designers choose to improve their product, they must choose where they are going to begin. The designers take steps back down the design process. Remember, we described the design process as a ladder. Each step of the process is a rung on the ladder. The designer can choose to revisit and redo any of the previous steps.

If the problem is a small issue, the designers may only need to go back a step or two. For example, imagine designers created a stapler. The problem with the product is that the staples do not fit inside of it. The designers may decide to go back to the step of

Standards for Technological Knowledge

9

Brainstorm
Have the students list products that must be designed to fit a standard size.

Research
Have the students identify the different types of standard screw sizes. Have them determine how the types are labeled and how to read the size.

Figure 18-12. Batteries come in standard sizes so they fit many different products. (Duracell)

TechnoFact: Batteries are designed in a series of different sizes. Most batteries are labeled with an American number, which is used to denote its size. For example, all AAA batteries are 10.5 mm in diameter and 44.5 mm in length. C batteries are 26 mm in diameter and 46 mm in length.

communicating designs. There, they may change the size of the stapler on the working drawings. The designers submit the new drawings to the manufacturing team. The improved staplers are then produced, and the problem is solved.

Unfortunately, it is not always this easy. There are times when the designers must go all the way back to the first step of the design process. If the product does not fit the users' needs or function correctly, the designers may have to begin with a new design brief. Other times, the designers are able to go back to a step in the middle of the process. Whichever step the designers go to is the point where they will start the process over. The designers must continue through the rest of the process. They cannot skip any of the steps just because they completed them when they designed the old product.

time will eventually become outdated. Designers have several reasons to improve their solutions. These reasons range from profit and market share to function and safety. Designers receive information about the need for product redesign from various sources. Customers and competitors are two sources of information. Other sources are government agencies and the company itself.

Whatever aspects the designers improve, they must go back and redo the steps of the design process. Some improvements require the designers to go back to the very beginning and redefine the problem. Other problems with the design can be solved by only going back a step or two. Innovative designers will often check their designs to see if improvements are needed. It is important that designs are improved. This is the only way our technology and knowledge will advance.

Summary

Design is a never ending process. Solutions can always be improved. A design that was a good solution at one time will eventually become outdated.

TechnoFact: Consultants are independent, "outside" experts, hired by a company to help improve its solutions. Consultants review the data the company gathers and the processes the company uses. They then make suggestions for improving the company. Consultants are helpful because they can offer a new perspective to the company.

Curricular Connections

Language Arts

Write a letter to a company regarding one of its products. In the letter, provide positive and negative feedback.

Science

Research the USDA and FDA regulations controlling the medicines and foods we use and eat. Choose two regulations from each agency and present the information to your class.

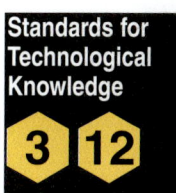
Activities

1. Choose a modern product and the original version of it. Research the improvements that have been made to keep the product up-to-date.

2. Gather several household products or toys. Develop a poster board showing different ways you can improve the products.

3. Interview a designer about the importance of improving designs.

TechnoFact: As products are improved, they are redesigned. However, sometimes consumers like the old designs better. For example, many digital phones today could be designed without an external antenna. However, such designs are slow to catch on because consumers feel there should be antennas on mobile phones.

Test Your Knowledge

1. Once a design reaches the improvements stage, it is complete. True or false.
2. Give three examples of products that have become obsolete.
3. _____ is the amount of money left after the bills are paid.
4. Define *competition*.
5. Recalls are rare in food and children's products. True or false.
6. Returns and complaints provide _____ feedback.
7. Competitors are great sources of information. True or false.
8. Standard parts are pieces that can be used to make _____ product.
9. Improving solutions is a process. True or false.
10. Why might a designer have to start over at the beginning of the design process?

Technology needs to be continually improved. The latest in communication technology connects us to the rest of the world and keeps us safe.

**Invention and
Innovation,
Textbook, page 426**
This activity is a
review of the history
of inventions. It
should be used in
the longer courses
covering Chapter 9:
"Invention and
Innovation." This
activity asks the stu-
dents to research an
invention and explain,
not only the inven-
tion's history, but also
the inventions that
led up to this inven-
tion. They will also
describe any inven-
tions that have devel-
oped from their
invention, as well as
the impacts of the
invention. In order for
the students to com-
plete this activity,
they must have
access to research
materials. The
Internet and basic
invention books, as
well as encyclo-
pedias, will all be
helpful to the
students.

Activity 3A

Invention and Innovation

Introduction

New inventions and innovations are created every day. You may never encounter some of them. There are many inventions, however, that most of us use all the time. These inventions have become part of our daily lives. Other inventions are outdated and would be impracti-cal to use today, like the telegraph, but it is important to understand that the invention of the telegraph led to other inventions we rely on today. In this activity, you will examine an invention used today. You will create a poster board explaining the invention and showing inven-tions that led up to it and inventions that have come from it.

Procedure

1. Select one of the following inventions: the telephone, television, printing press, automobile, airplane, telescope, laser, or SLR camera.

2. Using the library or Internet, research your invention in order to answer the following questions:
   a. Who invented the product?
   b. When was it invented?
   c. Why was it invented?
   d. What inventions led up to this invention?
   e. How does the invention work?
   f. What are the impacts (positive and negative) of the invention?
   g. What inventions have come from your invention?
   h. Was the invention patented?

3. Create a display (poster board) answering the research questions.

4. Present your display to the rest of the class.

Activity 3B

Communication Design

Introduction

Every day, new products, systems, and structures are designed. Some products are even designed to sell other products. Television commercials are an example of this type of design. Designers creating commercials use a design process including the following steps:

1. Identifying a problem.
2. Researching the problem.
3. Creating solutions.
4. Selecting and refining solutions.
5. Modeling the solution.
6. Testing the solution.
7. Communicating the solution.
8. Improving the solution.

This activity will give you an opportunity to complete the same steps, while designing a television commercial. The following design challenge has been developed for your use: Design and produce a 30 second commercial advertising the technology education course you are currently taking.

Materials and Supplies

➤ Pencil and paper
➤ Video cameras
➤ Video recording tape
➤ Activity sheets from the Student Activity Manual
➤ Costumes
➤ Props

Your teacher will divide you into design teams of three to four students. Each group will complete the following steps. Pay close attention to each step because some are to be done as a group, while others are to be done individually.

Communication Design, Textbook, pages 427–429, and Student Activity Manual, pages 55–58
This is a prescriptive activity challenging the students to design a television commercial to promote your class. It is organized around the design process outlined in Section 3. At several steps, the students are required to make design decisions. It is important that the students base their decisions on the design problem and design brief, instead of simply selecting the designs they like the most. This activity uses both group and individual work. When the students are designing the storyboards (or possible solutions in other activities), they should be encouraged to work individually. This will help each student gain a better understanding of the design process.

Identifying a Problem

1. Review the design challenge.
2. As a group, create a design brief for this challenge. The design brief should include the problem statement, criteria, and constraints.

Researching the Problem

3. As a group, create a survey that will help you gather information about different parts of the design. Ask questions to determine answers to the following questions:
   a. What information should be included in the video? (What projects or activities are important parts of the course?)
   b. Should there be teacher, student, or administration interviews?
   c. What types of music would make good background music for the video?
4. Distribute the survey to your classmates.
5. Collect the surveys and record the answers.
6. Make a list of the most popular responses.

Creating Solutions

7. Using the survey responses, each member of the group should create a storyboard for the video. The storyboard should include images of the different scenes. The images can be rough sketches. Also include the music or sound effects that will be used.
8. If you have several ideas for the video, you may create more than one storyboard.

Selecting and Refining Solutions

9. Gather as a group and review each of the storyboards.
10. Group members should explain their own storyboards.
11. Record your reactions (both positive and negative) to each of the storyboards.
12. When all storyboards have been presented, use your notes to help you select the one best fitting the design brief.
13. Make any changes necessary to the selected storyboard.

14. Have one student redraw the final solution.

15. The rest of the groups should begin writing a script.

Modeling the Solution

16. Gather the equipment needed to record your video.

17. Line up all the actors you will need.

18. Record your video.

Testing the Solution

19. As a group, watch the video you have recorded.

20. Make notes on things you like and dislike about the video.

21. Share your notes with the rest of the group.

22. Look back at your design brief. Answer the question "Does the video solve the problem and meet the criteria and constraints?"

23. If so, you are ready to share the video with the class. If not, you may need to go back and redesign the parts that do not solve the problem.

Communicate the Solution

24. Once the video passes your test, share it with your teacher and classmates.

25. View the rest of the videos your classmates made.

Improving the Solution

26. After viewing, write down suggestions that could have helped to improve your own solution.

Activity 3C

Design for Space

Design for Space, Textbook, pages 430–431
This is an open-ended design activity challenging the students to develop a model of a space station with living quarters, recreational areas, production facilities, office and managerial areas, mechanical support facilities, and a flight command center. To complete the activity, allow the students to work in groups of two to three students. Give each group a No. 10 tin can (a large can from the school cafeteria or local restaurant) and a supply of poster board, construction paper, a hollow tube or PVC pipe, and dowels. Then, have each group complete the following tasks:
1. Read the design brief in their textbooks.
2. Use the design process to design a space station.
3. Prepare a model of the station.
4. Present the model to the class.

Introduction

Technology has given us the life we have today. This life is very different from what our grandparents had. Likewise, life for future generations will be very different from life today. People may be living in deep space or undersea settlements. This activity will allow you to consider a life in space. You will design a self-contained space station.

Equipment and Supplies

➤ No. 10 tin cans (large cans found in most school cafeterias)
➤ Paper towel tube
➤ Five 6˝ cardboard or poster board squares
➤ 1½˝ wide strips of poster board
➤ Compass
➤ Ruler and pencil
➤ Scissors or small tin snips
➤ Glue
➤ Masking or transparent tape
➤ Wallpaper samples or scraps of cloth

Procedure

Your teacher will divide the class into groups of three to four students. Each group should perform the following steps:

1. Obtain the materials listed above.
2. See **Figure 3C-1** to determine the basic structure of the space station.

Figure 3C-1. Here is the basic structure of the model space station.

Paper towel tube (elevator shaft)

Cardboard discs (floors)

No. 10 can (space station)

3. List the basic areas needed for complete, comfortable living. These may include the following areas:

   a. Living quarters.

   b. Recreation areas.

   c. Agriculture, production, or factory areas (the reason for the space station).

   d. Mechanical support areas (such as areas for heating, air conditioning, and waste disposal).

   e. Office or managerial areas.

   f. Flight control areas.

   NOTE: Be sure to consider other types of facilities. This is not a complete list!

4. Assign each basic area to one of the four floors of the space station.

   Each member of the group should select one or two floors to develop. All group members should complete the following steps:

5. Cut a 5⅞″ disc from a square of cardboard or poster board.

6. Cut a 1½″ hole in the center of the disc.

7. Draw a floor plan on the disc. Be sure to consider the wise use of space, ease of movement from different areas, and grouping like activities together. Remember, 1″ = 8′.

8. Cut walls from the 1½″ strips of poster board.

9. Cut doorways in the wall sections.

10. Attach the walls to the floor.

11. Decorate the walls and floors with wallpaper samples and cloth to represent actual surface treatments.

   The group should complete their space station by completing these steps:

12. Cut openings in the elevator shaft (paper tube).

13. Attach the floors to the elevator shaft. See **Figure 3C-2.**

14. Insert the space station into the launch shell (tin can).

Figure 3C-2. This is an elevation view of the space station.

Section 4
Technological Contexts

In Chapter 2, you learned about systems. Now, you are about to study different systems. Each one is a complete, distinct system that is dependent on others. Some of them use many of the same resources, but they have different outputs.

These outputs may be new products, information, or energy. The output can also be a service for people. This section deals with the following seven systems and their processes:

➤ **Agriculture** plants, grows, and harvests food crops and fibers used for clothing.

➤ **Construction** transforms manufactured products and materials into structures we need for living, learning, and working.

➤ **Energy conversion** changes the forms of energy to make them more useful.

➤ **Communication** transforms information into organized designs, meaningful instructions, and purposeful skills.

➤ **Manufacturing** turns raw material into automobiles, clothing, toothpaste, and thousands of other items we use every day.

➤ **Medicine** promotes good health and treats illnesses and physical conditions.

➤ **Transportation** uses energy from chemicals, sunlight, and wind for power. This power is used to move people and materials.

19 Agricultural and Related Technology

20 Construction Technology

21 Energy Conversion Technology

22 Information and Communication Technology

23 Manufacturing Technology

24 Medical Technology

25 Transportation Technology

Technology Headline

Facial Recognition Systems

Facial recognition systems are computer software programs that compare pictures of faces in a crowd to images in a database. These systems can turn your face into a computer code so it can be compared to millions of other faces. This new technology can check 1,000 faces a second. It can also see through disguises such as hats, glasses, and facial hair.

This new technology has the ability to recognize faces and measure the features of each one. It measures the nodal points (the peaks and valleys making up your facial features) on your face. Some of these points include the distance between your eyes, the depth of your eye sockets, the width of your nose, your cheekbones, your jaw line, and your chin. Then, the system translates your measurements into a numerical code, called a *faceprint*. Your faceprint can be compared to a database of other faces.

The main users of facial recognition systems are law enforcement agencies. Police departments have begun using this technology to help find wanted criminals. These systems can also be used to eliminate voter fraud. They will recognize the face of someone who tries to register several times under different names. Another useful application of facial recognition systems is searching for missing children. The software can be used to scan public crowds, looking for matches to images of missing children on file.

Facial recognition systems can also be used to verify the identities of ATM users and people who want to cash a check. With the use of these systems, we will no longer need picture ID cards or personal identification numbers (PINs). You will also be able to secure your personal computer files with this software. Your face can become your password to get into your computer.

This new technology is part of a larger group of technologies, called *biometrics.* Biometrics uses biological information to distinguish one person from another. Some other types of biometrics are fingerprint scans, retina scans, and speech recognition. All of these technologies are expected to become more common in the near future.

Discussion Starter
Write the question "What are technological contexts?" on the board. Ask the students to volunteer definitions. You will probably get a variety of definitions. Discuss the differences.

Group Activity
Locate news articles discussing various technological systems and processes. Make enough copies of each article for a small group activity. Divide the class into small groups. Ask each group to read their article and present the highlights of it to the class. Ask the students to relate the article to the technology's context (agriculture, construction, energy conversion, information and communication, manufacturing, medicine, or transportation).

Writing Assignment
Ask the students to read the Technology Headline. Then have them write a few paragraphs about their feelings on technologies such as facial recognition systems. Do the students feel these technologies are infringements on their privacy? Do they see any other disadvantages to using such technologies? Can they think of any additional uses of these technologies that would benefit society?

Chapter 19
Agricultural and Related Technology

Did You Know?

➤ The following important events took place in the history of agriculture:

  ➤ Jethro Tull developed the first horse drawn seed drill in 1701.

  ➤ The first steam-powered plowing was done in 1830.

  ➤ The first successful combine harvester was developed in 1836.

  ➤ The first successful gasoline tractor was developed in 1892.

➤ The average American eats this amount of the following foods each year:

  ➤ About 300 pounds of fruit.

  ➤ About 420 pounds of vegetables.

  ➤ About 200 pounds of meat, poultry, and fish.

  ➤ About 10 pounds of nuts.

  ➤ About 200 pounds of flour and cereal products.

Objectives

The information given in this chapter will help you do the following:

➤ Explain what agriculture is.

➤ Name the two major types of agriculture.

➤ List the six major groups of crops grown on farms.

➤ Describe the seven main types of machines used in growing and harvesting crops.

➤ Label the five main types of livestock farms.

➤ Give examples of how technology is used in aquaculture.

➤ Summarize how biotechnology can be used in agriculture.

➤ Identify five methods of food preservation used today.

➤ Paraphrase the definition of *artificial ecosystem*.

Key Words

These words are used in this chapter. Do you know what they mean?

agricultural technology
agriculture
animal husbandry
animal science
aquaculture
artificial ecosystem
baler
biotechnology
combine
crop
cultivator
disc
drip irrigation
forestry
genetic engineering
gene splicing
grain drill
harrow
harvest
hydroponics
irrigation
pest control
pivot sprinkler
plant
plant science
plow
sprinkler
swather
tillage
tractor

Chapter Outline
Did You Know?
Objectives
Key Words
Types of Agriculture
Agricultural Crops
Technology in Agriculture
  Power, or Pulling, Equipment
  Tillage Equipment
  Planting Equipment
  Pest Control Equipment
  Irrigation Equipment
  Harvesting Equipment
  Storage Equipment
Raising Livestock
Special Types of Agriculture
Agriculture and Biotechnology
Food Production
  Food Processing
  Food Preservation
Artificial Ecosystems
Summary
Activities
Test Your Knowledge

Figure 19-1. Technology has greatly changed farming. This pea combine harvests green peas.

Reinforce
Review agriculture in a technological context (Chapter 3).

Research
Have the students find definitions for *plant science* and *animal science*.

Tech Terms
Agriculture—Using science and technology to grow plants and animals for human use.
Agricultural Technology—Developing and using machines and systems to grow plants and animals.

Early nomads are thought to have become the world's first farmers. They *planted* seeds of certain grasses for food. These grasses produced a new *crop* we now call *grain*. These people built villages as they waited for the seeds to grow and ripen. All this took place about ten thousand years ago. It happened in several places, including areas that are now Jordan, Iraq, and Turkey. The villagers tended the crops with crude hoes and bone sickles (cutters). This was some of the first technology people developed.

Today, modern farming uses both science and technology. Science is used in cross-pollinating crops and crossbreeding livestock. These types of science are called *plant science* and *animal science.* Likewise, science is used to describe the seasons. Scientific knowledge of weather helps guide planting and harvesting. Technology, however, has caused massive changes in farming. See **Figure 19-1.** New and modern machines and equipment allow fewer people to grow more food. Technological advances have helped people preserve and store food for later use. These and other advancements are part of *agricultural technology.*

What is *agriculture?* It is using science and technology in planting, growing, and harvesting crops and raising livestock. Agriculture includes using materials, information, and machines to produce the food and natural fibers needed to maintain life.

TechnoFact: Farming communities first appeared about twelve thousand years ago, following the end of the most recent ice age. Remnants of these communities can be found in widely diverse areas.

Types of Agriculture

Agriculture takes place on the farms and ranches of the world. See **Figure 19-2.** Agriculture is also practiced on small plots of land called *gardens.*

This type of activity has two main branches. Crop production grows plants for various uses. It provides human food, animal feed, and natural fibers to meet daily needs. Crop production produces ingredients for medicines and industrial processes. It provides plants for landscaping needs. Crop production produces trees for ornamental and wood product needs.

Animal husbandry involves breeding, feeding, and training animals. These animals are used for food and fiber for humans. In some cases, they are used to do physical work. Many animals are raised for pleasure. They are used in hobbies, for riding, and in racing.

Agricultural Crops

The crops raised today have evolved from specific regions of the world. Corn, beans, sweet potatoes, white potatoes, tomatoes, tobacco, peanuts, and sunflowers came from North and South America. China and

Extend
Compare gardening to farming and explain the differences.

Tech Terms
Crop Production—
Growing plants for food and fiber production.
Animal Husbandry—
Breeding, feeding, and training animals for food, fiber, work, and pleasure.

Figure 19-2. Crops and livestock are grown on farms and ranches. (U.S. Department of Agriculture)

TechnoFact: The primary goals of agriculture are to cause the land to produce more abundantly and to protect it from deterioration and misuse.

central Asia gave us peas, sugar cane, lettuce, onions, and soybeans. Sugar cane, rice, citrus fruits, and bananas came from Asia. The Middle East, southern Europe, and North Africa gave us the common grains. This region also was the home of sugar beets, alfalfa, and most grasses. Crops can be divided into several major groups:

➤ Grains. These are members of the grass family grown for their edible seeds. See **Figure 19-3.** This group includes wheat, rice, corn (maize), barley, oats, rye, and sorghum. Grains are the main food energy source for about 75 percent of the world's population. All the grains can be used in processed human food. Also, they can be used to produce oils, such as soybean, corn, and canola oils. Grains are widely used for animal feed.

➤ Vegetables. Plants grown for their edible leaves, stems, roots, and seeds are called *vegetables*. See

Figure 19-4. These watermelons are a vegetable crop. (U.S. Department of Agriculture)

Figure 19-4. These plants provide important vitamins and minerals for the daily diet. Vegetables include root crops, such as beets, carrots, radishes, and potatoes. They also

Figure 19-3. Grains are a major food crop. (Deere and Company)

TechnoFact: Modern wheat has developed over thousands of centuries. It started as a wild grain. About ten thousand years ago, a genetic mutation took place. The grain cross-bred with another grass to produce a more fertile hybrid called *emmer*. Then, a second mutation occurred eight to nine thousand years ago, producing a plump grain very similar to modern wheat.

Figure 19-5. These apples are almost ready to be harvested.

Tech Terms
Orchard—A farm or area on a farm on which fruit or nut trees are planted and cultivated.
Fruit—An edible part of a plant that is usually fleshy and contains seeds.
Nut—The shell-covered fruit of a plant. A nut usually has a hard outer shell and a fairly hard kernel, or inner part, which is the plant's seed.

include leaf crops, such as lettuce, spinach, and celery. Other vegetables provide food from their fruit and seeds. This group includes sweet corn, peas, beans, melons, squash, and tomatoes. Vegetables such as cabbage, cauliflower, and broccoli are called *cole plants.* Vegetables are widely grown on commercial farms called *truck farms.*

➤ Fruits. Other plants cultivated for their edible parts are called *fruits* and *berries.* See **Figure 19-5.** They include apples, peaches, pears, plums, and cherries, which are grown in temperate climates. Citrus fruits (oranges, lemons, limes, grapefruits, and tangerines), olives, and figs are grown in warmer climates. Tropical fruits include bananas, dates, and pineapples. Smaller fruits and berries include grapes, strawberries, blackberries, blueberries, raspberries, and cranberries. Fruits are grown on farms called *orchards* or *berry farms.*

➤ Nuts. Plants grown for their hard-shelled seeds are called *nuts.* They include walnuts, pecans, almonds, filberts (hazelnuts), coconuts, and peanuts. All, except peanuts, are grown on trees in orchards. Peanuts grow underground on plants grown on farms.

➤ Forage crops. Plants grown for animal food are classified as forage plants. These plants include the hay crops, such as alfalfa and clover.

TechnoFact: There is no clear botanical distinction between vegetables and fruits.

Reinforce
Review processes as parts of technological systems and apply this concept to the four agricultural processes (Chapter 2).

Tech Terms
Life Cycle—A series of changes in form and activity a living organism undergoes from its beginning to its maturity.
Equipment—Tools and machines needed to do a particular task or activity.

Figure 19-6. These are the processes in growing crops.

Also grasses used for pasture and hay are included in this group.

➤ Nonfood. Plants are grown for uses other than food. These plants include tobacco, cotton, and rubber. They also include nursery stock grown for landscape use and Christmas trees.

Technology in Agriculture

Crops, like all living things, have a life cycle. They are born when seeds germinate. Crops grow and reach maturity. They can be harvested or allowed to die. Farming takes advantage of this cycle through four processes. These include planting, growing, harvesting, and, in some cases, storing. See **Figure 19-6.** The edible parts of many crops are processed into food. For example, wheat can be processed into flour.

Flour can be further processed, with other ingredients, into bread.

At one time, growing crops was labor-intensive. It took many people to grow the food and fiber people needed. Today, technological advancements allow a few people to grow food for many people in a relatively short period of time. Farming has become equipment-intensive. New tools and machinery have been designed to make work easier and more productive. Fewer people are involved with producing food now, but more people are needed for processing, packaging, and distributing it.

Growing crops involves a number of technological devices and systems. These specialized pieces of equipment are used to improve the production of food, fiber, fuel, and other useful products. These devices can be divided into the following classifications:

➤ Power (pulling) equipment.
➤ Tillage equipment.

> Planting equipment.
> Pest control equipment.
> Irrigation equipment.
> Harvesting equipment.
> Storage equipment.

Power, or Pulling, Equipment

People tamed and trained animals to pull loads in the Stone Age. By 3,500 B.C., oxen were used to *plow* fields. Until the twentieth century, animals provided the majority of the pulling power needed in farming. See **Figure 19-7.** During the 1900s, the *tractor* replaced animal power on most farms. In the early 1900s, the modern all-purpose tractor was developed. By the 1950s, there were more than 3.5 million of these tractors in use. Today, the farm tractor can be found in all parts of the world. These devices provide the power to pull all types of farm equipment. There are two basic types of tractors: wheel tractors and track machines. See **Figure 19-8.** Both of these types of machines have the following features:

> A power source (engine).
> A way to transmit the power (transmission and drive train) for pulling a load.
> A method of controlling speed and direction.
> Traction devices (wheels or tracks).

Standards for Technological Knowledge

15

Brainstorm
Have the students list uses, other than farm applications, for tractors.

Tech Terms
Draft Animal—A domesticated animal used for work or transport.
Tractor—An agricultural vehicle used to pull such equipment as plows, cultivators, and mowers.

Figure 19-7. Up until the mid-1900s, horses, mules, and oxen provided pulling power needed on farms. (U.S. Department of Agriculture)

Standards for Technological Knowledge

15

Discussion
Discuss the advantages and disadvantages of minimum tillage and no-till farming.

Tech Terms
Tilling—Preparing land for growing crops.
Tillage Equipment—Machines designed to break and pulverize the soil.
Plow—An agricultural implement used to turn the soil in preparation for planting crops.

Figure 19-8. Wheel (left) and track (right) tractors are used on modern farms. Can you see the major parts of the tractor described in the text? (Deere and Company)

➤ An operator's area (seat, cab, and controls).

➤ A hitch onto which equipment can be fastened.

Tillage Equipment

Tillage equipment is designed to break and pulverize the soil. It develops a seedbed for the seeds and plants. The cornerstone of tillage is the moldboard plow. It can be traced back to tree branches and antlers used to prepare

Figure 19-9. A plow is used to cut, lift, and turn the soil. (Case IH)

the soil. In the 1800s, iron and steel plows were developed. They, like modern plows, had a blade-shaped plowshare that cut, lifted, and turned over the soil. See **Figure 19-9.**

Discs (or disc plows) are also used to prepare the seedbed. Sometimes these machines are used after the plow. At other times, the disc is used instead of a plow to start preparing a field for planting. A disc, like its name implies, is a series of curved discs on a shaft. See **Figure 19-10.** When the machine is pulled through the ground, it slices and crumbles the earth.

Generally, a *harrow* is used after plowing and discing are completed. A harrow is a frame with teeth. These teeth may be spikes or spring-shaped. The harrow is dragged over the ground to give the soil tilth. This means fine and crumbly soil.

Today, farmers are reducing the amount of tillage they do. This is especially true on soil that can erode easily. This technique is called *minimum tillage*. It uses small amounts of

TechnoFact: The early plow consisted of an iron-tipped wooden wedge attached to a single handle. It was pulled by men or oxen and was capable of breaking up the soil, but not turning it.

Standards for Technological Knowledge

Extend
Explain the role of fertilizers in agriculture.

Tech Terms
Fertilizer—Any natural or synthetic material applied to the soil to increase its ability to support plant growth.

Figure 19-10. A disc cuts and pulverizes the soil. Note the close-up of the discs on the right. (Deere and Company)

work to prepare the soil. Often, a set of chisels on a frame is used to open the soil. In some cases, no-tillage systems are used. In these cases, crops are planted without working the soil from the previous crop. Special planters slice open the soil and plant the seeds.

Planting Equipment

Once the soil is prepared, fertilizer must be applied. Fertilizer is a liquid, powder, or pellets containing important chemicals. It primarily delivers nitrogen, phosphorus, and potassium to the soil. Other nutrients are also in many fertilizers. Fertilizer can be applied before, during, or after planting seeds. It may be applied with special equipment or along with a seed planter. Often, dry fertilizer is scattered (broadcast) before planting. Liquid and gaseous (anhydrous ammonia) fertilizer is applied by injecting it into the soil. A machine with a series of knives is pulled over the ground. The liquid or gaseous fertilizer is injected into the trench the knives create.

TechnoFact: John Deere introduced the steel plow in 1837. It followed the development of a cast iron plow by Jethro Wood in 1819 and a steel blade plowshare by John Lane in 1833.

Figure Discussion
Describe how the farmer in Figure 19-13 is applying fertilizer while cultivating a crop.

Tech Terms
Cultivator—An agricultural implement for stirring and pulverizing the soil before planting or removing weeds and aerating and the soil after the crop has begun to grow.

Figure 19-11. This shows a grain drill in use. (Deere and Company)

The seeds must be planted to start the crop cycle. Over most of history, this was done by hand. In the early 1700s, however, a new machine was developed. It was called the *seed drill* or *grain drill.* See **Figure 19-11.** As it is pulled along, it opens a shallow trench, and then a seed is dropped. The trench is closed, covering the seed.

Other planting machines have been developed for potatoes and corn. See **Figure 19-12.** Specialized machines are used to plant vegetable plants, such as tomatoes and cabbage.

Pest Control Equipment

In nature, not all plants that sprout live to maturity. Diseases and insects kill some of the plants. Neighboring plants crowd out others. Farm crops face the same dangers. A number of machines have been developed to help control these pests.

Cultivators are used to remove weeds and open the soil for water. These machines are a series of hoe-shaped blades pulled through the ground. See **Figure 19-13.** The blades break the crust and allow rain and irrigation water to enter the soil. They

Figure 19-12. This farmer is planting corn with a six-row corn planter. Fertilizer is applied from the front hoppers. Corn seed is in the back hoppers. (Deere and Company)

Figure 19-13. This farmer is cultivating a row crop (corn). Notice the tank on the front of the tractor. Fertilizer and an herbicide are being applied as the plants are being cultivated. (U.S. Department of Agriculture)

TechnoFact: Henry Blair received patents for a seed planter in 1834 and a second patent in 1836 for a cotton planter. Blair was the second African-American to receive a patent.

Figure 19-14. A field is being sprayed from the air with chemicals protecting it from weeds and insects.

also cut off and pull out weeds. (A weed is any out of place plant.)

Sprays can be applied to control weeds and insects. Those controlling weeds are called *herbicides*. Insect sprays are called *pesticides*. Ground equipment or airplanes can apply both of these materials. The equipment has a tank, a pump, and spray nozzles on a boom. See **Figure 19-14.** As the plane or ground applicator crosses the field, a mist of *pest control* is applied.

Irrigation Equipment

In parts of the world, rainfall is sufficient to raise crops. Many places, however, are too dry for successful farming. In some of these areas, *irrigation* is used. This is artificial watering to maintain plant growth.

Irrigation systems can be traced back to 2100 B.C. The Egyptians developed systems using the water from the Nile to irrigate crops. Their systems, like all irrigation systems, contained these elements:

➤ A reliable source of water.

➤ Canals, ditches, or channels to move the water.

➤ A way to control and distribute the water.

The source of water is usually a lake, a river, or an underground source (an aquifer). A dam at its outlet often controls water in the lake. A dam backs up river water to form a reservoir. See **Figure 19-15.** A well and pump obtain underground water. A series of canals and pipes moves the water from the source to farm fields. One of three basic methods is used to apply the water to the land.

Flood irrigation is used on level fields and where there is a lot of water available. A sheet of water advances from a ditch across the field. Lateral ditches and pipes with holes along their

TechnoFact: Since prehistoric times, water has been diverted from waterways into ditches to irrigate fields. Later, devices to raise water were added to increase the acreage that could be irrigated. This use of a system of canals, dams, and reservoirs for the storage and distribution of irrigation water started in ancient Egypt.

Figure 19-15. This reservoir behind Grand Coulee Dam provides water for power generation and irrigation.

lengths supply the water to one side of the field. Gravity causes the water to flow across the field. Other ditches or pipes may carry off excess water. In row crops, furrows (small ditches) between rows of plants are used. They move the water from one end of the field to the other. See **Figure 19-16.** Pipes or tubes are used to control the water entering each furrow.

Sprinkler irrigation is used to better control water. It also uses less water. These systems involve a water source, a pump, main (distribution) lines, lateral (sprinkler) lines, valves, and sprinkler heads. The pump forces water into the main distribution lines. The water flows through them to pipes that have sprinkler heads attached at set intervals. Valves between the main and lateral lines can shut off or control the water flow. The water in the lateral lines enters the sprinkler heads, which spray water onto the land.

Sprinkler systems may have a number of straight sprinkler lines. Each line can apply water to a long, narrow band across the field. The

Figure 19-16. These strawberry plants are being flood irrigated. (U.S. Department of Agriculture)

TechnoFact: Areas of China, Egypt, Mesopotamia, and India have been continuously irrigated for thousands of years. In North America, Native Americans of the Southwest were the first to practice irrigation.

Figure 19-17. The pivot sprinkler system is irrigating crops. (U.S. Department of Agriculture)

Standards for Technological Knowledge

15

Demonstrate
Draw diagrams of the various irrigation systems and explain them to the students.

Extend
Ask the students which type of irrigation system is best suited for (a) a field of potatoes on level land, (b) an orchard, and (c) a large grain field on unleveled land.

Tech Terms
Harvesting—The procedure of gathering a crop.

water is allowed to run for a set time. This irrigates the bands on each side of the sprinkler lines. Then these lines are moved by hand or rolled to the next position. Here, they apply water to the next bands.

A large number of straight lines may be used to cover the entire field. When the lines are turned on, the entire field is irrigated at once. This eliminates the need to move individual lines. These lines are called *solid set sprinklers.*

Other sprinkler systems are called ***pivot sprinklers.*** See **Figure 19-17.** These systems use one long line. It is attached at one end to a water source. The line pivots around this point on large wheels electric motors power. The line is constantly moving very slowly

in a circle. Sprinkler or mist heads apply the water as the line pivots.

Drip irrigation is the third type of irrigation. The system uses main lines to bring water near the plants. Individual tubes or emitters bring water from the main lines to each plant. This ensures each plant is properly watered. It also reduces the amount of water lost to evaporation.

Harvesting Equipment

Once a crop reaches maturity, it must be *harvested.* Each type of crop has its own special harvesting equipment. *Combines* are used to harvest grains. See **Figure 19-18.** A combine is a combination of two early farm machines: the header (cuts heads from the grain) and thresher (removes grain

TechnoFact: Many modern irrigation projects also incorporate flood controls, provide drinking water, and supply hydroelectric power. India, China, and the United States are the three countries with the most irrigated land.

Standards for Technological Knowledge

Extend
Explain to the students why mechanical harvesters are an important part of agriculture.

Tech Terms
Combine—A machine that chops, threshes, and washes grain.
Picker—A person or machine that harvests fruit, nuts, berries, or other crops.

Figure 19-18. This combine is harvesting grain on a hillside. Notice that the header (cutting area) is at the same angle as the hillside. The machine itself, however, is kept level. This is necessary to make the grain separation efficient. (Deere and Company)

from chaff). The combine cuts off the tops of the plants containing the grain. The heads and straw move into the machine. A cylinder causes the grain to break away from the heads. Blasts of air and screens separate the grain from straw, chaff, and weed seeds. The grain is moved into storage hoppers on the machine. The unwanted materials are conveyed out the back of the machine and dropped onto the ground.

Other crops use different harvesting machines. A combine or a special corn-picking machine can harvest corn. Mechanical pickers are used to harvest almost all cotton grown in the United States. Vegetables may be harvested by special purpose machines or by hand. See **Figure 19-19.** Special machines dig and collect onions and

Figure 19-19. This special purpose machine harvests green beans. It is unloading a batch into a truck, which will move the beans to a freezing plant.

TechnoFact: Major improvements in farm implements included the development of the revolving disc harrow in 1847, the binder in 1850, the corn planter in 1853, and the combine harvester in 1860.

Figure 19-20. This harvesting machine is picking up onions that were machine dug and placed on top of the ground to dry. Workers of this machine separate clods and rocks from the onions.

Standards for Technological Knowledge

15

Research
Have the students research the plants that can be used as hay, or fodder, for animals.

Tech Terms
Hay—Grass or other plants that are cut, dried out, and then often used as fodder.
Bale—A large bundle or cube of hay that is tightly bound with string or wire to keep its shape while being transported or stored.

potatoes. See **Figure 19-20.** Fruits and grapes are usually picked by hand and placed in boxes. Special machines may, however, be used. They shake the trees, causing the fruit to fall into raised catching frames. Nuts are also harvested in this manner.

A series of machines harvests hay. A mover may cut the plants and let them fall on the ground. After the hay has dried for a day or more, a rake is used to gather it into windrows (bands of hay). In other cases, a windrower, or *swather,* may be used. This machine cuts and windrows the hay in one pass over the field. See **Figure 19-21.** After the hay has dried, it is usually baled. A hay *baler* picks up a windrow and conveys it into a baling chamber. There, the hay is compressed into a cube. Wire or twine is tied around the cube to maintain its shape. The finished bale is ejected out the back of the machine. See **Figure 19-22.** Special balers have been produced to make round bales.

Figure 19-21. The swather is cutting and windrowing hay. (Deere and Company)

TechnoFact: Cyrus McCormick invented the first successful reaping machine in 1831. Over time, he continued to improve the machine. McCormick patented the Self-Rake Reaper in 1858, and the Harvester and Binder in 1876.

Extend
Discuss why farms generally raise only one type of animal.

Tech Terms
Livestock—Farm animals raised for food or other products, or kept for use.
Dairy—A farm that processes milk and milk goods.
Poultry Farming—The raising of chickens, turkeys, ducks, and geese for their meat and eggs.

Figure 19-22. This picture shows a bale of hay leaving a baler. (Deere and Company)

Storage Equipment

Many crops are stored before they are sent to processing plants. Grain is stored in silos or buildings at grain elevators. See **Figure 19-23.** Hay is stored in hay barns that are roofs attached to long poles. These buildings generally do not have enclosed sides or ends. Many vegetables and fruits are stored in climate-controlled (cold storage) buildings. The crops will be transported to processing plants throughout the country and world, as demand requires.

Raising Livestock

Farmers and ranchers raise large numbers of livestock. These include cattle, sheep, goats, horses, swine (pigs), and poultry (chickens and turkeys). These animals are primarily raised to provide meat, milk, or materials for clothing.

In historical time, many farms raised a few of each of these animals. Today, most livestock are raised on single-purpose farms. These farms include the following:

➤ Cattle ranches. They raise beef cattle for meat and hides.

➤ Dairies. They raise dairy cattle primarily for milk.

➤ Swine farms. They raise hogs for meats and hides.

➤ Horse farms. They raise horses for pleasure riding and racing.

➤ Poultry farms. They raise turkeys for meat and chickens for meat and eggs.

A number of different technologies are involved in livestock raising.

Figure 19-23. These trucks are bringing grain to an elevator for storage.

TechnoFact: Since World War II, the poultry industry has become one of the most efficient producers of protein for human consumption. It raises animals in a much shorter time than beef and pork operations. A broiler can be produced in seven weeks, and a laying hen can be produced in five months.

Figure 19-24. These chickens are contained in a large house on a poultry farm.

Standards for Technological Knowledge

15

Extend
Discuss the rise of factory farms and large-scale feed lots.

Figure Discussion
Describe Figure 19-25 as a small-scale city (for example, providing housing, water, sewage disposal, food).

Tech Terms
Factory Farm—A large-scale farming operation where animals are raised using intensive methods and modern equipment.

Specialized equipment and practices are used in the care of animals. Many livestock operations require buildings to house animals and process feed. See **Figure 19-24.** These buildings are erected using construction technology. Livestock production also includes machines used to feed animals. Feed processing mills are required. They contain machines that grind and mix feed for the animals. Feed troughs or bunkers are required so the animals can eat grain and hay. See **Figure 19-25.** Water must be provided using manufactured pumps and tanks. Finally, machines and equipment are used to

Figure 19-25. Look at the cattle feedlot. There is a feed mill in the background. Fences contain the animals within a set area. Feed bunkers can be seen in the foreground.

TechnoFact: There is a difference between domesticated and tame animals. For example, many elephants are tamed and put to work. These animals are bred in the wild, captured, and trained. However, they remain like their wild brethren. A cow is a domesticated animal bred in captivity to serve human needs.

Standards for Technological Knowledge

⬡14 ⬡15

Research
Have the students research fish farming.

Tech Terms
Hydroponics—
Growing plants in nutrient solutions without soil.
Aquaculture—
Growing and harvesting fish, shellfish, and aquatic plants in controlled conditions.
Forestry—Growing trees for commercial uses, such as lumber and timber products, paper and pulp, and chips and fibers.

dispose of the animal waste (such as manure). See **Figure 19-26.**

Special Types of Agriculture

There are several unique types of agriculture practiced in North America. The three main types are hydroponics, aquaculture, and forestry:

➤ *Hydroponics.* This consists of growing plants in nutrient solutions without soil. Hydroponic systems supply nutrients in liquid solutions, while the plants grow in a porous material.

➤ *Aquaculture.* This is growing and harvesting fish, shellfish, and aquatic plants in controlled conditions. It uses ponds, instead of soil, to grow its crop. See **Figure 19-27.**

➤ *Forestry.* This includes growing trees for commercial uses, such as lumber and timber products, paper and pulp, and chips and fibers.

Figure 19-26. This hog farm contains pens in buildings and a waste containment lagoon (upper left of photo).

Figure 19-27. These workers are using equipment to harvest catfish at a fish farm.

Agriculture and Biotechnology

A specific technology that has greatly impacted agriculture is *biotechnology.* The term, *biotechnology,* is fairly new. The practice, however, can be traced into distant history. Evidence suggests the Babylonians used biotechnology to brew beer as early as 6000 B.C. The Egyptians used biotechnology to produce bread as far back as 4000 B.C. Both of these activities are directly related to agriculture.

Modern biotechnology can be traced back to at least World War I. Scientists used an additive to change the output of a yeast fermentation process. The result was glycerol instead of ethanol. The glycerol was a basic input to explosives manufacturing.

During World War II, the next stride in biotechnology took place. This involved the production of antibiotics (antibodies). These drugs are also the products of fermentation processes.

What is biotechnology? Specifically, biotechnology deals with using biological agents and principles in

TechnoFact: Bioconversion is a special type of biotechnology. It involves changing organic materials (plant or animal matter) into products or energy sources through biological agents or processes.

processes to produce commercial goods or services. The biological agents are generally microorganisms (very small living things), enzymes (a special group of proteins), or animal and plant cells. They are used as catalysts in the selected process. The word *catalyst* means they are used to cause a reaction. The catalyst does not, however, enter into the reaction itself.

Agricultural biotechnology is a type of biotechnology. It consists of techniques used to create, improve, or modify plants, animals, and microorganisms. This biotechnology can be used to improve many different activities impacting agriculture.

Biotechnology technology can be used to combat *diseases.* This was the first major use of biotechnology. For example, insulin used to treat diabetics can be produced less expensively using biotechnology. Also, enzymes reducing blood clots can be produced using biotechnology. Golden rice providing infants in developing countries with beta-carotene to fight blindness is a result of biotechnology. Biotechnology can be used to promote human health. The nutritional value of foods can be improved using biotechnological techniques.

Another use of biotechnology is fighting animal diseases. See **Figure 19-28.** It was used in developing a vaccine for shipping fever. This disease is a major factor in feedlot deaths. Biotechnology was also used in producing a vaccine protecting wild animals against rabies.

Standards for Technological Knowledge

14 **15**

Extend
Discuss the political problems associated with using biotechnology.

Tech Terms
Biotechnology— Using biological agents or processes to produce goods or services.
Catalyst— A substance causing a reaction, but not entering into the reaction itself.

Figure 19-28. Biotechnology can be used to fight diseases in animals.

TechnoFact: The techniques of genetic engineering (recombinant DNA technology) can be used to manipulate the genetic material (DNA) of bacteria and other organisms to manufacture biological products.

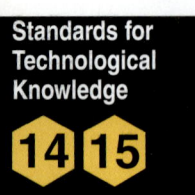
Tech Terms
Genetic Engineering— Altering an organism's genetic material to eliminate undesirable characteristics or to produce desirable new ones.
*Gene Splicing—*A genetic engineering technique in which a gene that produces a desired product or trait is inserted into DNA that did not previously contain the gene.

Biotechnology is a major factor in increasing crop yields. It has helped produce more food on the same number of acres. This factor has allowed farmers to feed more people using the same effort. For example, biotechnology was used to produce soybeans resistant to certain herbicides (weed sprays). Also, it was used to develop a cotton plant resistant to major pests.

It can be used to supplement the common techniques of selective breeding and pollinating. This specific application is often called *genetic engineering.* It enables people to move genes in ways they could not before. It allows people to develop plants and animals with desirable traits.

Gene splicing is based on a major discovery called *recombinant DNA.*

The structure of DNA is a double helix (spiral) structure. It consists of a jigsaw-like fit of biochemicals. The two strands have biochemical bonds between them.

The DNA molecule may be considered a set of plans for living organisms. It carries the genetic code determining the traits of living organisms. Scientists can use enzymes to cleanly cut the DNA chain at any point. The enzyme selected will determine where the chain is cut. Then two desirable parts can be spliced back together. This produces an organism with a new set of traits. The process is often called *gene splicing.*

This process allows scientists to engineer plants having specific characteristics. See **Figure 19-29.** For example,

Figure 19-29. Biotechnology can be used to develop crops yielding more food and resisting diseases and pests.

TechnoFact: Genetic engineering can be used to increase food production, produce vaccines and drugs, diagnose and treat disease, and dispose of industrial wastes and pollution.

Standards for Technological Knowledge

14 **15**

Figure 19-30. Foods must go through a lengthy transportation chain before you can buy them in your grocery store.

Brainstorm
Have the students list the type of workers involved in food production.

Tech Terms
Food—A plant or animal material containing essential nutrients, which is ingested to produce energy, stimulate growth, and maintain life.
Organic Food—Food that is raised without chemicals and is processed without additives.

resistance to specific diseases can be engineered into the plant. This could reduce the need for pesticides to control insect damage to crops.

This activity has received many headlines in newspapers and magazines. It is controversial. Some people think it will make life better. Others think we should not change the genetic structures of living things.

Food Production

People must eat to live. We eat many different kinds of food, prepared in a variety of ways. Few crops produced through agricultural technology are sold in their natural states.

Most foods have been processed or preserved in some way by the time we eat them. Food processing is one of the most important of all agricultural technology processes.

Food production is primarily the job of farmers. Many other people, however, handle the food the farmers produce before it reaches the consumers. After being harvested, most food products go to processing plants. Then, the food products must be transported to consumers. The transportation chain they follow includes truck, airfreight, and train personnel; food brokers; distributors; food wholesalers; and finally, food retailers, such as your local grocery store. See **Figure 19-30.**

TechnoFact: Dairy products include whole milk, low-fat milk, flavored milk, whole and nonfat dry milk, evaporated and condensed milk, butter, cheese, frozen dairy products, and fermented products, such as sour cream and yogurt.

Tech Terms

Food Processing— Converting harvested crops into foods that can be eaten.

Convenience Food—A food packaged before sale that can be prepared in a short time with minimal effort.

Fast Food— Reasonably priced food that can be prepared and served quickly.

Health Food—A food considered to be greatly advantageous to health.

Junk Food—A food high in calories, but low in nutritional value.

Meat—The edible portions of mammals, such as cattle, calves, sheep, lambs, and swine.

Food Processing

The conversion of harvested crops into food that is ready to eat is called *processing*. This is not just a single process, however; it includes many different processes, which vary depending on the food being processed. At the processing plants, skilled workers perform many jobs, including sorting, washing, peeling, slicing, roasting, grinding, canning, flash freezing, boxing, cooking, adding preservatives, and packaging.

Early humans ate their food exactly as they found it. They ate fruits, nuts, leaves, and roots as they gathered them, and they even ate the fish and game they hunted without cooking it. These people began cooking with the discovery of fire about 1 million years ago. This is the first way food was processed.

Cooking makes food softer and easier to digest. The heat kills bacteria that might cause the food to rot or make you sick. Most modern cooking appliances use heat from electricity or gas flames. Microwave ovens use radio waves to heat the food.

Some foods are sold ready to eat. Other foods come partly prepared. You must cook these types of foods, such as frozen pizzas, TV dinners, and cake mixes, before you eat them.

Grains are processed in mills. They can be ground, sifted, steamed, shredded, or toasted. Then they are made into products such as flour, bread, and cereal products. See **Figure 19-31.** Fruits and vegetables must be peeled before being processed, and sometimes fruits are crushed to make juice.

Milk is processed into whole milk, skim milk, condensed milk, cheese, butter, and cottage cheese.

Some processed foods contain additives. These are chemicals that improve the food or keep it from spoiling. Preservatives, flavorings, and dyes are all additives.

Food Preservation

Most food spoils after a while. Bacteria, mold, and insects feed on it. Because food is so important to survival, food preservation is one of the oldest technologies used by humans. Some of these technologies have been used for thousands of years. Over

Figure 19-31. Grains can be processed into many popular food items, including bagels.

TechnoFact: Gail Borden opened the first factory for the production of evaporated milk in 1860.

Figure 19-32. The most common form of food preservation is the household refrigerator.

Standards for Technological Knowledge

Research
Have the students research how a refrigerator works.

Tech Terms
Refrigerating—
Chilling food for preservation.
Freezing—
Preserving foods by keeping them in freezing temperatures.

time, people began to understand why these methods work. They learned how to create the conditions preventing the growth of microorganisms that spoil food. These are some of the techniques used today for preserving food:

➤ Refrigeration and freezing.

➤ Canning.

➤ Dehydration.

➤ Chemical preservation.

➤ Irradiation.

Other methods of food preservation include freeze-drying, pasteurization, fermentation, carbonation, smoking, and cheese making.

Refrigeration and Freezing

Refrigeration and freezing are the most popular forms of food preservation used today. See **Figure 19-32.** Long ago, people used to bury food in snow or ice to keep it fresh. This is because cold temperatures cause bacteria to stop growing. Refrigeration slows the growth of bacteria so food stays fresh longer—usually a week or two—instead of spoiling in just a few hours. Freezing stops bacterial growth all together. It keeps food fresher than refrigeration does. These methods are so popular because they have very little, if any, effect on the taste and texture of most foods.

TechnoFact: Thaddeus Lowe's compression ice machine, developed in 1865, and Carl Linde's ammonia compression refrigerator, developed in 1873, led to the use of mechanical refrigeration for preserving food products.

Standards for Technological Knowledge

15

Brainstorm
Ask the students when dehydrating is a more useful process than canning.

Tech Terms
Canning—
Preserving food by heating and sealing it in airtight containers.
Dehydration—
Preserving food by removing its moisture.

Bacteria stop growing below 14°F, but they do not die. To destroy all bacteria, food is steamed before freezing. In food preserved in this way, once the food is thawed, bacteria can grow again.

Vegetables, fish, and poultry are frozen by dipping the packaged foods into tanks of freezing salt water. The same process is used to freeze canned juices. A spray of liquid nitrogen with a temperature of −320°F freezes more expensive foods, like shrimp. Meat is frozen by traveling through a tunnel, as fans blow air at a temperature of −40°F on it. This process is called *blast freezing*.

Canning

Canning uses a cooking process to preserve food in glass jars or metal cans. It uses heat to remove oxygen from the container, kill microorganisms in the food, and destroy enzymes that could spoil the food. During the canning process, the can or jar is filled with food. The air is pumped out to form a vacuum, and the container is sealed. The food is heated, then cooled to prevent the food from becoming overcooked. This heating inactivates enzymes that could change the food's color, flavor, or texture. Canning is used to preserve many foods, including fruits, vegetables, jams and jellies, soups, and juices.

Dehydration

Dehydration is the process by which foods are dried to preserve them. During this process, most water is removed from the food. This increases the concentrations of salt and sugar, and these high concentrations kill any bacteria. Dried foods kept in airtight containers can last a relatively long time.

Early humans dried meat and fruit in the sunshine. Today, many foods, such as powdered milk, soup, potatoes, dried fruits and vegetables, beef jerky, pasta, instant rice, and orange juice, can be dehydrated. See **Figure 19-33.** You may eat them as they are sold, or you may need to add water. Drying often completely changes the taste and texture of foods, but many new foods created by dehydration have proven to be just as popular as the original forms.

Figure 19-33. We eat many foods, such as these dried cherries, that have been preserved through dehydration. (Cherry Marketing Institute)

TechnoFact: Nicholas Appert invented the canning process in France in 1809. Glass containers were used at first, but they were bulky, costly, and easily broken. The process became a more important food preservation method with the invention of the stamped can in 1847.

Chemical Preservation

People have used salt, sugar, and vinegar to preserve foods, especially meat, for thousands of years. When salt or sugar is used, the process is called *curing.* The salt or sugar dissolves in the water in the food and kills the bacteria. Today, this process is most often used to create foods such as "country ham," dried beef, corned beef, and pastrami. When vinegar is used, the process is called *pickling.* This process has been used to preserve meat, fruits, and vegetables, but today it is used almost exclusively for making pickles, or pickled cucumbers. See **Figure 19-34.**

Today, new chemicals, such as nitrites, benzoates, and sulphites, are used as preservatives. They either inhibit the growth of bacteria or kill them. These chemicals can be found on the ingredient lists of many different foods. For example, sulfur dioxide preserves fruits used to make jam, fruit juice, and dried fruits. All of these chemicals may be harmful if used in large amounts. The Food and Drug Administration (FDA) limits the amounts of these chemicals allowed in foods.

Irradiation

Irradiation is the process by which X-ray radiation is used to kill bacteria in food. This process kills bacteria without significantly changing the food. It can occur after foods have been packaged, which is a big advantage. If a food is sealed in plastic and then irradiated, it becomes sterile and can be stored on a shelf without refrigeration

Figure 19-34. One common form of chemical preservation turns cucumbers into pickles.

for a long time. The FDA has approved the irradiation of chicken and beef. The use of this technique could prevent many forms of food poisoning. Large amounts of radiation must be used, however, and many people are not sure whether or not it is safe. People generally do not like the term *radiation,* so this process is still not very common in the United States.

Artificial Ecosystems

Artificial ecosystems are human-made complexes reproducing some facets of the natural environment. They can be used to study agricultural processes and systems as they would

Standards for Technological Knowledge

Research
Have the students research how irradiation systems work.

Tech Terms
Irradiation— Exposing food to X-ray radiation to preserve it.
Ecosystem— Organisms living in a specific environment, such as a forest, and the physical parts of the environment affecting the organisms.

TechnoFact: Louis Pasteur's research into microbiology led to the development of pasteurization in 1861. This process kills bacteria in beverages and liquid foods by subjecting them to heat.

Technology Explained

landscape plants: Plants grown in a controlled environment. They provide flowers and vegetables for home gardeners and landscape contractors. Below is a picture story of how technology is used to produce these plants.

1. Seeding trays are selected.

3. Seeds are automatically placed in each cell of the trays.

5. The moist trays are placed in controlled environments until seeds sprout.

2. Planting mix (soil) is automatically placed in the trays.

4. The soil and seeds are sprayed with water.

6. The trays with new plants are placed in greenhouses.

Standards for Technological Knowledge

15

Figure Discussion
Explain why intensive agriculture is used to germinate and grow landscape plants.

Tech Terms
Ecology—The study of the relationship of plants and animals to their physical and biological environment.

be useful to biological ecosystems. Some examples of artificial ecosystems are terrariums and the hydroponics stations discussed earlier. They function as part of a larger closed system supporting living organisms.

A terrarium is used to nurture plant or animal life in an enclosed environment. It acts as a complete habitat using all the systems of life, such as food, water, shelter, and space. Terrariums can be used for decoration and enjoyment, but they can also be used to study the ways in which the elements of an ecosystem depend on one another. See **Figure 19-35.** Some

TechnoFact: Biosphere 2 is a privately funded ecological research project in which an artificial ecosystem is developed in a 3.15-acre structure. The structure, which is located near Tucson, Arizona, requires only sunshine and electricity from outside sources.

7. The plants are transplanted into trays with larger cells or into plastic retail packs.

9. Light and humidity are controlled to produce an ideal growing environment.

11. The finished plants are loaded onto carts and made ready for delivery to retailers.

8. Throughout the growing process, the plants receive automatic watering.

10. The plants are put into greenhouses when they are ready for sale.

12. The plants are transplanted into home gardens.

greenhouses can be considered large-scale terrariums, and they can be used to grow plants and animals in areas differing from their natural habitats.

A hydroponics station is used to grow plants without soil. Similar to terrariums, these stations are controlled environments supplying the light, humidity, food, and water the plants need for growth. They are an alternative to traditional agriculture for farmers in areas with poor soil. Using hydroponics stations, farmers can grow vegetables and other plants in the middle of a desert or in an alley in a crowded city.

Standards for Technological Knowledge

15

Research
Have the students research the way hydroponics systems operate.

Tech Terms
Environment—The external factors affecting an organism.
Hydroponics—Growing plants without soil, in water that has nutrients added.

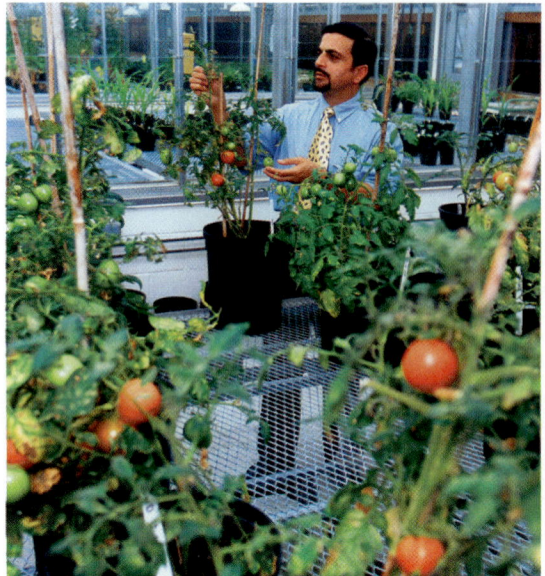

Figure 19-35. Artificial ecosystems, like this greenhouse, allow plants to grow in controlled environments. (Agricultural Research Service, U.S. Department of Agriculture)

Managing an artificial ecosystem entails collecting facts to plan, organize, and control processes, products, and systems. Operating such a system necessitates absolute control and cultivation. Temperature, nutrients, light, water, air, circulation, waste recycling, and monitoring of insects all require management in order for the system to function well.

Quality Control in Agriculture and Biotechnology

Quality control is critical in agriculture technology and biotechnology. The U.S. Food and Drug Administration (FDA) regulates food to ensure public health. Agricultural processes must adhere to FDA guidelines. Food products must be FDA approved before they can be sold.

There are many areas of quality control. In livestock farming, for example, the feed given to the animals must meet quality standards. The animals must be carefully checked to ensure that they have no harmful diseases. The preparation of meat for consumption must occur in a sanitary environment. In crop raising, crops are inspected for diseases or other signs of distress. Fertilizers must be approved for use on crops. Fruits and vegetables are inspected after they are harvested.

Food processing systems also have quality controls in place. Since these systems prepare food for human consumption, a breakdown in the quality control system could have health risks for the consumer. Therefore, the systems must be carefully designed to ensure that the food products do not become contaminated or spoiled. Food product is inspected as part of the system. All materials the food product contacts must be kept clean and sanitary.

Summary

Agriculture involves growing plants and animals on farms and ranches. It includes crop production and animal husbandry. It involves the use of many types of machines. It uses pulling, tilling, planting, pest control, irrigation, harvesting, and storage equipment. It also uses structures and equipment to feed and care for cattle, hogs, horses, and poultry. A special kind of agriculture is aquaculture. It

TechnoFact: Greenhouses are special glass-walled buildings with regulated temperature, humidity, and ventilation. They enable people to germinate, grow, and protect plants out of season.

involves raising fish and plants in controlled water environments.

Agriculture often employs biotechnology. This type of technology applies biological organisms to production processes. We use biotechnology to produce new strains of plants and drugs.

Technology is also used in food production. The food agriculture produces must go through many processes before reaching consumers. Technology is used to make sure the food stays fresh.

Natural agricultural environments can be recreated artificially. They may be used for enjoyment or for educational purposes. Terrariums and hydroponics stations are the most common types of these artificial ecosystems.

Standards for Technological Knowledge

3 **12** **13**

TechnoFact: Agriculture was the most important economic activity in America from the founding of the colonies in the early 1600s to about 1900. During the twentieth century, farming declined in economic importance. However, American agriculture continues to be the most efficient and productive in the world.

Curricular Connections

Social Studies

Develop a timeline for the invention of agricultural equipment. Select one machine and write a report on its inventor, operation, and importance.

Science

Research how cross-pollination and crossbreeding are used to improve animal and plant species.

Mathematics

Measure the school plot and convert the measurement to hectares and acres.

Science

Research DNA and its application to gene engineering.

Mathematics

Research the units of measurement used for agricultural crops. Develop a poster or other display presenting and contrasting these units of measurement.

Activities

1. Select a food you eat. Describe how its basic ingredient was planted, grown, and harvested.

2. Go to a local farm implement dealer or locate an appropriate site on the Internet. Gather information on how a specific piece of farm equipment operates.

3. Build a diorama (model) of a farm. Show the technology used on it.

Test Your Knowledge

Do not write in this book. Place your answers to this test on a separate sheet of paper.

1. What is agriculture, and why is it important?
2. What are the two branches of agriculture?
3. Members of the grass family that have edible seeds are called _____.
4. Grass and hay crops grown for animal feed are called _____ crops.
5. The four processes in growing crops using technology are _____, _____, _____, and _____.
6. Growing and harvesting fish in controlled conditions is called _____.
7. Using biological agents to produce goods is called _____.
8. Producing a new organism by cutting and joining genes is called _____.
9. Label the following types of food preservation:
   A. Used to preserve pasta, instant rice, beef jerky, and powdered soup. _____
   B. Can be used to preserve meat, but people are concerned about its safeness. _____
   C. Used to make pickles and corned beef. _____
   D. The most common form of food preservation today. _____
10. Summarize the purpose of artificial ecosystems.

Answers to *Test Your Knowledge* Questions
1. Agriculture is growing and harvesting crops and raising livestock. It is important because it produces the food and natural fibers needed to maintain life.
2. Crop production and animal husbandry.
3. grains
4. forage
5. planting, growing, harvesting, storing
6. aquaculture
7. biotechnology
8. gene splicing
9. A. Dehydration. B. Irradiation. C. Chemical preservation. D. Refrigeration and freezing.
10. Artificial ecosystems reproduce some facets of the natural environment so scientists can study agricultural processes and systems as they would be useful to biological ecosystems.

Chapter 20
Construction Technology

Did You Know?

➤ The pyramids were built to protect the tombs of the early pharaohs. These pharaohs were the ancient kings of Egypt. There were about 100 pyramids built near present-day Cairo.

➤ The Appian Way was the first major Roman road. It was built in 312 B.C. The Romans built roads throughout their empire. The empire stretched from Scotland to the Sahara Desert and from Spain to Syria.

➤ Eleanor Coade developed a new building material in 1722. It was called *Coade stone.* This material was a type of concrete molded to look like stone.

➤ William Jenney developed a new method of construction. This method was used to build the Home Insurance Building in Chicago. It gave birth to a new type of building—the skyscraper.

➤ Rural settlements may be called *villages* or *hamlets.* Urban settlements are often called *towns, cities,* or *metropolises.* Villages and hamlets usually have a few houses, one or two churches, and some stores. Cities have many houses, stores, churches, factories, and schools. They also have hospitals; water, power, and transportation systems; banks; post offices; and recreational centers.

Objectives

The information given in this chapter will help you do the following:

➤ Describe how early humans developed shelters.

➤ Name the five main types of buildings.

➤ Give examples of civil structures and discuss their uses.

➤ Cite and describe the steps in the construction process.

Key Words

These words are used in this chapter. Do you know what they mean?

apartment	joist
boundary	mechanical system
civil structure	public building
commercial building	rafter
condominium	religious building
construction	residential building
contractor	roof
flooring	specification
foundation	stud
industrial building	superstructure
infrastructure	survey
insulation	utility

Chapter Outline
Did You Know?
Objectives
Key Words
Buildings
  Residential Buildings
  Commercial Buildings
  Industrial Structures
  Public Buildings
  Religious Buildings
Civil Structures
Constructing a Structure
  Preparing to Build
  Preparing the Site
  Setting Foundations
  Building the Framework
  Enclosing the Structure
  Finishing the Interior
  Finishing the Site
Summary
Curricular Connections
Activities
Test Your Knowledge

Standards for Technological Knowledge

7

Research
Have the students research which type of columns the ancient Greeks and Romans used in their buildings.

Tech Terms
Structure—A building, bridge, framework, or other object assembled from a number of parts.

Throughout history, people have sought places to live. Our early ancestors lived in natural features of nature, such as in caves and under trees. As they developed farming techniques, people started living in villages. These villages contained houses for a number of families. They also had buildings in which to store grains and other foods. As civilization progressed, people built more structures. Their villages started to become small cities. See **Figure 20-1.** There were simple homes for most of the people, but the cities also had fancy houses for leaders and rich people. These cities had buildings for businesses and government activities. In or around the cities, people built temples to their gods. See **Figure 20-2.** People built roads to travel on and bridges with which to cross rivers. They also built dams to contain a water supply and aqueducts to move water from one place to another. There are remains of

Figure 20-2. These are the remains of a temple in Herculaneum, an ancient city in Italy.

many of these ancient structures throughout the world. You can see the pyramids in Egypt, the Colosseum in Rome, and Mayan temples in Central and South America.

The kind of structures people built reflected their lifestyles. People who moved in search of food needed simple, portable structures. The early plains Indians used tents for their villages. Today, nomads in Central Asia still

Figure 20-1. This photograph shows the remains of Pompeii. Pompeii was a city built in what is now Italy. In 79 a.d., an eruption on Mount Vesuvius destroyed it.

TechnoFact: Ancient humans constructed many types of tunnels. These types included catacombs and crypts, in which to bury the dead, and many transportation tunnels. Over twenty-five hundred years ago, the Greeks constructed a 3,500' long, 6' square tunnel to carry water. Over two thousand years ago, the Romans built a one mile long, 30' wide highway tunnel connecting two towns.

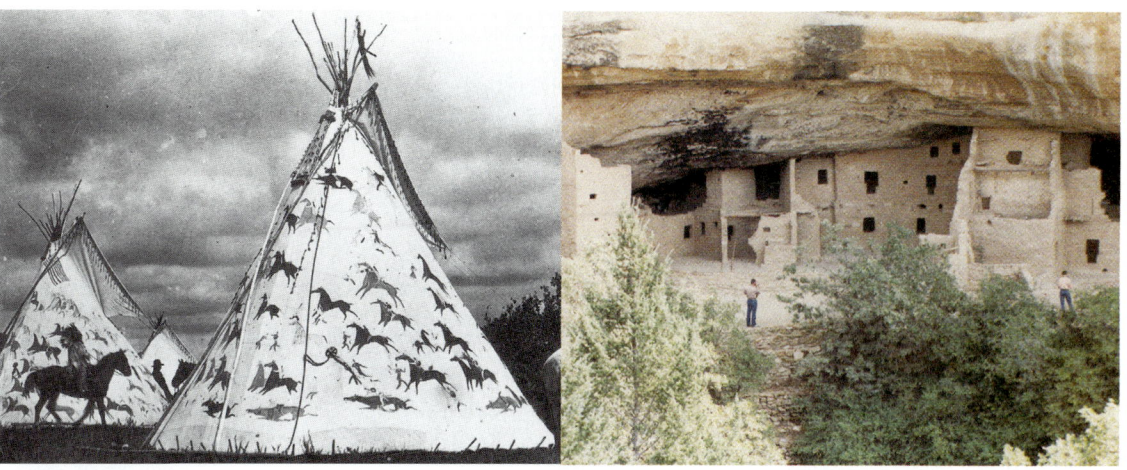

Figure 20-3. These are examples of early buildings. On the left is a Sioux dwelling of the western plains. (Smithsonian Institution). On the right is a cliff dwelling built by ancient Pueblo Indians who lived in Arizona, New Mexico, Utah, and southern Colorado.

Standards for Technological Knowledge

7

Reinforce
Discuss construction as a technological system (Chapter 3).

Tech Terms
Construction—The act of building structures on a site.
Build—To produce a structure by combining resources or parts.

use this type of structure. People who farmed or lived in an area with ample food supplies built more permanent structures. See **Figure 20-3.** These people developed wood or stone structures.

Today, people continue to build structures. Many of us live in neighborhoods made up of things people have erected. We call these buildings, roads, and monuments *constructed works.* They are structures erected on the sites where they are to be used. The act of building these structures is called *construction.* The design and construction of structures for service and convenience have developed from the advancement of methods for measurement, controlling systems, and the knowledge of spatial relationships.

Construction is a series of carefully planned events. This area of technology uses materials, work, processes, and equipment to build a structure on a site. The structures people design and build meet many different needs.

These include buildings and civil structures. See **Figure 20-4.**

Buildings

Some of the earliest constructed works were buildings. Buildings are structures providing protection and safety for humans and their possessions. They serve a variety of purposes. Today, there are at least five different types of buildings:

➤ Residential.
➤ Commercial.
➤ Industrial.
➤ Public.
➤ Religious.

Residential Buildings

Shelter is one of our basic needs. It protects us from the weather and outside threats. Homes make us safer and more comfortable. They contain

TechnoFact: Pompeii was an ancient Roman city near Naples, Italy. It was a busy port and popular resort. In 63 A.D., the city was severely damaged by an earthquake. Sixteen years later, an eruption of nearby Mount Vesuvius buried the city under cinders and ash, preserving the ruins of the city and the bodies of many of its inhabitants.

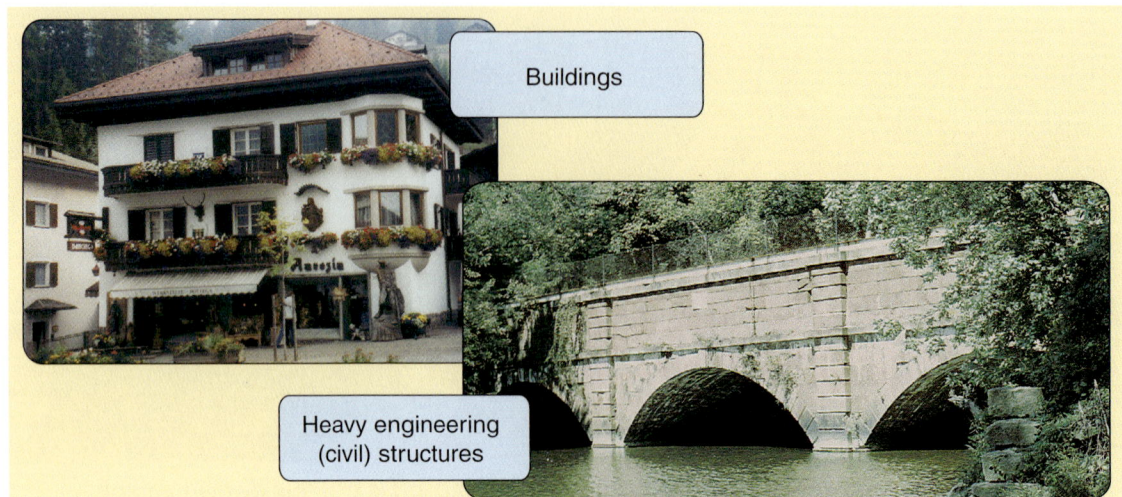

Figure 20-4. Construction technology is used to erect buildings and civil structures.

the basic spaces people need. Most *residential buildings* (homes) have living, food preparation, entertainment, sleeping, storage, and sanitary (bathrooms and laundry) spaces. They may have attached or detached automobile storage. Several different types of homes are built. See **Figure 20-5.** Single family dwellings stand alone on a plot of land. These buildings are designed for one family to occupy. Duplexes are buildings containing separate spaces for two families. These spaces are often side-by-side in the same building. Condominiums and apartment buildings have several living units joined together. In *condominiums,* separate families own each living unit. A company or one person generally owns an *apartment* building. The units are rented to the people living in them. Apartment buildings and many condominiums are multistoried (have more than one floor or level) buildings. A single, large building may have many separate dwelling units.

Figure 20-5. People live in different kinds of structures. These include single family dwellings (left), apartments, and condominiums (right).

TechnoFact: William Levitt addressed the huge demand for housing that developed after World War II by building Levittown, New York, in 1947. This was the first major planned community. The Long Island community had 17,500 two-bedroom homes that sold for about seven thousand dollars each.

Figure 20-6. People take care of daily business and shopping in commercial buildings.

Commercial Buildings

Each of us does some business in a building. See **Figure 20-6.** We buy groceries or clothing in a building. Each of us receives medical or dental care in a building. We may buy an airline ticket at a travel agent's office. Any of us may have our car repaired in a garage. We may stay in a motel or hotel as we travel. All these activities are part of commerce. They take place in *commercial buildings.* These buildings may be professional offices, shopping centers, supermarkets, lodging establishments, or repair facilities. The businesses occupying these buildings provide us with services or sell us goods.

Industrial Structures

All societies produce goods and services they need. They make products in factories. Societies generate electricity in power plants. They develop news and entertainment programs in radio and television studios. These activities are called *industrial or productive activities.* They require special structures we call *industrial buildings.* See **Figure 20-7.** These structures provide workspace and shelter for people, materials, and equipment. They generally contain office space, production or service areas, storage areas, and worker support areas (such as restrooms, locker areas, and cafeterias).

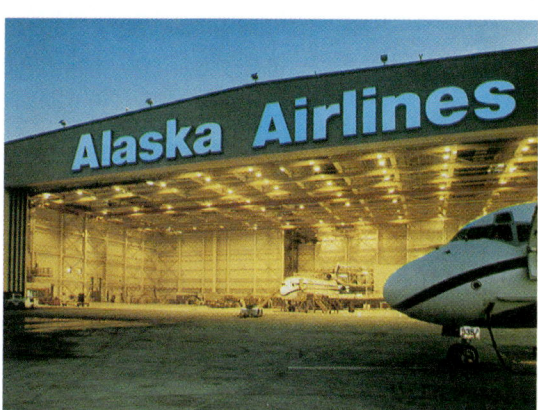

Figure 20-7. Industries use buildings in their business activities. (Alaska Airlines)

TechnoFact: The nation of Zimbabwe takes its name from an ancient masonry city. In the city center was an outdoor amphitheater or temple structure called a *Zimbabwe.*

Figure 20-8. Churches, synagogues, and mosques are religious buildings. This is a church in Northern Italy.

Public Buildings

People use their tax money to erect special purpose buildings. These buildings are designed to meet public needs. They provide areas for administration, police and security, fire protection, health care, education, and other government functions. Government agencies, such as cities and towns, school districts, states, and the federal government, build them. Most *public buildings* are paid for with tax money. These buildings may be schools, government office buildings, courthouses, jails, and monuments.

Religious Buildings

People erect buildings in which they can practice their religious activities. See **Figure 20-8.** *Religious buildings* are called a number of different names. They may be called *churches, cathedrals, synagogues,* or *mosques.* These buildings are used for worship, fellowship, education, and other activities associated with religion. In many cases, community service activities are also provided in the buildings. These activities may include counseling, collecting food and clothing for the needy, providing shelter for the homeless, and conducting community interest programs (such as music and speakers).

Civil Structures

Civil structures are constructed works supporting the public interest or other technological activities. They are built for the convenience of all the people who live in an area or nation. Civil structures include constructed works that are not buildings. Examples of civil structures are streets and roads, bridges, tunnels, railroad lines, airport runways, sewers, pipelines, dams, ponds and reservoirs, communication towers, sports fields, monuments, and water towers. See **Figure 20-9.**

Bridges and overpasses take people and vehicles over rivers, canals, and other obstacles. Overpasses cross obstacles, such as other roadways or railroad tracks. Airport runways provide landing strips for airliners and other airplanes. Pipelines move petroleum from wells to refineries.

Sewers carry waste away from homes and other buildings. Water towers and reservoirs store water for drinking, firefighting, and industrial and commercial activities. Pipes and cables supply water, fuel, and electricity.

TechnoFact: A major type of civil structure is a bridge. One of the most famous is the Golden Gate Bridge. Its bold design included a 4,200' central span that made it longer than any suspension bridge ever built before that time. The suspension cables that hold the bridge deck are 3' in diameter and woven from over twenty-seven thousand individual strands. The bridge was opened to vehicles in 1938 and has been in continuous use since then.

Figure 20-9. Earth is moved and compacted to make the foundation for a road. Then the asphalt superstructure is laid down. (Natchez Trace Parkway, National Park System)

Resource
Reproducible Master 20-2: Steps in Constructing a Structure

Research
Have the students identify a local or state monument and indicate the person or event it honors.

Tech Terms
Construction Process—The series of actions completed to build a structure.
Monument—A building or place paying public tribute to a person or an event.

Communication towers broadcast radio, television, and telephone signals. Power transmission lines move electricity from generators to homes, factories, and businesses.

Monuments are structures built to pay honor to people, ideas, and events. See **Figure 20-10.** Almost everyone can recognize a picture of the Statue of Liberty in New York. This statue was a gift from the French people to the United States. Likewise, most people can recognize the Washington Monument and the Lincoln Memorial in Washington, D.C.

Constructing a Structure

Construction is a technological activity. It requires a series of actions that have to be done in the right order. These actions are part of a technical process called the *construction process.*

This process generally follows these eight steps:

1. Preparing to build.
2. Preparing the site.
3. Setting foundations.
4. Building the framework.
5. Enclosing the structure.
6. Installing utilities.
7. Finishing the interior.
8. Finishing the site.

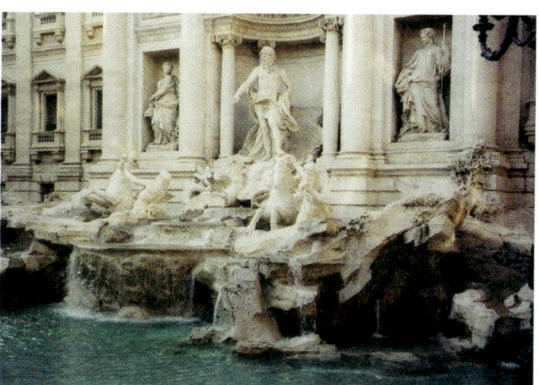

Figure 20-10. This monument in Rome is one type of public structure.

TechnoFact: The Washington Monument is 555′ high and weighs over ninety thousand tons. This marble-faced granite structure is the tallest masonry structure ever built and approaches the maximum theoretical height for such structures. The Washington Monument was finished in 1885, the same year the first steel-frame skyscraper (the nine-story Chicago Home Insurance Building) was completed.

Technology Explained

tower crane: A self-raising crane used to lift materials on high-rise building projects.

Building high-rise structures offers a number of unique challenges. One challenge involves lifting the structural steel members that will be part of the building into place. This is often done with tower cranes, like those shown in **Figure A.**

A tower crane has several parts: the tower sections, main jib (cross arm), slewing (turning) gear, and cab. A truck-mounted crane lifts the base unit of the tower crane into place. The height of the crane needs to be increased as the building is constructed. The

Figure A. This tower crane is used to lift building materials into place.

crane rises under its own power. As shown in **Figure B,** several steps are involved.

When additional height is needed, a device called a *climbing frame* is used. The

Standards for Technological Knowledge

20

Resource
Reproducible Master 20-3: Planning and Management of Construction Projects

Extend
Explain the types of information drawings can provide to builders.

Tech Terms
Architectural Drawings— Drawings used to convey the size, shape, and space arrangements of constructed structures.

Preparing to Build

A structure starts with an identified need. A family needs a new home. A community needs a new courthouse. A business needs a new store. A doctor needs a new office. The list could go on and on.

The need must be changed into a design. The selection of designs for structures is based on many factors. Builders need to consider style, convenience, cost, climate, and function. They also must pay attention to building laws and codes, which are typically part of the city or county regulations for construction. Once all of these factors have been considered, a suitable design can be created.

The design is described with a set of plans. The new structure is shown on architectural drawings and specifications. The drawings show the shape and size of the proposed structure and the arrangement of spaces within the structure. They also show the location of features, such as windows and doors, and how foundations, floors, walls, and roofs are to be constructed. Electrical, heating, and plumbing systems

TechnoFact: Architecture is the art and technology of designing buildings in which human requirements and construction materials are related to one another to produce practical and aesthetically pleasing structures.

frame is positioned beneath the cab. Hydraulic cylinders lift the jib and cab unit above the tower. Then, as shown in the middle drawing of **Figure B,** the crane lifts another section and swings it into the opening in the climbing frame. The new section is bolted in place to produce a stable tower. This procedure is repeated as many times as necessary. The climbing frame is generally removed when it is no longer needed.

The tower crane uses a trolley, cables, and a hoist (a power-driven drum) to lift loads. The trolley is a frame with pulleys that can be moved along the main jib. A cable extends from the cab out to the trolley and down to the

load. It is wound around the hoist drum to lift the load. When the load is at the right height, the main jib turns, and the trolley moves along the jib to position the load over the structure. Then, the load is lowered into position. When the project is completed, the tower crane is disassembled and moved to the next construction site.

A typical tower crane can have a maximum unsupported height of about 265´ (80 m). It can be much taller, if it is tied to the building as the building rises around the crane. This type of crane can have a maximum reach of about 230´ (70 m) and maximum lifting power of about 20 tons (18 metric tons).

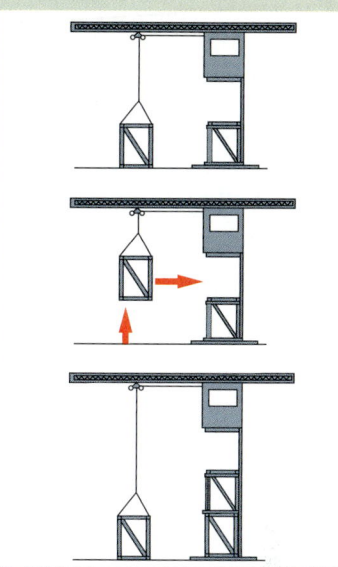

Figure B. Here is a drawing showing how a tower crane increases its height. The climbing frame lifts the cab and jib, and a new section is put in place.

are shown. The plans often include additional information sheets called *specifications.* These information sheets are descriptions telling people how the work must be done. They also explain what quality of materials to use. The plans and specifications tell the builder what the structure will look like and how to build it.

Once the plans are developed, the future owner must obtain the money to build the project. Public projects get funds from taxes. Government officials may take the money from existing tax revenue. In some cases, they may have

to levy a new tax to pay for the project. Individuals pay for private projects. Some people may already have the money in the bank. Many people, however, have to borrow the funds from a bank to build the structure.

With the money secured, someone must locate and obtain a building site. A dam or a bridge will be built on a river. There will be few choices of location. A house, store, office, or factory is different. There are many places where these structures can be built. It is important to choose the right place. A house should be in a residential area.

Research
Ask the students to research various sources of money spent for building structures.

Tech Terms
Specifications—
Descriptions of characteristics of materials or the procedures to be used in building a structure.
*Loan—*Money borrowed that is to be paid back at a particular interest rate.
*Site—*A parcel of land on which a structure will be located.

TechnoFact: The money spent on construction comes from many sources. Home construction is frequently financed through banks, lending institutions, or personal savings. Governments may issue bonds to finance construction projects.

Standards for Technological Knowledge

20

Figure Discussion
Refer to Figure 20-11 as you discuss with the students why businesses, homes, and factories should be built in separate areas.

Brainstorm
Have the students develop a list of their ideas about what makes a good residential area.

Tech Terms
Infrastructure—Essential services and systems required for a neighborhood or society to operate, such as roads, phone lines, power systems, water lines, sewage management facilities, and schools.
Boundary—The border or limit of a parcel of land.
Contractor—A person or company agreeing to supply resources or perform services for a particular fee.

Figure 20-11. Homes should be built in residential neighborhoods.

See **Figure 20-11.** The house should also be close to things the owner needs. Being near shopping areas, schools, parks, and jobs may be important. In contrast, an industrial building needs to be away from residential areas. It needs to be located near transportation, communication, water, sewage, and power systems. These elements are called the *infrastructure.*

When land is bought, the new owner must know where the property begins and ends. The defining lines are known as *boundaries.* They are established by doing a survey. Also, soil tests are often needed. Builders need to know how well soil will support a structure.

Also, a *contractor* or several contractors must be selected. On many building projects, a general contractor is hired to oversee all the work. A general contractor manages materials, other contractors, and workers.

Contractors are responsible for hiring workers. They also get materials and equipment needed to construct the structure. The contractors must inspect the quality of the work and materials. They see that the work meets the standards set in the specifications.

There are several quality controls in construction systems. Structures are designed based on building code requirements, which generally focus on the safety aspects of the structure. The building specifications list quality requirements for materials. Standard construction materials, such as steel bolts and plywood panels, are rated before arriving at the job site. Materials produced on-site, such as concrete, must be tested in the field. Throughout the construction process, inspectors review the work to ensure that the contractor is meeting the building code requirements.

The last step before building can start is getting a permit. Most communities control what is built in their area. Factories should not go up in a residential neighborhood.

Officials of a city or county review the plans. They want to know that the building meets local building codes and zoning ordinances. When the plans are approved, a building permit

TechnoFact: Suburbs are not new phenomena. They have existed in one form or another throughout history. In ancient times, many cities were surrounded by walls. The villages outside of the city walls were smaller and often considered lower in the social hierarchy. Now, suburbs are often considered the opposite of that: upscale places for upper- and middle-income families. This "suburb" may have appeared over twenty-six hundred years ago in Babylon. The word *suburb* comes from the Latin *suburbani,* which Roman Emperor Cicero used in the first century B.C. to describe luxurious estates outside Rome.

Figure 20-12. This hillside lot is being prepared as a building site.

is issued. This document allows the contractor to start work.

Preparing the Site

Few building sites are ready for a new structure. There may be old buildings that must be torn down. Brush and trees may need to be removed. Rocks and debris must be hauled away. The ground may need to be leveled. See **Figure 20-12.** Often the topsoil is scraped away and stored for later use. Sometimes, temporary buildings and service roads are built to aid the construction of permanent structures. All these activities help prepare the site for the new structure.

Once the site is prepared, a structure can be located on it. It must be kept a certain distance from the property of others. Local restrictions dictate these distances. Work crews must *survey* the site. See **Figure 20-13.** These crews

measure distances from the boundary lines of the property. Stakes are driven to show where to place the structure's foundation.

Setting Foundations

A constructed structure has two major parts. See **Figure 20-14.** There is the *substructure* below the ground. It is often called the *foundation.* The substructure connects the structure to the earth. Also, it spreads out the weight of the structure so the structure does not sink into the ground. The other part of the structure is the *superstructure,* which is usually seen above the ground. It is the part of the structure that is used. The superstructure is the reason the structure is built.

Figure 20-13. This surveyor is surveying a building site. (U.S. Department of Agriculture)

Standards for Technological Knowledge

20

Resource
Reproducible Master 20-4: The Many Tasks in Building a Structure

Demonstrate
Show the students how a site can be surveyed by laying out a boundary.

Tech Terms
Surveying—
Measuring distances, elevations, and directions of landmasses to locate property boundaries, lay out construction, and produce maps.
*Foundation—*The part of a structure on which it stands or by which it is supported.

TechnoFact: Concrete is a common material used in foundations. It is a structural material made by mixing gravel, sand, cement, and water. The Romans used a material very similar to modern concrete about twenty-two hundred years ago for constructing roads, buildings, and other public works, such as their aqueducts.

Tech Terms
Pile—A heavy timber, steel beam, or column of concrete set into the earth as a foundation or support for a structure.
Roadbed—The foundation on which railways are constructed.
Riprap—A foundation made of loose broken stones, used in water or on soft ground.

Railroad

Highway or airport runway   Dam   Home   High-rise building

Figure 20-14. Here are samples of superstructures and substructures.

Structures rest on foundations. A foundation can be constructed of such materials as packed earth, gravel, wood, concrete, or steel. Wood or steel poles, called *pilings*, may be sunk into the ground. These poles support the structure when the soil is not firm.

There are several different types of foundations used for buildings. The type of structure determines the type of foundation needed. Three common types are spread, raft, and pile foundations. See **Figure 20-15.**

Foundations start with excavating. This process prepares the ground for the foundation. Holes or trenches (ditches) are dug in the ground. It is important to place the foundation on solid soil or rock. Also, the foundation footings must be below the frost line. The frost line is the deepest level at which the ground freezes during winter. Otherwise, freezing and thawing of the earth will damage the foundation and the superstructure.

One of the most popular materials for foundations is concrete. It is always used for raft and pile foundations. Concrete is a mixture of cement, gravel, and water. Freshly mixed concrete will not hold its shape. Therefore, temporary structures are needed to aid in the construction of these foundations. The

Figure 20-15. These are some kinds of substructures and superstructures for houses.

TechnoFact: The Leaning Tower of Pisa is a marvel to see, but it also shows what a poor understanding of foundations can lead to. Construction was started in 1173 and completed about two hundred years later. The tower is over 180' tall and sits on soft soil. Because of its poor foundation, the tower started to tilt, even as it was being built. It has continued to sink and tilt over the last eight hundred years. Efforts are under way to stabilize the tower.

Figure 20-16. This is a set of forms with the concrete curing between them.

concrete must be poured into forms until it sets. See **Figure 20-16.** After the concrete becomes solid, the forms can be removed. If a footing is needed for the foundation, the footing is poured first. Then the foundation form is erected, and the foundation is poured.

Building the Framework

A framework or superstructure is the part of the structure built on top of the foundation. It can be made of steel, concrete, adobe (straw and mud blocks), wood, or other material. There are several types of frameworks. The most common is called a *framed superstructure.* See **Figure 20-17.** Other types include solid (for example, a dam) and air supported (for example, a sports dome) frameworks.

Let us explore common frame construction, as it is used in a house. The superstructure of the house is built on top of the foundation. It has three main parts:

➤ Floor.

➤ Wall.

➤ Roof.

The superstructure must be securely fastened to the foundation. Bolts or straps are embedded in the top of the foundation wall. Wood pieces are attached to the bolts or

Standards for Technological Knowledge

20

Demonstrate
Show the parts of a structure using pictures or a model.

Brainstorm
Have the students develop theories on the purpose of the superstructure.

Tech Terms
Superstructure—
The part of a building built above the foundation.

Figure 20-17. Can you see the framework, or superstructure, of the bridge (left) and the office building (right)?

TechnoFact: Construction on the Taj Mahal in India was started around 1630 by Emperor Shah Jahan as a tomb for his wife, Mumtaz Mahal. The white marble monument has a dome that rises 240' and four minarets that are 140' tall. Over twenty thousand people worked twenty-two years to build it.

Research
Have the students investigate when balloon framing was developed and for what reasons.

Tech Terms
Joist—Parallel wood, metal, or concrete beams supporting a floor or ceiling.
Girder—A large, strong beam providing the main spanning and supporting member in a floor system.

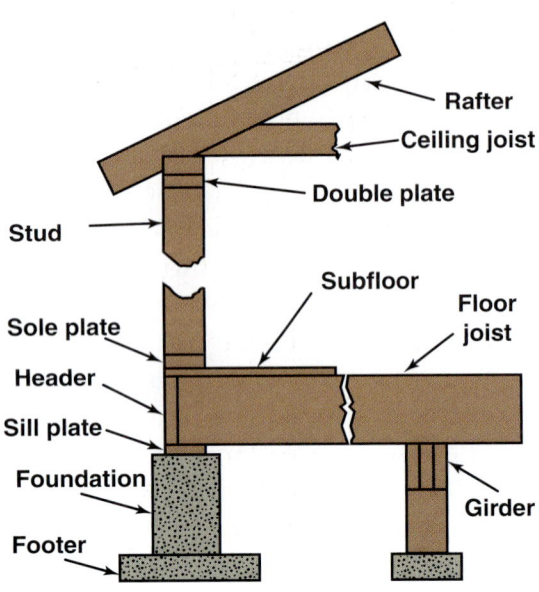

Figure 20-18. Here are the parts of the framework of a frame-built house.

straps. These parts make up the *sill plate.* See **Figure 20-18.**

Floor Framing

The floor frame is fastened on top of the sill plate. **Joists** extend from one wall to the other. See **Figure 20-19.** The joists are placed on an edge and nailed to a header. They must carry the weight of the floor, furniture, appliances, and people that will be on them.

Sometimes the foundation walls are far apart. The joists cannot reach from wall to wall. Also, the span may be too great for them to carry the weight that will be in the house. Then an extra support is placed between the two walls. This steel or wood support is called a *girder.*

Figure 20-19. This is an example of floor framing.

TechnoFact: Joseph Monier patented a reinforced concrete beam in 1878. Later, other people used reinforced concrete for pipes, aqueducts, bridges, tunnels, and buildings.

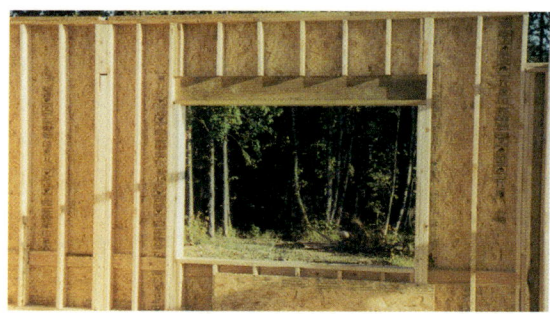

Figure 20-20. Here is a framed wall for a new home. Can you see the plates, studs, and headers?

Wall Framing

In most houses, walls are made of wood framing. Recently, steel framing is being used in some homes. Wall frames are designed to be light and strong. They include plates, studs, and headers. See **Figure 20-20.**

Plates are the horizontal parts at the top and bottom of the wall. They hold the *studs,* which are the vertical parts. Studs are 2 x 4 or 2 x 6 boards or metal shapes spaced 16″ or 24″ apart. The spaces between the studs are later filled with insulation. *Headers* are beams placed above door and window openings. They support the weight over these openings.

Roof Framing

Buildings may have flat or pitched *roofs.* A pitched roof is attached to the walls. It rises several feet to the center of the house. The high part of the roof is called the *ridge.* The low ends are called the *eaves.*

The roof framing members are called *rafters.* They are generally spaced 16″ to 24″ apart. At the ridge, the rafters are fastened to a *ridge board.* In some cases, trusses are used for the

roof framing. These are manufactured elements that include the ceiling joists and rafters in one unit.

Enclosing the Structure

Buildings are designed to serve many purposes:

➤ Keep out the weather (rain, snow, wind, heat, and cold).

➤ Protect people and goods.

➤ Give privacy.

To accomplish these tasks, the building must be enclosed. Builders will perform many tasks to close in the structure. They will cover the frame with a sheet material, such as plywood, flake board, oriented strand board (OSB), or insulating board. The material closes the structure and makes the frame more rigid. When this material is applied to outside walls, it is called *sheathing.* Sheathing used on roofs is known as *decking.*

Enclosing Wall Materials

After the sheathing is applied, windows and outside doors are installed. See **Figure 20-21.** The windows and

Figure 20-21. This worker is installing vinyl windows in a wall of a house.

Standards for Technological Knowledge

Extend
Have the students sketch and name the parts of a wall, floor, and roof system.

Tech Terms
Plate—Horizontal members on the top and bottom of a wall.
Stud—Vertical members in a wall.
Rafter—Sloping roof members extending from the ridge beam of a roof to its edge.
Roof—The framework and covering of the top of a building.
Sheathing—A layer of materials applied to the studs, beams, and supports of a structure to reinforce the building and act as a base for an exterior cladding.

TechnoFact: Before the mid-1800s, houses were built with heavy interlocking beams. A construction technique was developed in Chicago in the 1830s in which a house was framed with 2″ × 4″ or 2″ × 6″ studs, crisscrossed with pieces of wood. This technique was called *balloon framing* and remained popular until about World War II. Now, most houses are framed with a technique called *platform framing,* in which each floor or level serves as a platform for the next level.

Figure 20-22. Shingles are used on many homes. Notice how they overlap to provide a waterproof surface. (Asphalt Roofing Manufacturers Association)

doors enclose important openings giving people sunshine and giving air access to the structure.

Sheathing encloses and strengthens the frame. This is usually 4' x 8' sheet material, such as OSB, plywood, or a similar material. The sheathing, however, is not very attractive. To improve the beauty of the building, attractive weatherproof materials cover the sheathing. There are many different types of exterior building materials. The material may be brick, stone, wood siding, aluminum, vinyl (plastic), brick veneer, glass, wood panels, rocks, logs, or shingles.

Enclosing the Roof

Roofs are more likely to leak than walls are. Special waterproof materials must be used to keep the building dry. Shingles are normally used on all pitched roofs.

A shingled roof is installed by attaching many small pieces of waterproof material. This material must withstand many kinds of weather conditions. As each piece is installed, it overlaps the one beneath it. See **Figure 20-22.** Shingles are made of many different materials, such as wood, slate, ceramic, aluminum, fiberglass, or asphalt.

Installing Utilities

Buildings usually contain a variety of subsystems, including electrical, water, waste disposal, climate control, communication, and structural subsystems. Most of these subsystems are referred to as utilities. *Utilities* are services coming into a building. They are supplied from pipes and cables. Most buildings have electrical wiring. They also have a plumbing system to bring in water and carry away wastes. In many homes, an air-conditioning system supplies cool air. Telephone and television cables support entertainment and communication systems. Sometimes, pipes bring in natural gas for heating. All these systems are hooked up to outside utility lines. The community or private companies supply these lines.

In a building, these systems are usually called the *mechanical systems.* They must be installed before the inside of the house is finished. Installing mechanical systems is called *roughing in.* Usually, you cannot see the systems. Mechanical systems are placed inside the walls and under floors. If you go into a basement or under a house, you may see the pipes, ducts, and electrical wiring.

TechnoFact: The first steel-frame, glass-curtain-wall building was completed in 1918. By the 1960s, this technique, pioneered by L. Mies van der Rohe, was used in most high-rise buildings.

The steps of roughing in are usually carried out in a special order. Ductwork or pipes for the heating system are installed first. They carry warm or cool air from a furnace or air conditioner. These pipes go to each room of the house. They are fitted between floor joists and wall studs.

Plumbing systems are installed next. See **Figure 20-23.** This system of pipes carries liquids and gases. Plumbing in a building is needed to accomplish several purposes:

➤ Provide fresh, pure water.

➤ Remove wastewater.

➤ Carry fuel to furnaces, water heaters, and stoves.

Like the ductwork, plumbing is installed in the open spaces between the wall studs and floor joists.

Electrical systems distribute electrical power. They also are used for communication. Electric power runs appliances and is used for lighting. Telephones and intercom systems use electrical impulses to carry speech. Some homes have home theater systems requiring wiring to various

Figure 20-23. Plumbing system pipes supply pure water and carry away wastewater.

Figure 20-24. Insulation is made from materials that do not conduct heat. Placing it in walls and ceilings prevents heated air from passing in or out of a building. (Owens Corning)

speakers. Since electrical wiring is small, it is installed last.

Finishing the Interior

Once the utilities are roughed in, the interior can be finished. The walls and ceilings can be insulated. Also, they can be covered with protective and decorating materials. Finally, appliances, cabinets, and fixtures can be installed.

Insulating

Outside walls and ceilings of the house must be insulated. *Insulation* is a material that resists heat passage. It keeps heat inside during cold weather. In hot weather, it keeps the heat outside. Insulation is very light, soft material made in several forms. Loose insulation can be blown into ceilings. Long strips of insulation, called *batts,* can be placed between studs and joists. See **Figure 20-24.**

TechnoFact: Some people are building modern houses out of straw. They start with a concrete foundation with vertical reinforcing bars extending out of the foundation. Then, bales of straw are forced over the steel rods. This process continues, much like building a house from a set of blocks. Finally, stucco is applied to the outside and plaster is applied to the inside. The thick straw walls make excellent thermal insulators.

Tech Terms
Floor—The base of a room on which one stands, or the lower surface of a building.
Flooring—A material used to make floors.
Paint—A liquid composed of a solid dye in a liquid medium, used as an ornamental covering or a protective finish.

Figure 20-25. These workers are attaching drywall to the interior walls and ceilings. (U.S. Department of Agriculture)

Enclosing Interior Walls and Ceilings

Once the insulation is in, workers can enclose the interior. First, walls and ceilings are covered to provide a smooth surface. This is generally done with drywall or wallboard. See **Figure 20-25.** Drywall is manufactured by putting a layer of gypsum (a chalky substance) between two layers of heavy paper. Drywallers attach large sheets of drywall with glue or nails. Seams and nail heads are concealed with a special tape and filler.

Finishing the Inside

Finishing includes jobs making the interior of the building attractive. These tasks generally follow a certain order:

1. Painting and decorating.
2. Installing finish flooring.
3. Installing window, door, and baseboard trim.
4. Installing electrical and plumbing fixtures and accessories.
5. Cleaning up.

Painting and decorating are usually done first. They are done to beautify the interior. Paint protects wood and drywall surfaces. Wallpaper, wood paneling, and ceramic tile are other choices for wall coverings.

Finish *flooring* is usually installed after the painting and decorating are done. Installing it at this point keeps the floors from being damaged during the painting process. Flooring materials are

TechnoFact: Early humans decorated their surroundings much like we decorate our homes. The technology of painting was part of the Stone Age. There are remnants of Stone Age cave paintings in many areas, including northern Spain and southwestern France.

designed to wear well and look attractive. Many different materials make good finish flooring. Wood, carpeting, linoleum, and ceramic tile are among the most commonly used. See **Figure 20-26.** These materials are fastened to the subfloor or underlayment with special nails or adhesives (glues).

The final steps involve trimming the structure and installing accessories. Trim consists of the decorative wood or plastic strips covering joints. Joints appear where floors, walls, and ceilings meet. They are also where window and door frames meet walls.

Next, room doors are carefully hung on their hinges. Closet doors are installed. Edges are dressed (planed) so they fit openings.

Then, cabinets are installed in kitchens, bathrooms, and other rooms. These cabinets provide storage space

Figure 20-26. This ceramic tile is a good entry floor covering.

and support countertops. During this installation, the countertops and kitchen appliances are installed.

Finally, hardware and accessories are installed. Hardware includes doorknobs, latches, catches, and brackets. Other accessories include closet shelving and towel bars. Plumbing faucets and electrical fixtures are among the final items installed.

Finishing the Site

Construction is not complete until the building site is finished. During construction, the site becomes cluttered. Scraps of building materials are everywhere. There may be piles of dirt. Several things must still be done:

1. Clearing the site.
2. Leveling and grading the ground.
3. Creating walks and drives.
4. Adding landscaping.

Clearing the Site

Clearing of the site may need to be done before any other finishing steps are taken. Some of the dirt may not be needed. It must be hauled away. Rocks, trash, and scraps of building materials must be removed. Temporary buildings and fences are taken down.

Leveling and Grading the Ground

Earth may have to be moved. Holes may have been dug for foundations. Some of the earth is pushed back to fill in around the foundation. This

Extend
Discuss the ways in which people decorate their homes.

Brainstorm
Have the students discuss floor coverings and then determine what they think is the best for (a) kitchens, (b) living rooms, (c) bathrooms, and (d) bedrooms.

Tech Terms
Grading—Flattening or leveling to a preferred surface or incline.
Backfilling—Filling the areas around the foundation of a building, typically with dirt or gravel.

TechnoFact: Some people have compared the Statue of Liberty to the Colossus at Rhodes, which is one of the Seven Wonders of the Ancient World. The Colossus of Rhodes is said to have been over 100′ tall. The Statue of Liberty is over 150′ tall. With her pedestal height included, she rises over 300′ above New York Harbor.

Figure 20-27. This natural stone makes for a decorative walkway.

is called *backfilling*. The earth is shaped around the structure. Soil may be moved from one spot and placed in another, so it is more pleasing. Topsoil may be returned to areas that will have plants and lawn.

Creating Walks and Drives

Walks and drives give users access to a building. Drives must have a heavy base of gravel. The surface may be finer gravel, concrete, or asphalt. Sidewalks can be constructed of concrete, natural stone, wood, or masonry units. See **Figure 20-27.**

Adding Landscaping

Landscaping is a way of making the site more attractive. It includes planting trees, shrubs, and flowers. Landscaping also often includes planting grass or putting down sod. Ground cover, such as bark and rock, is sometimes used where grass is not wanted. It keeps soil from washing away and covers unattractive soil.

Summary

Construction is building a structure at the spot where it will be used. Our ancestors knew little about putting up buildings. They lived in caves or built crude shelters that did not last. In time, people began to settle in one place and erect permanent buildings.

Structures of today are built for several purposes. Residences and commercial buildings shelter people where they live and work. Industrial buildings house companies making products and providing important services. Public buildings provide space for tax-supported activities. Religious buildings provide places for worship and fellowship. Civil structures help move people and materials or provide some other benefit to the public. Construction processes include acquiring land, preparing the site, putting in foundations, and building the frame or superstructure. They also include installing mechanical systems and finishing the structure and site.

Curricular Connections

Standards for Technological Knowledge

3 **13**

Social Studies

Research the types of housing people developed (1) in one region or country at different points in history and (2) in different cultures or countries at the same point in history.

Science

List the major materials used in construction. Investigate their properties (such as strength, hardness, and color).

Mathematics

Measure the rooms in a home. Develop a chart comparing the sizes of these rooms using square footage.

Social Studies

Describe how (1) climate, (2) natural materials available, and (3) lifestyle affect the type of housing a society develops.

Science

Design and conduct a test on the wear resistance of several different flooring materials.

Activities

1. On your way to school, look for construction projects. Make a list of these projects. Try to separate the projects into one of the following groups: residential buildings, commercial buildings, public buildings, and religious buildings.

2. Locate a map of your city. Mark where residential, industrial, and commercial areas are located. Explain why you think they are located where they are.

3. List the reasons your school is located where it is.

TechnoFact: Landscape gardening is an ancient art. The Egyptians used formal walled gardens, while the Mesopotamians constructed private parks and terraced gardens as landscape elements. The Persians developed the use of water for decorative effects. The Japanese art of bonsai gave rise to the unique miniature gardens and dish gardens.

Test Your Knowledge

Do not write in this book. Place your answers to this test on a separate sheet of paper.

1. A building, road, or monument is called a _____.

2. Match the types of buildings with the right description(s):

    _____ May be used as a grocery or clothing store.

    _____ Contains spaces for living, food preparation, and sleeping.

    _____ May be called a church, synagogue, or mosque.

    _____ May be used as a jail or firehouse.

    _____ May be apartments.

    _____ May be used as factories.

    _____ Used by businesses providing services or selling goods.

    A. Residential.

    B. Commercial.

    C. Industrial.

    D. Public.

    E. Religious.

3. Paraphrase the definition of *civil structure* .

4. Building a structure begins with recognizing a _____.

5. The part of the structure connected to the earth is called the _____.

6. The frame of the house includes what three major parts?

7. Explain the three main purposes buildings serve.

8. Inside a building, the heating ducts, electrical wiring, and plumbing pipes are known as the _____.

9. Materials used to keep heat in or out of a building are called _____.

10. Planting the lawn and shrubs around a home is called _____.

Modular Activity

This activity develops the skills used in TSA's Structural Challenge event.

Structural Challenge

Activity Overview

In this activity, you will create a balsa-wood bridge and determine its failure weight (load at which the bridge breaks).

Materials

➤ Grid paper
➤ 20´ of ⅛´´ x ⅛´´ balsa wood
➤ 3´´ x 5´´ note card
➤ Glue

Background Information

General. There are several types of bridges: beam, truss, cantilever, suspension, and cable-stayed. The type of bridge used in a particular situation is generally determined by the length of the span and available materials. For this activity, a truss design is considered the most efficient.

Truss bridges. The truss bridge design is based on the assumption that the structural members carry loads along their axes in compression or tension. The members along the bottom of the bridge carry a tensile load. The members along the top of the truss carry a compressive load. The members connecting the top and bottom chords (members) can be in tension or compression.

Gussets. Gussets are plates connected to members at joints to add strength. Gussets are normally used in steel construction. The structural steel members are welded or bolted to the gusset. When designing your bridge, include a gusset at each joint, if possible.

Wood properties. Due to its molecular structure, wood can normally carry a larger load in tension than it can in compression. Also, a shorter member can carry a greater compressive load than a longer member.

Guidelines

➤ You must create a scale sketch of the bridge before building.

➤ Pieces of balsa wood can be glued together along lengthwise surfaces. No more than *two* pieces of balsa can be glued together. You cannot use an excessive amount of glue.

➤ Gussets cut from the 3″ x 5″ card can be no larger than the diameter of a US quarter dollar coin. A gusset cannot touch another gusset. A gusset cannot be "sandwiched" between two pieces of balsa wood.

➤ The bridge design must take into account the loading device. Your teacher will provide specific guidelines for bridge length, width, and required details for attachment of loading device.

➤ Your bridge will be weighed before a load is applied.

Evaluation Criteria

Your project will be evaluated using the following criteria:

➤ Accuracy of sketch compared to completed bridge

➤ Conformance to guidelines

➤ Efficiency (failure weight ÷ bridge weight)

With the use of construction technology, we can build many structures that make our lives easier.

Chapter 21
Energy Conversion Technology

Did You Know?

➤ One-fourth of the world's population consumes three-fourths of the world's energy production. This group of energy consumers lives in the industrialized countries.

➤ The Grand Coulee Dam on the Columbia River in Washington is 5,223' long and 550' high. It was made from 11,975,000 yds³ of concrete. This dam has 33 generators that can produce 6,809 megawatts of electrical power.

➤ On the average, the yearly cost of operating a clothes dryer is over $47. A personal computer and monitor cost over $24 a year to operate.

➤ Petroleum and natural gas provide over 50 percent of the energy used in the United States. Alternate (such as solar, wind, and geothermal) energy sources provide less than 5 percent.

Objectives

The information given in this chapter will help you do the following:

➤ Explain what energy conversion is.

➤ Summarize how an electric generator works.

➤ List and describe several types of heat engines.

➤ Explain how a battery works.

➤ Describe a fuel cell.

➤ Give examples of the common types of solar converters.

➤ Describe how wind and water turbines are used to generate electricity.

➤ Identify and explain the major ways energy and power can be transmitted.

➤ Summarize the importance of energy conservation.

Key Words

These words are used in this chapter. Do you know what they mean?

active solar system
anode
cathode
chemical converter
conductor
conservation
electricity
electromagnetic induction
energy converter
external combustion engine
fluid converter
four cycle engine
fuel cell
gas turbine
heat engine
internal combustion engine
jet engine
mechanical converter
passive solar system
rocket engine
solar cell
solar converter
step-down transformer
step-up transformer
thermal converter
Thrombe wall
transformer
transmission
turbine
water turbine

Chapter Outline
Did You Know?
Objectives
Key Words
Conversion and Converters
  Mechanical Converters
  Thermal Converters
  Chemical Converters
  Solar Converters
  Fluid Converters
Energy and Power Transmission
  Mechanical Transmission
  Fluid Transmission
  Electrical Transmission
Using and Conserving Energy
Summary
Activities
Test Your Knowledge

Standards for Technological Knowledge

16

Resource
Transparency 21-1: Energy Converters

Research
Have the students investigate which energy converters led the Industrial Revolution.

Tech Terms
Energy—The ability to do work or produce change.
Conversion—Changing something from one form or function to another.

Figure 21-1. This early photograph shows a group of farmers next to a steam engine.

As you learned in Chapter 6, energy is the ability to do work. Using energy is essential for meeting basic human needs. It helps us have a better standard of living. Using energy contributes to people living longer and healthier lives.

People have learned to use energy over thousands of years. This started with human beings learning to make fire. As history progressed, farmers used animals as a source of energy to do work. Later, people used mechanical energy to power machines. They harnessed wind power and water-power for their workshops and mills. The Industrial Revolution added new ways to power devices. Coal and steam power laid the foundation for new ways to do work. See **Figure 21-1.** Today, society uses more recent developments in power generation. These developments include the internal combustion engine and large-scale electricity generation.

Conversion and Converters

Throughout history, people have depended on energy. They have used biological (human and animal), chemical, mechanical, solar, and hydraulic sources. People developed ways to convert this energy from one form to another, and they have used it to do many different types of work.

Energy converts easily from one form to another. This allows us to make energy more usable. Power systems are used to propel and provide force to other technological products and systems. For example, when a person rides a bicycle, chemical energy stored in the body is changed into mechanical

TechnoFact: Energy converters have been around for centuries. The first reaction engine, called the *aeolipile*, was invented by Hero of Alexandria more than two thousand years ago.

TechnoFact: Wind energy is a promising source of electrical power because it is a clean and inexhaustible resource. However, because wind speeds vary by time of day, season, and even from one year to the next, wind energy is an intermittent resource. At frequently windy sites, it is common for wind turbines to operate only 60 percent of the year.

Figure 21-2. A streetcar converts electrical energy into mechanical energy so it can move.

Standards for Technological Knowledge

Demonstrate
Demonstrate how a battery in a flashlight is an energy converter.

Brainstorm
Have the students identify all the energy converters in the classroom. Then, ask students to identify the type of converter for each device.

Tech Terms
Energy Converter—A device that converts energy from one form to another.

energy (pedals moving a chain that moves the back wheel). An automobile engine converts heat energy into mechanical energy so the automobile can be moved. A streetcar converts electrical energy into mechanical energy (motion). See **Figure 21-2.**

People have developed many types of converters. See **Figure 21-3.** An *energy converter* is a device that changes one type of energy into a different energy form. (Refer back to Chapter 6 to review the types of energy.) There are five main types of converters:

➤ *Mechanical converters.* These convert kinetic energy to another form of energy.

➤ *Thermal converters.* These convert heat energy to another form.

➤ *Chemical converters.* These convert energy in the molecular structure of substances to another energy form.

➤ *Solar converters.* These convert energy from the Sun to another energy form.

➤ *Fluid converters.* These convert moving fluids, such as air and water, to another form of energy.

Energy may be converted several times before it is used. Consider generating electricity. We change the heat in steam to mechanical energy in the motion of a generator. This motion is changed into electrical energy. If the electricity is used to light a room, the lightbulb converts the electricity into light and heat energy. Even animals and plants can be considered energy converters. They take in food and change it

TechnoFact: One of the oldest energy converters is the waterwheel, which harnesses the power of flowing or falling water. The Norse wheel is the oldest known type of waterwheel. This waterwheel, which originated in the Middle East, has a vertical shaft. The top of the shaft is directly connected to the work device, such as a millstone. The bottom of the shaft extends into the moving water. Vanes or paddles attached to the lower part of the shaft harness the water's movement and, in turn, spin the shaft.

Standards for Technological Knowledge

16

Resource
Reproducible Master 21-1: Energy Conversion

Extend
Discuss how a turbine operates.

Tech Terms
Mechanical—
Involved with or done by devices that use physical forces.
Mechanical Converters—
Devices that convert mechanical energy to another form of energy.
*Generator—*A device that changes mechanical energy into electrical energy.
*Turbine—*A machine that uses a moving fluid (liquid or gas) striking rotor blades to produce rotary motion that can be transformed to electrical or mechanical power

	Chemical	Electrical	Heat	Light	Mechanical	Acoustical (Sound)
Chemical	Foods Plants	Battery Fuel cell	Fire Food Hot water boiler Steam boiler	Lamp Gas lantern Candle Firefly	Gas engine Human muscle Animal muscle	Smoke alarm Exploding matter
Electrical	Batteries Electrolytes	Diodes Transistors Transformers	Electric blanket Hairdryer Toaster	Lightbulb TV screen Lighting	Electric motor Relay (type of magnet)	Horn Loudspeaker Thunder
Heat	Distilling Vaporizing Gasifying	Thermocouple Thermopile	Heat exchanger Heat pump Solar panel	Fire Lightbulb	External and internal combustion engines Turbine	Explosions Flame tube
Light	Camera film Plant growth through sunlight	Photovoltaic (solar) cell Photoelectric cell	Heat lamp Laser	Laser Reflector	Photoelectric door opener	Sound track of movie Video disc
Mechanical	Gunpowder	Alternator Generator	Brake Friction	Flint Spark	Flywheel Pendulum Water stored in tower	Wind instrument Voice
Acoustical (Sound)	Hearing	Hearing Microphone Telephone	Sound absorption	Color organ	Ultrasonic cleaner	Megaphone

Figure 21-3. Energy can be changed from one form to another. The chart shows you what is used to make each kind of change. For example, to change electrical energy to heat, you could use a hair dryer.

into either chemical or mechanical energy. Let us look at some examples of common energy converters.

Mechanical Converters

Nearly one-fourth of all energy used in the United States and Canada is converted into electricity. This energy is then used in different ways. Some energy provides light for homes and offices. See **Figure 21-4.** Some energy is used to power electric motors. Another portion is used for heat in buildings and industrial processes. Still more is used to run various electrical and electronic devices.

A machine called an *electric generator,* or *alternator,* provides most electrical energy. It converts mechanical energy (machine motion) into electrical energy. A *turbine,* or an engine, usually turns it at high speeds.

The operation of an electric generator is based on an important scientific principle. Michael Faraday discovered this principle in 1831. He found that if a conductor is moved through a magnetic

TechnoFact: About twenty-one hundred years ago, waterwheels with horizontal shafts came into use, supplementing the vertical shaft machines. Two main types of horizontal shaft waterwheels were developed. The overshot wheel used water falling over the top of the wheel to spin the shaft. The undershot wheel used water moving below the wheel to spin the shaft.

Figure 21-4. Electricity is used in many ways. In this photograph, you can see electricity used to light the area and power the elevator.

field, an electrical current is set up (induced) in the conductor. This process is called *electromagnetic induction.*

The wire is part of a path along which electricity can flow. This path is called *circuit.* Imagine that you can see inside a very simple electric generator. This generator is a loop of wire turning in the space between the poles of a magnet. See **Figure 21-5.** As the loop spins, it moves through the magnetic field. This makes the electricity move in one direction through the wire. As the loop continues to rotate, it becomes parallel to the magnetic field. The flow stops for a moment because wires are not cutting through any lines of magnetic force. As the loop continues, current starts up

again. Now it moves in the opposite direction through the wire loop. This type of electricity is called *alternating current.* Electricity in North America changes direction 120 times a second. It is called *60 Hz alternating current.* In other areas, it may be 50 Hz.

In reality, an electric power station is simply a large generator. Homes, businesses, and factories are parts of the circuit, as are streetlights, emergency sirens, and other devices. They make up what is called the *electric power system.*

Thermal Converters

Heat engines are thermal converters. An engine is a machine that converts energy into motion. See **Figure 21-6.** Oil, gasoline, steam, or electricity usually supplies the energy. The output is almost always rotary motion of a shaft.

Figure 21-5. A simple generator consists of a magnet and a loop of wire. The wire loop rotates through the magnetic field.

Standards for Technological Knowledge

Demonstrate
Use a child's whirligig (propeller on a stick) to demonstrate the operation of a windmill or turbine.

Tech Terms
Direct Current— Electrical current in which the electrons flow in one direction.
*Alternating Current—*Electrical current in which the current flows first in one direction, then in the opposite direction.
Thermal Converters— Devices that convert heat energy to another form of energy.
*Engine—*A device that changes energy into motion.
*Heat Engine—*A machine that converts heat energy into mechanical energy (motion).

TechnoFact: An early use of steam engines was to pump water into a pond. The pond had a spillway through which water exited the pond. Water from the spillway turned a waterwheel. However, the steam engine would quickly replace the waterwheel.

Tech Terms
Intake Stroke—The part of a four cycle engine's cycle in which the piston moves down to bring a fuel-air mixture into the cylinder.
Compression Stroke—The part of a four cycle engine's cycle in which the piston moves up to compress the fuel-air mixture.
Power Stroke—The part of a four cycle engine's cycle in which the fuel is ignited, and the burning gases propel the piston down.
Exhaust Stroke—The part of a four cycle engine's cycle in which the piston moves up to force exhaust gases out of the cylinder.

Figure 21-6. An internal combustion engine is used to power this racecar.

Engines can be classified in a number of ways:

➤ Use for which the engine is designed: automobile, locomotive, or aircraft.

➤ Form of energy used: steam, compressed air, or gasoline.

➤ Type of cycle: otto (in ordinary gasoline engines) or diesel.

➤ Type of motion produced: reciprocating or rotary.

➤ Place where chemical energy is converted into heat energy: internal or external combustion.

➤ Cylinder position: V, in-line, or radial.

➤ Number of piston strokes needed to complete a cycle: two or four.

➤ Cooling method used: air or water cooled.

Engines are sometimes called *motors*. The term *motor*, however, is generally used to describe devices that transform electrical energy into mechanical energy.

Internal Combustion Engines

In an *internal combustion engine,* the fuel is first drawn into the engine. Then the fuel is ignited and burned. The heated gases expand so rapidly they appear to explode. This provides the power to create the desired motion.

All engines operate in much the same way. A sealed chamber takes in a mixture of fuel and air. It burns there. See **Figure 21-7.** There are four steps or strokes for a typical piston engine:

1. Taking in a fuel air mixture (intake stroke).

2. Compressing the mixture (compression stroke).

3. Burning the fuel (power stroke).

4. Exhausting the gas (exhaust stroke).

This design is known as a *four cycle engine.* Another type is the two cycle engine. Unlike the four cycle engine, it burns a fuel charge every time the piston is at the top of the cylinder. The intake and compression occur in one

Figure 21-7. This is how a four cycle engine operates. A–As the piston moves downward, it pulls fuel into the combustion chamber. (This chamber is the space above the piston.) B–The piston squeezes fuel charges into a small space. C–The fuel charges burn. Hot gases force the piston downward. This is the power that does work. D–Burned gases are pushed out of the combustion chamber.

TechnoFact: The first person to experiment with an internal combustion engine was Christian Huygens during the 1680s. In 1859, J. J. Étienne Lenoir developed the first effective gasoline-powered engine, a spark ignition engine that could be operated continuously. In 1862, Alphonse Beau de Rochas patented the first four stroke (cycle) engine. However, he never built the engine. Nikolaus A. Otto built a successful four stroke engine in the mid-1870s, known as the "Otto cycle" engine.

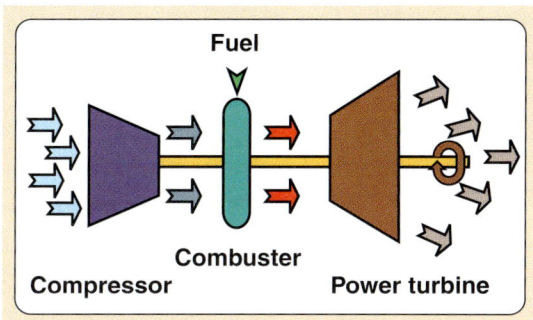

Figure 21-8. Gas turbine engines have three parts: (1) a compressor to compress the incoming air, (2) a combustion area burning fuel and producing high-pressure, high velocity gas, and (3) a power turbine that uses the energy from the gas flowing from the combustion chamber.

cycle (upward motion). The power and exhaust are in the other cycle (downward cycle).

Jet Engines and Gas Turbines

Jet engines, gas turbines, and rockets are jet propulsion devices. They burn a mixture of fuel and air. Jet propulsion devices are used primarily in high-speed aircraft, missiles, and spacecraft. They are also used in portable electric generating stations and experimental cars.

In a *jet engine,* the compressor draws in air. See **Figure 21-8.** The compressor is a cone-shaped cylinder with rows of small fan blades. As the air is forced through the compression stage, its pressure rises significantly. In some engines, the air pressure rises to 30 times the normal pressure. This high-pressure air enters the combustion area. Here, fuel injectors inject a steady stream of kerosene, jet fuel, propane, or natural gas. A "flame holder" ignites the fuel, and the hot gases enter the

turbine section. One set of turbines drives the compressor. Another set of turbines drives the output shaft. The final set of turbines spins freely without any connection to the rest of the engine.

If you go to an airport, you will see the commercial jets that carry many people. Huge engines power these jets. Turbofan engines power most commercial jets. These engines are a type of engine called *gas turbines.* These engines create power from high velocity gases leaving the engine.

Rocket Engines

When most people think about engines, they think about rotation. This rotation is used to drive other devices. *Rocket engines,* however, are different. They are reaction engines based on a principle of physics. This principle suggests that for every action, there is an equal and opposite reaction.

An inflated balloon can show the principle of a reaction engine. See **Figure 21-9.** In a closed balloon, the air pressure is equal in all directions. The balloon remains in place. If the air is allowed to quickly escape from the balloon, motion is created. The internal air is "thrown" out the nozzle. This creates an equal and opposite reaction. The balloon moves in the opposite direction.

Like a balloon, a rocket engine is throwing mass. It burns solid or liquid fuel to create hot gases. These gases leave the rocket engine nozzle at a high speed in one direction. This causes the rocket to move in the opposite direction. Unlike a jet engine, a rocket carries its own oxygen. This

Figure Discussion
Figure 21-8 shows a schematic of a gas turbine. Explain how each stage of the engine operates.

Tech Terms
Gas Turbine—A machine that uses hot gases striking rotor blades to produce rotary motion.
Rocket Engine—A device that carries both fuel and oxidizer and produces thrust by expelling expanding hot gases through a nozzle.

TechnoFact: The first rockets were made from gunpowder packed into sections of bamboo. These were developed in China about one thousand years ago. Later, improved gunpowder was developed in Europe, which increased the range and effectiveness of rocket-propelled weapons.

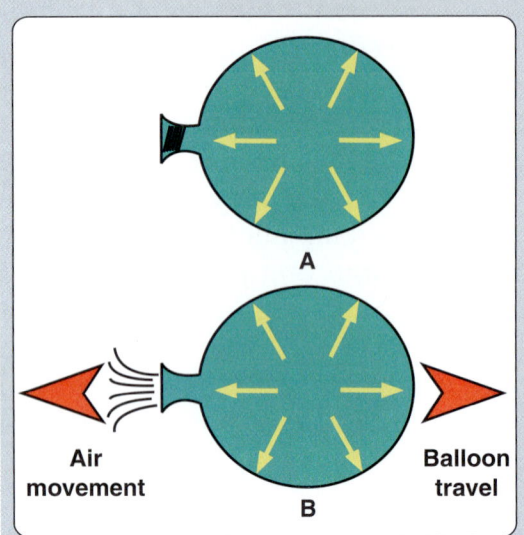

Figure 21-9. A balloon will show us the principle of reaction engines. A–Air inside the balloon pushes on all parts of the balloon equally. The balloon does not move. B–Air escapes from the balloon when the neck is open. This flow creates an equal and opposite reaction causing the balloon to move.

allows the engine to operate in deep space, where no oxygen exists.

There are two major types of rocket engines: solid fuel and liquid fuel. Solid fuel rocket engines were invented hundreds of years ago in China. The principle of a solid fuel rocket is simple. See **Figure 21-10.** This rocket uses a fuel that burns very quickly, but does not explode. The fuel is generally in a cylindrical form with a hole down the middle. When the fuel is ignited, it burns from the middle outward. This creates gases that travel out the nozzle and propel the rocket. Fireworks use solid fuel rockets to send them into the air. Also, model rockets are lifted with solid fuel rockets.

Robert Goddard developed the first *liquid fuel rocket engine* in 1926. The

operation of a liquid fuel rocket is fairly simple. See **Figure 21-11.** This rocket uses a fuel and an oxidizer. For example, Goddard used gasoline and liquid oxygen. The fuel and oxidizer are pumped into a combustion chamber. They are burned in the chamber to create a high-pressure, high velocity

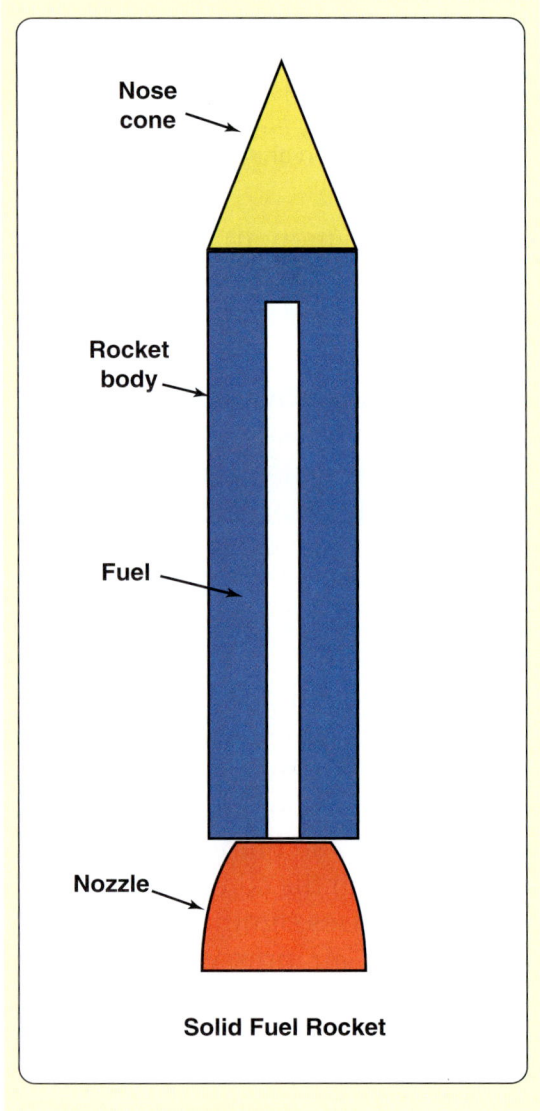

Figure 21-10. This is a cross-sectional view of a solid fuel rocket.

TechnoFact: Konstantin E. Tsiolkovsky suggested the use of rockets in space exploration at the beginning of the twentieth century. In 1926, Robert H. Goddard launched the first liquid fuel rocket and demonstrated the rocket's potential for carrying payloads into space.

stream of gases. These gases flow through a nozzle accelerating them further. The gases leave the nozzle at speeds of up to ten thousand mph. This stream of gases propels the rocket.

External Combustion Engines

An *external combustion engine* burns fuel outside of the engine chambers. Only the expanding fluid enters the engine. The steam engine is an example of an external combustion engine. See **Figure 21-12.** Coal, wood, or some other fuel is burned outside

Figure 21-12. This stationary steam engine was used in logging operations in the early 1900s.

the engine to create steam. The steam is introduced into a cylinder chamber. The steam moves a piston, which turns a crankshaft. Cooled steam is exhausted from the engine. Another charge of live steam is introduced, and the cycle is repeated.

Steam engines are used very little today. Some historical railroads and collectors maintain steam engines. See **Figure 21-13.**

Chemical Converters

Chemical converters change the energy found in molecular structures into another form of energy. Batteries and fuel cells are the most common chemical energy converters. We will look closer at each of these converters to see how they work.

Batteries

The most common chemical energy converter is the household battery. It changes chemical energy into electrical

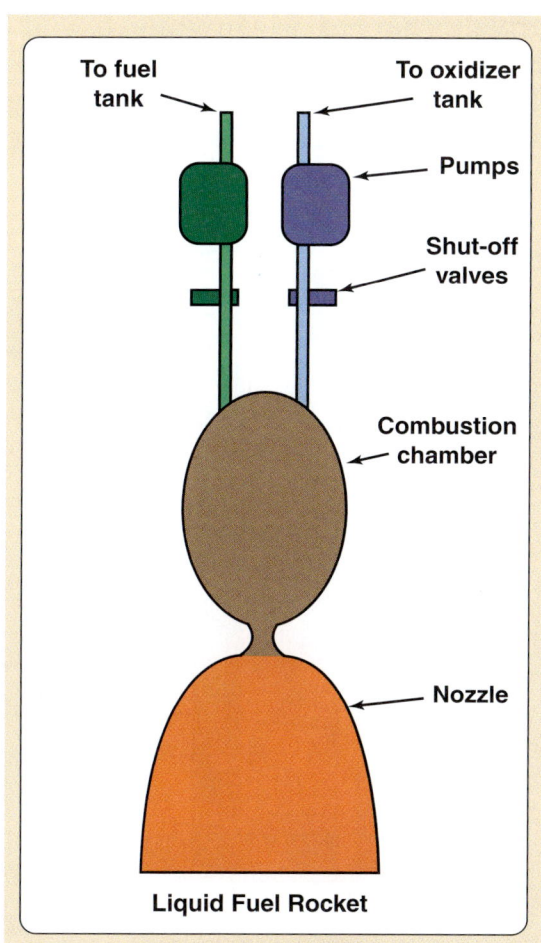

Figure 21-11. Here is a cross-sectional view of a liquid fuel rocket.

TechnoFact: In early steam engines, steam pushed the piston out of the cylinder. The cylinder was then cooled, the steam condensed, and a vacuum was formed. The vacuum drew the piston back into the cylinder, and the cycle began again. These engines were very large because they used low-pressure steam. Steam engines developed after 1800 used high-pressure steam, which allowed a reduction in the size of engines. This allowed the steam engine to be adapted to transportation needs.

Standards for Technological Knowledge

16

Tech Terms

Anode—The negative terminal of a battery.
Cathode—The positive terminal of a battery.
Fuel Cell—A device that converts the chemical energy of a fuel and an oxidant to electric energy.

Figure 21-13. This historic railroad in Colorado still uses a steam engine.

energy. All batteries have a *cathode* (the positive side) and an *anode* (the negative side). See **Figure 21-14.**

The cathode is made from a mixture of chemicals. In an alkaline battery, the cathode is made from manganese dioxide, graphite, and an electrolyte. This mixture is compacted into a hollow cylinder. The cylinder is inserted into a steel container. The steel can and the compacted chemicals become the cathode. They make up the positive charge of the battery.

The cathode and anode cannot come into contact with one another. Therefore, a separator is placed next to the cathode. The anode is made from zinc powder and several other materials. It is produced as a gel. This gel is inserted into the steel can against the separator. Finally, a brass nail (acting as the current collector), a plastic seal, and a metal end cap are attached. The brass nail collects the current the battery produces.

Fuel Cells

Another chemical energy converter is the *fuel cell.* It looks like an electrical storage battery. See **Figure 21-15.** It produces electricity from a fuel and oxygen. It does not, however, burn the fuel.

A fuel cell has a container to hold an electrolyte, such as phosphoric acid. Two carbon plates are placed into the electrolyte. These are the terminals that allow electricity to flow through them.

To start the cell, oxygen and hydrogen (or another fuel) are fed into it. The fuel loses electrons to one of the carbon plates. The plate becomes negatively charged. Meanwhile, oxygen is fed to the other plate or terminal. The oxygen collects electrons from the second plate. This leaves the plate with a positive charge.

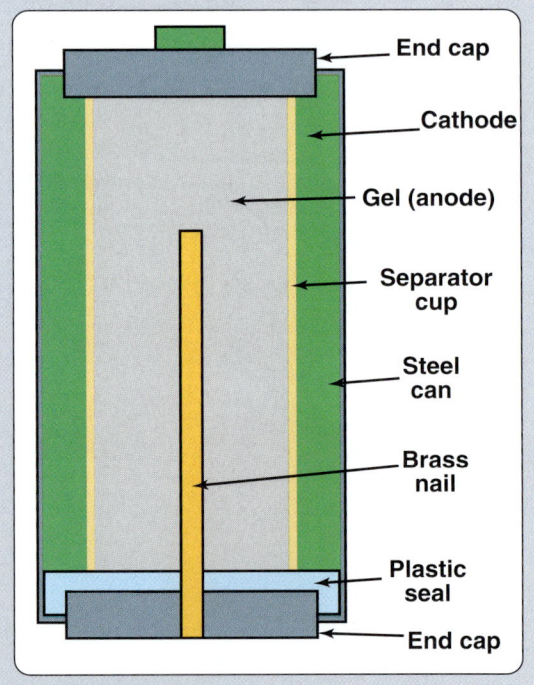

Figure 21-14. Here is a cross-sectional view of a simple battery.

TechnoFact: William Grove developed the first fuel cell in 1839. His device combined hydrogen and oxygen to form water and release electricity.

The fuel cell can now provide electric current. This current becomes active when a load, such as a light, is connected into the circuit. Free electrons on the negative plate travel through the conducting wire. They flow through the lightbulb and back into the fuel cell through the other plate. In the electrolyte, ions of the hydrogen combine with oxygen ions to form water.

Solar Converters

Many electrical devices never need batteries. Some devices use solar converters instead. These converters change energy from the Sun into another form of energy.

Solar Cells

There are calculators that never use batteries. Some emergency road signs or call boxes can be far away from electrical lines. Satellites have unique power systems. There are buoys in harbors and lakes not powered by

Figure 21-16. This is a photovoltaic cell in use.

batteries. *Solar cells* power these devices. See **Figure 21-16.**

Solar cells used for these and other devices are photovoltaic cells. The cells get their name from two terms: *photo* means "light," and *voltaic* means "electricity." Therefore, the term *photovoltaic* means "obtaining electricity from light."

Photovoltaic cells are made from a special material called a *semiconductor*. See **Figure 21-17.** The most common material for these cells is silicon. When light strikes the silicon semiconductor, this material absorbs it. The light energy causes electrons in the semiconductor to become loose. The electrons can flow freely within the material. This flow of electrons is called *current.* If metal contacts are placed on the top and bottom of the cell, current can be drawn off for use.

Solar Heating Systems

Many homes and some businesses use solar systems for heat. These systems are another useful type of solar converter. Homes and businesses may use one of two types of systems:

➤ *Passive solar systems* use no moving parts to capture and use solar energy.

Figure 21-15. The fuel cell makes electricity directly from fuel.

Research
Have the students investigate current uses of solar cells for traffic signals, construction barricades, and other similar applications.

Brainstorm
Ask the students to list ways in which solar power (converters) can be used in a home.

Tech Terms
Solar Energy—Any form of energy radiated by the sun, including visible light, radio waves, and ultraviolet radiation.
Solar Converters—Devices that convert energy from the sun to another energy form.

Standards for Technological Knowledge

Demonstrate
Use a model to show how a passive solar system works.

Discussion
Conduct a discussion session with the students on why a greenhouse is a passive solar system.

Tech Terms
Passive Solar Systems—Systems that capture and use solar energy without the use of moving parts.
Active Solar Systems—Systems that capture and use solar energy with the use of moving parts, such as pumps.

Figure 21-17. Here is a close-up view of a solar cell.

➤ *Active solar systems* use moving parts to capture and use solar energy.

Passive Solar Systems

There are three major types of passive solar energy systems: direct gain, indirect gain, and isolated. Direct gain is the simplest passive solar energy system. See **Figure 21-18.** Sunlight enters the house through collectors. These collectors are usually windows facing south. The sunlight then strikes masonry floors or walls. These materials absorb and store the solar heat. The surfaces of the masonry are generally a dark color because dark colors absorb more heat than light colors do. At night, the heat stored in the masonry heats the room.

An indirect gain passive solar system has thermal storage between the windows and living spaces. See **Figure 21-19.** This storage is usually called a *Thrombe wall.* It is an 8″–16″ thick masonry wall with a layer of glass mounted about 1″ in front of the wall. The wall's dark colored outside surface absorbs solar heat. This stored heat radiates into the living space.

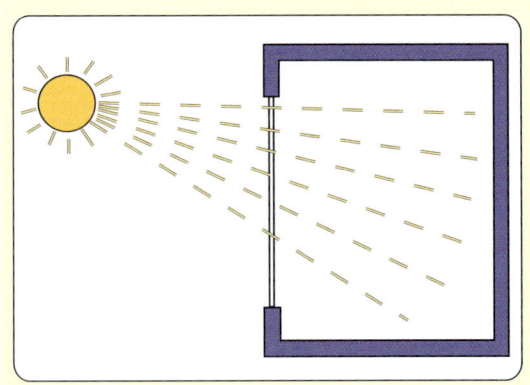

Figure 21-18. This is a direct gain solar heating system.

TechnoFact: Solar heating systems can supply a home with heat and hot water. In one method, special collectors on the roof capture the heat and route it to a containment area filled with rocks or water, where it is stored for later use.

Figure 21-19. Here is an indirect gain solar heating system. It gathers solar energy during the day and releases the energy at night.

A sunspace, or isolated solar system, is often called a *solar room*, or *solarium*. The simplest sunspace design uses vertical windows with no overhead glazing (windows). Sunspaces experience high heat gain and high heat loss through their windows. Because of this fact, many sunspaces are separated from the home with doors or windows. The heat generated in the sunroom may be moved to other parts of the building through vents or ducting.

Active Solar Systems

Active solar systems can be used to provide hot water and heating for homes. These systems are either hot air or water systems. In the hot air system, solar energy is used to heat air. See **Figure 21-20.** The heating system

operates like any forced air heating system. Fans and ducts circulate the air. Some of the air is used to heat the home during the day. Additional hot air is used to heat a rock storage area. During the night and on cloudy days, the hot rocks are a heat source. Air is

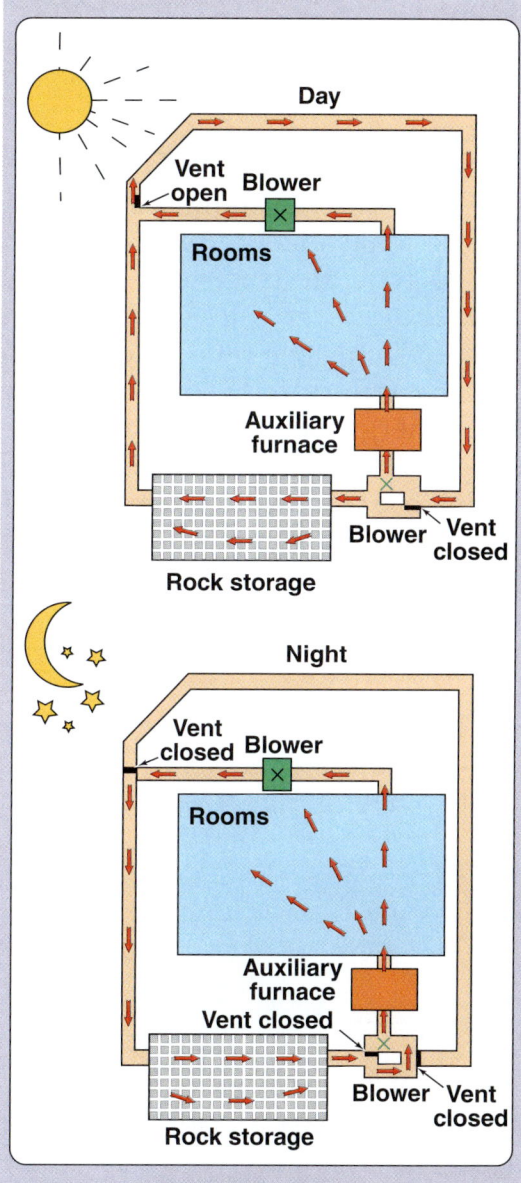

Figure 21-20. Here is an active solar hot air system.

Standards for Technological Knowledge

16

Figure Discussion
Have the students follow along in their books as you explain the flow of warm and cool air through the system shown in Figure 21-20.

Research
Have the students investigate the advantages and disadvantages of passive and active solar energy systems. Then, discuss which system is most appropriate for your area.

Tech Terms
Solarium—A room built for enjoying sunlight and collecting solar heat.

TechnoFact: A number of devices and systems use solar energy. Solar batteries use light striking photosensitive substances to generate electricity. Thermoelectric generators convert the heat generated by solar energy directly into electricity. Solar energy is used in air-drying processes, distillation, and the extraction of salt from seawater. Solar heating systems supply heat and hot water. Solar stoves focus the sun's heat directly onto a cooking surface.

Standards for Technological Knowledge

16

Research
Have the students trace the historical development of devices that harness wind energy.

Tech Terms
Fluid Converters— Devices that convert moving fluids, such as air and water, to another form of energy.
*Wind—*The effect produced by the flow of air across Earth.

blown over the rocks to heat the home. In most of these systems, an auxiliary furnace provides heat during long, cold or cloudy periods.

Hot water systems allow the solar energy to heat water. The hot water is circulated through the house. It enters heat exchangers, allowing it to heat air in the various rooms. The hot water can also be used to heat tap water for household use.

Concentrated Power Systems

Active solar systems can also be used to make electricity. A common way to generate electricity is called *concentrated solar power technologies.* These systems use reflective materials, such as mirrors, to concentrate the Sun's energy. See **Figure 21-21.** The energy is focused on a specific point. The intense rays are used to turn water into steam. The steam is used to drive turbines generating electricity.

Fluid Converters

Fluid converters are yet another way to change energy into a usable form. They convert energy from

Figure 21-21. The tower in the center of this photo contains many mirrors. These mirrors concentrate solar energy on the collectors on the ground.

Figure 21-22. This windmill is used to pump water for cattle in Nebraska.

moving fluids into other forms of energy. Wind and water turbines are the most common fluid converters.

Wind Turbines

Wind is a free, but somewhat undependable, energy source. When it blows, it provides a great deal of energy. On a calm day, however, there is little or no wind to capture.

The energy from wind is generally changed into mechanical or electrical energy. A sail changes wind energy into mechanical energy. The energy is used to move the boat through water. Likewise, farmers have used wind energy to do work for many years. They have used windmills to pump water for household and livestock use. See **Figure 21-22.** In Holland,

TechnoFact: Wind energy is contained in the force of the winds blowing across Earth's surface. A sail can harness wind energy to move a boat. Wind energy converters can be attached to machines to produce mechanical energy for pumping water, sawing lumber, and grinding grain. Egyptians may have been the first to use wind energy. They used it with their newly developed sails to move boats up the Nile River beginning about twenty-five hundred years ago. Persians developed the first windmill about thirteen hundred years ago.

windmills are used to pump water from low-lying areas. See **Figure 21-23.**

Today, the wind is becoming a more common way to make electricity. See **Figure 21-24.** Wind is used to spin the blades of a wind turbine. To be effective, wind speeds must be above 12–14 mph. Speeds below this will not turn the turbines fast enough to generate electricity. The blades turn a shaft attached to a gear transmission box. The transmission turns a high-speed shaft, which turns an electric generator. Commonly used turbines produce about 50–300 kilowatts of electricity each. A kilowatt is one thousand watts. Therefore, a 50-kilowatt (50,000 watts) wind turbine can produce enough electricity to light 500 100-watt lightbulbs.

Water Turbines

Water turbines use the energy of moving water to make electricity. Possible sources of water are fast flowing

Figure 21-23. This is a replica of a Dutch windmill.

Figure 21-24. A wind turbine is used to generate electricity.

rivers and reservoirs behind dams. The water flows from the river or reservoir through a pipe called a *penstock.* See **Figure 21-25.** The rushing water pushes against blades in a water turbine. The turbine spins a generator to produce electricity.

Energy and Power Transmission

Energy is not always used where it is produced. Many times it must be moved to where it is needed. Moving it to where it performs work is called *transmission.* Three major types of transmission are mechanical, fluid, and electrical.

Figure 21-25. Note the penstock (pipes) leading from the dam to the powerhouse.

Mechanical Transmission

Mechanical transmission uses physical objects to move power and energy from one place to another. This can be done in a number of ways. The three simplest techniques are shafts, pulleys, and gears. See **Figure 21-26.**

Shafts are rods that can be used to transmit power. An example is a camshaft in an automobile. It transmits rotary power along its length. Also, the cams change the rotary power to reciprocating (up and down) motion.

Belts are another way of transmitting mechanical power. The system includes at least two pulleys. One pulley is on some source of power. The

Figure 21-26. Here are some examples of mechanical means of power transmission.

other one is on the load. The belt transmits the power from one pulley to the other. A factory machine operated by an electric motor might use a belt to transmit the mechanical power of the electric motor.

Gears transmit mechanical power with teeth meshing with other gears. Sprockets are special types of gears that are toothed wheels. A chain transmitting power links two sprocket wheels. A bicycle is a good example of this kind of transmission. See **Figure 21-27.**

Figure 21-27. Bicycle sprockets and chains move power from the pedals to the rear wheel.

Fluid Transmission

If a power source exerts force on a fluid (liquid or air), the fluid will transmit the power to where it is needed. The fluids must be contained so the force exerted on them travels to where the force is needed. Usually, the fluids are held inside pipes and cylinders. A common example of fluids transmitting energy is the brake system of an automobile. See **Figure 21-28.** The driver's foot pressing down on the brake pedal provides the energy. A piston attached to the brake pedal presses on fluid. The pressure forces the fluid through lines

TechnoFact: An early use of mechanical transmission of power was in the water clock. The first water clocks (clepsydras) were developed thirty-five hundred to four thousand years ago. They were vessels with small openings, through which the water dripped. As the water level dropped, it exposed marks on the walls of the vessel indicating the time that had elapsed since the vessel was full. Later water clocks used a float in a steadily filling or draining tank to tell time. The movement of the float upward and downward drove gears that moved hands around a dial.

(tubing) to each wheel. Other pistons are attached to the brake shoes.

Electrical Transmission

Electricity is the movement of electrons through materials called **conductors**. Electricity is easy to transmit. Lines carry it from the power plant to wherever it is to be used. The lines are conductors made of either copper or aluminum.

Electric current leaving the generating station is fed into transformers. These *transformers* are devices increasing or decreasing the force (voltage) of electric current. (A transformer increasing the voltage is a *step-up transformer*.) These transformers boost the voltage from around 13,800 volts to as high as 700,000 volts. The current is then carried to the point of use. There, *step-down transformers* reduce the voltage.

Quality Control in Energy and Power Systems

Quality control in electrical and power systems focuses on monitoring

Figure 21-28. Automobile brakes are an example of a system transmitting energy through fluids. (EIS Div., Parker-Hannifin)

system operation and output. The equipment and machines within the system are produced using the inspection and design guidelines typical of manufacturing systems. This equipment is subjected to periodic maintenance and inspection. Most electrical and power systems have a built-in monitoring system. When the operation or output is beyond specified limits, the system may signal an alert to an operator, who can then make required adjustments. In some cases, the system may shut down automatically.

Using and Conserving Energy

Energy and power technology has many benefits. It makes life better. This technology promotes global transportation and communication.

Business and industry rely on energy and power technologies to provide electricity so machines, lights, computers, and equipment can operate. Mechanical power transmission is critical in the manufacturing and automotive industries. Nearly all communication systems and devices need electrical power, sometimes provided by batteries, to operate. Businesses' dependence on electricity is such that the operations of nearly all businesses are greatly reduced if the business loses electrical power.

Our technological society requires a great deal of energy, and the rate at which energy is being used in the world is increasing. Natural resources may be depleted in the future before other energy resources are available to replace them.

TechnoFact: Over one-half of all hydroelectric power is generated in the United States, Canada, Brazil, China, and Russia. North America and Europe lead the world in the development of hydroelectric power generation. Nations in Africa, Asia, and South America have the greatest potential for additional hydroelectric power development.

Technology Explained

turbine: A device that transforms energy from a flowing stream of fluid into rotating mechanical energy.

Humans have harnessed the energy of nature to do work for centuries. Early efforts used animal power. Later, the forces of the wind and flowing water were harnessed. The first efforts in this arena involved the windmill and waterwheel, **Figure A**. Both of these devices are effective, but provide limited amounts of power.

The device that has served best in capturing the energy of flowing fluids is the turbine. Turbines can be used to convert energy from wind, steam,

Figure A. This wind turbine was derived from an early windmill.

or running water into rotating motion. One common use of turbines is in hydroelectric

power plants. These turbines work at about 90 percent efficiency. This means they convert 90 percent of the energy from the rushing water into rotating mechanical energy.

The first water turbine was developed in France in 1827. Benoit Fourneyron invented it. A modern water turbine is the Francis turbine, **Figure B**. In this device, water is piped into the turbine. The water swirls through the area between the casing and blade unit. The casing of the turbine includes a series of vanes that catch the rushing water and direct it onto the blades. These vanes are set so the force of the water is tangent to the rotational plane of the blades. This angle captures the maximum amount

Standards for Technological Knowledge

16

Discussion
Discuss how energy is used, conserved, and wasted in your school.

Brainstorm
Ask the students to list how they can conserve energy in their daily activities.

We can, however, do our part to slow down the rate at which energy is being used. Much of the energy used in our environment is not used efficiently. *Conservation* is the act of making better use of energy. Some efforts are already being made to save energy:

➤ Manufacturers are building more energy efficient products. For example, automobiles are being made with more fuel efficient engines. Furnaces are more energy efficient. Home appliances use less energy.

➤ Buildings are better insulated. Higher insulation standards are being used in new construction. This means thicker insulation in ceilings and walls. Triple glazing (three panes of glass) is available in windows.

➤ People are beginning to recycle more materials. It takes large amounts of energy to make a ton of aluminum from bauxite ore. Converting scrap aluminum requires much less energy.

People have personal roles in conserving energy. We have formed

TechnoFact: Blowing wind does not necessarily mean a wind turbine will be producing electricity. There must be sufficient wind power for a wind turbine to convert into electricity. The typical output of a wind turbine may be as low as 25 to 35 percent of its full capacity. In comparison, the power output of coal-fired power plants may be as high as 75 to 85 percent of full capacity.

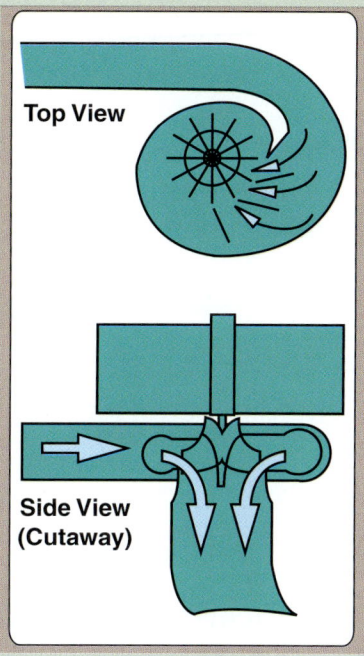

Top View

Side View
(Cutaway)

Figure B. Shown are two views of the Francis water turbine.

of energy from the water. The turbine shaft is usually connected directly to a generator.

Another common type of turbine is the steam turbine, **Figure C.** The steam turbine uses a series of blade units similar to those in a jet engine. The hot steam enters at one end of the turbine. As the steam moves through the unit, it forces the blades to turn. The steam cools and expands as it travels across the blade units. To account for this, each succeeding set of blades is slightly larger in diameter. The steam finally condenses into water at the back of the turbine and is returned to the boiler for reheating. Steam turbines are used in coal-fired, natural gas,

and nuclear power plants. They also power many ocean-going ships.

Figure C. This is a cutaway model showing the inside of a steam turbine.

habits that waste energy. Consider how we use electricity. How many times have you left radios and lights on when leaving a room?

Our standard of living is the highest in the world. We have many benefits that are the results of modern technology, and we also have a responsibility to do what we can to conserve natural resources for future generations. What are some of the things you and your family can do to conserve? Is there anything that can be done in your school to use less energy? What can governments do about promoting energy conservation?

Summary

We use energy to do all kinds of work. In many cases, energy is transformed from one form to a more usable form. Energy is converted and used through an energy system. Such a system will have an energy source, a converter, a means of transmission, controls, and a load. Much of our available energy is wasted. Throughout the process, concern for energy conservation must be shown.

Standards for Technological Knowledge

16

Brainstorm
Conduct a brainstorming session with the students to develop ideas for maintaining a high standard of living, while using less energy.

Curricular Connections

Mathematics

Make a graph showing the energy consumption of major home appliances.

Social Studies

Compare the energy use of various countries in the world. Contrast the energy sources used. Map the locations of major energy resources in the country and the world.

Science

Explain the scientific principles of distillation used in petroleum refining. Describe the scientific principles used in electricity generation and use (for example, motors and lightbulbs).

Mathematics

Prepare a display explaining the measurement systems used for various energy uses (such as volts, amperes, joules, and kilowatts).

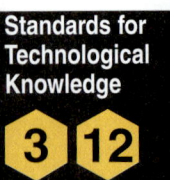

Standards for Technological Knowledge

3 12

Activities

1. Visit an electrical power plant to learn how it operates. Write a report on what you see. Use a computer to prepare the report, if you can.

2. Build a model of a windmill or waterwheel. Explain how it works.

3. Make a list of devices you see in your community using energy in any form. Place them in the order of their importance to the well-being of the community.

4. Prepare a list of the energy using devices you use each day. Describe how you can conserve energy while using them.

TechnoFact: *Renewable energy* is a term used a great deal in popular literature. Renewable energy includes using direct solar energy to heat water and power steam boilers and solar cells, which convert solar energy to electricity. It also includes moving water, which powers hydroelectric turbines, and wind, which powers windmills. In addition, renewable energy may include engines that take energy from the warmer surface of the ocean and discharge it into the colder water below.

Test Your Knowledge

Do not write in this book. Place your answers to this test on a separate sheet of paper.

1. Energy is the ability to do _____.
2. Match the types of converters with the right description(s) and give an example of each:

   _____ Converts liquids into another form of energy.

   _____ Converts the Sun's energy into another form.

   _____ Converts heat energy into another form.

   _____ Converts kinetic energy into another form.

   _____ Converts energy in the molecular structure of a material into another form.

   A. Mechanical converter.
   B. Thermal converter.
   C. Chemical converter.
   D. Solar converter.
   E. Fluid converter.

3. A _____ converts mechanical energy into electrical energy.
4. Internal combustion and steam engines are _____ engines.
5. The two types of rocket engines are _____ and _____.
6. Summarize how batteries work.
7. Explain how fuel cells work.
8. Another name for a solar cell is a _____ cell.
9. The two types of solar heating systems are _____ and _____.
10. Direct gain and indirect gain are examples of _____ solar heating systems.
11. _____ are used to change wind energy into electricity.
12. _____ change the energy in flowing water into electricity.
13. Three major mechanical means used to transmit power are _____, _____, and _____.
14. Transformers used to increase electrical voltage are called _____ transformers.
15. Why is conserving energy important?

15. Conserving energy is important because the rate at which energy is being used in the world is increasing and most of the fuels we use are limited in supply.

Chapter 22
Information and Communication Technology

Did You Know?

➤ The Bible was the first book printed with Johannes Gutenberg's movable type. It was, therefore, the first printed book.

➤ The first machine with a keyboard and levers was the tachygrapher. It was invented in 1823 and is the ancestor of the typewriter.

➤ The first pictures made from a camera were called *daguerreotypes*. They were named after the inventor Louis-Jacques-Mandé Daguerre.

➤ The first camera with a lens was invented in 1840.

➤ The first American president to appear on television was Franklin Delano Roosevelt. The black-and-white image was broadcast from the 1939 New York World's Fair.

➤ Charles Babbage invented an "analytical engine" in 1835. It did calculations and stored data. This engine is an ancestor of the modern computer.

Objectives

The information given in this chapter will help you do the following:

➤ State the definition of *communication.*

➤ Describe several ways in which people communicate.

➤ Define *communication technology.*

➤ Identify and give examples of the four types of communication systems.

➤ Diagram and explain the communication process.

➤ Cite and describe the four steps in the process for producing a message.

➤ Explain the difference between computer hardware and software.

Chapter Outline
Did You Know?
Objectives
Key Words
Communication and Information
How We Communicate
Communication Technology
  Graphic Communication
  Wave (Electronic) Communication
  Graphic vs. Electronic Communication
Types of Communication Systems
  People-to-People Communication
  People-to-Machine Communication
  Machine-to-People Communication
  Machine-to-Machine Communication
The Communication Process
  Encoding
  Transmitting
  Receiving
  Decoding
  Storage and Retrieval
Producing Communication Messages
  Producing Published Messages
  Producing Film Messages
  Producing Broadcast Messages
Information Technologies
  Computer Hardware
  Computer Software
Summary
Activities
Test Your Knowledge

Key Words

These words are used in this chapter. Do you know what they mean?

audience assessment
audio recording
broadcast message
carrier
central processing unit (CPU)
communication
communication satellite
communication technology
compact disc read-only memory (CD-ROM)
comprehensive
computer
decode
digital videodisc (DVD)
digitized
downlink
electromagnetic radiation
encode
film message
flexography
format
graphic communication
hardware

hypertext markup language (HTML)
information
input device
layout
medium
memory
monitor
operating system
output device
photographic system
printing system
published message
receive
receiver
rough
storyboard
telecommunication system
transmit
transmitter
uplink
videocassette recorder (VCR)
video recording
wave communication
World Wide Web (WWW)

The modern world is vastly different from the past because of technology. Probably the most important technologies that have changed our life are communication and information processing technologies. These technologies started with the development of tools to support speech. The written word was developed. It started with hieroglyphics, which the Egyptians developed. A number of developments followed this, including movable type, printing presses, telephones, radios, televisions, computers, and cellular telephones.

Today, humans communicate constantly. They talk to one another. People read what other humans write. They listen to voices on radios. Humans view and listen to television programs. They are in constant communication through electronic media. In fact, humans cannot avoid communication. Anytime people send or receive information, they are communicating and processing information. See **Figure 22-1.** Think about how many times you receive and send messages in one day.

Communication and Information

People often use the terms *information* and *communication* together. These terms are related, but each means something different. Let's review two words you learned in Chapter 7. *Data* includes individual facts, statistics (numerical data), and ideas. These facts and ideas are not sorted or arranged in any manner. *Information*

Figure 22-1. We receive information in a number of ways. These geologists are reviewing a satellite image, looking for petroleum deposits. (American Petroleum Institute)

TechnoFact: Written communication marked the beginning of recorded history. The rise of book, magazine, and newspaper publications created avenues for widespread information sharing.

is data that has been sorted and arranged. It consists of organized facts and opinions people receive during daily life. Changing data into information is called *data processing* or *information processing*. It involves gathering, organizing, and reporting data so it is useful to people. It is often done using *information technology.*

Data and information are of little use, however, unless they are shared. This is what communication does. *Communication* is the act of exchanging ideas, information, and opinions. It involves a message someone wants someone else or something to receive. When we use technical equipment in communicating, it is called *communication technology.*

How We Communicate

People must process information before they can communicate. You may know something. The knowledge is in your mind. It belongs to you. No one has access to it. There are a number of ways you can share your knowledge. See **Figure 22-2.** You can use your ability to speak and tell someone what you know, or you can write a paper containing the information you have. A person can read the paper to learn what you know. You can draw a picture communicating the information. The picture shows a person what you know. You can take a photograph communicating what you know, use a symbol, or develop a sign representing the information. The use of drawings, symbols, and measurements promotes clear communication by offering a general language with which to convey ideas. Technical systems use specific symbols and terms.

So far, we have discussed people directly communicating with other people. This is face-to-face communication. People show pictures and symbols, or they use formal language. These are common types of communication, but they use little technology. There are no technical means used. Machines and equipment are not involved in the communication. A communication system is not present.

Standards for Technological Knowledge

Demonstrate
Use words, such as school, bus, or stairs, to show the difference between speaking, writing, drawings, photos, and symbols. Refer to Figure 22-2.

Tech Terms
Information Processing (Data Processing)— Organizing and manipulating information (data) into a form that is more usable for a specific need.
Information Technology— Using technical means, usually computer systems, to manage and process information.
Communication— Exchanging ideas, information, and opinions.
Communication Technology— Use of technical means (equipment and systems) to exchange ideas, information, and opinions.

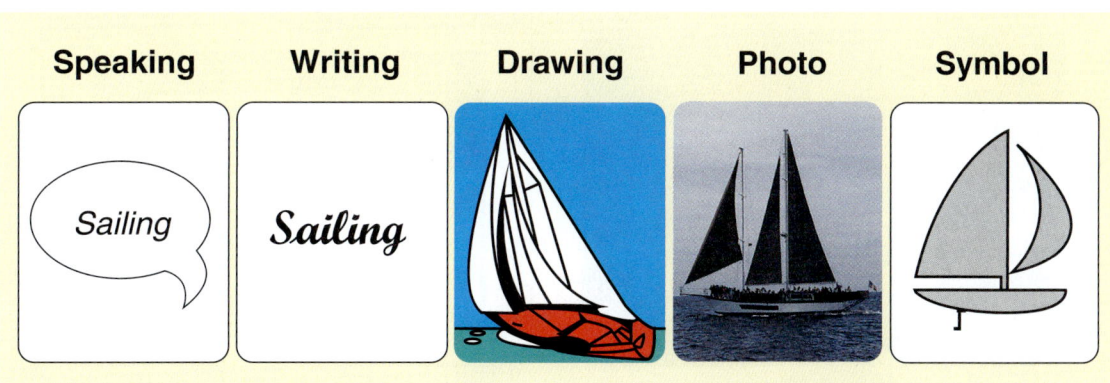

Speaking	Writing	Drawing	Photo	Symbol
Sailing	*Sailing*			

Figure 22-2. People can communicate with pictures, symbols, writing, or speech.

TechnoFact: A major advancement in writing technology was made by the Egyptians. They replaced the commonly used, bulky clay tablet with a new writing material made from papyrus plants. They wrote their hieroglyphic language on this paper-like material.

Figure 22-3. Communication technology uses equipment and systems.

Communication Technology

Communication technology uses equipment and systems to send and receive information. See **Figure 22-3.** This technology communicates information using *graphic* and *wave* systems. See **Figure 22-4.** *Graphic* comes from a word meaning "to draw or write, as on paper." *Wave* refers to radio waves, a kind of energy.

Graphic Communication

Information may be communicated using drawings, pictures, graphs, photographs, or words on flat surfaces. See **Figure 22-5.** These types of messages are called *graphic communication.* In graphic communication, paper or film carries the message. Graphic communication systems produce printed graphic and photographic *media.* They communicate through drawings, printed words, and pictures. The systems also use photographic prints and transparencies (slides and motion pictures). These messages are visible at all points as they move from the sender to the receiver.

Wave (Electronic) Communication

Wave communication systems depend on an energy source called *electromagnetic radiation.* *Electromagnetic radiation* is energy moving through

Figure 22-4. The two major communication systems are graphic and electronic (wave) systems. (Bren Instruments Inc., TSA)

TechnoFact: The basic forms of communication are signs (sight) and sounds (hearing and language).

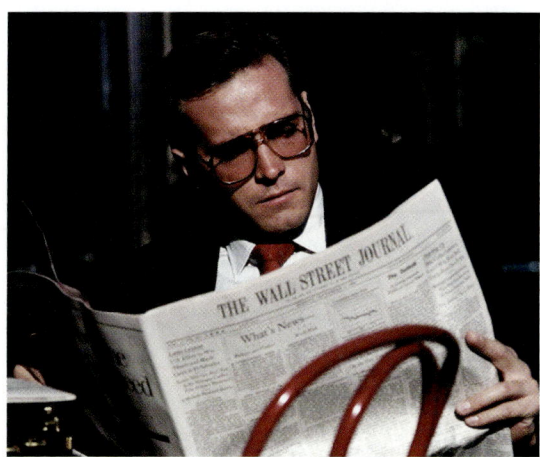

Figure 22-5. A newspaper is an example of a graphic communication medium. (Gannett Co.)

space in waves. It travels at the speed of light—186 thousand miles per second. See **Figure 22-6.**

Sound waves are much slower than electromagnetic waves. They travel at different speeds in different media. In air, they travel at about 750 mph at sea level. They travel about four times faster in water.

People can use light, sound, or electrical waves to send information. The information is coded at the source. It is then transmitted (sent) to the receiver. There, the code must be changed back to information. This type of communication is often called *electronic communication.* See **Figure 22-7.** These systems often use electronic equipment to code and transmit the information.

For example, suppose your favorite sports figures are interviewed. They speak into microphones. Speech is changed into pulses of electromagnetic energy. This energy is transmitted from the radio station's broadcast tower. Your radio's antenna picks up some of these pulses of energy. Various stages of the radio change the energy. Finally, the radio speaker changes electrical pulses back into *audible* sound. You hear the voices, but between the microphones and the speaker, no one can hear or see the messages.

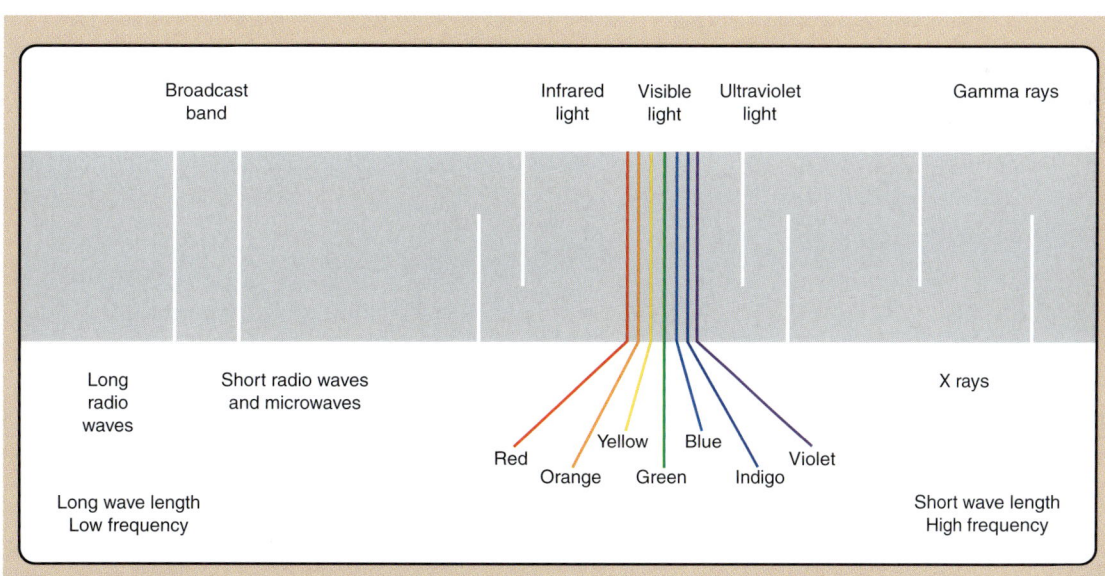

Figure 22-6. This is the electromagnetic spectrum.

TechnoFact: Edwin Armstrong researched the behavior and application of radio. He conducted many experiments that led to the improvement of the young radio field. Eventually, he developed the first practical frequency modulation (FM) radio in 1933. FM broadcasts were far superior to AM broadcasts, and FM would become the standard for TV sound broadcasting.

Figure 22-7. Here is a typical communication system. A—Cameras capture pictures and sound. B—Complex equipment processes the signals. C—The signals are then sent through the air. D—The viewer gets the message on a receiver (TV set). (Westinghouse Electric Corp.)

Graphic vs. Electronic Communication

Electronic communication has many advantages over graphic communication. A book can hold only so much information. Readable type can be only so small. A book can be only so thick and still be useful, but wave communication carries vast amounts of information in almost no space. A space satellite communicates a library of information in a matter of seconds. Fiber-optic systems are said to carry encyclopedias of information per second. A compact disc (CD) can hold many books of information on a single disc. A *digital videodisc (DVD)* can hold an entire movie on a disc, with room to spare.

You cannot, however, take wave media everywhere with you. There isn't always a television set next to you. A signal from a radio station can only reach so far. A cell phone will work only when a tower is nearby, but books and magazines are portable. Books and magazines can also be available at any time or place. They capture ideas and visual images. On the other hand, they cannot provide sound, like music.

Therefore, we need different types of communication systems for different

TechnoFact: Peter Goldmark and a team at RCA developed the long-playing (LP) record in 1948. This record spun at a rate of 33 1/3 rpms, which allowed much longer recordings than the 78 rpm records of the day. The compact disc, introduced in the early 1980s, has almost completely replaced phonographic records.

People-to-Machine

People-to-People

Machine-to-People

Machine-to-Machine

Figure 22-8. These are the types of communication. (Hirdinge, Inc.)

uses. Wave communication is quick and inexpensive. Current events reach us almost as they happen. Electronic communication has made our world smaller. It has caused us to live in what many people call a *global village.* Print communication media is slower, but can be used almost everywhere. It can be easily selected, used, and stored. This media will last a lifetime.

Types of Communication Systems

Communication technology is used for four distinct types of communication:

➤ People-to-people communication.

➤ People-to-machine communication.

➤ Machine-to-people communication.

➤ Machine-to-machine communication.

See **Figure 22-8.** Each of these information and communication systems affects our daily lives. We come into contact with each type as we interact with our environment. People create information and communication technology systems to gather facts, manipulate data, and communicate information more effectively. Information is transmitted and received using various systems.

All the information and communication systems have similar components. See **Figure 22-9.** They all have a device to develop and send the message. This device is often called a *transmitter.* All systems must have a

Standards for Technological Knowledge

17

Resource
Reproducible Master 22-1: Types of Communication Systems

Figure Discussion
Give several examples of each communication system shown in Figure 22-8.

Tech Terms
Global Village—The entire world as a single community served by electronic media and information technology.
Transmitter—The part of a broadcasting system producing a radio frequency wave, altering it to convey a meaningful signal, and radiating the signal from an antenna.

Figure Discussion
Explain each part of the communication system shown in Figure 22-9.

Tech Terms
Receiver—The part of a broadcasting system that receives incoming signals and converts them to light or sound.
Carrier—The part of a broadcasting system that moves the encoded message from the transmitter to the receiver.

Transmitter

Receiver

Figure 22-9. This is a simple communication system.

channel to carry the message. This channel is called the *carrier*. It can be the airways, a wire, or optical fibers. Finally, each system has a device to gather and process the message. This device is generally called a *receiver*.

People-to-People Communication

Everyone communicates with other people in one way or another. Humans have developed complex

TechnoFact: Telecommunication signals can be transmitted through the air, wire and cables, or optic fibers. In 1956, Narinder Kapany coined the term *fiber optics* to describe a bundle of flexible glass fibers coated with a reflective material.

A ·—	S ···		
B —···	T —		
C —·—·	U ··—		
D —··	V ···—		
E ·	W ·——		
F ··—·	X —··—		
G ——·	Y —·——		
H ····	Z ——··		
I ··	1 ·————		
J ·———	2 ··———		
K —·—	3 ···——		
L ·—··	4 ····—		
M ——	5 ·····		
N —·	6 —····		
O ———	7 ——···		
P ·——·	8 ———··		
Q ——·—	9 ————·		
R ·—·	0 —————		

Sending station   Receiving station

Standards for Technological Knowledge

17

Demonstrate
Construct and demonstrate the operation of a simple telegraph system.

Tech Terms
Telecommuni-cation—Using electronic means to exchange information over a distance.

Figure 22-10. An early telegraph is shown at the top left. A schematic of a closed loop is at the bottom left.

technological systems to improve their communication. Machines and devices have been produced to help us communicate better and easier.

People-to-people communication includes a number of basic systems:

➤ Telecommunication systems.
➤ Audio and video recording systems.
➤ Computer (Internet) systems.
➤ Printing systems.
➤ Photographic systems.
➤ Drafting systems.

Each of these systems is in wide use today. All of them have their places and serve specific purposes.

Telecommunication Systems

Communication systems exchanging information over a distance are called *telecommunication systems.*

There are two major types of telecommunication systems. These are individual and mass telecommunication systems.

Individual telecommunication systems have been designed to let one person communicate to another individual. These were the earliest telecommunication technologies. Two systems of this type were developed in the 1800s.

The first was the work of Samuel Morse in the 1830s. It was the *telegraph.* This system depended on an electrical circuit (complete pathway through which electricity flows) to carry the message. (The circuit is the carrier.) An operator uses a special code. See **Figure 22-10.** A series of dots (short electrical pulses) and dashes (long pulses) represent letters and numbers. The operator produces the pulses by

TechnoFact: The telegraph came into general use after Samuel F. B. Morse developed an operational system. His system used a single overhead wire as one conductor in a circuit and the earth as the other conductor. With this system, only one message could be sent at a time on the one wire. In 1872, J. B. Stearns invented a way to send two messages over a single wire at one time. This became known as duplex telegraphy. In 1874, Thomas A. Edison invented quadruplex telegraphy, which allowed four messages over the same wire at one time.

Figure 22-11. Here is a communication relay tower.

transmitter that changes sound waves into electrical pulses. It also contains a receiver that changes the electrical pulses back into sound.

Often, the two telephones are connected to different central offices. Then the message is transmitted first to the central office on wires. From there, the signal moves between the two offices. This can be done on wires, by microwaves (high frequency waves) between towers or satellites, or by glass fiber cables (fiber optics). See **Figure 22-11.** Today's cellular telephone operates in a similar manner. The main difference is that this telephone is really a low power radio transmitter. The cellular telephone directly broadcasts its signal to a relay tower. The tower connects the signal to a Mobile Telephone Switching Office (MTSO). This office connects the signal to normal telephone landlines. From there, the signal travels like a normal telephone call.

Two *mass telecommunication* methods have been designed to reach large groups of people. These are radio and television. See **Figure 22-12.**

They operate very much alike. Radios and televisions change sound

pressing a key (the transmitter). At the other end of the circuit, a sounder (the receiver) is activated. It changes the pulses into clicks read as dots or dashes.

The second personal telecommunication system was the *telephone.* Alexander Graham Bell invented this device in 1876. The telephone uses an electrical circuit to connect the two communicators' phones with a central office. The handset contains a

Figure 22-12. The control room (left) is the nerve center of a television broadcasting activity (right). (Viacom International, Inc.)

TechnoFact: Alexander Graham Bell filed for patents in 1876 and 1877 on what would evolve into the modern telephone. Many people were working on similar devices around the same time as Bell, but he is credited as the inventor of the telephone. Since so many people were working on similar devices at about the same time, there are disputes about who is the actual inventor of the telephone.

Research
Have the students trace the history of communication satellites.

Tech Terms
VHF (Very High Frequency)—The range of radio frequencies falling between 30 and 300 megahertz.
UHF (Ultrahigh Frequency)—The range of radio frequencies from 300 to 3,000 megahertz.
Communication Satellites—Artificial satellites that are in Earth orbit and serve as part of a global communication network.

Figure 22-13. Communication satellites high in space can receive and send signals from Earth (left). Here is a space communication platform of the future (right). A powerful antenna will be able to link up with small Earth stations. You could use it for your own personal communication link! (NASA)

and light waves into high frequency signals. These systems use microphones and cameras to change sound and light into electrical pulses. Transmitters process these pulses into high frequency waves. Signals are carried to a broadcast tower. They are *radiated* (projected) over a large area. The atmosphere is the carrier for the signal. A receiving antenna captures these waves, and the receiver separates a signal from other waves. Speakers and television picture tubes change the signal back into sound and light. Today, television signals are also broadcast by satellite (dish) systems and over cables. These systems replace local broadcast towers. Signals are carried over cables connected to each home or

beamed from a satellite to a personal dish receiver.

Radio and television broadcast systems use two types of signals. Radio and most television stations use the very high frequency (VHF) bands (groups of signal frequencies). Some television stations use the ultra high frequency (UHF) bands.

The advent of space exploration ushered in a new phase of telecommunications. This allowed the use of *communication satellites.* These satellites are broadcasting stations in outer space. They can receive messages from one place on Earth. Then, they relay the messages back to Earth at a distant place. See **Figure 22-13.**

TechnoFact: Communication satellites are an important part of the global communication network. *Echo 1,* which was launched in 1960, was the first communication satellite. It was basically a balloon that reflected radio signals back to Earth. Later satellites contained electronic devices for receiving, amplifying, and rebroadcasting signals to Earth.

Technology Explained

instrument landing system (ILS): A navigational aid used to help pilots guide aircraft onto runways for safe landings.

Weather often does not allow pilots clear visibility while approaching an airport. A modern navigational aid is the instrument landing system (ILS). It is found at almost all commercial airports, **Figure A.**

An ILS produces two radio beams. One beam is in the aircraft's horizontal plane, and the other beam is in the aircraft's vertical plane. The horizontal beam is a rather narrow beam giving the pilots the proper glide slope (angle of descent) for a safe landing. The other

Figure A. Modern aircraft use many landing aids. (Delta Air Lines, Inc.)

beam, which is wider, radiates at a right angle to the glide slope beam. This beam allows pilots to align the airplane with the center of the runway.

The pilots have display screens indicating the aircraft's alignment with the two beams. Two lines are shown on the display telling the pilots how

Standards for Technological Knowledge

17

Discussion
Discuss the use of communication satellites for commercial and military purposes.

Brainstorm
Have the students develop a list of several reasons why they think communication satellites are important to global communication.

A narrow beam of microwaves carries the message. The beam has to be carefully aimed to hit the satellite. The signals are beamed back to Earth at many different angles and to more than one Earth station.

Because of the need to aim the narrow beam, the satellite must always be in the same position above the Earth. It must complete one orbit every 24 hours. The satellite travels at the same speed as Earth's rotation. For this reason, its orbit is called *geosynchronous* (at the same time as Earth). If it were

to travel slower or faster, the microwave link (contact) would be broken.

Communication satellites are placed about 22,300 miles above Earth. There, they can transmit to over 40 percent of Earth's surface. (A message beamed to a satellite from Boston or Montreal can be received in Germany.)

Earth stations transmit the microwave signals through bowl-like antennae. The upward signal is called the *uplink.* The signal relayed back is called the *downlink.* Signals can also be relayed from one satellite to another.

TechnoFact: A satellite in orbit above Earth's atmosphere can remain in orbit indefinitely. When a satellite is placed in an orbit 22,300 miles above Earth, it will make one revolution around Earth every twenty-four hours. This is the exact time Earth takes to rotate once on its axis. In this orbit, the satellite can remain over the same point on Earth's surface. This is called a *geosynchronous orbit* and allows for communication satellites.

the plane is aligned with the runway. When the plane is making a proper approach, these lines will be in alignment with vertical and horizontal index lines on the screen. An improper approach causes the lines the ILS produces to be displayed away from the index lines, **Figure B** (upper left and lower right).

Arrayed along the center beam are three additional vertically broadcast beams. The first beam is known as the outer marker beacon. The beam closest to the runway is called the *boundary beacon.* These markers tell the pilots the distance they are from the runway, as the plane approaches for a touchdown.

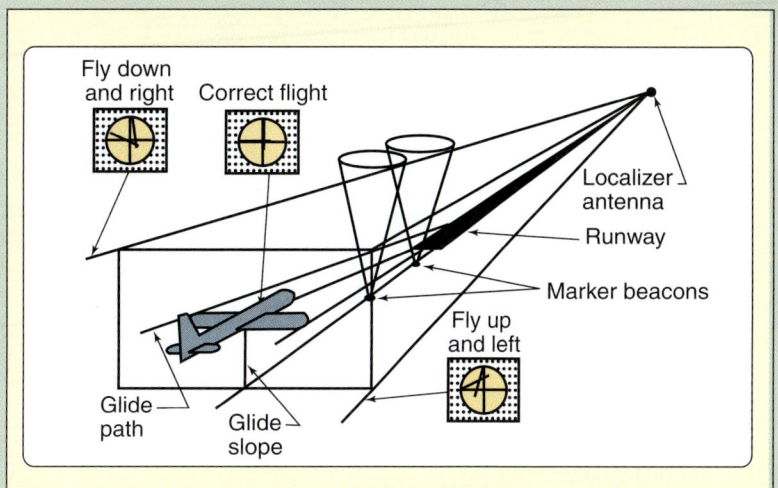

Figure B. Shown are the beam paths of an ILS.

The ILS guides the pilots to about 200´ above the runway. The actual landing is completed visually. Used with the lights marking the boundaries of a runway and verbal help from air traffic controllers, the ILS provides an effective aid in landing.

Audio and Video Recording Systems

Audio recordings and *video recordings* are extensions of radio and television communications. They record the same type of information radios and televisions broadcast. These recordings can be seen as warehouses for audio and video information.

These communication systems can be traced to Thomas Edison's invention of the phonograph. This 1877 invention allowed people new freedom. No longer did they have to listen to a speaker or a concert when someone else scheduled it. They could select the time and place they wanted to listen to the speech or music. See **Figure 22-14.** Audio and video recordings use one of three technologies:

➤ Grooves in discs. The first audio records used this recording technology. Records were produced with wavy grooves on their surface. A *stylus* (needle) vibrated as it followed the grooves on a moving record. This vibration produces a small electrical signal. This signal is

TechnoFact: Audio (sound) recording is a process converting acoustic energy into a form that can be permanently stored and reproduced at any time. In 1855, Leon Scott developed a recording device called a *phonautograph.* However, Thomas Edison invented the first machine to audibly reproduce recorded sound. He patented this talking machine (phonograph) in 1878. The quality of sound recording was greatly improved with the adoption of electromechanical systems using electronic amplifiers.

Extend
Explain the advantages of DVD recordings over tape and disc recordings.

Tech Terms
Downlink—The transmission path used to send a communication signal and data from an aircraft, a spacecraft, or a communication satellite to Earth.
Audio Recording—A device or system that stores and reproduces sound.
Video Recording—A device or system that stores and reproduces visual images and often sound.
VCR—Videocassette recorder.
CD-ROM—Compact disc read-only memory.
Digital Data—Data expressed using a system of discreet numbers, such as ones and zeros.
Analog Data—Data expressed using a variable physical quantity, such as voltage.

Figure 22-14. Sound and pictures have been recorded using grooved discs, tapes, and optical recorded disks.

amplified (made stronger) and changed into sound waves. Few people use this type of recording today.

➤ Magnetic charges on a tape. To produce a magnetic recording, coated plastic ribbons are fed through a recording unit. The unit produces a pattern of magnetic charges in the coating. The playback unit "reads" the charges. The charges are then changed into sound (audio) and light (video) messages. Cassette audiotapes and *videocassette recorder (VCR)* tapes are produced in this way.

➤ Digital codes on a disc. This is the technology used for audio, video, and *compact discs read-only memory (CD-ROMs).* Sound and light waves to be recorded are *digitized* (numbered). Each frequency is assigned a specific number. The number is recorded on the disc on microscopic pits and flats. The playback unit reads the code with a laser beam. The digital readout is converted back to very accurate pictures or sound.

Each of these systems has advantages and disadvantages. Disc recordings were manufactured inexpensively, scratched easily, and lost quality quickly. Tape recordings are small and unbreakable and can have new material recorded over a used tape. They lose quality over time and sometimes jam in the player. CDs, CD-ROMs, and DVDs produce high quality reproduction, but are fairly expensive.

Computer (Internet) Systems

Computers are the main tools used for networking information and communication technologies. A development that has become a key

TechnoFact: What would become the World Wide Web started as a text-only interface in the 1980s and was introduced to the public in 1990. However, it was not until Marc Andreessen developed a graphical browser that the Web began to be accepted by the public. By 1994, UNIX, Windows, and Macintosh versions of Andreessen's browser were available. The continued development of improved browsers and the advent of powerful search engines have helped make the Internet a vital communication medium.

Figure 22-15. Computers are now a part of global communication, using the Internet.

communication medium is the Internet. This system allows people to communicate through their *computers.* The U.S. Department of Defense started the Internet in 1966. Designed to allow researchers to use their computers to communicate, the system divides messages into packets (small parts). These packets each have an address for a destination. Each packet does not have to travel the same route to its destination. When all the packets arrive, the message is reassembled. The system showed so much promise that other people also developed networks. In 1982, the networks were merged, or inter-networked, into one.

The amount of information available on the Internet is huge. Until recently, however, this information was hard to find and use. An innovation that has helped solve the problem is the *World Wide Web (WWW).* Documents that are accessible on the web are formatted in a language called *hypertext markup language (HTML).* This language allows users to select words, known as *hot links,* to jump from one document to another. For example, the word *airplane* may be highlighted in a document. If a person clicks on the word, additional documents about airplanes will be made available.

A second advancement was *Mosaic.* This system was the first set of programs called *browsers.* Browsers are computer software programs. They allow you to use key words, like *airplane,* to search for related documents on the Internet. People with a computer containing the proper software and Internet access can have information from around the world at their fingertips. See **Figure 22-15.**

Printing Systems

Printing systems are the third type of people-to-people communication. They include all systems used to produce letters and pictures on non-treated paper. See **Figure 22-16.**

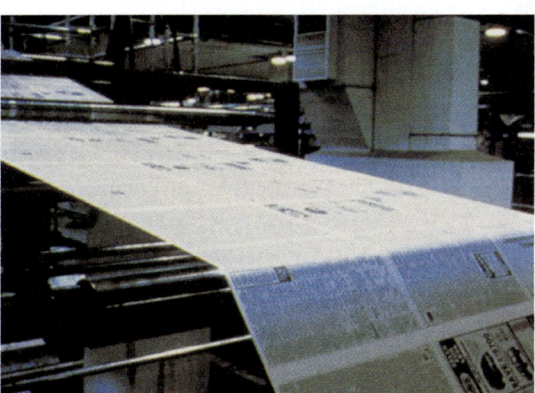

Figure 22-16. High-speed printing presses produce thousands of newspapers each day. (The Chicago Tribune Co.)

Standards for Technological Knowledge

Extend
Differentiate between browsers, documents, websites, and the Internet.

Tech Terms
Computer—A device that electronically stores, processes, and outputs data according to instructions provided by the user through programs or manual entry.
Printing—The process of producing text and images on various media, such as paper.

TechnoFact: The first mechanical reproduction of illustrations was achieved by means of wooden blocks. First, a picture was drawn on the smooth surface of a wooden block. Then, the wood on either side of the lines was carved away. The resulting relief image was inked. Parchment or paper was pressed onto the block, thus transferring the image. The process could be repeated again and again, producing many identical pictures from a single block.

Figure 22-17. Here is letterpress printing (left). This is the image it produces on paper (center). A raised surface makes the image (right). (Graphic Arts Technical Foundation)

The printing revolution started with Johannes Gutenberg about 1450. Until that time, scribes produced most books by copying manuscripts one at a time. This process was slow and open to many errors. Gutenberg developed movable type (individual letters on pieces of lead). This development allowed the printer to arrange the type for an entire page. Then, a number of copies of the page could be produced. This printing from raised letters on type is called *letterpress printing* or *relief printing.* See **Figure 22-17.** Other more efficient processes are rapidly replacing it. One similar process is called *flexography.* This process uses raised type similar to letterpress printing. Flexography, however, uses a synthetic rubberlike sheet, much like a rubber stamp. This sheet is used to print the desired message.

Another more efficient printing process is called *offset lithography.* It uses a negative (reverse image) of the page to be printed. The negative is placed on an offset plate. The plate has a photographic coating bright light can expose. The negative allows light to expose the plate where the letters are to be. When the plate is developed, the type to be printed is on the plate.

The plate is placed on a special press. This press coats the plate with a liquid (fountain solution). This liquid will not stick where the letters are on the plate. Then ink is applied. The ink will not stick where the fountain solution is. Therefore, ink is only on the type portion of the plate. The plate then presses against a blanket that picks up the ink. The blanket transfers the ink image to the paper. See **Figure 22-18.**

Other printing processes include intaglio, screen printing, and electrostatic reproduction. *Intaglio* prints from a recessed image. It is often called an *etching* and is used for high quality printing. Stamps and paper money are printed by this process.

Screen printing forces ink through a fabric screen to produce an image. See **Figure 22-19.** Fabrics, posters, and T-shirts are screen printed.

Electrostatic reproduction is rapidly becoming a major printing process. The office photocopier is an example

TechnoFact: Early printing depended on type similar to the ancient seals and stamps used to make impressions in clay. The first movable type made from metal molds was developed in Korea in the thirteenth century. Independently, Johann Gutenberg developed a similar system in Europe. His Mazarin Bible is believed to be the first European book printed with movable type and was completed in Mainz in the mid-1450s.

Figure 22-18. The offset printing process (left) produces an image on paper (center) from a flat printing surface (right). (Graphic Arts Technical Foundation)

of an electrostatic machine. High-speed machines of this type can print several thousand copies an hour, rather inexpensively.

Photographic Systems

Directly related to printing are *photographic systems.* These systems use photographs to convey messages.

Figure 22-19. The screen printing process produces an image by forcing ink through openings in a screen carrier. (Graphic Arts Technical Foundation)

TechnoFact: The first printing press in North America was set up at the Massachusetts Bay Colony's new college, Harvard, in 1638. It was operated by a Mrs. Glover, the widow of Reverend Jose Glover. The first items published were an almanac, "The Freeman's Oath," and the now famous *Bay Psalm Book,* which was the first book printed in the colonies. The press later was owned by Samuel Green, who printed fifteen hundred copies of a Bible in the Algonquian language.

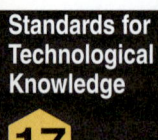

Standards for Technological Knowledge

17

Extend
Give examples of people-to-machine and machine-to-people communication.

Tech Terms
Machine-to-People Communication— Information and data provided through mechanical means, such as meters, dials, lights, and switches, to people.

People on an airliner review emergency procedures with a series of pictures. The message was carefully designed and produced. The pictures tell a story.

This use of photography is very much different from personal snapshots. Personal photographs are used to capture moments of time. They generally are not planned to communicate a message. Therefore, most personal photography is not done for communication purposes. It is done to record a bit of history or an important occasion.

Drafting Systems

The last type of person-to-person communication is drafting. This communication technique uses lines and symbols to communicate designs. It includes engineering and architectural drawing. Using sketching and drawing techniques to communicate designs was covered in Section 3 of this book.

People-to-Machine Communication

People communicate to machines daily. We set controls "telling" machines how to operate. See **Figure 22-20.** We set the thermostat to communicate the room temperature we want to a heating and cooling system. To communicate the speed at which we want to travel, we set a speed control system in a car.

Also, we write computer programs to tell computers what to do. We can tell computers to print letters, draw lines, calculate costs, or perform hundreds of other acts. To ensure the clock will signal a time to go to bed or begin a new day, we set time on digital clocks and timers. The timer will tell us when food is done in a microwave or oven. Symbols or icons are used on many computers, elevators, and telephones to stand for ideas and to communicate what should be done when the symbol is pressed or used.

Machine-to-People Communication

Directly related to people-to-machine communication is machine-to-people communication. The machine we communicate to, through switches and dials, responds. It will present us with meter readings, flashing lights, or alarms.

Pilots have many machine-to-people communication systems on the aircraft flight deck. Lights tell them if the engines are running properly.

Figure 22-20. This worker is communicating to a machine using a control panel. (Ford Motor Company)

TechnoFact: George Eastman is credited for making photography an everyday activity. He invented a dry-plate process and, in 1880, built a factory to make the plates. In 1888, he invented roll film and the Kodak® camera to use it. In 1928, he developed a process for color photography.

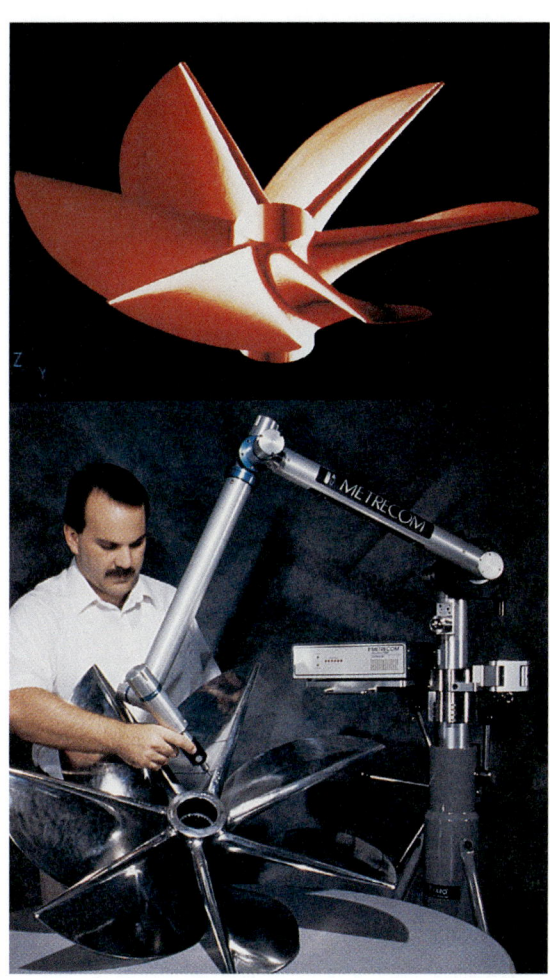

Figure 22-21. This computer-aided design (CAD) system (top) was used to design this industrial part (below). (Faro Technologies, Inc.)

Alarms sound as the plane approaches stall speed. Video screens display the information the radar system gathers.

In our automobiles, gauges and lights are also used to communicate. The fuel gauge tells the driver when to buy more gasoline. The oil light warns the operator of low oil pressure. A blue light tells the driver the headlights are on high beam. A flashing light tells the driver the turn signals are operating.

Machine-to-Machine Communication

The most recent communication systems have machines providing information to machines. Computer-aided design (CAD) systems help people design parts. See **Figure 22-21.** These systems do drafting. They can direct machines to produce parts.

Computer-aided manufacturing (CAM) uses computers to directly control machine operations. See **Figure 22-22.** The computer can direct

Figure 22-22. This robotic production line is an example of computer-aided manufacturing (CAM). (AMP, Inc.)

Brainstorm
Have the students develop a list of ways in which machine-to-machine communication impacts their life every day.

Tech Terms
Computer-Aided Design (CAD)—
Using computer systems to facilitate the design of mechanical and architectural components and the production of their drawings.
*Computer-Aided Manufacturing (CAM)—*Using computer systems to operate and control manufacturing processes.

TechnoFact: A common machine-to-people communication device is a speedometer. This device is an instrument indicating the speed at which a vehicle is traveling. Some speedometers are mechanically linked to the drivetrain, while others use electronic sensors to determine speed. In both cases, a visual display communicates the vehicle's speed to the driver.

Figure 22-23. Here is the communication process.

the machine to run at specific speeds. It can set material feed rates. The computer can cause the machine to change cutting tools. All this occurs without human action. Even more complex systems totally merge the design and manufacturing activities. The complex systems are called *computer-integrated manufacturing (CIM)*. These are but a few examples of machines communicating to other machines.

The Communication Process

You have now read about many communication systems. These systems are made up of a source, an encoder, a transmitter, a receiver, a decoder, and a destination. A communication system is similar to other systems in that it includes input, processes, outputs, and sometimes feedback. Each of these systems follows a basic communication process. This process has four major steps. See **Figure 22-23.** These steps move the information from the sender to the receiver:

1. Encoding.
2. Transmitting.
3. Receiving.
4. Decoding.

At any point along the process, the information can be stored. Later, it can be retrieved from storage.

Encoding

Communication involves exchanging information between a sender and a receiver. The sender must first decide what the receiver wants to know. See **Figure 22-24.** Then, the

Figure 22-24. This designer is developing the message and layout for an advertisement. (Ohio Art Co.)

TechnoFact: The development of mass media in the twentieth century has played a major role in changing social, economic, political, and educational institutions.

message must be designed. It must attract the attention of the receiver. Finally, the message must be communicated in a form the receiver will understand. All of these tasks are parts of encoding. To *encode* means "to change the form of a message."

Messages may be encoded in a number of ways. See **Figure 22-25.** Information can be encoded using symbols and graphics. Symbols may be used to convey the message. A red, octagonal street sign means "stop!" Two people shaking hands mean agreement. A skull and crossbones mean poison. All these are symbols carrying meaning.

Other messages may be written. The information may be communicated using language. This book is communicating a message in this manner.

Figure 22-25. Highway signs use shapes, words, and symbols to communicate.

Figure 22-26. This scene shows the first stage in transmitting a message using electronic media.

Transmitting

Once the message is encoded, it must be delivered to the receiver. Remember, we are talking about communication technology. Therefore, a technical means (machine or equipment) must be used. Switching circuits allow signals to be sent back and forth in the communication process. A network is a system connected by communication lines to transfer information from one device to another.

We have already presented the basic ways of *transmitting* (sending) the message. Transmission entails sending signals in a form that can travel over a distance. Graphic means and wave transmission are the basic forms of transmission. The message may be printed on paper or carried on a series of photographs. These are the graphic communication technologies. Electronic (or wave) messages may be carried on sound, light, or radio waves. See **Figure 22-26.** We may use radio or television transmitters to send our information, or the light a laser produces may carry our telephone

Figure Discussion
Decode each sign shown in Figure 22-25.

Tech Terms
Encode—To put a message into code that can be transmitted.
Transmit—To send a message from one point to another.

TechnoFact: Telecommunication has been defined by international agreement as any emission, transmission, or reception of signs, signals, sounds, and writing.

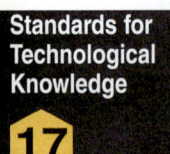

Research
Ask the students to investigate how a radio receiver works.

Tech Terms
Receive—To pick up communication signals.
Decode—To convert a received message into a meaningful form.

conversations. A public address system may use sound waves to broadcast our ideas.

Receiving

The transmitted message must be *received.* See **Figure 22-27.** The message must arrive at a desired location. It must be available to the receiver. The message the transmitter sends is often in a special form. It may be a series of electrical pulses on a wire. The message may be radio waves varying in strength (AM radio) or frequency (FM radio). It may be in digital code.

Often, receivers are electronic devices. They change the transmitted code back into a form people can understand. They may change radio waves into sound waves. They also can change digital code into printed words.

Decoding

The final act in the communication process is putting meaning to the message, or *decoding.* Information must be decoded in order to be understood by the reader. Decoding is the opposite of encoding, with data being changed back to symbols and graphics. The people receiving the message must take action. They must read the printed word, listen to the broadcast, or watch the television program. This is not, however, enough. The people must place meaning on the message.

For example, you might hear someone shout "fore!" First, you must decide if the word is *fore* or *four.* Then, you must put it in context (compare it with the situation). If you are on a golf course, you should become alert. "Fore!" means someone is hitting a golf

Figure 22-27. The platemaking machine is receiving messages from the operator at the keyboard. (R. R. Donnelley & Sons Company)

TechnoFact: Valdemar Poulsen patented a device he called the *telegraphone,* in the 1890s. This device is the ancestor to the telephone answering machine, which was first introduced commercially in the 1960s.

ball. If you are cooking at a fast-food outlet, you may take other action. The manager may have told you to make four hamburgers.

A message is only effective if it is received. Time can be wasted designing, producing, and sending messages that are never received. How many tornado, flood, and hurricane warnings are transmitted, but ignored? Many "No smoking" signs go unnoticed. Highway speed limit signs are often viewed with indifference. Health warnings on cigarette packages are sometimes disregarded.

Storage and Retrieval

At any point in the communication process, the message may be stored. It can be recorded on video or audiotape. The message can be stored on CDs, DVDs, CD-ROMs, or computer

discs. The printed message can be placed in a warehouse.

Later, the message is taken out of storage. It reenters the communication process. Most television programs are taped (stored) for later broadcast (retrieved). Books are stored in libraries. The reader must retrieve and decode the messages.

Producing Communication Messages

Formal communication messages are carefully planned and produced. See **Figure 22-28.** The production of these messages involves four major steps:

1. Designing the message.
2. Preparing to produce the message.
3. Producing the message.
4. Delivering the message.

These steps apply to all specific activities: publishing, filmmaking, and broadcasting. You will see there are some activities each group does alike. Other activities belong only to a specific industry.

Producing Published Messages

Publishing includes all activities producing a *published message.* This includes newspapers, magazines, books, greeting cards, and flyers. Publishing is part of a larger activity called *printing.* In addition to publishing,

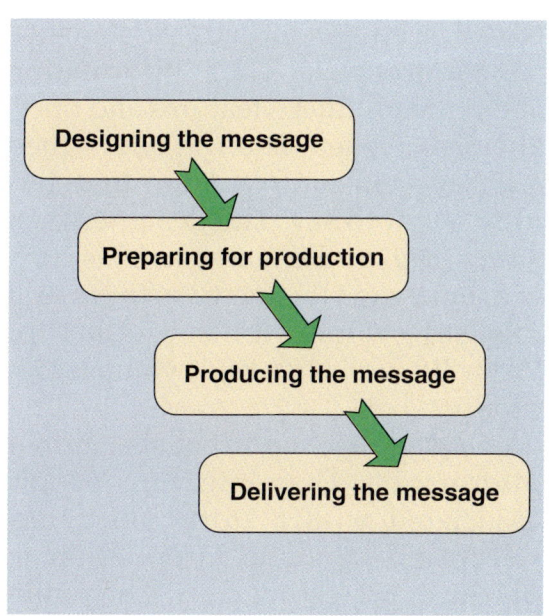

Figure 22-28. Here are the steps in designing a communication message.

Standards for Technological Knowledge

Resource
Reproducible Master 22-3: Creating Media Messages

Resource
Reproducible Master 22-4: Steps Used to Produce a Message for Print Communication

Figure Discussion
Explain the importance of each step shown in Figure 22-28.

Tech Terms
Message—The intended meaning of the information conveyed in communication.
Published Message—Information or ideas prepared for public distribution or sale using print or electronic means.

TechnoFact: The first magnetic disk drive was developed in the early 1950s by Reynold Johnson for IBM and first commercially available in 1956. This drive was a large unit consisting of fifty spinning 24″ magnetic disks.

printing includes the production of forms, stationery, and other "nonmessage" materials.

Designing Published Messages

Mass communication messages are designed to reach large audiences. They are planned to inform people, entertain people, or cause people to take action, but there are many audiences for published materials. Therefore, the designers must gather information about a specific audience they want to reach. In particular, the following questions must be answered:

➤ Who is the audience? (for example, young people, sports fans, business leaders, or senior citizens)

➤ What gets the audience's attention? (for example, words, pictures, or comic presentations)

➤ What does the audience value? (for example, status, financial security, fun, freedom, or power)

Factors such as the intended purpose and nature of the message also influence the message's design. All these factors should be taken into account when the message is created and transmitted to a specific audience. Gathering this information is called *audience assessment.* The information provides the base for designing the message. It gives the designer guidance and direction to complete several specific tasks.

The first design task involves selecting a *format.* This is planning for the physical size and shape of the message carrier. The carrier may be an 8½″ x 11″

Figure 22-29. These are examples of advertisements designed to sell products.

flyer. It can be a large billboard on a busy highway. A newspaper can be a tabloid (smaller size) or regular format. A magazine can be any size. The design can cover an entire page or part of a page.

Next, the designer gathers information. The message may be designed to sell products (an advertisement). See **Figure 22-29.** Then, information about the product's features and operation are needed. This can be matched with audience information to determine what to say. The designer then *writes copy* for the message. This copy is the message the media is to carry. It may say "You need this product" or "Your life will change, if you use this product."

Newspaper reporters also gather information. They interview people associated with a story. Then they select and combine information to produce interesting copy. The publication may be about a movie star or sports figure. Then biographical and

Figure 22-30. This photograph of an archeological dig can be used in a story about discovering our past.

Standards for Technological Knowledge

17

Resource
Reproducible Master 22-6: Preparing to Produce a Message

Resource
Transparency 22-2: Roughs

Resource
Transparency 22-3: Comprehensives

Research
Have the students compare the layout of two dissimilar publications, such as *Time* and *People* magazines. Then, have them attempt to explain why the publications are arranged as they are.

Tech Terms
Roughs—Initial layout for a publication
Comprehensives—Final layout for a publication.

current information about the person must be gathered.

Finally, the designer selects illustrations to support the copy. These are drawings and photographs attracting attention and adding interest to the message. See **Figure 22-30.** These illustrations may help the news story convey information. They may show a famous figure at work and play.

Preparing to Produce Published Messages

The format, copy, and illustrations must be brought together. They must be placed in a pleasing arrangement. In designing advertisements, designers first sketch out their ideas. *Roughs* (initial sketches) are prepared to show some basic ideas. See **Figure 22-31.**

The better ideas are converted into *refined* sketches. These sketches show more detail and thought. Finally, *comprehensives* are prepared. These

are more complete sketches for the message. They describe the arrangement of the elements, including type, illustration, and white space. Type size and style are called out (written on the sketch).

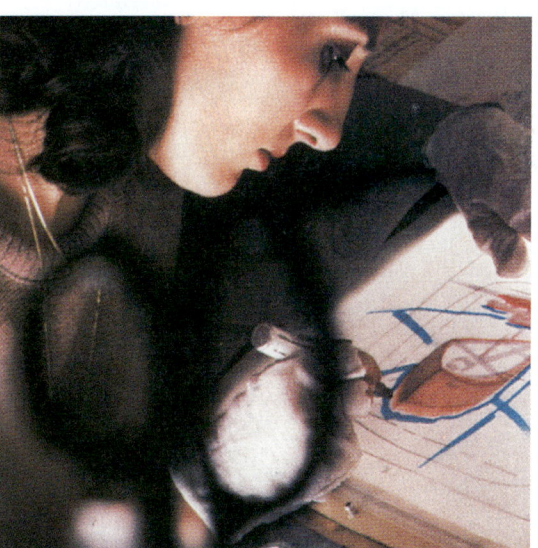

Figure 22-31. This artist is preparing roughs for an advertisement. (American Petroleum Institute)

TechnoFact: Printing from handset type was the common method until William Ged developed a new way to produce a printing plate in 1725. He molded a plate of type in plaster that he used to cast a lead copy of the plate. This allowed him to avoid the slow process of handsetting duplicate plates. He called his process *stereotype*.

Likewise, newspapers have layouts designed for their audiences. *USA Today* and other newspapers use colorful formats with many photographs, charts, and graphs. *Wall Street Journal* caters to its audience with a large amount of text.

Producing Published Messages

The message is now ready to be produced. The advertisement, magazine, or newspaper is ready to be printed. The first major task is to schedule the production through the plant. Resources must be allocated to each production task.

Then, the comprehensive is used to set type, size illustrations, and prepare photographs. These components are then attached to a large sheet of paper. This is called a *layout,* or paste-up. See **Figure 22-32.**

The pasteup is used to make printing plates or silk screens for the production process. These items are called *image carriers.* They carry the image from the press to the paper.

Once the large carriers are produced, the communication product must be printed. See **Figure 22-33.** Ink must be applied to the substrate (such as paper, plastic, or foil). As described earlier, this can be done by one of five basic processes:

➤ Letterpress (or relief) printing. This is printing from a raised surface.

➤ Offset lithography. This is printing from a flat surface.

Figure 22-32. This person is preparing a layout for a printed product. (The Chicago Tribune Co.)

TechnoFact: Charles Stanhope invented the first iron printing press. Like Gutenberg's press, it used a hand-operated screw to press inked type against paper. However, it allowed much faster printing. The Stanhope Press opened the door for later inventions, including the high-speed presses that made books inexpensive and readily available in the nineteenth century.

Figure 22-33. The printed word brings us information from around the world. (Gannett Co.)

➤ Intaglio (or gravure) printing. This is printing from a recessed surface.

➤ Screen printing. This is printing by forcing ink through openings in a screen.

➤ Xerography. This is printing using electrostatic means.

Delivering Published Messages

Delivering a printed message uses some standard distribution methods. Subscribers may receive the message in newspapers and magazines. Books are available in stores and libraries. Billboards, posters in store windows, and mailed flyers are still other distribution techniques.

The task is to select a distribution method that will reach the identified audience. Using *Sports Illustrated* magazine to reach large numbers of older women would be unwise. A message in *Seventeen* magazine, however, will reach many young women.

Producing Film Messages

Film messages are photographs and transparencies. Transparencies include movies, slides, and filmstrips. These messages are designed to present information, change attitudes, or entertain.

Designing Film Messages

Like published works, film messages are based on the results of audience assessments. Moviemakers have a good idea about what teenagers will go to see. These movies are greatly different from what grandparents feel are "good" movies.

The first design decision concerns format. Will a motion picture or a series of photographs present the message? How long should the presentation be? Should it be color or black-and-white? Answering these questions will establish the format.

Next, research must be done. Information about the subject must be developed. Educational films require much research. A good example of this type of film is the *National Geographic* video series of programs.

Entertainment films often take less research. Sometimes a film is based on a novel. After reading the novel, the author has already done most of the research.

The third step is *scriptwriting*. Most films use one or more actors to portray a story or present information. The actions and words of the actors must be described in writing. This writing is called the *script*. See **Figure 22-34.** The script carefully describes all events in the film.

Stage sets are designed and built to support these actions. See **Figure 22-35.** Also, many times on-site filming locations are selected and used. The sets and locations provide the realism for the message.

Preparing to Produce Film Messages

The production of film messages must be scheduled. The efficient use of resources must be planned. Then, the cast and production crew must be hired.

Directors must stage the various scenes. They decide how each scene will be shot. Placements of cameras and lights are considered. Also, directors determine how to use extras (people

Tech Terms

Script—The copy for a theatrical production, radio program, television show, or film.

Shooting Script—The final version of a script with the scenes set in the order they will be recorded or put on tape.

Screenplay—The script for a motion picture, including the dialog, descriptions of settings, and camera directions.

Set—Arrangement of furniture, scenery, and backdrops on a stage.

Various Voices off: *H'ya Buck!...Howdy, Buck!...How's things in Bisbee, Buck? Have a good trip?*

Meanwhile the SHOTGUN GUARD, who has guarded the treasure box from Bisbee, jumps down to the sidewalk.

SHOTGUN GUARD: *So long, Buck!*

Men begin unhitching the horses. BUCK acknowledges the cheery greetings as the WELLS FARGO AGENT in Tonto pushes his way through the crowd.

WELLS FARGO AGENT: *Howdy, Buck. Got that payroll for the mining company?*

Buck kicks the box which is under his seat.

BUCK: *She's right here in this box.*

The WELLS FARGO AGENT climbs up to the top of the coach, calling to a colleague as he does so.

WELLS FARGO AGENT: *Give us a hand with this box, Jim.*

BUCK: *Jim, I'll pay you that $2.50 when I get through.*

JIM: *Okay.*

The two agents get the box down and carry it off between them—BUCK looks over his shoulder to the other side of the coach.

BUCK: *Now you kids, get away from them wheels!*

He starts to get down and calls out to the men who are leading the horses away.

BUCK: *Well...sir, we ran into a little snow up there, quite bad, so you fellers better prepare for a good frost.* He jumps down and disappears round the side of the coach. The Tonto Hotel is seen on the other side of the road.

Medium shot of the stagecoach as BUCK comes round to open the coach door.

Figure 22-34. This is a page from the script for "Stagecoach," a John Ford and Dudley Nichols film. Lorimer Publishing, in London, printed the script.

TechnoFact: The motion picture camera rapidly captures a series of single images at a rate of eighteen or twenty-four frames per second. When projected at this same rate, the images produce the illusion of movement.

who add interest to the scene). The camera, lighting, wardrobe, and set crews work closely with directors.

The cast learns lines and movements from the script. Then, they come together to *rehearse*. Changes are made in the script and staging, until the director, cast, and crew are happy with the production.

Producing Film Messages

After the final rehearsal, filming can start. The actors complete each scene they rehearsed. (The scenes are seldom shot in the order outlined in the script.) Wise use of resources (such as stage time, natural light, and location availability) dictates the proper order in which to shoot the scenes.

Crew and cast may shoot each scene a number of times to develop several different effects. Later, the film is cut and spliced to combine the various scenes. These scenes are edited into a final product.

Distributing Film Messages

Most movies are first distributed to theaters. Special companies schedule the films and collect royalties for the film producers. Other films are made for television. They are distributed to the television stations or cable networks. After they are shown on these outlets, many films are converted into VCR tapes. These tapes are distributed through rental businesses or for purchase in stores.

Producing Broadcast Messages

Broadcast messages are communications radio stations, television stations, and cable television systems carry. These include entertainment, information, news, sports events, and advertisements. Regular programming follows the basic steps outlined for films. Scripts are prepared, crews and actors are hired, the production is

Brainstorm
Ask the students to develop a list of possible reasons why different types of programs are broadcast at different times and on different stations.

Tech Terms
Cinematography— The art or science of making motion pictures.
Rehearse—To practice the lines and movements needed for a play or film.
Broadcast—To communicate by way of radio or television.
Broadcast Messages— Information and ideas delivered by radio or television.

Figure 22-35. This outdoor set can be used in making movies and television shows. Notice the walkways at the top of the "buildings."

TechnoFact: There were many pioneers in filmmaking. Georges Méliès and Edwin S. Porter found that the speed of motion was related to the speed at which motion was filmed. Filming at a speed slower than the projection speed speeds up the motion. On the other hand, D. W. Griffith discovered people could easily understand lengths of dissimilar footage that had been edited or cut together.

Figure 22-36. This is an example of a storyboard for a television commercial. (Clorox Co.)

rehearsed and filmed, and the product is edited into its final form.

The major difference is in the way these messages reach viewers. Film producers expect people to pay to see their programs in a theater. Broadcast programming sells advertising time to enable "free" delivery of the product. The word *free* is in quotes because it is *not* free. We pay for radio and television programming every time we purchase an advertised product. Likewise, cable and satellite television channels exist through advertising and payments the cable and satellite companies make. When we sign up for cable or satellite services, we subscribe to certain channels. These channels are sold in packages of several channels per subscription tier. In a similar manner, we pay for part of the production cost of magazines and newspapers with our subscriptions, but we pay the rest of the costs through the advertisers.

Broadcast advertisements follow the design steps used for print advertising described earlier. The only difference is in the layout. Radio advertisement uses a script to "lay out" the ad. Television uses a *storyboard.* See **Figure 22-36.** The storyboard shows each shot for the advertisement. Also, the script is included.

Producing broadcast advertising follows the film model. The director and actors rehearse the advertisement.

TechnoFact: Advertising is used to make consumers aware of a product's features, price, or benefits. It also creates a brand image for the product. Consumers frequently buy the product, not only for its price and characteristics, but also because of the image associated with the brand.

Figure 22-37. A computer receives, processes, and stores information. (Tandy Co.)

Filming and editing then take place. Finally, time is purchased to present the advertisement on the broadcast station.

Most 60 second advertisements will be shot a number of times (sometimes into the hundreds) to get just the right effect. The ads are sponsors' ways to convince people to buy their products. A great deal of money is spent to produce and air the short messages. Therefore, sponsors want their messages to be nearly perfect.

Information Technologies

Many people call the times we live in the *Information Age*. This is because of the vast amount of information available. Today, we have access to more information than at any other time in history. The challenge is to access and use this information.

The primary technology allowing us to deal with large amounts of information is computer technology. It is based on the developments of the late 1900s. The computer is a processing machine. See **Figure 22-37.** This machine uses electronic parts and circuits to process information.

Computer systems have two basic elements: hardware and software. The *hardware* is the equipment used. It includes the computer, printers, scanners, and data storage devices. Software includes the instructions causing the computer to do specific tasks.

Computer Hardware

Computer systems include a number of different types of hardware. See **Figure 22-38.** There is the computer itself. Many computers are general-purpose information processing devices. We often call them *personal*

Discussion
Discuss the impact of information processing on people and society.

Brainstorm
Ask the students to develop a list of reasons why they think the present period in history is called the *Information Age.*

Tech Terms
Computer Hardware—The equipment making up a computer system.
Personal Computer—A computer and software used by one person at any given time.

TechnoFact: There are three basic types of information storage and retrieval systems. Document retrieval systems store entire documents; database systems store the information as a series of discrete records divided into discrete fields; and reference retrieval systems store references to documents rather than the documents themselves.

Resource
Reproducible Master 22-9: Parts of a Computer

Demonstrate
Show a computer system and identify its major parts.

Tech Terms
Global Positioning System (GPS)—A worldwide navigation system that uses information received from orbiting satellites to determine location.
Input Device—Devices that allow data and information to be entered into a computer.
Central Processing Unit—The part of a computer that performs operations and executes software commands.

Figure 22-38. Today, it is important to be able to use computers as tools. Computers are used to communicate, process data, and control machines. (FMC Corporation)

computers (PCs). These computers take information from a person, another device, or a network and process the information. The processed information is shown on a monitor, stored on a device, or sent to another location on the network.

In addition to PCs, there are mainframe computers that do the same tasks on networks. They process vast quantities of information for businesses, banks, and industrial companies. Also, there are specialized computers that do only one task. For example, a global positioning satellite (GPS) unit can only handle signals from GPS satellites. A computer in a washing machine can only control that

machine. A Gameboy® system is a specialized computer for playing games. See **Figure 22-39.** A computer has four basic parts:

➤ An *input device.* These devices allow for inputting information into the computer. They include keyboards, scanners, and network connectors. Input devices take the information and change it into electrical signals the computer can understand.

➤ A *central processing unit (CPU).* This is the "brain" processing the information. It follows the program (set of instructions) fed into it. The CPU oversees everything a computer does.

TechnoFact: The integrated circuit was developed in 1959. Robert Noyce developed a silicon-based circuit. Jack Kilby developed a circuit that used germanium for the base material. These scientists were working independently.

Figure 22-39. Here is a diagram of a computer system.

Figure Discussion
Discuss the computer system elements shown in Figure 22-39 and elaborate on each component.

Tech Terms
Memory—A location for the storage of data and information.
Random-Access Memory (RAM)—Computer memory that temporarily stores information.
Read-Only Memory (ROM)—Computer memory that permanently stores information.
Output Devices—Devices that provide information to the user or an external device.

➤ A *memory.* This is the place where the information is stored. A memory will also store the program of instructions. There are several specific types of memory in a computer. Two types are more important than the others:

➤ Random-access memory (RAM). This type of memory temporarily stores information with which the computer is currently working.

➤ Read-only memory (ROM). This is a permanent type of memory storage. It stores important data that does not change.

➤ *Output devices.* They receive and act upon information the computer provides. These devices include printers, monitors, and modems. A *monitor* is the primary device for displaying information from the computer. A modem is the standard method of connecting to the Internet. Other output devices are removable storage devices. They are used to store data. These devices include disc, CD-ROM, and DVD-ROM drives. Still other output devices include speakers, machines connected to the computer, and motors.

Computer Software

Computers need instructions to operate. For some computers, the instructions are simple and built into the system. For example, the computer

TechnoFact: Ted Hoff led a team that, in 1971, developed the first microprocessor, which is the entire computer central processing unit on one chip. The first chip had the same power as the original thirty-ton ENIAC computer that was built in the 1940s.

controlling a microwave has one set of tasks to perform. It uses simple input and output methods. A keypad is used to input instructions. A liquid crystal display (LCD) shows the output.

Operating Systems

Many computers, however, perform a variety of tasks. They need varying instructions and control. Programs, or software, provide these instructions. All desktop computers have a basic *operating system* overseeing all operations. Common operating systems are Windows® computer services, Linux® software, and MacOS® software. There are hundreds of other operating systems available for special purpose applications, such as mainframe computers, manufacturing applications, and robotics. An operating system does two things in a computer system:

➤ The operating system manages the computer system. This system controls the hardware and software of the computer system. It manages such things as the processor, memory, and disc space.

➤ The operating system provides a consistent way for application software to deal with the hardware.

Application Software

The instructions for the computer are called *programs* and *subroutines*. Another name for all of the programs and subroutines is *software*. The software tells the computer what to do and how to do it. The computer follows your instructions exactly. In doing so, it does something useful. This can include drawing a line,

implementing a word processor, checking a paragraph for spelling and grammatical errors, or balancing a checkbook. Without software, the computer cannot do complex tasks or solve problems.

Quality Control in Communication Systems

Quality control in communication systems ensures that the intended message reaches the receiver. In graphic communication, the production of the message follows acceptable standards for text and image preparation. Messages are often "proofed" prior to being sent to the receiver. Printing systems are monitored to ensure that the duplicates match the original message. Electronic communication uses industry-accepted standards and protocols to ensure that messages can be transmitted and received by appropriate devices.

Summary

Communication and information technologies are part of every person's life. We use graphic and electronic communication media daily. This media allows us to send or receive information. Information technology allows us to access and use vast quantities of data. Communication and information technologies inform us or cause us to take action. Without communication and information technologies, we would know little about the world around us.

Demonstrate

Show how software allows a computer to do specific tasks.

Tech Terms

*Software—*Computer programs and applications that can be run on a particular computer system.

*Operating System—*The master control program in a computer that maintains files, runs applications, and operates specific devices.

Figure 23-5. These are secondary manufacturing processes.

process is called *casting,* or *molding.* See **Figure 23-6.**

All casting and molding activities follow some common steps:

1. A mold is produced.
2. The material is made liquid.
3. The material is put into the mold.
4. The material hardens.
5. The finished item is removed from the mold.

If you have ever made an ice cube, you have produced a casting. The first step is to make or buy a mold. This mold has a cavity in it. The shape of

Figure 23-6. Here are casting and molding processes. Liquid material is poured into a mold, where it hardens.

TechnoFact: Early machines were fairly crude and produced inaccurate parts, causing failures and accidents. This led to a new type of machine called *machine tools.* These were the machines needed to build other machines.

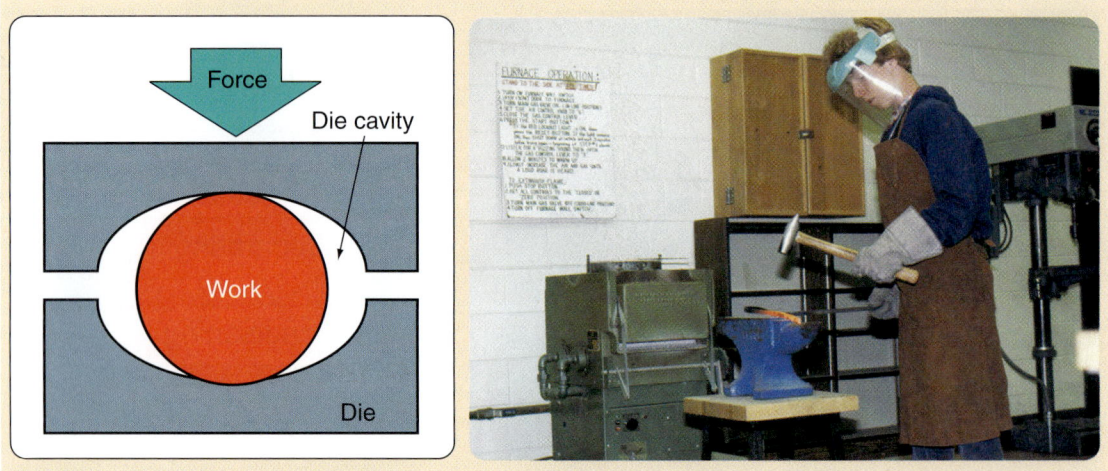

Figure 23-7. This is a forming process. Left—Here is a simplified drawing of a die. Right—Blacksmithing is a forming process.

the cavity gives the finished product its shape.

Some casting processes use a new mold for each product. These molds are usually made of sand or plaster. Such processes are called *one-shot mold casting,* or *expendable mold casting.*

Other casting processes use metal or plaster of paris (ceramic cement made from gypsum) molds. These molds can be used a number of times. This type of casting is called *permanent mold casting.*

Most materials cast are solids. They must be made liquid before they can be cast. Often, they are melted. Other materials, such as clays, are suspended in (mixed with) water.

The liquid material is then put into the mold. In some cases, the material is poured. Gravity draws the material into the mold cavity. In other cases, machines force the material into the cavity.

Once in the cavity, the material must solidify. Hot materials must be cooled to become hard. With other materials, the mold absorbs water to create a solid part. Some materials harden with a chemical action.

Finally, the hardened material must be removed from the mold. One-shot molds are broken away from the cast part. Permanent molds are opened. The finished casting is ejected from the mold.

Casting processes can form many metallic, plastic, and ceramic materials. Typical examples are automobile engine blocks, parts for plastic models, ceramic bathtubs and lavatories, and plaster wall decorations. Candy and other food products can also be cast into shapes.

Forming

Sometimes industrial materials are shaped using force. The materials are squeezed or stretched into the desired shape. These processes are called *forming.* See **Figure 23-7.** Forming includes bending, shaping, stamping,

TechnoFact: Many new machine tools were developed during the Industrial Revolution. These included boring machines in 1774, wood planers in 1802, gear cutters in 1820, grinders in 1834, metal shapers in 1836, and screw-cutting machines in 1880.

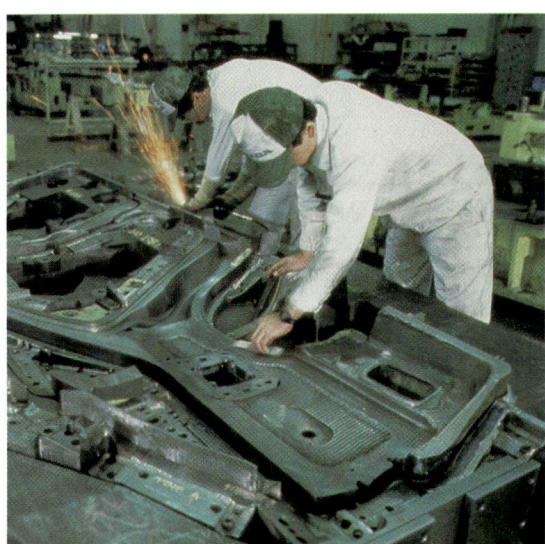

Figure 23-8. These workers are using forming dies to make automobile parts. (Honda)

and crushing. All forming processes require two things. They must have a shaping device and an applied force.

One forming process uses a shaping device called a *die.* This die is usually a set of metal blocks. Cavities are machined in them. Hot material is placed between the die halves. A hammer or press applies force. One die half moves toward the other half. As they close, the dies apply force on the material. This causes the material to flow into the die cavities. See **Figure 23-8.** This process is called *forging.* It is used to make hand tools, automotive parts, and other products requiring a high level of strength.

Another forming process uses a single shaped die or mold. The material may be forced into the die cavity, or it may be drawn around a mold. An example of this type of process is *thermoforming.* A plastic sheet is placed above a mold. The

sheet is heated. It is then lowered onto the mold. A vacuum is pulled in the cavity or around the mold. This causes the hot plastic to draw tightly to the sides of the mold. Thermoforming produces plastic parts of all shapes.

Rolls are also used to form materials. The material is fed between rotating rolls. This action stretches and squeezes the material into a new shape. This process, called *roll forming,* is used to make corrugated roofing and large tank parts. These are just three forming processes. There are many more that use force and shaping devices to form materials.

Separating

Some manufacturing processes shape material by removing excess stock. See **Figure 23-9.** The extra material is cut, sheared, burned, or torn away. These processes are called *separation.* They separate or remove the unwanted portion of the workpiece. This leaves a properly shaped part.

One type of separation is called *machining.* See **Figure 23-10.** This process uses a tool to cut chips of material from the workpiece. All machining requires motion between the tool and workpiece.

The tool may spin to make the chip. Drilling and many sawing operations use a rotating tool. In other cases, the work is rotated against a solid tool. Lathes use this action. In still other cases, the tool is drawn across the stationary work. The band saw and scroll saw (jigsaw) move the tool across the work to make the cut.

Standards for Technological Knowledge

19

Resource
Transparency 23-4: Separating

Extend
Have the students make an object with modeling clay. Present the activity as a forming process.

Tech Terms
Die—A device used in forming processes to cut out, form, or stamp a material.
Forging—Using pressure or hammering force to shape a material.
Separating—Manufacturing processes in which the material is given shape by removing excess material.
Machining—Separating processes that use cutting tools to remove material from a workpiece.

TechnoFact: The factory system provides large increases in per capita output through division of labor, automation, and high volume production. At the same time, it limits the creativity of individual workers.

Standards for Technological Knowledge

Demonstrate
Show the students how materials can be shaped using machining and shearing processes.

Figure Discussion
Use Figure 23-9 to present machining as chip removal.

Tech Terms
Shearing—
Manufacturing processes in which opposing edges or blades fracture a material to give it shape.
Flame Cutting—
Manufacturing processes in which the heat from burning gases melt away parts of a workpiece to give it shape.

Figure 23-9. These are separating processes. Saws, knives, or other tools cut away unwanted material.

A second separation action is called *shearing.* This process uses blades, which move against each other. The material is placed between the blades. Then, the moving blades apply force. This force fractures or breaks the material into two parts. Scissors, tin snips, and shears are shearing tools.

The third type of separation process is called *flame cutting.* Burning gases are used to melt away unwanted materials. Oxyacetylene cutting is an example of flame cutting.

Newer separating processes use beams of light, sound waves, electric sparks, and even jets of water to cut

Figure 23-10. This machining process incorporates the use of electromagnets to hold the workpiece in place. (Technomagnete)

TechnoFact: Elias Howe invented the lockstitch sewing machine in 1846. This was followed by Isaac M. Singer's 1851 invention of the first practical household sewing machine, which also became the first major consumer appliance.

Standards for
Technological
Knowledge

19

Resource
Transparency 23-5:
Conditioning

Resource
Transparency 23-6:
Assembling

Research
Have the students
research heat-
treating.

Extend
Differentiate
between hardening,
tempering, and
annealing as heat
treatment
processes.

Tech Terms
Conditioning—
Manufacturing
processes that use
heat, chemicals, or
mechanical force to
change the internal
structure of a mate-
rial, imparting it with
new properties.
Firing—
Conditioning
processes that use
heat to change the
properties of a
material.
Assembling—
Manufacturing
processes that use
adhesion, cohesion,
or fasteners to hold
parts together.

Figure 23-11. This scientist is experimenting with a laser that can be used for cutting materials. (Hewlett Packard)

away unwanted materials. Laser machining, ultrasonic (high sound) machining, and electrical discharge machining are examples of separating. See **Figure 23-11.**

Conditioning

A fourth type of secondary processing is *conditioning.* These processes alter and improve the internal structure of materials. This action will change the properties of the material. Conditioning may be done by heating or cooling, or it can be done with mechanical forces or chemical action. See **Figure 23-12.**

The most common conditioning activity is thermal (heat) conditioning. The material being conditioned is heated to make it harder, softer, or easier to use. Three major types of heat treatments are used on metals. These are hardening, annealing (softening), and tempering (removing internal stress).

Heat can also condition ceramic materials. This process is called *firing.* It involves slowly heating material to a very high temperature. The item is then allowed to cool slowly. During the process, a glass-like ingredient in the ceramic material melts. This coats the clay particles in the material. As the material cools, it becomes solid. The solid, glass-like materials bond the clay particles into a rigid structure. The result is a very hard, brittle product.

Assembling

Nails, screws, baseball bats, and combs are all one-part products, but most products are made up of several parts. The act of putting parts together is called *assembling,* or combining. Parts may be held together using

TechnoFact: France was the first country to manufacture parts that could be interchanged in factory assembly. A French gunsmith was the first to make muskets with interchangeable parts, eighteen years before Eli Whitney made his interchangeable-parts muskets.

Figure Discussion
Use Figure 23-13 to differentiate between permanent assembly (fusion) and temporary assembly (fasteners).

Brainstorm
Ask the students to list the products they use that were partly or wholly assembled by welding processes.

Tech Terms
Fastener—A mechanical device that holds parts together.
Welding—Using heat or pressure to fuse parts together.
Adhesion—Using a tacky substance to bond parts together.
Cohesion—Using the force of attraction between molecules to bond parts together.

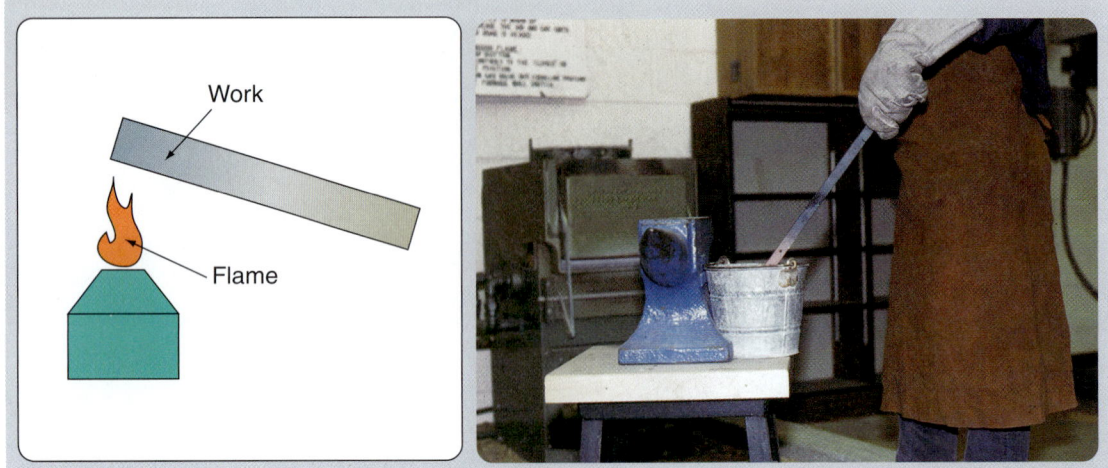

Figure 23-12. Here is a conditioning process. Metal is heated, then cooled, to make it harder.

mechanical fastening or bonding. See **Figure 23-13.**

Mechanical fasteners grip parts and hold them in place. Typical mechanical fasteners are nails, screws, rivets, nuts, bolts, staples, and stitches. Bonding permanently assembles parts together. This may be done either by fusion or by adhesive bonding.

Fusion uses cohesion. It uses the same forces that hold the molecules of the material together. The parts are melted at the joint area. This causes the materials to flow together. When the material cools, a bond is formed. The two parts become one. This assembling process is usually called *welding.* See **Figure 23-14.**

The second bonding technique uses an adhesive. An adhesive is a sticky substance holding the parts together. It is first applied to the separate parts. The adhesive must be able to attach itself to these parts. When the

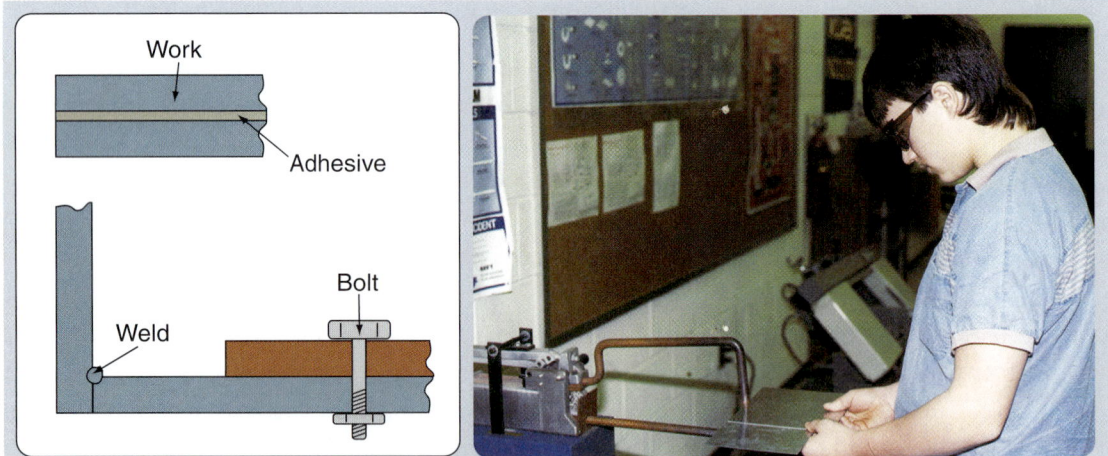

Figure 23-13. The assembling process fastens parts together.

TechnoFact: Although French arms manufacturers had already developed the use of interchangeable parts in mass production, the credit for first using interchangeable parts in manufacturing went to Eli Whitney in 1800. The system became known as "the American system."

Figure 23-14. The worker on the left is welding two parts together. The robot on the right is welding body parts of a car together. (Coachman Industries and Ford Motor Company)

Standards for Technological Knowledge

19

Resource
Transparency 23-7: Finishing

Extend
Discuss the difference between organic and inorganic finishes.

Research
Have the students identify some typical organic finishing materials and their properties and uses.

Tech Terms
Finishing—
Converting a material's surface or applying a coating to protect or beautify the material.
Electroplating—
Using electrochemical processes to deposit a metallic coating on a material.
Dipping—
Immersing a part in a finishing material to form a protective coating.

parts are placed together, an adhesive "bridge" is formed between them. The parts are fastened together.

Different materials require different adhesives. A wide range of adhesives is available to adhere almost any two materials together. Plastic trim parts are adhered to automobile bodies. The surface covering of aircraft wings is adhered to the wing structure. Wallpaper is glued to walls of homes and apartments.

Finishing

The final group of secondary manufacturing processes is *finishing*. This area includes all activities protecting and beautifying the surface of a material. See **Figure 23-15.**

A finish is usually a surface coat applied to the material. This coating can be an organic, metallic, or ceramic material. Organic finishes are plastic materials suspended in a solvent.

They are applied by brushing, spraying, rolling, or dip coating.

The finish dries when the solvent (thinner) evaporates. As this happens, the plastic material changes into more complex molecules. This new form produces a hard, uniform coat. The coat keeps water, oil, and other environmental elements from the base material. See **Figure 23-16.**

Organic finishes are called paints, enamels, varnishes, and lacquers. They provide an attractive, protective coating for metals and woods. Metals can be applied as a finish. In this process, a base material is coated with the metal. This may be done in several ways. The following are the most common:

➤ Electroplating. This is using electricity to deposit the metal on the part.

➤ Dipping. This is suspending the part in a vat of molten metal.

TechnoFact: One common process of coating metal to protect and beautify it is known as *silver plating*. It was developed in 1742 by a silversmith in Sheffield, England. This silversmith discovered that copper and silver would bond when pressed together between two rollers. Starting in 1762, factories were built to produce Sheffield plate objects. However, in 1840, the process of electroplating was patented, and the Sheffield plate process soon became obsolete.

Research
Have the students research the way steel is galvanized.

Extend
Discuss the advantages of coating steel with zinc, tin, or chromium.

Tech Terms
Galvanized Steel— Steel that has a zinc coating.
Tin Plate— Steel that has a coating of tin.

Figure 23-15. This is a finishing process. Paint can be sprayed onto a product. It protects the product and improves its appearance.

Chromium is a common metal coating material. The metal protects and provides a shiny surface. It is generally applied by electroplating the part.

Several other metals are applied by dip coating. Zinc coatings protect steel barn siding and roofing, garbage cans, and other steel products. The steel is dipped into molten zinc and then allowed to cool. This process is called *galvanizing.*

A similar process applies tin to steel. The result is a tin coated steel sheet. This material is widely used to make "tin cans."

Ceramic materials also make good coatings. Porcelain and glaze (a glass-like material) are often used. The coating material is applied to the part while it is cold. The part is then heated to melt and fuse the finish to the product.

Figure 23-16. The finish on this prototype automobile both protects and beautifies the automobile's surface. (Ford Motor Company)

TechnoFact: The Industrial Revolution made a major leap forward into a period called the *Age of Steel.* Steel is an alloy of iron that has a precise amount of carbon in it. It is stronger than cast iron, but was only made in small quantities until 1870. A new furnace, the Bessemer Converter, allowed large quantities of steel to be made relatively quickly and cheaply. The furnace removed excess carbon by blasting air into the molten iron.

Many ceramic products are finished with glaze. This material provides a colorful and water-resistant coating for dishes, planters, and other products. Porcelain is often used to coat metals and ceramic products. Some kitchen appliances are coated with porcelain enamel. Many bathroom fixtures also have a porcelain coating.

Manufacturing Systems

Secondary manufacturing processes are used to make products for everyday use, but they must be organized to be effective. They must be used as part of a manufacturing system. See **Figure 23-17.** There are four major types of manufacturing systems:

➤ Custom manufacturing.

➤ Intermittent and batch manufacturing.

➤ Continuous manufacturing.

➤ Flexible manufacturing.

Each of these systems is used today to make products. These systems all have advantages and disadvantages.

Custom Manufacturing

Custom manufacturing is the oldest system. In early history, one person made an entire product. This person had all the skills needed to process materials into products.

Most products of Colonial times were custom-made. Silversmiths made silver bowls and candleholders.

Figure 23-17. Here are the four major types of manufacturing.

Cobblers made shoes. Weavers made cloth. Tailors made clothing.

Later, custom manufacturing systems were used to make very special products. These products were designed for the customer. Only a few products were built to fill a specific need.

Today, spacecraft, ships, some cabinets and furniture, and clothing for special needs are custom manufactured. One person may make simple products, while many people work on complex custom-made products.

Intermittent Manufacturing

As the nation grew, custom manufacturing could not meet customer demand. There was a growing population and increasing wealth. People wanted more and better products. The

TechnoFact: A factory is a production facility characterized by standardized goods, interchangeable parts, wage-earning labor, the large-scale use of machinery, and the division of labor.

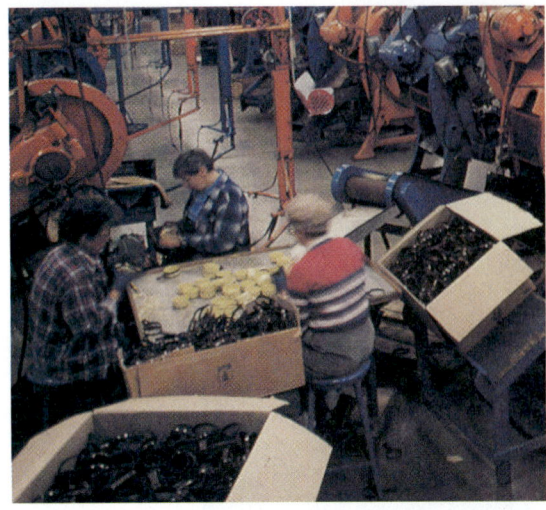

Figure 23-18. The metal stampings in the boxes were produced in job lots. (Ohio Art Co.)

skilled craft workers could not produce products fast enough.

Small factories were started. Products started to be made in small batches. A dozen or more candlesticks were made at a time. Several pairs of shoes were made in the same size. This system was called *intermittent manufacturing.*

Intermittent manufacturing is widely used today. See **Figure 23-18.** In this system, the parts for a product travel in a lot or batch. For example, suppose 100 birdhouses are needed. One part is the front. First, workers select lumber to make the fronts. Next, they move the boards to a saw. Here 100 birdhouse fronts are cut to length. These parts are put in a tray. The tray moves to a drill press. Workers drill the entry hole for the bird in all 100 parts. The tray of parts travels to another drill press. Perch holes are drilled in each piece. The parts finally move to another saw. Here, the roof peak is cut on all 100 parts. In the example, the parts moved from operation to operation in a batch.

Continuous Manufacturing

When many products are needed, *continuous manufacturing* is generally used. The parts move down a manufacturing line. At each station on the line, a worker completes a specific operation. Workers at each station are trained to do the job quickly. The product takes shape as it moves along the line. Completed parts flow to an assembly line. At this step, the parts are put together to form the finished product.

Flexible Manufacturing

A new system of manufacturing is called *flexible manufacturing.* See **Figure 23-19.** This system uses complex machines and computers for control. Flexible manufacturing can produce small lots like intermittent

Figure 23-19. This flexible manufacturing system uses robots performing several functions. (Kalb)

TechnoFact: An assembly line is a manufacturing technique used to efficiently assemble large numbers of uniform products. With the assembly line technique, a product is carried by mechanized means from station to station, where various assembly operations are performed. Henry Ford is credited with establishing the first assembly line, for his Model T® automobiles.

manufacturing, but it uses continuous manufacturing actions. Thus, flexible manufacturing is the way many modern products are built. It produces low cost products, as the products are needed.

Developing Manufacturing Systems

Manufacturing systems are developed for one purpose. They produce products to meet people's needs and wants. These systems have significantly increased the number of products available, while improving quality and cutting costs. The manufacturing process includes the designing, development, making, and servicing of products and systems. It includes the use of materials, hand tools, human operated machines, and automated machines. The development of a manufacturing system involves several actions:

1. Selecting operations needed to make the product.
2. Putting the operations in a logical order.
3. Selecting equipment to make the product.
4. Arranging the equipment for efficient use.
5. Designing special devices to help build the product.
6. Developing ways to control product quality.
7. Testing the manufacturing system.

Each of these elements contributes to efficient production of products. These elements help us use technology wisely.

Selecting Operations

Most decisions about manufacturing system design are based on product drawings. These documents describe the product that will be built. One of the first system design steps is to decide which operations are needed. This may sound easy. A hole is needed. What could be simpler? There are many options, however; should the hole be drilled, punched, cut with a laser, or produced with an electrical discharge machine?

Each feature of the product is studied. Tasks to be performed are listed. Then, methods for doing each task are selected. The result of this activity can be a set of operation sheets. Each sheet lists all the operations needed to make a part.

Sequencing Operations

After the operations are selected, a planner must put them in the proper order. The product must be built efficiently. Also, moving the product from workstation to workstation must be considered. Inspections must be scheduled. Plans must be made for storing parts and products until they are needed or sold.

Remember our example of the birdhouse fronts? The material for the front was first *cut* to length. Then, parts were *moved* to a drill press. A

Standards for Technological Knowledge

Extend
Discuss the selecting and sequencing operations for a simple task, such as changing a lightbulb.

Tech Terms
Manufacturing System—People, procedures, and equipment integrated to produce products efficiently.
Operation—A process or series of actions taken to complete a task.

TechnoFact: As industry moved into the Industrial Age, new technologies were developed, and new skills were required. The field of engineering included a number of specific skills that were growing in importance. One branch of engineering is industrial, or manufacturing, engineering, which designs and implements manufacturing systems.

Demonstrate
Work with the class to develop an operation (flow process) chart for making a simple product (for example, a paper airplane or a kite).

Extend
Discuss the importance of planning. Proper planning prevents poor performance.

Tech Terms
Operation—Actions taken to change the shape or size of the material.
Transportation—Actions taken to move parts or products from station to station.
Inspection—Actions taken to determine the quality of a part or product.
Storage—Actions taken to place a part or product in a safe place until needed.

Product Name Bird House - End		Flow begins Standard stock	Flow ends Finished part	Date 10-17
Prepared by: R.T. Wright	Section: R&D		Approved by: Deb	
Process symbols and no. used	◯ Operations 4 ▢ Inspections 2 ▭ Delays 0 ◻ Transportations 5 ▽ Storages 1			

Task No.	Process Symbols	Description of Task	Machine Required	Tooling Required
	◯●▷▢▽	Move material to saw	stock cart	
	●◯▷▢▽	Cut to length	radial saw	Stop #301
	◯●▷▢▽	Move to drill press	conveyor	
	●◯▷▢▽	Drill large hole	drill press #1	drilling jig 208
	◯●▷▢▽	Move to drill press	conveyor	
	●◯▷▢▽	Drill small hole	drill press #2	drilling jig 308
	◯▷◻●▽	Inspect		
	◯●▷▢▽	Move to circular saw	conveyor	
	●◯▷▢▽	Cut gable	circular saw	sawing fixture
	◯▷◻●▽	Inspect		
	◯●▷▢▽	Move to storage	stock cart	storage tray
	◯▷◻▢●	Store		
	◯▷◻▢▽			
	◯▷◻▢▽			
	◯▷◻▢▽			

Figure 23-20. This operation, or flow process, chart has been filled in for the sample birdhouse front.

hole was *drilled.* Again, the part *moved* to another drill press. There, another hole was *drilled.* At this point, the part was *inspected.* The quality and location of the holes needed to be checked. Then, the parts *moved* to another saw. There, the gable (pointed) end was *cut.* The finished part was *inspected.* Finally, the parts *moved* to a storage area. There, they wait for other parts so assembly can start. This simple example includes the following:

➤ Four operations (changing the shape or size of the material).

➤ Four transportations (moving parts from station to station).

➤ Two inspections (checking the quality of the part).

➤ One storage (placing the part in a safe place until needed).

All these words describe the order of operations, transportations, inspections, and storage acts, but sometimes words are hard to follow. Charts communicate better. *Flow process charts* contain the same information. See **Figure 23-20.** These charts are often used to design new manufacturing activities. They are also used to study old procedures. Studying them can result in finding a better way to make the part.

Selecting Equipment

Each operation requires equipment. See **Figure 23-21.** Saws cut lumber. Drill presses drill holes. Conveyors move materials between workstations. Racks are used to store materials and parts.

For each manufacturing system, equipment must be provided. The

Figure 23-21. Look at the equipment on this production line. You can see material handling equipment (conveyors). There is processing equipment (the shaping machine) on the left, with power controls.

TechnoFact: Different types of workers are needed in a manufacturing facility. These include skilled workers, semiskilled workers, and unskilled workers. Skilled workers, such as toolmakers, millwrights, or carpenters, possess a broad, recognized skill in a specific area. Semiskilled workers possess a high degree of skill, but in a narrow range of work. Unskilled workers do routine work that requires little training and experience.

Figure 23-22. This is a plant layout drawing for the birdhouse front production line. See how handy both the beginning and the end of the line are to the stockroom.

flow process chart helps engineers determine the equipment they need for the process. Some equipment may already be owned. Other items will be purchased.

Arranging Equipment

The equipment must be arranged for production. Sometimes the machines will be used for a number of different products. This is true with intermittent manufacturing activities. In this type of manufacturing, like equipment is grouped together. A roughing department may be formed to contain all saws, jointers, and surfacers. A finishing department may be equipped to paint several different products.

In other cases, the equipment is set up to make only one product. A continuous manufacturing system is designed. In this system, the equipment is arranged to make the selected

product. In our example, the line to make our birdhouse fronts would contain the following equipment: a saw, a drill press, another drill press, and another saw. See **Figure 23-22.**

Designing Tooling

Many times, special devices make manufacturing more efficient. These devices may hold a part so the part can be machined. See **Figure 23-23.** Some devices may hold several parts for welding, or they may be special dies for forming the material. All these items are called *tooling.*

These items are designed to make operation more efficient. They should make the operation faster, easier to complete, and safer. For our birdhouse front, several pieces of tooling can be used. They might include devices to do the following:

➤ Hold the part so the entry hole is always drilled in the same place.

➤ Hold the part so the perch hole is correctly placed.

➤ Hold the part so the gable ends are accurately cut.

Controlling Quality

Product *quality* is a major concern in manufacturing. Everyone wants products that work well and look good. Therefore, parts and products must be checked for quality. This action is called *inspection.* See **Figure 23-24.** The parts are compared to the drawings. Each part and product must meet the standards designers and engineers set.

Figure Discussion
Have the students suggest ways to rearrange the equipment in Figure 23-22 so three people can do all the work.

Tech Terms
Tooling—A special holding device or tool that increases the speed, accuracy, or safety of an operation.
Quality—The degree of fineness built into the product.
Inspection—Assessing the quality of materials, parts, and products.

Brainstorm
Have the students think of ways to encourage workers to take a personal interest in product quality.

Tech Terms
Scrap—Material and parts that fail to meet quality standards and cannot be reworked or reused.
Rework—Altering parts and products that failed to meet quality standards so they meet those standards.

Figure 23-23. Note the simple drilling tooling on the left and the assembly fixture on the right.

Parts failing the inspection may be scrapped. Other parts might be reworked. The defect may be removed from the part. Scrapping and reworking parts add cost to the product. Therefore, every effort is made to make the product right the first time.

Quality control efforts also encourage workers to build good products.

These efforts reduce scrap and waste. They also help ensure customers receive quality products.

Testing the System

The last step in manufacturing system design is testing the system. The parts of the system must be put

Figure 23-24. These technicians are testing electrical wiring components. (AMP, Inc.)

TechnoFact: Among the number of different quality control methods and approaches, the Taguchi method, developed by Dr. Genichi Taguchi in the 1980s, has become one of the more popular. This system employs quality engineering techniques, including both statistical process control (SPC) and quality-centered management techniques.

Figure 23-25. Can you identify the resources being used in this manufacturing line?

in place. The machines must be positioned. Conveyors and other material handling devices must be installed. Tooling must be attached to the machines. Then, the system can run.

People produce test products using the manufacturing system. This is called a *pilot run.* Engineers check to see that the operations are working correctly. They observe the flow of material. Also, product quality is carefully checked.

A pilot run is important. It shows where changes are needed. This run may identify tooling needing to be improved. It may indicate that the equipment should be reorganized. The pilot run may suggest that the workstations need to be relocated.

After the changes are made, engineers make another pilot run. Changes continue until the system is operating properly. Only then will full-scale production start.

Producing Products

Manufacturing of products requires all the major types of resources. People use information and machines, which are powered by energy, to change materials into products. See **Figure 23-25.** In general, machines, many of which are computer controlled, are capable of making higher quality goods than a skilled craftsperson can do alone.

Most manufacturing is managed. A group of people sees that the system

Standards for Technological Knowledge

1 19

Extend
Have the students develop their own set of Burma-Shave® shaving cream signs. See the TechnoFact at the bottom of this page.

Tech Terms
Planning—Setting goals and determining a course of action to reach them.
Organizing—Dividing major tasks into individual jobs.
Actuating (Directing)—Assigning jobs to workers and supervising their performance.
Controlling—Comparing the results from the system to the original plan.
Marketing—The process by which a product or service is priced, promoted, and distributed to customers.
Advertising—A mass communication message designed to promote an idea or the sale of a product or service.

runs properly. These people are called *managers*. They make up management.

Managers do not make the products. They organize the systems so the products are made efficiently. In doing this, they complete the four basic functions:

1. Plan. The managers set goals.
2. Organize. The managers divide tasks into jobs.
3. Actuate. The managers assign jobs and supervise workers.
4. Control. The managers compare the results to the plan.

Good managers get work done through other people. They provide direction and support. If managers do their jobs well, workers can more easily make products.

Marketing Products

Today, there are thousands of products available to each of us. These products must be marketed. Potential customers must be told about them. Generally, *advertising* and marketing do this task. See **Figure 23-26.**

When the idea for a product or service is being developed, the inventors create a marketing plan. The marketing plan specifies the intended customers, expected sales, and possible advertising and sales methods.

Figure 23-26. Magazines and catalogs are good places to advertise.

TechnoFact: Advertising takes all forms. Burma-Shave shaving cream signs were one form that was dear to the heart of motorists. These were six signs along the road that had a cute saying and ended with the product name: Burma-Shave. They first appeared in 1927 and were last seen in 1963. At their height of popularity, there were seven thousand Burma-Shave shaving cream signs throughout America. This is an example of a set of Burma-Shave shaving cream signs:
"A guy...Who drives...A car wide open...Is not thinkin'...He's just hopin'...Burma-Shave."

Marketing a product means telling the public about it, as well as assisting in selling and distributing it. It involves gauging what the public wants and then advertising and selling products to consumers.

Ads are delivered in several ways. Some are placed in magazines and newspapers. Others are aired on radio and television. Still others are delivered through the mail or displayed on billboards and signs.

Advertisements make people act. They bring customers to stores. There, the sales effort takes place. Salespeople encourage customers to buy the product.

Salespeople sell expensive products, like automobiles and computers. They present the value of the products to the customer. Less expensive products (such as toothpaste and colas) are simply displayed. The customers' actions are often based on the advertising effort, the appearance of the package, or previous experience.

After a product or system is sold or leased, it sometimes needs to be serviced. Servicing, or providing support after a sale has been made, is an important part of the manufacturing process. These services can include installing, troubleshooting, maintaining, and repairing.

Summary

Manufacturing provides all the products we use. The food we eat, the clothes we wear, and the vehicles we travel in have all been manufactured.

Manufacturing generally includes primary and secondary processing. Primary processes convert raw materials into industrial materials. Secondary processes change industrial materials into usable products.

To meet the large demand for products, complex manufacturing systems have been developed. They let people efficiently make things we need and want. The major types of manufacturing systems are custom manufacturing, intermittent and batch manufacturing, continuous manufacturing, and flexible manufacturing. Custom manufacturing is used to make specialized products designed for a specific customer. In intermittent manufacturing, the parts of a product travel from operation to operation in a batch, so more products can be produced in a short amount of time. Continuous manufacturing uses an assembly line to produce many products quickly. Flexible manufacturing uses continuous manufacturing actions, but it can be used to produce small batches of a product, much like intermittent manufacturing.

There are several steps involved in developing a manufacturing system. To produce products efficiently, operations must be selected and sequenced, equipment must be selected and arranged, tooling must be designed, quality must be controlled, and the system must be tested. Managers assist in efficient production by providing direction and support. After the products are produced, they must be marketed to potential customers.

Standards for Technological Knowledge

Extend
Show the students a number of advertisements and discuss the audience and message for each one.

Tech Terms
Sales—The process of exchanging products for money.
Distribution—Moving the product from the point of manufacture to the point of sale.

Curricular Connections

Social Studies

Investigate the products manufactured in your city or state. Try to determine why these, and not other products, are made there.

Mathematics

Select a simple product. Determine the geometric shapes it contains. Measure and record these shapes in standard and metric units.

Science

Select a manufacturing process, such as sawing or injection molding. Investigate the scientific principles used in the process. Prepare a poster display explaining these principles.

Social Studies

Select a manufacturing job and determine the education or training needed for it.

Standards for Technological Knowledge

3

Activities

1. Select a simple product, then list the types of processes used to make it (for example, casting and molding, forming, separating, conditioning, finishing, and assembling).

2. Select a simple product, such as a bookend, then list the following:
   A. The operations you would use to make it.
   B. The equipment or tools you think you would need.
   C. The points you would check for quality.
   D. Any special tooling you think you would need.

3. Suppose you were to drill a hole in the center of a 4″ x 4″ piece of wood. Sketch the piece of tooling you would use to make the hole accurately in 100 parts.

4. Working in a small group, prepare a marketing plan for an idea, a product, or a service of your choice. Developed advertising flyers using desktop publishing software, and develop a web page for your product.

TechnoFact: Advertising can turn an inventor's name into a household word. Consider King C. Gillette, who invented the safety razor with throwaway blades in 1895, or J. Schick, who invented the electric razor in 1928.

Test Your Knowledge

Do not write in this book. Place your answers to this test on a separate sheet of paper.

1. Manufacturing changes _____ into useful _____.

2. There are two types of manufacturing:

   A. Changing raw materials into industrial materials is called _____.

   B. Changing industrial materials into usable products is called _____.

3. Match the terms and descriptions. Give an example of each.

   _____ Cuts materials.　　　　　　　　　A. *Chemical processing.*

   _____ Heats materials to change them.　B. *Mechanical processing.*

   _____ Uses chemicals to make new materials.　C. *Thermal processing.*

4. Name the six secondary manufacturing processes.

5. Identify the processes described below:

   A. Uses hollow forms to shape liquid materials. _____

   B. Forces or squeezes materials into new shapes. _____

   C. Cuts away excess stock by burning, shearing, or cutting. _____

   D. Alters interior structures of materials. _____

   E. Fastens parts together by any means. _____

   F. Protects or beautifies the outsides of products. _____

6. Describe each type of manufacturing system:

   A. Custom.　　　　　　　C. Continuous.

   B. Intermittent.　　　　　D. Flexible.

7. Flow process charts are sometimes used to show the order of operations. True or false.

8. Tooling is (select all correct answers):

   A. A device that holds a part while it is being manufactured.

   B. Designed to make manufacturing more efficient.

   C. Sometimes a special shape for forming the material.

9. Label the following basic managerial functions:

   _____ Setting goals.

   _____ Dividing tasks into jobs.

   _____ Assigning jobs and supervising workers.

   _____ Comparing the results to the plan.

10. Finding customers and telling them about a new product is known as _____.

---

8. A. A device that holds a part while it is being manufactured.
B. Designed to make manufacturing more efficient.
C. Sometimes a special shape for forming the material.

9. Planning
Organizing
Actuating
Controlling

10. advertising

Answers to *Test Your Knowledge* Questions

1. raw materials, products

2. A. primary processing
B. secondary processing

3. B. *Mechanical processing.*
C. *Thermal processing.*
A. *Chemical processing.*
Examples will vary. [cutting lumber (mechanical), making copper (thermal), and refining aluminum (chemical)]

4. Casting and molding, forming, separating, conditioning, assembling, and finishing.

5. A. Casting and molding
B. Forming
C. Separating
D. Conditioning
E. Assembling
F. Finishing

6. A. An entire product is made by one person, or a few very special products are designed for the customer.
B. Products are made in small batches, and they travel from operation to operation in a batch.
C. Parts move down a manufacturing line to an assembly line, where they are put together to form the finished products.
D. Small lots are produced, but continuous manufacturing actions are used.

7. True

![TSA] Modular Activity

This activity develops the skills used in TSA's Manufacturing Challenge event.

Manufacturing Challenge

Activity Overview

In this activity, you will obtain discarded materials from local business or industry, and design products that can be developed with those materials. You will design a manufacturing process to produce the product, perform market research for the product, and prepare a comprehensive report.

Materials

➤ Discarded materials from local business or industry (will vary)
➤ Paper
➤ Pencil
➤ Three-ring binder
➤ Computer with CAD and word processing software

Background Information

Obtaining materials. Contact local businesses to obtain scrap material. You can write letters, make telephone calls, or visit businesses in person. Your local telephone book will provide many leads.

Design. After obtaining materials, use brainstorming techniques to develop a list of potential products. Create rough design sketches for some of the products. Consider the available tools and materials when selecting the best idea.

Drawings. Create working drawings using a CAD system. Use multiview drawings with as many views as needed to fully describe the product. If your product is small enough, create the drawings at full scale.

Prototype. Produce a prototype of the product. As you develop the prototype, consider the most effective manufacturing sequence. Are there design changes that could improve the manufacturing process? If design changes are made as you develop the prototype, be sure to update the drawings to reflect those design changes.

Production plan flowchart. Develop a production plan flowchart that illustrates the manufacturing sequence for the product.

Manufacturing. Manufacture several units of your product using the production process. Make a trial run first, and adjust the flowchart as needed to improve efficiency. Take photographs to document each step of production.

Marketing. Develop a marketing plan, including an advertisement. Who may want to purchase the product? What is a reasonable price for the product?

Guidelines

In the course of this project, you must develop a report containing the following items:

➤ Cover page
➤ Contents
➤ Written description of product
➤ Print advertisement for product
➤ Design sketches
➤ Working drawings
➤ Materials list
➤ Tools and machine list
➤ Production flow chart
➤ Photographs of the manufacturing process with written explanations for each image
➤ Letters of donation from the businesses supplying materials

Evaluation Criteria

Your project will be evaluated using the following criteria:

➤ Report
➤ Quality of product
➤ Creativity in design and use of materials

Chapter 24
Medical Technology

Did You Know?

➤ One of the most important diagnostic tools in medicine is the stethoscope. It is an instrument physicians use to detect sounds in the body, such as a heartbeat. This tool was invented in 1819.

➤ The first completely artificial heart was installed in a person in 1982. The device was the Jarvik-7™ heart.

➤ The first inoculation for smallpox was developed in China in about 600 B.C.

➤ A British physician, Edward Jenner, made the first major advancement in modern times for preventing smallpox. In 1796, he discovered people could be immunized by inoculating them with material from cowpox sores.

➤ Louis Pasteur created the first successful vaccine against rabies in 1885.

➤ In 1954, Jonas Salk introduced an injectable vaccine for poliomyelitis (polio). Later, Albert Sabin developed an oral vaccine for the disease.

Objectives

The information given in this chapter will help you do the following:

➤ Relate some major points in the history of medicine and health care.

➤ List and describe the three major goals of health and medicine.

➤ Explain the two major thrusts of illness and injury prevention.

➤ Describe the role of technology in prevention programs.

➤ Summarize the function of vaccines in illness prevention.

➤ Explain the two major things health care professionals can do to help ill or injured people.

➤ Give examples of ways technological devices are used in diagnosing illnesses and physical conditions.

➤ Summarize how technological devices are used to treat illnesses and physical conditions.

Key Words

These words are used in this chapter. Do you know what they mean?

clinical testing

computerized tomography (CT) scanner

diagnose

disease

drug

electrocardiograph (EKG)

endoscope

immunization

inoculate

intensive care

magnetic resonance imaging (MRI)

pathologist

prevent

radiology

surgery

treat

ultrasonic

vaccination

vaccine

wellness

X-ray

Chapter Outline
Did You Know?
Objectives
Key Words
Goals of Health and Medicine
  Prevention Programs
  Dealing with Illness and Injury
  Using Technology in Medicine
Summary
Curricular Connections
Activities
Test Your Knowledge

Standards for Technological Knowledge

14

Research
Have the students research the ways Egyptians treated illnesses.

Tech Terms
Medicine—The science and technology used in treating and preventing disease.

Diseases have been the concern of people over the ages. Early humans blamed diseases on demons. Treatments were based on magic and folk remedies. As societies progressed, treatments became mixes of magic and rational approaches. Treatments for some diseases of the skin and eyes were developed first because these problems were visible. Internal disorders continued to be treated with magic-based approaches.

By about 2600 B.C., physicians were a recognized part of society. These people practiced an early form of medical science. They used simple technological devices and instruments to do their work. Over the years, the field of medicine slowly developed. Hospitals started to be built and used during the Middle Ages in Europe. They treated people with many diseases. Hospitals were especially important during the large epidemics of bubonic plague, leprosy, and smallpox that swept the continent.

An important milestone in medical history happened in the seventeenth century. At that time, scientists discovered that blood circulates in the human body. An English physician named William Harvey established that the heart pumps the blood in continuous circulation. Marcello Malpighi's discovery of tiny blood vessels, called *capillaries*, followed Harvey's discovery. Other important work investigated the brain and nervous system. Also, new knowledge of the liver, muscles, and heart was discovered. These and many other discoveries led to a new age in medicine.

Great advances in diagnosis and treatment of diseases were made during this age. The discovery of germs and their role in diseases was established. This led to the discovery of the causes for major diseases, such as anthrax, diphtheria, tuberculosis, and plague. Techniques to stop these diseases through vaccines were developed.

New surgical methods were also developed. Aseptic (sterile) surgery became common. Physicians started using sterilized instruments and techniques to avoid infecting patients. By the mid-1800s, anesthesia was being used in surgery.

By the end of the 1800s, new tools, such as X rays and ultraviolet lamps, were developed to diagnose and treat illnesses. Huge advances in medicine have occurred since 1900. They have helped to greatly increase the average person's life expectancy. Longer life has given medicine new challenges. People who live longer suffer from higher rates of heart disease, cancer, and stroke. These conditions have replaced infectious diseases as the leading causes of death. This has given rise to new treatments and the need for new technologies. Progress and improvements in medical technologies are used to improve health care. The use of new medicine and technology helps people live healthier and less painful lives.

Goals of Health and Medicine

There are three major goals of health and medical programs. See **Figure 24-1.** First, people use technology in health and medicine to help

TechnoFact: Prior to 1300, medicine was basically an academic discipline. Doctors used the logic methods proposed by Aristotle to deal with illness. This all changed with a cruel turn of history: the Black Death. Bubonic plague was killing huge numbers of people and could not be treated with logic and traditional medicine. New answers were sought through empirical observation and trial and error treatments. Medical scientists turned to dissecting the dead to gain insight into how the body functions.

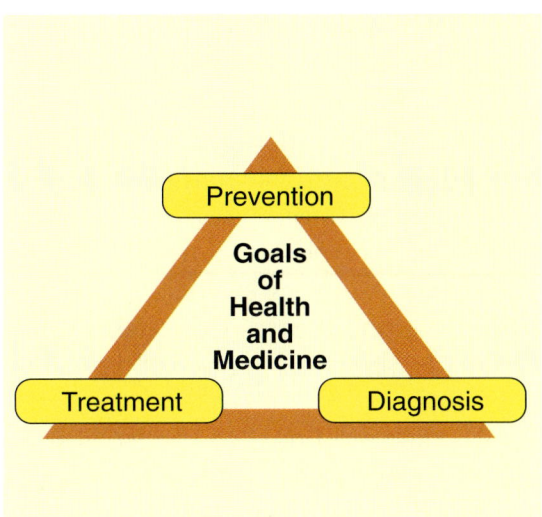

Figure 24-1. These are the goals of medical science and technology.

prevent illness and injury. Second, they use technology to diagnose diseases and injuries they think they have. Finally, they use technology to treat illnesses and injuries they could not prevent.

Prevention Programs

There is an old saying that "an ounce of prevention is worth a pound of cure." This means it is much better to *prevent* an injury or disease than to try to cure one after it occurs. There are two major ways people use technological products to help prevent illnesses and injuries. The first is wellness, and the second is vaccination, or *inoculation.*

Wellness Programs

People are considered being well when their bodies are in good health. Therefore, *wellness* can be described as a *state of personal well-being.* Wellness programs have people do things that help keep their bodies healthy. These actions can be considered preventative medicine. They involve at least four major factors. See **Figure 24-2.** These factors are nutrition and diet, environment, stress management, and physical fitness.

Wellness programs stress that people must be concerned with what they eat. These programs emphasize proper nutrition as an important factor in personal health. Wellness programs focus on the quantity and value of the food people eat. They are concerned with the intake of vitamins, fats, proteins, and other life-sustaining food components. Technology is used to preserve and improve the nutritional value of foods. It is used to process farm products into food we eat. See **Figure 24-3.**

These programs stress that people must be aware of the environment in which they live. They emphasize the need to control and improve the quality of the air we breathe. Wellness programs indicate the hazards associated with excessive exposure to

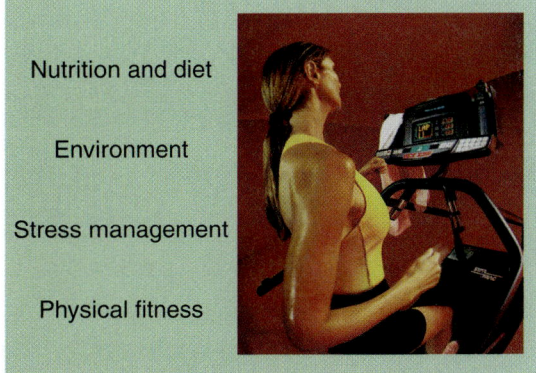

Figure 24-2. Here are the four areas of wellness programs. (StarTrac).

Nutrition and diet

Environment

Stress management

Physical fitness

Resource
Reproducible Master 24-2: Wellness

Figure Discussion
Explain how the three goals presented in Figure 24-1 can be applied to a person.

Tech Terms
Wellness—A state of well-being.
Wellness Programs—Programs that encourage well-being through nutrition and diet, stress management, physical fitness, and a healthy physical environment.

TechnoFact: The Black Death was a disease that caused major change in society. By the time the plague began to recede in 1430, Europe had lost three-quarters of its population. The plague changed life for those who survived. It undid much of the feudal system that had oppressed people, and it left behind many wealthy survivors. Because of the smaller population, manual labor became precious, wages rose rapidly, and the mechanical clock was developed to keep more precise track of time.

direct sunlight. Technology can be used to deal with health hazards in the environment. It can be used to purify water and clean the air. Technology has been used to develop sunscreen lotions and sunlight filtering clothing.

A wellness approach suggests that people must be aware of the health damage emotional stress can cause. People should keep their bodies fit through activity and exercise. Technology has also been applied to stress management and physical fitness through two major areas: exercise and sports.

Technology, Wellness, and Exercise

Exercise requires that people exert their bodies to improve their health. People do two major types of exercises.

First, there is anaerobic exercise, which involves heavy work using only a few muscles. Examples of anaerobic exercise are weight lifting and sprinting. This type of exercise is maintained for short intervals of time. It is done to increase strength and muscle mass. Anaerobic exercises have limited benefits, however, to cardiovascular health, which is the major goal of wellness exercises.

The second type of exercise is aerobic exercise. Typical aerobic exercises are walking, jogging, and swimming. These exercises use a large number of different muscles. They use a lot of oxygen to keep the muscles moving continuously. Aerobic exercises place demands on the cardiovascular and respiratory systems to supply oxygen to the working muscles. Aerobic exercise

Figure 24-3. Technology is used to process the food we eat. This food helps us stay healthy. (U.S. Department of Agriculture)

TechnoFact: Cortez established the first European-style hospital in North America in 1521. This hospital was located in Mexico City and was available to Spaniards and Indians alike. The University of Mexico was founded thirty years later, and by 1579, it had started a medical school. This school graduated medical scholars long before French or English settlers arrived in the New World.

Figure 24-4. This is a typical treadmill used in exercise programs. (SportsArt)

➤ Stationary bikes. These are non-moving bicycles allowing people to obtain the benefits of bicycling without leaving home. The handlebars move and provide resistance to provide an upper body workout, as well as the lower body workout. See **Figure 24-5.**

➤ Stair climbers. These machines are devices allowing people to obtain the benefits of climbing without using stairs. They provide lower body workouts. See **Figure 24-6.**

➤ Rowing machines. These machines simulate actual rowboats with oars. They allow people to simulate rowing a boat with oars, using their arms and legs.

reduces the risk of heart disease and increases endurance.

Aerobic exercise can be done without special equipment. People can walk or jog to improve their health. They can swim laps in a pool or lake. Even these activities, however, often require technology. Special shoes and clothing have been designed and produced to help in these activities. Swimming pools and walking paths in parks are built using construction technology.

Also, many people use exercise equipment to improve their health and well-being. This equipment is a result of technological design and production activities. Typical exercise equipment includes the following:

➤ Treadmills. These are moving belts allowing people to walk or jog in place to provide an aerobic fitness workout. See **Figure 24-4.**

Figure 24-5. This stationary bike can work many muscle groups. (StarTrac)

TechnoFact: Most early medicine was crude by present standards. It was said that many hospitals did more harm than good. Before the development of anesthetics, surgeons rushed through operations because the patient was in great pain. Also, many people died of blood poisoning because of unsterile operating conditions.

Tech Terms
Home Gym—A multi-station exercise machine allowing people to work on many different muscle groups.
Sports—Games or contests (played individually or in teams) involving physical strength, skill, and endurance.
Contact Sport—A game involving bodily contact among players as part of the usual play.
Venue—The location of a sport meet or event.

Figure 24-6. This climber is an exercise machine simulating the work required to climb stairs. (StarTrac)

➤ Home gyms. These are multi-station exercise machines allowing people to work on many different muscle groups.

All these machines have sensors and other technological devices to help people stay fit. These devices measure and display information, such as speed and heart rate. They may graph the amount of energy used and the effects on the body. See **Figure 24-7.**

Technology, Wellness, and Sports

Sports are another way to promote wellness. They are games or contests involving skill, physical strength, and

endurance. Sports may be played as an economic activity, where players are paid, and fans pay admission fees. This discussion, however, will deal with sports for the sake of fitness. It will explore sports played for personal health and enjoyment.

All sports require technological products. These products may be constructed playing fields, or *venues*. These venues are part of our built environment. See **Figure 24-8.** The playing fields are the results of construction technology, which was presented in Chapter 17. For example, if you use softball for a fitness activity, it requires a specific playing venue. Softball needs a field with bases and probably a backstop. In contrast, if golf is the fitness sport, a course with holes and traps is required.

Often, technology is used to improve the natural playing surfaces for sports. For example, special grasses have been developed for playgrounds and playing fields. Fertilizers have been developed and manufactured to

Figure 24-7. Many exercise machines have technological displays providing information about the effects of exercising. (StarTrac)

TechnoFact: Holistic medicine is a health care approach promoting health and wellness through a whole body approach. In the holistic approach, a person's body, mind, spirit, and emotions must be in balance with the environment to maintain wellness. The holistic approach encourages people to take control of their own well-being by developing a healthy lifestyle: eating properly, avoiding excessive alcohol consumption, reducing stress, exercising, and not smoking.

Figure 24-8. A basketball court is an example of a typical venue for sports.

encourage grass to grow. Special lawn grooming and moving equipment has been developed to maintain the fields. Likewise, snow grooming equipment prepares and renews the surfaces of ski runs. Snowmaking equipment has been developed to supplement natural snow for skiing. These are just a few examples showing how technology has helped develop and maintain sport venues.

Most sports require manufactured *equipment* and *personal protection*. Each game or contest, however, uses specific technological products. The game equipment for baseball (such as a baseball, bats, and gloves) is different from the game equipment for golf (for example, clubs and golf balls). Similarly, the personal clothing and protection equipment for baseball is different from the clothing and protection equipment for golf.

Players in many sports wear special clothing and protective gear. Special shoes may be developed to provide foot support. They may absorb the shock running on hard surfaces causes. The clothing may be made from fabrics shedding rain and snow. It may maintain body heat or wick away perspiration. Participants in some sports wear protective gear to reduce the chance of injury. For example, baseball players wear batting helmets to protect their heads from wild pitches. The game equipment and protective equipment are technological products. They are designed and manufactured for specific purposes.

TechnoFact: Physical fitness is a combination of health and physical development. The goals of a physical fitness program are to make participants healthier and stronger and to improve their physical skills and endurance. Physical fitness programs use diet, exercise, sufficient sleep, and medical supervision to accomplish these goals.

Research
Have the students determine the diseases for which vaccines have been developed. Have them research the way technology is used in developing vaccines.

Tech Terms
Vaccine—A preparation of a diluted or destroyed pathogen that starts antibody production, but will not contaminate the receiver.
Vaccination—Protection from a particular disease through the administration of a vaccine.
Booster Shot—An additional dose of vaccine administered a predetermined time after the initial vaccination to maintain the patient's immunity.

Figure 24-9. Inoculations are given using technological devices.

Prevention through Immunization

Healthy bodies can go only so far in preventing illnesses. Another approach is called *immunization,* or *vaccination.* *Immunization* is the process of systematically vaccinating people through a series of shots to prevent disease. See **Figure 24-9.** This process is the same as inoculation and is the common method used to promote natural resistance to specific diseases. Over history, people have developed two different approaches to immunization. The first is passive immunity. This approach injects blood from an actively immunized person or animal into a patient. The other system is active immunization. In this system, a disease-causing microorganism or a product of the microorganism is injected into the body. This approach uses modified or killed bacteria or viruses to create protection.

Vaccines are special drugs created to prevent diseases from affecting people. They do not cause disease. Instead, they cause the body's immune system to build a defense against a disease. This defense allows the body's immune system to immediately respond to a particular disease.

More than 50 vaccines for preventable diseases have been developed. They are designed to begin with birth and are available throughout a person's lifetime. How they are developed will be discussed later in the chapter.

The vaccines developed for use in immunization require specific equipment to sustain settings in which ample quantities of the vaccines can be created. The technological system designed to produce the right environment in which a vaccine may be cultured is vital to the success of the

TechnoFact: In 1952, Jonas Salk developed the first successful polio vaccine. At the time, over fifty-two thousand Americans contracted the disease each year, and about three thousand died from it. In 1961, the oral live virus vaccine developed by Albert Sabin was recommended for combating the disease.

large amount of the vaccine required for immunization. Increasing the production of a vaccine requires understanding how an organism is modified to produce a vaccine and how a vaccine works. It also requires addressing the quantity needed for all concerned and providing sufficient resources for proper production of the vaccine.

Dealing with Illness and Injury

The second focus of health and medicine deals with people having diseases or injuries. This area of medicine involves diagnosing and treating these diseases and injuries. Its goal is to reduce human suffering and physical disability.

A *disease* can be described as any change interfering with the normal functioning of the body. Treating diseases and injuries requires a number of different health care professionals. These professionals include *physicians,* who diagnose and treat diseases and injuries. They also include *nurses,* who help physicians in their work, and *medical technologists,* who gather and analyze specimens to assist physicians in diagnosis and treatment. *Dentists* are health care professionals who diagnose, treat, and help prevent diseases of the teeth and gums. *Dental hygienists* assist dentists in surgery and clean teeth. Also on the health care team are *pharmacists,* who dispense prescription drugs and advise people on the drugs' uses.

This team of health care professionals uses technology to make their work more effective. Many people seek medical care because they are ill or injured. To help these people, health care professionals do two major things:

➤ *Diagnose.* The professionals determine medical problems. They try to establish why the people are ill or what injuries they have. Diagnosis is done using interviews, physical examinations, and medical tests. See **Figure 24-10.**

➤ *Treat.* Health care professionals use medical procedures to cure diseases, heal injuries, and ease symptoms. Treatment can involve the use of surgery, drugs, or other procedures.

Using Technology in Medicine

Diagnosing and treating illnesses and injuries involve tools and equipment. Physicians and dentists use

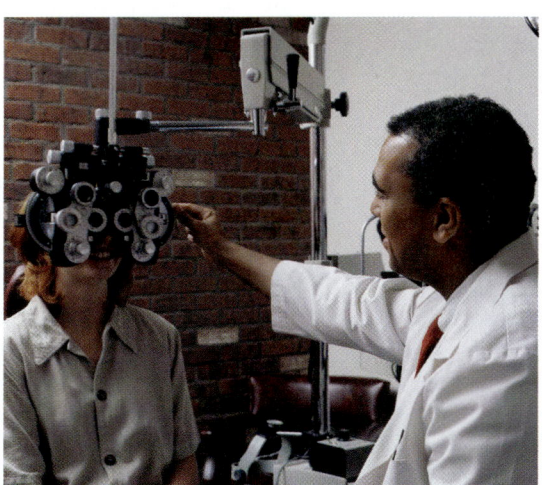

Figure 24-10. Physicians diagnose medical conditions through interviews and examinations. This doctor is examining a patient's eyes using special equipment.

Standards for Technological Knowledge

14

Tech Terms

Disease—Any change interfering with the normal functioning of the body or any of its parts.
Physician—A health care professional who diagnoses and treats diseases and injuries.
Nurse—A health care professional who helps physicians in their work.
Medical Technologist—A health care professional who gathers and analyzes specimens to assist physicians in diagnosis and treatment.
Dentist—A health care professional who diagnoses, treats, and helps prevent diseases of the teeth and gums.
Dental Hygienist—A health care professional who assists dentists in surgery and cleans teeth.
Pharmacist—A health care professional who dispenses prescription drugs and advises people on their use.
Diagnosis—The process of determining the medical problem.
Treatment—The process that uses medical procedures to cure diseases, heal injuries, or ease symptoms.

TechnoFact: Folk medicine attempts to cure and prevent illness and injury with a number of unconventional treatments. Folk medicine remedies are rooted in tradition or superstition and may include the application of healing objects and animal parts, magical or spiritual ceremonies, and the ingestion or application of herbs. Herbal medicine, which uses natural plant substances to treat and prevent illness, is closely related to folk medicine.

technology to extend the potential to deal with medical problems. Many different types of technology have been developed over the years.

Technology and Diagnosis

In times past, physicians had to depend upon people to describe their symptoms. Often, these descriptions were not accurate, and many were hard to understand. They were used, however, to plan treatments. At times, this led to an inability to cure the diseases or treat the injuries. To deal with these problems, people saw a need for diagnostic equipment. For this discussion, routine, noninvasive, and invasive diagnostic equipment will be discussed.

Routine diagnostic equipment is used to gather basic information about the

patient's condition and general health. See **Figure 24-11.** This equipment provides a baseline of general information about the patient. It often includes the following:

➤ Thermometers to determine body temperature.

➤ Scales to measure body weight.

➤ Devices to measure blood pressure.

➤ Stethoscopes to listen to heartbeats and lung conditions.

Noninvasive diagnostic equipment is used to gather information about the patient without entering the body. A typical example of this type of diagnostic is called ***radiology.*** It uses electromagnetic waves and high frequency sounds, or ***ultrasonics,*** to diagnose diseases and injuries. Diagnostic radiology uses special equipment

Resource
Reproducible Master 24-1: Technology and Medicine

Research
Have the students determine how the device medical professionals use to check blood pressure works.

Brainstorm
Ask the students what technological devices they have seen at the doctor's or dentist's office.

Tech Terms
Radiology—Using equipment that generates electromagnetic waves to diagnose and treat medical conditions.
Ultrasound (Ultrasonics)—Using equipment that generates high-frequency sounds to diagnose diseases and injuries.

Figure 24-11. Doctors use general diagnostic equipment to start finding the cause for the medical condition. This photo shows a stethoscope.

TechnoFact: W. C. Roentgen discovered X rays in 1895. He called them *X rays* because he didn't understand their nature. X-ray machines are widely used in medicine, but they have other applications. Researchers use them to peer into the very structure of matter; industry uses them to detect minute flaws in metal parts, and airports around the world use them to scan bags and cargo for contraband and weapons.

called *body scanners,* which produce images of the body without entering it. Examples of body scanning equipment include the following:

➤ *X-ray* machines. An X-ray machine is essentially a camera. See **Figure 24-12.** It uses X rays, instead of visible light, to expose the film. These X rays are short electromagnetic waves that can pass through solid materials, such as human tissue. Denser materials, however, such as human bones, absorb some or all of the waves. Assume someone puts a piece of film under your hand, then passed X rays through your hand. Your skin and tissue would let most of the X rays pass through them. The film behind that part of your hand would be almost completely exposed. The bones in your hand, however, would absorb most of the X rays. The film behind them would not be exposed completely. When the film is developed, an image of the bones in the hand will appear. Any fractures or joint deformities will be shown.

➤ *Computerized tomography (CT) scanner,* or *computerized axial tomography (CAT) scanner.* A major disadvantage of X rays is that the image is flat. The image shows the body in two dimensions. To deal with this shortcoming, CAT scanners have been developed. The scanner sends a thin X-ray beam as it rotates around the patient's body. See **Figure 24-13.** Crystals opposite the beam pick up and record the absorption rates of the bone and tissue. A computer processes the data and creates a cross-sectional image of the part of the body being scanned.

➤ *Magnetic resonance imaging (MRI).* X rays can cause damage to body parts. To deal with this hazard, a

Extend
Compare and contrast X-ray, CT, MRI, and ultrasound imaging.

Tech Terms
X Rays—Invisible, highly penetrating electromagnetic radiation of much shorter wavelength than visible light.
X-Ray Imaging—Using high-frequency (invisible) light of exposed photographic film to produce a flat image of the body.
CT (Computerized Axial Tomography) or CAT Scanning—Using high-frequency (invisible) light to generate digital data that can be converted by a computer system into a three-dimensional image of the body.
MRI (Magnetic Resonance Imaging)—Using magnetic waves to generate digital data that can be converted by a computer system into a three-dimensional image of the body.

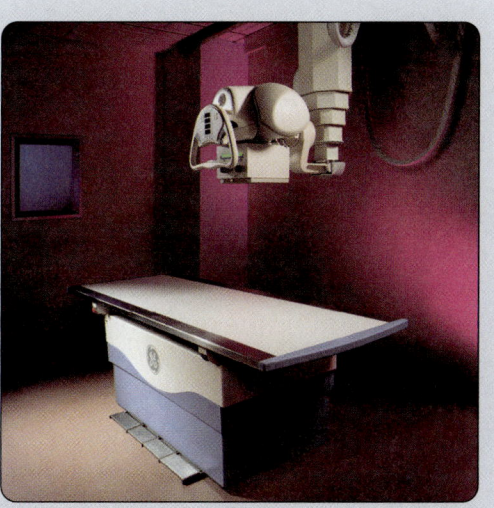

Figure 24-12. The X-ray machine (right) uses electromagnetic waves to create an image of a body part. The drawing on the left shows the machine's operation. (GE Medical Systems)

TechnoFact: In 1977, Dr. Raymond Damadian, Dr. Larry Minkoff, and Dr. Michael Goldsmith performed the first MRI exam of a human being. The process was the result of seven years of developmental work. Generation of the first image was slow by today's standards, taking nearly five hours. The new imaging process was equally slow to catch on; five years later only a handful of machines were in use. However, today, there are thousands in use, and they take only a few seconds to produce an image.

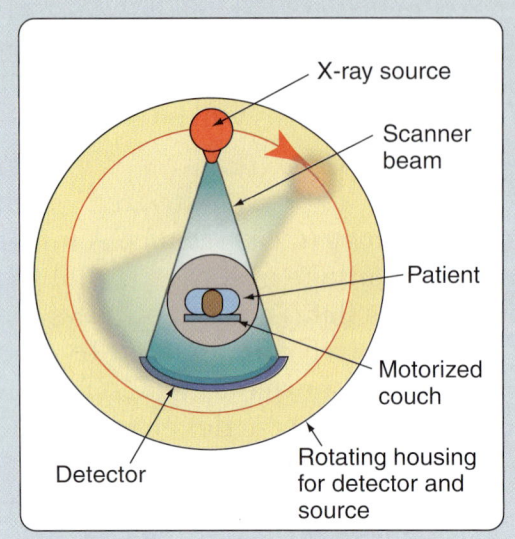

Figure 24-13. This illustrates how a CT machine works.

new imaging technique was developed. This is called *MRI*. See **Figure 24-14.** This technique uses magnetic waves, rather than X rays, to create the image. It can produce a computer-developed cross-sectional image of any part of the body very quickly.

➤ Ultrasound. This technique uses high frequency sound waves and their echoes to develop an image of the body. The ultrasound machine subjects the body to high frequency sound pulses, using a probe. The sound waves travel into your body. There, they hit a boundary between tissue and bone or tissue and fluid. Some of the sound waves are reflected back to the probe. Others travel further, until they reach another boundary and get reflected. A computer processes the reflected sound waves to produce still or moving images.

Not all diagnostic activities involve imaging equipment. There are other technological devices used in diagnosis. One important nonimaging diagnostic device is the *electrocardiograph (EKG)* machine. See **Figure 24-15.** This device is used to produce a visual record of the heart's electrical activity. As the heart works, it sends off very small electrical signals. These signals can be detected on the skin. Electrodes are attached to selected locations on the body. These electrodes capture the signals. The EKG machine amplifies the signals, and it produces a graph of their values. A physician can read this graph to determine how the heart is functioning.

Another important diagnostic device is the endoscope. It allows a physician to actually look inside the body. An

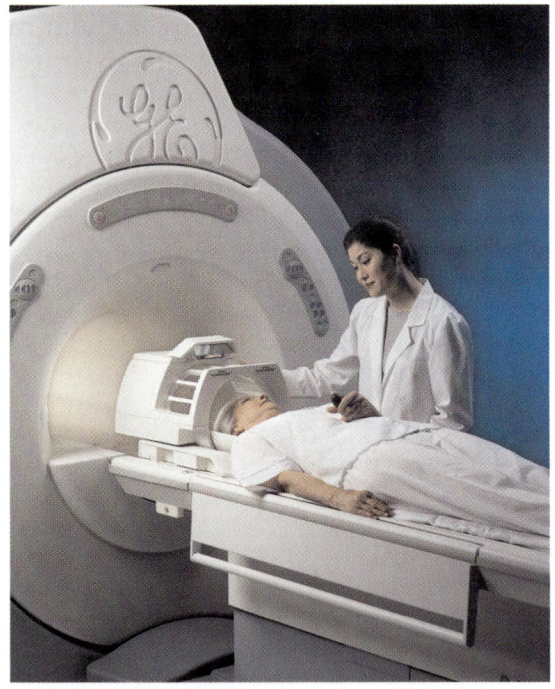

Figure 24-14. Here is an MRI machine in use. (GE Medical Systems)

TechnoFact: Hippocrates, who was born about twenty-five hundred years ago, produced a great body of medical writings, building on the ideas of others. His clinical and thoughtful approach to medicine replaced the superstitious medicine of the earlier Greek doctors. Followers of his approach were fairly good at diagnosing illnesses. However, because they had very few instruments to check bodily functions, their diagnoses depended heavily on observation and the human senses. They knew little about the human body because Greek custom required the dead to be buried quickly. Dissection would have been seen as a desecration of the body.

endoscope is a narrow, flexible tube containing a number of fiber-optic fibers smaller in diameter than a human hair. The tube can be threaded through a natural opening, such as the throat, or through a small incision. Light is sent through the fibers. This light shines on an interior part of the body and is reflected back through the fibers. This reflected light forms a series of dots. Each fiber in the tube produces one dot. These dots form a picture of an internal organ or other part of the body.

These examples are just a few of the many devices designed and built to help diagnose illnesses and physical conditions. They are examples of the dramatic uses of technology to help reduce human suffering.

Invasive diagnostic technologies involve removing tissue or fluids from the body for analysis. These approaches may include drawing and testing blood samples. They may also include taking

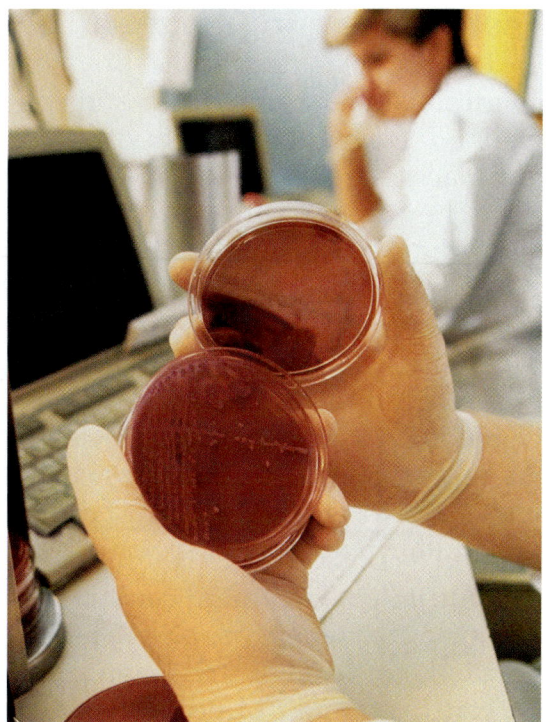

Figure 24-16. This technician is examining a culture that can identify an illness.

tissue samples (biopsies) for laboratory examination.

A blood test is a chemical analysis of a sample of blood. The sample is tested in a laboratory, using a number of different technological procedures. These procedures identify the composition of the blood. The tests can determine the presence of specific chemicals associated with a disease. They can also detect imbalances in the chemical composition of the blood. This data provides health care professionals information needed to treat illnesses or physical conditions.

Many medical conditions, including cancer, are diagnosed by removing a sample of tissue. A *pathologist* examines this tissue to determine if it is normal or diseased. See **Figure 24-16.**

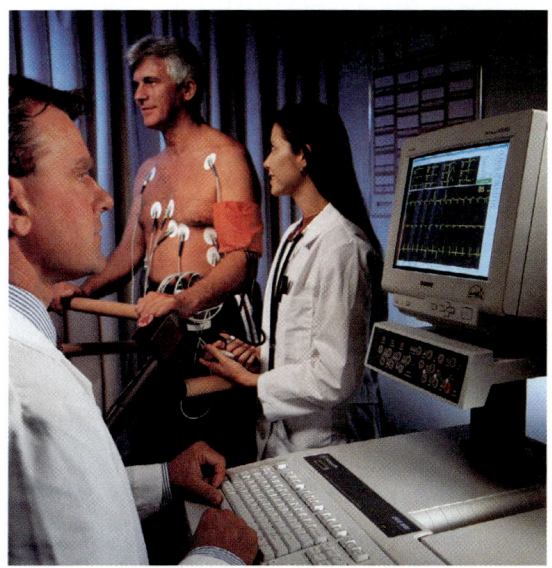

Figure 24-15. An EKG machine senses and records heart actions. (GE Medical Systems)

TechnoFact: Ultrasound, or ultrasonography, is a fairly new medical imaging technique that uses high-frequency sound waves and their echoes. The sound is called *ultra* because of its high frequency. The human ear can hear sound waves having a frequency from 20 to 20,000 Hz. Ultrasound waves have a frequency higher than 20,000 Hz and are, therefore, outside our hearing range. Ultrasound imaging is similar to the sonar location techniques used by bats, whales, dolphins, and even submarines.

Tech Terms
Hospital—A public or private institution for the care of the sick or injured.
Teaching Hospital—An institution that treats patients, educates health care professionals, and performs research on the causes and cures of diseases.
Hospital Ship—A ship serving as a hospital, often used to treat the wounded in wartime.

Technology and Treatment

Healing of illnesses and physical conditions can require drugs and specialized equipment. Both of these approaches are the results of technology. They are the products of design and production actions.

Treatment may be performed in a physician's office, at an immediate care (emergency) center, or in a hospital. A hospital is a facility where trained professionals use knowledge and equipment to treat illnesses and injuries. See **Figure 24-17.** Most hospitals contain a number of special areas:

➤ Outpatient treatment center. This is the area where people are scheduled to receive treatment and return home. Minor surgeries, medical tests and examinations, and other routine treatments are performed here.

➤ Emergency room. This is the area where seriously ill or injured people enter the hospital. Ambulances or family members usually bring these people here.

➤ Operating room. This is the area where surgeries are performed. It includes preoperation (pre-op) areas, operating theaters, and postoperation (post-op) recovery areas.

➤ Medical and surgical floors. These are areas of general care for people who are ill or have had surgery.

➤ *Intensive care* unit. This is the area where seriously ill people receive constant care and monitoring. Often, heart attack and stroke victims and people in serious accidents are placed in intensive care.

➤ Pediatric floor. This is the area where sick and injured children receive general care.

➤ Maternity ward. This is the area where mothers give birth. It has sections for caring for both the mothers and newborns.

➤ Physical therapy room. This is the area where people receive treatment to strengthen muscles. Treatment for loss of mobility and pain is received here.

➤ Pharmacy. This is the area for preparing and storing all drugs used in the hospital.

➤ Radiology unit. This is the area where various imaging equipment is used. In this area, the images are also read.

➤ Pathology unit. This is the area where blood and tissue samples are taken and analyzed.

Each of these areas in a typical hospital uses modern equipment and techniques. These areas all depend on technology to treat patients.

Treatment with Medical Equipment

The equipment used to treat patients comes in many shapes and forms. See **Figure 24-18.** This equipment includes life-support equipment, such as cardiac pacemakers, defibrillators, and artificial kidneys. It includes computer systems to monitor patients during surgery and in intensive care. The equipment includes instruments and devices, such as laser systems for eye surgery and catheters to open

TechnoFact: Godfrey Hounsfield invented Computed Tomography (CT) imaging in 1972. The original system could only image a patient's head. In 1976, systems were introduced that could allow the whole patient to be imaged. The first CT scanner took several hours to scan the patient and days to create an image from the collected data. The latest multi-slice CT systems can image a patient's chest in five to ten seconds.

Figure labels: Emergency Room, Laboratory, Maternity, Medical/Surgery, Physical Therapy, Pediatrics, Surgery, Pharmacy, Out-Patient, Radiology

Standards for Technological Knowledge

14

Tech Terms

Emergency Room—An area where seriously ill or injured people enter the hospital.

Laboratory—Area where blood and tissue samples are examined.

Maternity Area—Area where mothers give birth to children, and mothers and newborn children receive care.

Medical/Surgical Floors—Areas of general care for people who are ill or have had surgery.

Physical Therapy Room—Area where people receive treatment to strengthen muscles.

Pediatrics Floor—Area where sick and injured children receive general care.

Pharmacy—Area where drugs are prepared and stored.

Operating Room—Area where surgeries are performed.

Outpatient Treatment Center—An area where people receive treatment and return home.

Radiology Unit—An area where various imaging equipment is used.

Figure 24-17. These and other areas of a hospital use technology to help patients. Look for all the examples of technology you can see.

TechnoFact: James Earl "Jimmy" Carter, the thirty-ninth president of the United States, was the first American president to be born in a hospital. He was born at the Wise Clinic in Plains, Georgia on October 1, 1924.

Standards for Technological Knowledge

Extend
Discuss the equipment used in surgeries.

Tech Terms
Radiology Treatment—
Therapeutic radiology using high-energy radiation to destroy cancer cells.
Surgery—Medical treatment used to remove or replace diseased organs, implant medical devices, repair broken bones and internal injuries, and stop bleeding.
Plastic Surgery—
Surgical procedures used to alter, restore, or repair body parts, usually through the relocation of tissue.

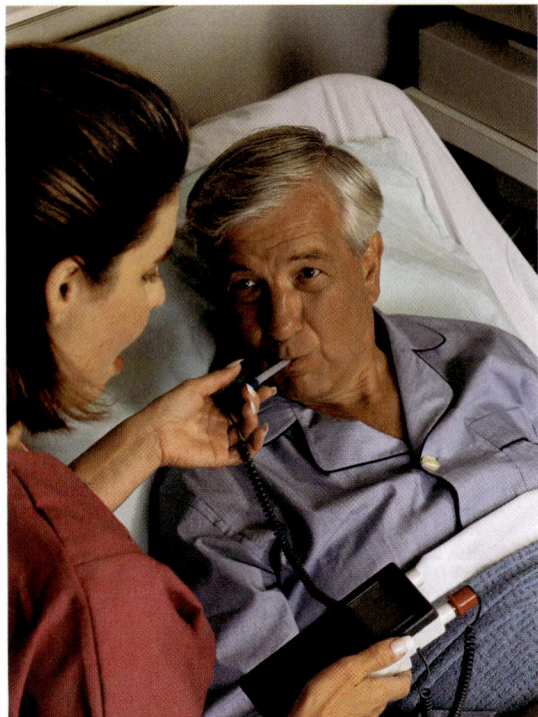

Figure 24-18. This device measures the temperature of the patient.

blocked blood vessels. It includes radiology treatment systems for cancers and other growths.

A description of all the technological devices used to treat diseases would fill books. For this chapter, two major devices will be discussed. These will provide a quick view of technology as it is applied to treating injuries and illnesses.

Radiology treatment is used for types of cancer. This treatment is called *therapeutic radiology.* It uses high-energy radiation to destroy the cancer cells' ability to reproduce. This radiology works because normal cells can recover from the effects of radiation better than cancer cells can.

Sometimes, radiation therapy is only part of a patient's treatment.

Patients can be treated with radiation therapy and chemotherapy (chemicals or drugs). Surgery may follow these treatments.

Radiology is also used as a nonsurgical treatment for a number of ailments. This use is called *intervention radiology.* The images the radiology equipment produces allow the physician to guide catheters (hollow, flexible tubes), balloons, and other tiny instruments through blood vessels and organs. An example of this approach is balloon angioplasty, which uses a balloon to open blocked arteries.

Surgery is a very common way to treat diseases and injuries. It can be used to remove diseased organs, repair broken bones, and stop bleeding. See **Figure 24-19.** Most surgery involves manually removing diseased

Figure 24-19. Surgery removes diseased tissues or repairs body damage.

TechnoFact: Surgery is not a new medical procedure. Prehistoric people performed surgical operations with the tools of the time: sharpened flints. The level of skill of the early Greek and Roman surgeons and the cleanliness of the conditions under which they worked made infection a relatively rare occurrence. As the Greek and Roman civilizations declined, their knowledge of surgical techniques and sanitation were lost. In Europe, during the Middle Ages, the task of surgery fell to unskilled and uneducated barber-surgeons. Death from surgery and infection became common. Surgery once again rose to the professional level in the 1700s.

tissue and organs. New technologies, however, are being used for many types of surgery.

High frequency sound waves can be used to break up kidney stones. Lasers use a beam of light to vaporize or destroy tissue. Transplant surgery allows organs removed from one person to be implanted into another person. Also, devices, such as pacemakers, can be implanted.

To prevent diseases from spreading, it is important for all medical equipment to be kept sanitary. Sanitation processes used in the disposal of medical products help to guard people from dangerous organisms and illnesses. These processes shape

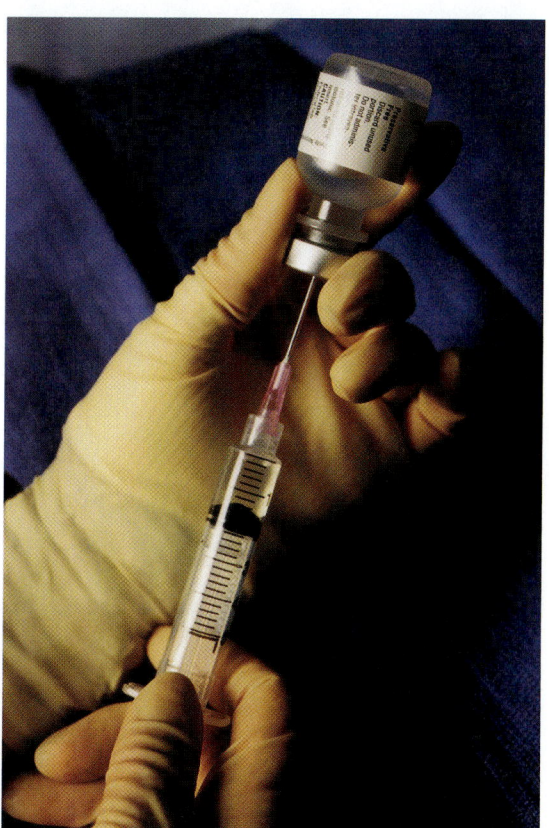

Figure 24-20. A vaccine helps prevent illness.

the principles of medical safety. Appropriate use and management of harmful materials help to shield people from avoidable harm and also help to ensure safe environments.

Treatment with Drugs and Vaccines

Humans have always experimented with substances to treat pain and illness and restore health. These substances are called *drugs.* They are any substances used to prevent, diagnose, or treat diseases. Drugs can also be used to prolong the lives of patients with incurable conditions. A special type of drug is called a *vaccine.* See **Figure 24-20.** Vaccines are substances administered to stimulate the immune system to produce antibodies against a disease. These substances have helped eliminate diseases, such as measles, whooping cough, and mumps.

Most modern drugs are the products of chemical laboratories. See **Figure 24-21.** These drugs are called *synthetic drugs.* A number of new drugs have been developed by using gene splicing, or recombinant DNA. Genetic engineering entails altering the structure of DNA to create new genetic makeups. This approach joins the DNA of a selected human cell to the DNA of a second organism, such as a harmless bacterium. The new organism can produce the disease-fighting substance the original human cell could produce. This new substance is extracted from the bacterium and processed into a drug. Genetic engineering is done in a laboratory, using reagents and other tools allowing researchers to make

TechnoFact: The first medical use of a recombinant DNA protein was approved in 1982. It was for the hormone insulin, which is used to treat diabetes. By inserting the human insulin gene into the *Escherichia coli (E. coli)* bacteria, scientists were able to produce great quantities of insulin. Before this breakthrough, insulin was only available in limited quantities from hogs.

Figure 24-21. Drugs are developed through chemistry and technology.

controlled variations in genetic information and structure.

The drug development process generally starts with a need to treat a disease or physical condition. Researchers start looking for a chemical substance that may have some medical value. These researchers may work with thousands of different substances before they find one that can serve as a drug.

Once a substance that may have medical value is discovered, an extensive testing program starts. See **Figure 24-22.** The first tests are performed on small animals, such as rats and mice. If the tests are successful,

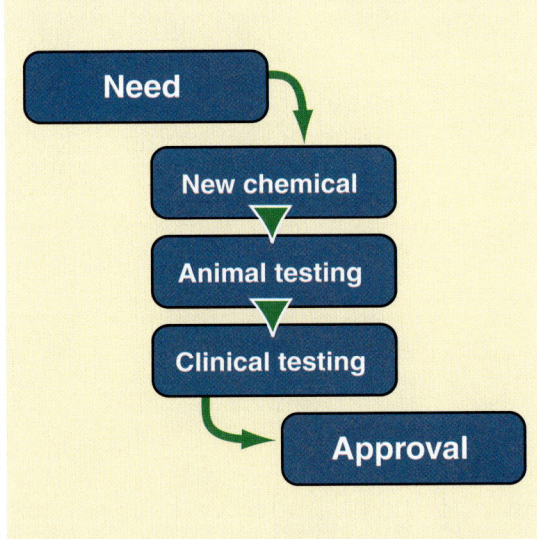

Figure 24-22. These are the steps in testing a new drug.

TechnoFact: Herbal medicine has been practiced since prehistoric times and is still in use in many parts of the world. It is estimated that over 75 percent of the world's population uses this approach as their primary form of medicine. There are over eighty thousand species of plants throughout the world used for medical purposes. For over twenty-five hundred years, herbal medicine and acupuncture have been the primary health care approaches in China.

additional tests are conducted on larger animals, such as dogs and monkeys.

The tests are carefully evaluated to see if the drug treats the disease or physical condition. Its toxicity (capability to poison a person) is also evaluated. Obviously, a drug must be effective and have low toxicity before it can be used on people. At this point, a request is made to the Food and Drug Administration (FDA) to conduct *clinical tests.* These tests are generally conducted in three phases that take many months to complete.

During the first phase, the drug is given to a small number of healthy individuals. These tests are designed to determine the drug's effects on people. If the drug passes this test, it moves into the next phase.

In the second phase, the drug is given to a small number of people who have the disease or physical condition the drug is intended to treat. The test subjects are divided into two groups. The first group is given the drug, while the other group is given an inert substance, such as sugar. This inactive compound is called a *placebo.* The test will determine if the group receiving the drug fares better than the group not receiving the drug (the placebo group).

During the final phase of testing, the drug is given to a much larger group of people. This test is used to determine dose levels, side effects, and interactions with other drugs. The results of all the tests are submitted to the FDA for approval. The agency weighs the drug's benefits against any risks that may be present. It must decide if the drug is effective and safe. If the FDA determines that the drug meets its criteria, it approves the drug for use.

Summary

Medicine and medical professionals treat ill and injured people. These professionals provide illness prevention, diagnosis, and treatment programs. They help people prevent illness by promoting good diet and proper exercise. Medical professionals diagnose illnesses and injury using imaging and other equipment. They treat illnesses using equipment and drugs developed through technological actions. By using technology, health care professionals help people live better lives with reduced levels of pain, injury, and illness.

Curricular Connections

Social Studies

Investigate a major disease and describe the drugs and technological devices that have been developed to treat it. Try to determine by whom and where each development was made.

Mathematics

Do a series of physical exercises while you are wearing a heart rate monitor. Graph the results of the exercise over time.

Science

Select a technological device used to treat illnesses. Describe the scientific principles (such as optics or radiation) the device uses.

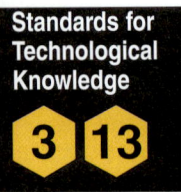

Standards for
Technological
Knowledge

Activities

1. List five exercises or sports you can do. List the equipment (technological products) required for each.

2. Visit a hospital, nursing home, or retirement center. List all the technological devices you see used to treat illnesses or physical conditions.

3. Prepare a display or poster explaining how a major medical device works. Include the inventor, the operating principles, and its uses in the presentation.

TechnoFact: Alternative medicine attempts to prevent and cure illness through means the Western medical community deems as having little or no scientific basis. Internal medicine attempts to cure diseases of the internal organs without the use of surgical techniques.

Test Your Knowledge

Do not write in this book. Place your answers to this test on a separate sheet of paper.

1. What are capillaries?
2. Construct a simple timeline of milestones in the history of medical technology.
3. The three goals of medicine are _____, _____, and _____ of diseases and injuries.
4. Paraphrase the saying "An ounce of prevention is worth a pound of cure."
5. Give two examples of ways to prevent illness and injury.
6. List five types of exercise machines or devices.
7. Technology is used in sports to construct game _____, _____ equipment and clothing, and playing fields, or _____.
8. Describe the procedure of immunization.
9. Drugs promoting natural resistance to diseases are called _____.
10. Determining the cause of a medical problem is called _____.
11. Procedures curing medical conditions are called _____.
12. _____ use short electromagnetic waves to expose a film.
13. An MRI machine uses _____ waves to create an image of a body part.
14. Removing diseased tissue is called _____.
15. _____ are substances used to treat pain and illness.

Answers to *Test Your Knowledge* Questions

1. Tiny blood vessels.
2. Answers will vary.
3. prevention, diagnosis, treatment
4. It is better to prevent an injury or disease than to try to cure one after it occurs.
5. Answers will vary. (examples: wellness programs, vaccination)
6. Answers will vary. (examples: treadmills, stationary bikes, stair climbers, rowing machines, home gyms)
7. equipment, personal protection, venues
8. People are systematically vaccinated through a series of shots to prevent disease.
9. vaccines
10. diagnosis
11. treatments
12. X rays
13. magnetic
14. surgery
15. Drugs

Chapter 25
Transportation Technology

This chapter covers the benchmark topics for the following
Standards for Technological Literacy:

Did You Know?

➤ The first human-powered aircraft was the Gossamer Condor™ airplane. It flew a figure eight course in 1977.

➤ The tunnel connecting England with the rest of Europe is designed for trains only. Special trains are designed to carry cars and trucks. Passenger lounges carry people.

➤ In Europe, trains that have hydraulic tilting mechanisms help them make curves at high speed. In Italy, the tilting train, called the *Pendelino*, can travel at 156 mph (250 kilometers per hour).

➤ Synthetic automobile fuel made from coal is produced in South Africa.

Objectives

The information given in this chapter will help you do the following:

➤ Describe some important events in the history of transportation.

➤ Define *transportation*.

➤ List and summarize the three types of transportation systems.

➤ Identify and explain the six processes of transportation.

➤ Give examples of the four modes of travel.

➤ Summarize the two main parts of a transportation system.

➤ Name and describe the five vehicular systems.

➤ Explain the functions of support facilities in transportation systems.

Key Words

These words are used in this chapter. Do you know what they mean?

air transportation
cargo
control
conveyor
ferry
freighter
guidance
intermodal transportation
land transportation
load
ocean liner
pipeline
propulsion
roadway
route
schedule
shipping lane
space transportation
storage
structure
suspension
tanker
terminal
transportation
tugboat
unit train
unload
water transportation

Chapter Outline
Did You Know?
Objectives
Key Words
What Is Transportation?
Types of Transportation
Transportation Processes
Modes of Transportation
  Land Transportation
  Water Transportation
  Air Transportation
  Space Transportation
  Intermodal Transportation
Transportation Systems
  Vehicle Systems
Summary
Curricular Connections
Activities
Test Your Knowledge

Figure 25-1. This sled, or stoneboat, is based on the same principle as the earliest sleds. It was used to move rocks from farmers' fields.

Many people think nothing of jumping into a car and going to the supermarket or mall. Other people catch planes to faraway places. Transportation was not always so easy. In the earliest times, people had to walk everywhere they went. To improve transportation, people started to develop new ways to move loads. They developed sleds, rafts, and other crude devices. See **Figure 25-1.** These people started to use pack animals to carry loads. Later, they developed canoes and other boats. By 4000 B.C., the sailboat

was in use. About 5,000 years ago, the wheel was invented. See **Figure 25-2.** These devices laid the foundation for new land transportation systems. People developed ships to carry themselves on the water and wagons to move loads on land. By the 1400s, sailing ships were used to explore the world. A steam power carriage was developed in 1769. The Montgolfier brothers flew the first hot air balloon in 1783. The first railroad locomotive was developed in the early 1800s. Rail lines were used for streetcars during the same period. In 1885, Karl Benz built the first practical automobile. The Wright brothers flew their airplane in 1903. In 1926, Robert Goddard flew the first liquid fuel rocket. The first commercial jet, the *Comet,* flew in 1949. The Russians launched *Sputnik,* the first human-made satellite, in 1957. Neil Armstrong became the first man to walk on the Moon in 1969. The space shuttle started flying in 1981. As you can see, transportation has advanced greatly over time. Much of this change has come in the last 250 years.

Figure 25-2. This wheel was made from a cross section of a log (left). An iron band was placed around it. The wheel was used on an early logging wagon in Oregon (right).

TechnoFact: The first record of water transportation can be found in early Egyptian drawings. They show bundles of papyrus (reeds) tied together to form rafts. These vessels were used on the Nile River about nine thousand years ago. These were followed by more traditional boat-shaped craft. Sailing ships with square sails were used for voyages on the Mediterranean Sea as early as five thousand years ago.

Figure 25-3. Notice the load carriers (the train cars) and the pathway (the rails).

What Is Transportation?

We all use *transportation* devices and systems. Transportation systems move people or *cargo* from an origin (starting point) to a destination (ending point). Most transportation systems include a way to carry a load and a dedicated pathway. See **Figure 25-3.** The load is carried in a vehicle (such as a car or plane) or fixed container (such as a pipeline). The path may be a permanent structure (for example, a roadway or railroad) or an invisible path (for example, an air route). Transporting people and goods requires a combination of individuals and vehicles.

While carrying a load, the carrier (vehicle) has degrees of freedom to move people and cargo. See **Figure 25-4.** For example, a pipeline or railroad track has one degree of freedom. It travels in a fixed path from the origin to the destination. Once the material is entered into the system, this kind of vehicle can only go one way. By contrast, cars, boats, and bicycles have two degrees of freedom. They can go forward or backward, or they can turn to the left or right. The operators can decide to travel a number of different routes on the land or water. These vehicles cannot, however, go up or down like an airplane can. Aircraft and spacecraft have three degrees of freedom: forward or back,

Standards for Technological Knowledge

Resource
Transparency 25-1: Transportation

Demonstrate
Use models to demonstrate the degrees of freedom available to transportation vehicles.

Tech Terms
Transportation—Moving something from an origin to a destination.
Transportation Technology—Using technical means (equipment) to move people or cargo to a destination.
Degrees of Freedom—The number of planes in which a vehicle can travel.

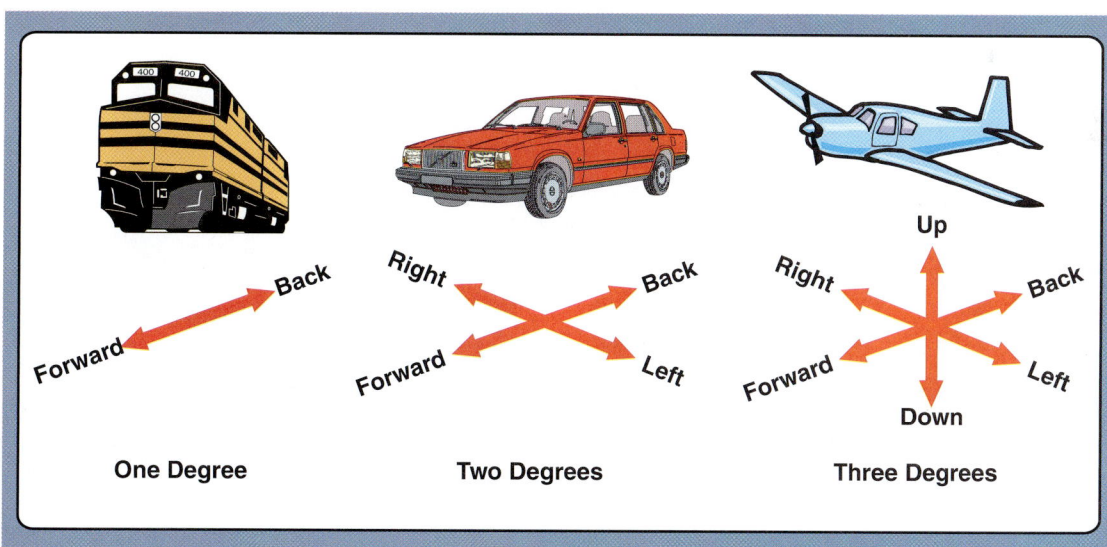

Figure 25-4. Here are the degrees of freedom to operate for various vehicles.

TechnoFact: Land transportation vehicles can be traced back to the wheel and axle. The wheel was not invented by a single person in a single place. It probably started out as log rollers to move heavy objects. Then, shorter, lightweight rollers were found to work equally well. The final leap was to put two narrow rollers on an axle.

TechnoFact: Once the wheel and axle was invented, wagons soon followed. In one of the earliest developments, a sledge was placed on a single wheel and axle assembly to make a cart for moving objects. Later, wheel and axle assemblies were placed at each end of a box to form a wagon.

left or right, and up or down. They can travel in all directions and many altitudes as they move between points.

Types of Transportation

Over the years, three types of transportation have evolved. There are personal, public, and commercial transportation systems. See **Figure 25-5.** *Personal transportation* involves people or families traveling in their own vehicles. Driving to work and riding a bicycle to the store are examples of personal transportation. *Public transportation* involves systems governmental agencies operate. This includes riding on buses and commuter trains built with public (tax) money. Profit-making companies operate *commercial transportation* systems. These systems provide transportation services in the hopes of making money. Airlines, trucking companies, and taxi services are examples of commercial transportation systems.

Transportation Processes

Transportation involves a number of different processes or actions. These processes include routing, scheduling, loading, receiving, moving, unloading, delivering, storing, evaluating, marketing, managing, communicating, and using conventions. See **Figure 25-6.** These processes are vital for the whole transportation system to operate efficiently. They may be used alone or in a

Figure 25-5. These are types of transportation systems.

TechnoFact: One of the first types of public mass transportation developed was the subway. It is an underground rail system designed to move people rapidly in a city. The oldest such system is the London Underground, which went into service in 1863. Boston installed the first subway in the United States in 1898. Today, there are nearly one hundred subways in about sixty countries. The London system has the most track (256 miles), and the Moscow system carries the most passengers annually (3.1 billion).

Routing
Scheduling
Loading
Moving
Unloading
Storing

Figure 25-6. Here are the processes of transportation.

Extend
Trace mail through the transportation processes.

variety of combinations to transfer cargo and people. We will look closer at some of the more important processes.

Routing determines the path the load will travel. For example, people may decide the path they will take on a vacation from Chicago to Los Angeles. An airline develops a route map for its operations. It plans the origin and destinations for each flight in its daily schedule.

Scheduling is assigning a time for the travel. Again, the person on vacation will decide how many miles to travel each day. Meal and lodging stops will be determined. In the airline example, each flight will be assigned a departure time and an arrival time.

Loading is the physical placement of cargo and people on a vehicle or allowing people to enter the vehicle. Loading may take place in a terminal or at some other site. In the vacation example, the family climbs into the family car. The airline loads its passengers at the airport terminal.

Moving involves transporting the people or cargo to the destination. See **Figure 25-7.** In the vacation example,

the driver guides the family car along streets and highways until it reaches the destination. The pilot of the airplane guides the airliner through takeoff, flight, and landing.

Unloading is the opposite action to loading. It allows passengers to disembark at the destination. See **Figure 25-8.** Cargo is removed from the vehicle. The family on vacation gets out of the car at the vacation spot. The airline passengers leave the airplane at the destination terminal.

Figure 25-7. This gondolier propels the craft with cargo along a route.

Tech Terms
Routing—
Determining the path the people or cargo will travel.
Scheduling—
Assigning a time for the travel.
Loading—
Physically placing cargo on a vehicle or allowing travelers to board the vehicle.
Moving—
Transporting the people or cargo to the destination.
Unloading—
Removing cargo from a vehicle or allowing passengers to disembark at the destination.
*Storage—*Placing cargo in a secure location away from the weather.

TechnoFact: Mass transit refers to public transportation open to all passengers without a reservation. Because of the development of industrialization and urbanization, the distance between people's homes and their workplaces grew. Mass transit systems were needed to shuttle people back and forth. By 1900, most major cities, including London, New York, and Paris, had subway systems and trolley lines. By the 1920s, buses were common in these and other cities.

Figure 25-8. These students are unloading from a school bus (a public transportation vehicle).

Storage is required in some transportation acts. A container may be loaded and stored until it is scheduled to leave a port. Bulk materials may be stored until a ship arrives to receive the cargo. Products may be placed in a warehouse so they are secure.

Modes of Transportation

There are several ways of moving people and goods from one point to another. For example, you can move a load of computers on a truck. This load can be sent on a ship, or it can be flown to the destination. The way the computers are moved is called the *mode of travel*. The modes of travel include the following:

➤ Land. This is using transportation means on the surface or below the earth to move people and cargo.

➤ Water. This is using transportation means on or below the water to move people and cargo.

➤ Air. This is using transportation means traveling though the atmosphere to move people and cargo.

➤ Space. This is using transportation means traveling beyond the atmosphere to move people and cargo.

Land Transportation

Transportation systems operating on or beneath the earth's surface are known as *land transportation.* They move over or through constructed pathways. These systems include the following types:

➤ Fixed path systems. These systems have one degree of freedom. They move from one origin to one destination. Fixed path systems include railroads, pipelines, and conveyors. They also include on-site systems, such as elevators, moving sidewalks, and escalators.

➤ Variable path systems. These systems use vehicles that can be guided through two degrees of freedom. These vehicles include automobiles, bicycles, buses, trucks, forklifts, and motorcycles.

Fixed Path Transportation

Fixed path systems provide a single path from the origin point to the destination. These systems include railroads, pipelines, and conveyors. *Railroads* are efficient systems for moving people and goods. They can carry many passengers and huge amounts of cargo. The vehicles and their tracks are built to carry heavy loads. Because schedules control the number of trains

TechnoFact: In the mid-1500s, a mining railway was developed that used cars on wooden poles. The wooden rails were replaced with iron rails in the late 1700s. In 1804, Richard Trevithick developed a locomotive that could pull loads up to 20 tons. Steam-powered freight and passenger service started in England in 1825.

Figure 25-9. This modern passenger train can move people between major cities quickly and comfortably.

on a track, the trains move cargo quickly, without traffic jams.

Cargo is carried on two major types of trains. *Freight trains* are made up of individual cars carrying many different products and materials. A freight train traveling along its route will drop off cars and pick up others along the way. This is usually done in central locations called *freight yards*. A **unit train** is used when large quantities of a single material are being hauled. This train moves the cargo from the same origin to the same destination, time after time. For example, coal is moved from Wyoming coal mines to midwestern power plants on unit trains. These trains also move grain from America's heartland to ports along the coasts.

People are moved on two types of trains. Commuter and light-rail trains take people to and from their jobs.

Long distance travelers may ride on standard railroad passenger trains, such as Amtrak® railway cars. See **Figure 25-9.**

Some materials are moved on systems that do not use vehicles. These systems include pipelines or conveyors. In pipelines, pumps suck or push material through the system. Petroleum and petroleum products, natural gas, and coal are often moved this way. Belt and bucket conveyors are used to move grain, gravel, and wood chips.

Pipelines are generally buried, which conserves valuable land. The cargo they carry encounters little danger from the weather and thieves. It is not likely the materials being transported will be damaged or contaminated.

Conveyors are stationary, built-in structures that move materials and products. They are often used in

Research
Have the students select a high-speed train and research its development, features, and speed.

Tech Terms
Freight Train—A train made up of individual cars carrying many different products and materials.
Unit Train—A train made up of a single type of car designed to carry a single commodity quickly from the same origin to the same destination, time after time.
Pipelines—Transportation systems constructed of pipes that move liquid or gaseous materials.
Conveyors—Stationary, built-in structures that move materials and products over a fixed path.

TechnoFact: The first railroad in the United States was built in Massachusetts in 1826. It used horse-drawn wagons that followed iron rails. Three years later, two locomotives were imported from England, but they were too heavy for the tracks. Therefore, locomotives started being produced in America. The first "long-haul" railroad was the Baltimore & Ohio Railroad (B&O). It started operation in 1828 with horse-drawn cars, but soon experimented with steam-powered locomotives. Its successful run of the *Tom Thumb* locomotive, built by Peter Cooper in 1830, paved the way for steam-powered locomotives on all railroads.

Extend
Discuss the advantages and disadvantages of fixed path and variable path systems.

Brainstorm
Ask the students why most people prefer their personal, variable transportation vehicle (automobile).

Tech Terms
Elevator—A platform or area that is moved up and down in an upright passage to convey individuals or cargo.
Escalator—A moving staircase consisting of steps attached to a constantly circulating belt that is used to transfer people from one floor to another.

Figure 25-10. Notice the conveyors transporting tires behind the worker. (Goodyear Tire and Rubber Company)

manufacturing. See **Figure 25-10.** These structures move materials along manufacturing lines. They are used to transmit materials from mine shafts or pits to processing operations. Conveyors lift grain in elevators. Special conveyors are also *people movers.* They transport people from one part of a large building to another on moving sidewalks.

Variable Path Transportation

Cargo and people can be moved in variable path (steerable) vehicles. These vehicles travel on highways, streets, and even inside buildings. See **Figure 25-11.** People using these vehicles have flexible schedules and routes. Drivers can move people and cargo day or night. They can take a direct route or an alternate route to miss traffic. Variable path vehicles include automobiles, trucks, buses, and taxis. In addition, forklifts and tractors are variable path vehicles used in factories, on construction sites, and at airports.

Water Transportation

Waterways provide another type of transportation. These waterways may be natural oceans, lakes, or rivers. Some waterways are human-made

Figure 25-11. Variable path vehicles move cargo and people (A) on streets, (B) in the open countryside, and (C) inside buildings. (Caterpillar, Inc.)

TechnoFact: The railroad played a significant role in passenger and freight movement during the 1800s and early 1900s. However, the automobile and the aircraft reduced the importance of rail travel. To counter this trend, high-speed rail systems, which travel as fast as 185 miles per hour on high-quality track, have been developed in many countries. Japan opened the era of high-speed rail service with the introduction of the Shinkansen, or "bullet train," in 1964. The French introduced the high-speed train to Europe in 1981, followed by the United Kingdom in 1984, Italy in 1988, Germany in 1991, and Spain in 1992. In 2000, the United States introduced its first high-speed train, the *Acela Express,* which runs between Washington, D.C. and Boston.

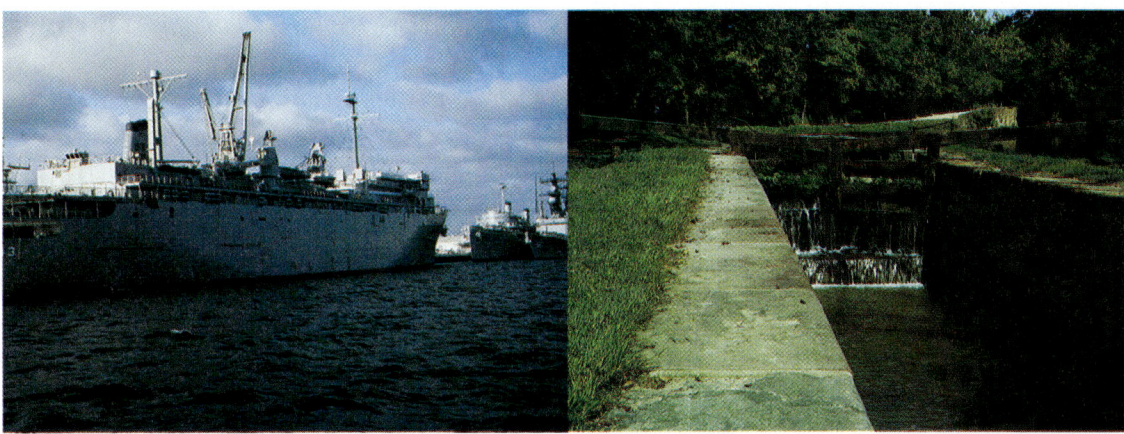

Figure 25-12. Waterways may be natural features, like oceans, lakes, or rivers (left), or human-made, like the historic B & O Canal (right).

channels, such as the Suez and Panama Canals. See **Figure 25-12.**

Water transportation is generally cheaper than land transportation. It can be used, however, only where rivers, lakes, and other bodies of water are *navigable.* The bodies of water must be wide enough and deep enough for heavily loaded watercraft to travel on them.

There are two major types of waterways used for transportation. These are oceans and inland waterways. Oceans and seas are the masses of water separating continents and major landmasses. Inland waterways are the rivers and lakes within a landmass.

Ocean transportation carries freight and people on ships. Ships carrying people are called *ocean liners.* *Freighters* carry products and solid materials. Some freighters carry cargo in containers that are easily loaded and unloaded. See **Figure 25-13.** *Tankers* carry liquids like petroleum and chemicals.

There are no constructed paths or highways on the open seas. Ships do, however, keep to regular routes over

the water. These routes are known as *sea-lanes,* or *shipping lanes.* Ships travel along these routes.

A number of vehicles are used for inland shipping. *Barges* are large floating cargo boxes without engines. See **Figure 25-14.** *Tugboats,* or towboats, are small vessels that move the barges over the water. They can be seen as the "locomotives" of the water. *Ferries* are used to move people and vehicles across bodies of water.

Figure 25-13. This freighter is loaded with containers. (U.S. Department of Agriculture)

Research
Have the students select a type of ship and research its general size, construction, and use.

Tech Terms
Oceans and Seas—Masses of water separating continents and major landmasses.
Freighters—Ships designed to carry products and solid materials.
Tankers—Ships designed to carry liquids.
Sea Lanes or Shipping Lanes—Regular routes over the water on which ships travel.
Barges—Large floating cargo boxes without engines.
Tugboats—Small vessels that move the barges over the water.

TechnoFact: The Marquis Claude de Jouffroy d'Abbans created the first steamboat, a paddle-wheel steamer, in France in 1783. John Fitch launched a water-jet–propelled steamboat on the Delaware in 1787. John Stevens designed the first successful screw-propeller steamboat in 1802. However, it was Robert Fulton who introduced the first commercially successful steamboat, the *Clermont,* which sailed from New York to Albany in 1807.

Figure 25-14. Barge transportation is a familiar sight on canals and rivers. (U.S. Park Service, Natchez Trace Parkway)

Air Transportation

A mode of transportation developed in the last century is *air transportation.* It uses aircraft to move people and cargo to their destinations. Aircraft include all vehicles traveling within Earth's atmosphere.

Airplanes are the most commonly used aircraft. These airplanes may be commercial airliners, company planes, or private aircraft. Helicopters can be used for air transportation. They are usually used for short, commuter-type hops.

Airliners are used to move large numbers of people along set routes. Airlines (air transportation companies) operate them. Businesses operate company planes to move their employees and cargo. Private aircraft are individually owned airplanes used for personal use. They are called *general aviation craft.* See **Figure 25-15.**

Like the oceans, the sky has no highways. To be safe, aircraft follow established routes. In addition, air lanes are at various heights. For example, planes flying west may travel at a different height than planes flying east.

Space Transportation

Space transportation is a very new transportation mode. It includes unmanned and manned flights. Unmanned flights have rockets traveling far into outer space. They are used to explore the universe. Cameras aboard the space vehicles send back photographs of unexplored space.

Manned flights have taken astronauts to the Moon. The space shuttle ferries humans between space stations and Earth. See **Figure 25-16.**

Intermodal Transportation

Often, moving people and goods from one point to another requires more than one transportation mode. For example, travelers may travel from

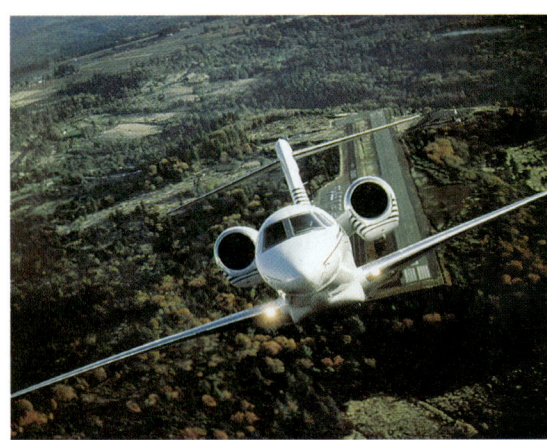

Figure 25-15. Smaller aircraft are used for private and company travel. (Cessna Aircraft)

TechnoFact: Early airplanes were light and slow. Because the airplanes did not require smooth or paved surfaces to land and take off, they required only crude airfields that usually had no designated runways. Multiengine airplanes that could carry up to 75 passengers were introduced in the 1930s, requiring airport improvements. The 1959 introduction of commercial jet aircraft that could seat more than 150 passengers required new, even larger airports. Runways were longer, and the terminals were larger to handle the increased number of passengers.

Figure 25-16. Space travel has opened up many new technologies.

home to the airport in a taxi or shuttle. They use land transportation. These people may fly to a port city by using air transportation. There, they may board a cruise ship for a holiday vacation. In all, they used three modes of transportation. Likewise, a shipment of apples may be loaded in a refrigerated container. The container may be transported to the port on a train. There, it is loaded on a ship bound for Japan. This use of more than one mode of transportation is called *intermodal transportation.* See **Figure 25-17.**

Transportation Systems

A transportation system includes several elements. There is a vehicle or mechanism to carry people or cargo. Also, there are support facilities. These support components include a constructed or assigned pathway, one or more terminals, and subsystems to provide life, legal, operational, maintenance, and economic support.

Vehicle Systems

Most transportation systems include vehicles. These vehicles are designed to move people and cargo safely. Passengers must be protected and comfortable. The vehicles must be able to operate in one or more of the modes described earlier. See **Figure 25-18.** Transportation vehicles are made up of subsystems that must operate together for a system to work efficiently. Every vehicle has five subsystems:

➤ Structure. This provides a rigid framework to protect the vehicle's contents and support other systems.

Figure 25-17. Intermodal transportation means use of more than one type of transportation. Three are shown here: ship and tug (water transportation), train (fixed path land transportation), and truck (variable path land transportation).

TechnoFact: To speed up shipping, Malcolm McLean adapted semitrailers so they could be moved on trains and cargo ships. This development was called *containerized shipping.*

Figure 25-18. Here are the systems in a vehicle.

➤ Propulsion. This provides the means to move the vehicle.

➤ Suspension. This maintains the vehicle on the pathway.

➤ Guidance. This receives information needed to operate the vehicle.

➤ Control. This enables the vehicle to change speed and direction.

Structure

You ride in many vehicles each year. Probably, you have ridden in your family car. You may ride in a bus to school. To get to a faraway place, you may ride on a train or in a plane.

Each of these vehicles has a *structure.* Structural systems are the framework and body of a transportation vehicle or system giving it shape. This structure provides a number of important areas. There is a power area containing the engines, transmission, and other systems. In addition, there is an area for the driver to operate the vehicle. This area contains a place to sit or stand and a series of operating controls. Finally, there is a place for the load. This area may contain and protect cargo. It will keep the cargo away from the weather. This area will keep unwanted people from stealing or damaging the cargo. It may keep the cargo from freezing or keep the cargo cool so it does not spoil. Railroad boxcars, semi truck trailers, and shipping containers are examples of load-carrying structures.

If passengers are transported, the structural system contains seats. See **Figure 25-19.** There are lighting and climate controls (heating and air-conditioning). The structure has windows and doors. Railroad cars and the passenger compartments of airplanes are examples of people-carrying structures. The operator and passenger areas are located together in automobiles and buses.

Figure 25-19. Vehicles transporting people have attractive and comfortable areas in which passengers can ride.

TechnoFact: All vehicles have structure to protect people. An early example is the carriage, which is a passenger vehicle drawn by animals. The first carriages appeared in the Bronze Age as two-wheeled carts and four-wheeled wagons for transporting goods. An early passenger carriage was the chariot.

Figure 25-20. Diesel engines power this car and passenger ferry.

Propulsion

Transportation vehicles must move from place to place. To do this, they require a *propulsion* (power) system. This system includes a power source, energy converter, power transmission system, and drive mechanism. For example, a car has an engine (the power source), a transmission or transaxle (the power transmission), and wheels and tires (the drive mechanism).

Heat engines are the most common way to power vehicles. These engines, as discussed in Chapter 21, are gasoline, diesel, gas turbine, and rocket engines. Electric motors are also used to power some vehicles.

Gasoline engines use a piston and cylinder to generate the power. Fuel and air are drawn or injected into the cylinder. The piston compresses the fuel and air mixture. A spark ignites the fuel. The rapidly burning fuel is turned into a gas. The expanding gas

drives the cylinder down, creating power. Gasoline engines are used in smaller vehicles, like cars, delivery trucks, small airplanes, and pleasure boats.

Diesel engines operate like gasoline engines. They have a piston and cylinder moving through fueling, compression, power, and exhaust strokes. The engines differ in the way the fuel is ignited. In gasoline engines, a spark plug causes the fuel to burn. A diesel engine uses the heat that compressing the fuel generates. Diesel engines operate at higher compression ratios (up to 20 to 1) and produce more power. They are used in large vehicles, such as heavy-duty trucks, railroad locomotives, and ships. See **Figure 25-20.** Also, because of the better fuel economy, these engines are used in some automobiles and delivery trucks.

The gas turbine and jet engines pass air through a compressor. The

TechnoFact: Karl Benz produced a prototype of the automobile using an Otto four stroke cycle internal combustion engine in 1885. In same year, Gottlieb Daimler patented his gasoline engine. In 1892, Rudolf Diesel patented the engine named for him.

Standards for Technological Knowledge

18

Resource
Reproducible Master 25-2: Vehicle Systems: Propulsion

Research
Ask the students how a gasoline engine compares with a diesel engine in terms of operation, power generated, cost to operate, and pollution emitted.

Reinforce
Review the energy conversion machines (engines and motors) presented in Chapter 7.

Tech Terms
Internal Combustion Engine—An engine in which the combustion of fuel takes place in a confined space to produce expanding gases used to provide mechanical power.
Diesel Engine—An internal combustion engine differing from the gasoline engine in that the fuel is ignited by heat generated through compression of air in its cylinders instead of by a spark.
Steam Engine—An engine converting heat energy into mechanical energy, using steam as a working fluid.
Jet Engine—An engine producing forward thrust from the rearward discharge of combustion gases.

Standards for Technological Knowledge

18

Resource
Reproducible Master 25-3: Vehicle Systems: Suspension

Research
Have the students research how hybrid cars work.

Brainstorm
Ask the students to think of reasons manufacturers and car buyers have been slow to adopt the diesel engine.

Tech Terms
Rocket Engine—An engine producing thrust by burning a fuel and an oxidizer in a combustion chamber and expelling the expanding hot gases through a nozzle.
Electric Motor—A machine converting energy from electricity into mechanical energy.

compressed air is sent into a combustion area, where fuel is added and ignites. The burning fuel becomes hot gas, which passes through turbine blades. The gases producing the power spin these blades. Turbine and jet engines are used primarily in aircraft.

Rocket engines use solid or liquid fuels to create power. The burning fuels become gases, which are then ejected through a nozzle. The force of the gas leaving the nozzle creates the power needed to move the vehicle. Rocket engines are primarily used to power spacecraft.

Finally, electric motors can be used to power vehicles. The motors are used to turn the wheels on commuter trains, streetcars, and some automobiles. See **Figure 25-21.**

Figure 25-21. The mechanism shown in this picture picks up electricity to power a commuter train.

A combination of more than one power type is used in some vehicles. For example, a railroad locomotive is called a *diesel-electric locomotive.* It uses a diesel engine to drive an electric generator. The electricity produced powers electric motors driving the wheels of the locomotive. Nuclear-powered ships use nuclear reactors to produce steam. The steam powers turbines driving the ship.

Suspension

Vehicles must move along a path from an origin to a destination. To keep them on the path, they use *suspension* systems. These systems connect or associate vehicles with their surroundings. The systems use direct contact, lift, or buoyancy to suspend the vehicles. Wheels on land vehicles are in direct contact with *roadways* and rail lines. They suspend the vehicle as it follows the path.

Ships and boats use buoyancy to suspend themselves in water. This principle of science tells us that a force equal to the weight of the water it displaces sustains an object in water. See **Figure 25-22.** If a boat weighs 1,000 lbs., it will sink into the water until it displaces 1,000 lbs. of water. It will sink if its shape will not displace 1,000 lbs.

Aircraft and hydrofoils use lift to create suspension. See **Figure 25-23.** The airfoil has a relatively flat bottom and curved top. When air flows over the shape, the air on top has farther to travel. This causes it to increase speed. The faster moving air has less pressure. The difference in pressure gives the wing lift.

TechnoFact: George Stephenson built his first locomotive in 1814. In 1829, the *Rocket,* which Stephenson designed with his son Robert, won a competition for locomotive design. The Stephenson design set the standard for future locomotive developments.

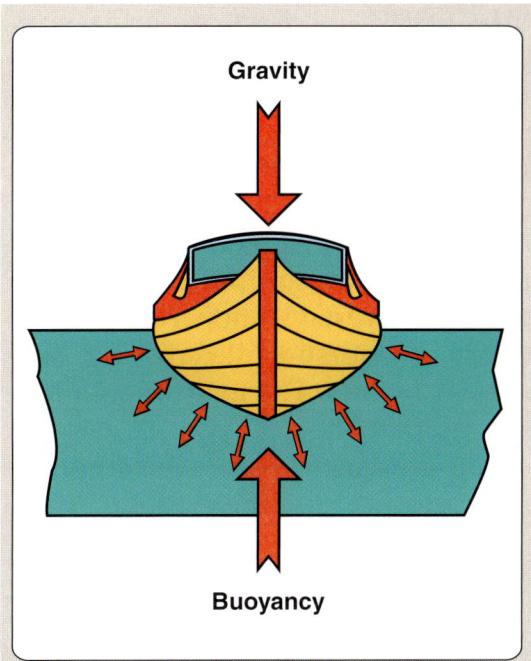

Figure 25-22. A boat will float if the force of the displaced water equals the force of gravity.

There are less common suspension systems. One of these is known as a *ground-effect machine, or hovercraft*. A fan in the vehicle creates a thin cushion of high-pressure air. The machine moves about on this cushion of air. See **Figure 25-24.**

Another system uses magnetic forces to suspend the vehicle. It is called *magnetic levitation (maglev)*. The vehicle has a set of lifting magnets causing it to rise above the pathway. Another set of propelling magnets causes the vehicle to move forward and backward.

In all vehicles, there are forces trying to stop forward movement. For example, wind resistance hampers forward motion. The flatter the surface, the more wind resistance it has. This resistance is called *drag*. Semi trucks and buses have more wind resistance than automobiles have. Their flat front surfaces provide large areas of wind resistance. Many vehicles are designed to overcome this wind resistance. They are said to be aerodynamically designed. See **Figure 25-25.**

Standards for Technological Knowledge

Extend
Explain or demonstrate to the class how a hovercraft works.

Research
Have the students research how maglev transportation systems operate.

Tech Terms
Air-Cushion Vehicle (Ground-Effect Machine or Hovercraft)—A craft designed to travel just above the ground or water.
Magnetic Levitation—Using magnets to support and propel vehicles.

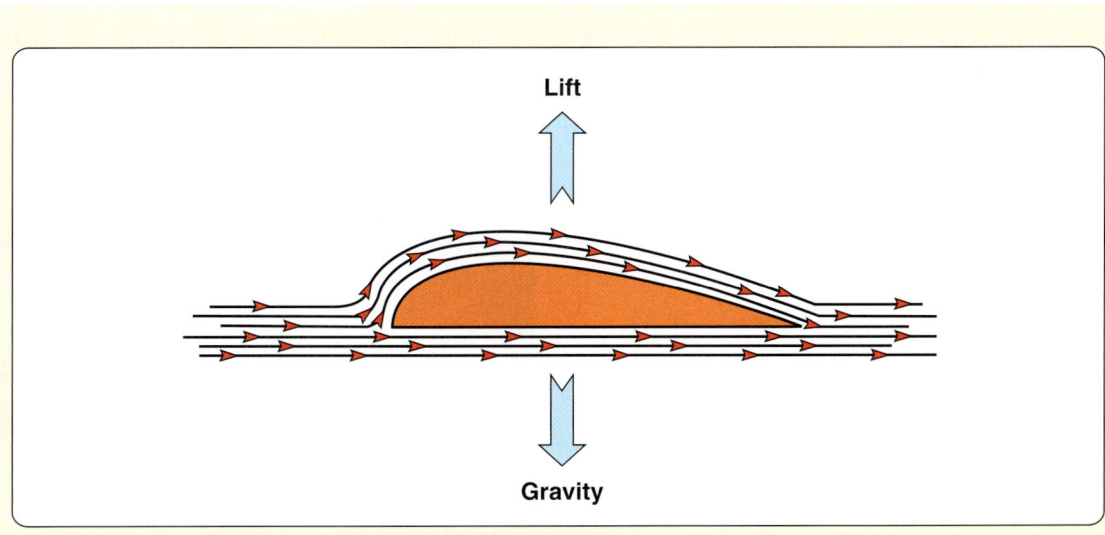

Figure 25-23. The faster flowing, lower-pressure air across the top of a wing creates lift.

TechnoFact: Emmanual Swedenborg designed the first air-cushion vehicle in 1716. The craft was never built because there was no available power source adequate to operate it. In the 1870s, Sir John Thornycroft worked with air-cushion effects and filed patents involving air-lubricated hulls. In 1955, Christopher Cockerell obtained a patent for a vehicle that was "neither an airplane, nor a boat, nor a wheeled land craft;" however, few people showed interest in the designs. In 1962, a British air-cushion vehicle became the first commercial hovercraft, servicing a 19-mile long ferry run.

Figure 25-24. A—A hovercraft floats on a high-pressure air bubble. A fan at the top compresses air and sends it to the bottom of the craft. B—The U.S. Navy developed this hovercraft, which was tested in the Arctic. Someday it may also be used for oil exploration there. (Standard Oil of Ohio)

Also, gravity pulling down on the vehicle reduces forward motion. Heavier vehicles require more power to move than lighter ones. Therefore, the heavier the vehicle, the larger power source it requires. Trucks require larger engines than do minivans.

Friction is another factor reducing forward motion. There is friction within the vehicle. The vehicle's moving parts resist movement. Also, land vehicles experience rolling resistance. There is friction between the tires or wheels and the roadway or rail line. These forces use power.

Guidance

Guidance systems provide information to the vehicle's operator. They help the person operate the vehicle safely and correctly. See **Figure 25-26.** Guidance may be onboard or outside the vehicle. Many vehicles use electronic guidance equipment. Ships, for example, have navigational instruments to stay on course while crossing large bodies of water. The instrument panel of an automobile has many gauges providing guidance. Future automobiles may carry electronic maps.

Outside guidance systems include the control towers giving pilots instructions about takeoff and landing. Railroads have signals along their tracks. Highways have traffic signals and travel information on road signs.

Control

Control systems allow operators to direct and regulate a vehicle's travel. They obtain information from the guidance system to determine the variation in speed, direction, or altitude of

Figure 25-25. The aerodynamics of this vehicle design are being tested. (Ford Motor Company)

TechnoFact: The term *dead reckoning* refers to the method of calculating a current position based on the last known latitude and longitude, the speed and direction of the vessel, and the elapsed time.

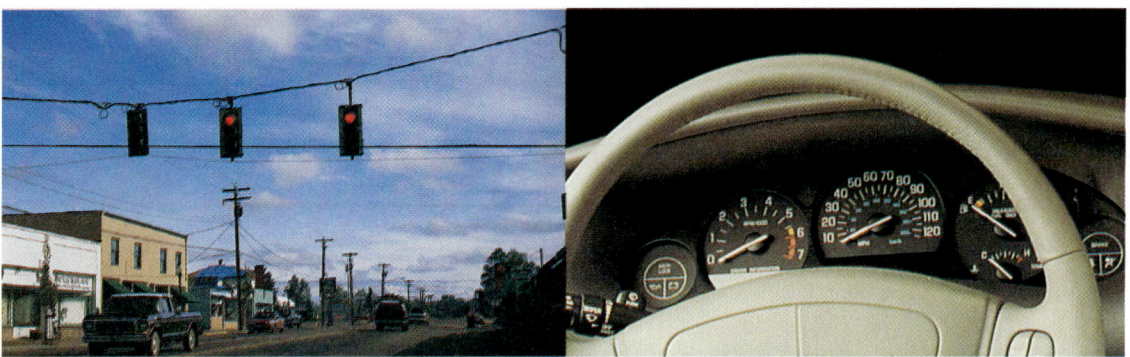

Figure 25-26. Traffic lights (left) and the gauges on an automobile dash (right) provide guidance information. (Buick)

a vehicle. The operator must be able to start, stop, speed up, slow down, and turn the vehicle. These acts are managed through systems of control. Control systems vary somewhat from vehicle to vehicle.

Starting and stopping a vehicle are done with *braking systems.* Trains, highway vehicles, and airplanes have wheel brakes. Air or hydraulic pressure pushes pads or blocks against the wheels. Friction produces drag on the wheels to slow or stop the vehicle. Airplanes also have wing flaps. These flaps produce air drag and help slow the plane. Aircraft engines are also used to slow the plane after landing. Reversing the propellers slows or stops ships.

Acceleration and deceleration controls are used to vary a vehicle's speed. Controlling the amount of fuel delivered to the engine varies the speed. These controls are called *throttles.* The gas pedal in a car is a throttle.

Vehicles also must have directional control. Variable path land vehicles have wheels that turn left and right. Ships and airplanes have rudders for left and right movement. Airplanes also control up and down movement with an elevator. The elevator is part of the tail assembly.

Most vehicles have other controls. These include switches to turn on lights, windshield wipers, windshield washers, defrosters, heaters, and radios. These are usually simple electrical switches. They control electrical current to the devices.

Support Systems

A transportation system is more than vehicles moving people and cargo. It also includes a number of support facilities. Support systems include vehicle pathways and terminals. They also supply life, legal, operational, maintenance, and economic support for safe and efficient operation.

Transportation systems require pathways. There are roadways, bridges, and overpasses on highways. See **Figure 25-27.** Tunnels allow roads and rail lines to pass through mountains. Towers support pipelines as they cross ravines. These facilities must be maintained so the transportation system is efficient. See **Figure 25-28.**

TechnoFact: Airports differ in size and layout, depending on the aircraft they are designed to accommodate. The three types of airports are military, general aviation, and commercial airports. Military airports have 10,000'–15,000' runways and are used only by military aircraft. General aviation airports have 3,000'–5,000' runways and are used by small civilian aircraft. Commercial airports have 6,000'–12,000' runways and are used by airlines.

Technology Explained

flight simulator: A computer-controlled device allowing pilots to practice flying an airplane without actually using an airplane.

The first time you rode a bicycle, you did not hop on and ride away. You had to learn how to balance the bicycle and pedal. Until you mastered these skills, you may have fallen over a few times. You made mistakes as you practiced how to ride. Making mistakes with a bicycle was not a big deal. You got up, dusted yourself off, picked up your bike, and tried again.

Pilots who are learning to fly also need to practice. Making a mistake with an airplane,

Figure A. This simulator is used to train pilots to fly the Boeing 767 aircraft.

however, can be dangerous and costly. In order to practice without risk, pilots use *simulators*. Simulators range from simple training devices for small aircraft to sophisticated units for large airliners, **Figure A.**

We will discuss large simulators here.

A simulator mimics the sounds, motions, and views from an airplane in flight. Pilots can improve their skills and learn to fly new models of aircraft. A trainee pilot can develop the ability to take directions from flight controllers and practice takeoffs, cruising, and landings. Emergency procedures can be practiced over and over. The safety of the pilot, the public, and a very expensive aircraft are not put at risk.

A simulator has several major systems. The first is the control system. The instructor uses the control system's computer to set up a desired flying condition. The second is a capsule

Also, most systems have stations or terminals. See **Figure 25-29.** Most *terminals* provide offices for employees of the transportation system.

Figure 25-27. Roadways, bridges, and other structures support transportation systems.

Figure 25-28. This crew is using equipment to repair a railroad track.

There may be areas to receive cargo or sell tickets to passengers. Usually, there is a place to secure the cargo or allow passengers to wait. Other services may be provided, including food areas, rest rooms, and shops.

TechnoFact: Denver International Airport, opened in 1995, is the newest major commercial airport in the United States. The airport covers 53 square miles (nearly 34,000 acres). It has five 12,000' runways that are at least 4,300' apart. This arrangement allows two or three streams of aircraft to land simultaneously in all types of weather. The terminal and concourses are more than 6 million square feet and include eighty-nine airline gates.

motion system, **Figure B.** This is a series of hydraulic cylinders moving the capsule. The movements of the capsule imitate the motions of an aircraft in a particular situation. The next part is an audiovisual system, which uses projectors, screens, and speakers. This system conveys to the pilots what they would see and hear if they were flying real airplanes. Finally, the simulator has a set of controls and instruments just like those in a real aircraft. The pilot uses them to "fly" the simulator. They are connected to the control system computer.

The flight simulator uses a closed loop control system. The computer sets up a situation. It directs the audiovisual system to show the pilots what they would see and hear in a real situation. The computer also directs the motion system to provide the physical feel of the situation. The computer causes the instruments to show readings that would be expected in the situation. Then, when the pilot reacts, the computer makes the simulator act like an airplane. The computer will change the view, sounds, instrument readings, and motions to *simulate* the actions of an aircraft. If the pilot makes a mistake, the computer can simulate an emergency. The pilot can practice how to get the aircraft back to a safe condition. No one is injured, and the pilot can find out how to avoid a crisis.

Figure B. The capsule is suspended on hydraulic cylinders and moves to mimic the motions of an aircraft. It contains a complete set of instruments and controls (inset).

Stations and terminals are needed to make intermodal travel convenient. Buses and automobiles bring passengers to and from airports. There must be facilities for unloading. Dockside facilities provide needed services for cargo handling. Vehicles and cranes move seagoing containers from ships. Containers are transferred to and from land transportation. Escalators, elevators, moving sidewalks, stairs, and conveyors are used at terminals. They help travelers reach airplanes, buses, and other modes of transportation.

Many terminals provide space for vehicle service. See **Figure 25-30.** Trucks, airplanes, ships, and other vehicles need routine attention. This service keeps them in good operating condition. Lubricants must be changed. Vehicles need to be refueled. Tires may need replacing. The interiors are cleaned, and the outsides are washed.

Areas for vehicle storage are provided at many terminals. Park and ride lots let people park their cars and take public transportation (buses and commuter trains). Airports and railroad stations must have space for

Standards for Technological Knowledge

Extend
Discuss terminals as the keys to intermodal transportation.

Tech Terms
Terminal—Buildings that provide spaces to receive, retain (waiting or storage), and load people and cargo.
Parking Lot—An area near a transportation terminal to store personal vehicles.

TechnoFact: Ports are the terminals for water transportation systems. They include facilities to connect to other transportation modes (rail lines, highways, and pipelines), wharves, storage, and cargo loading and unloading machinery.

Figure 25-29. Passenger terminals provide services for the traveler.

long-term parking facilities. Most terminals also have space to store cargo. Airports have baggage sorting and storage areas. Warehouses are built next to docks and airports.

Besides providing pathways and terminals for transportation systems, the support systems also must provide support for safe and efficient operation. Providing legal support is one way they can do this. Governmental regulations often affect the design and operation of transportation systems. State organizations control the use of highway systems, establish speed limits, and monitor other operating conditions. The Federal Aviation Administration regulates airspace and air safety and issues licenses to pilots.

Quality Control in Transportation Systems

Quality control in transportation involves inspection and maintenance of three separate components: transportation vehicles, transportation structures, and transportation processes.

Transportation vehicles such as automobiles, locomotive engines, and boats are manufactured using a quality

TechnoFact: Tents may be the oldest human constructed shelter. Now they are being used to build large buildings called *membrane structures*. Membrane structures are commonly used in transportation terminals. The Hajj Terminal, in Jeddah, Saudi Arabia has a roof area of more than 120 acres and is the largest membrane structure in the world. Horst Berger, who designed the Hajj Terminal, also designed the roof for the new Denver airport's central terminal.

Figure 25-30. A plane is serviced and refueled at the boarding gate.

Figure Discussion
Discuss the work of a ground crew in preparing an aircraft for service.

control process that produces an acceptable product. These vehicles also require periodic maintenance to ensure proper operation.

The construction of transportation structures includes quality control methods typical of the construction industry. These include material testing and on-site inspections. Following construction, structures such as roads, bridges, tunnels, and railways are continually inspected and repaired as needed.

The quality of a transportation process is based on the effectiveness of the system in transporting goods or people in a timely and safe manner to the intended destination. The process is controlled by monitoring the system and making adjustments as needed. For example, overcrowded roads can produce unacceptable delays and unsafe conditions. To remedy this situation, the city planner may recommend adding more lanes to the road, improving traffic control systems, developing additional routes, or providing mass-transit options.

Summary

Transportation is the moving of people and goods from one place to another. The three main types of transportation are personal, public, and commercial transportation. Transportation involves many processes, including routing, scheduling, loading, moving, unloading, and storing. Modern transportation includes land, water, air, and space travel. When more than one system is employed, it is called *intermodal transportation*.

A transportation system is made up of structures and support systems. The structures are buildings and vehicles. Vehicles are made up of several subsystems. The subsystems include structure, propulsion, suspension, guidance, and control systems. Support systems include pathways, terminals, and other forms of support for safe and efficient operation.

Curricular Connections

Social Studies

Research the types of transportation systems used at a specific time in history. Explain the reasons they were used.

Science

Develop a display explaining the scientific principles of (1) lift, (2) buoyancy, and (3) drag.

Mathematics

Select a city in your state you would like to visit. Calculate the distance to this town. Determine how long it would take to travel there, using set speed limits. Calculate the cost of gasoline for the trip, using a predetermined fuel economy figure.

Standards for Technological Knowledge

3

Activities

1. Select a vehicle. Describe its subsystems (structure, propulsion, suspension, guidance, and control).
2. Obtain a map of your state. Select a city you would like to visit. From the map, determine (1) the most direct route and (2) an alternate, more scenic route.
3. Interview a parent or older friend about travel experiences as a youth. Write a report on the interview.

TechnoFact: The Industrial Revolution that swept England in the 1700s created a transportation problem. Goods had to be shipped from the new factories to market. But the English had not been road builders. Their road system had changed little from the time the Romans had ruled the island fifteen hundred years earlier. The new manufacturers addressed their problem with a canal system and used horse-drawn rail cars to move the goods between canals.

Test Your Knowledge

Do not write in this book. Place your answers to this test on a separate sheet of paper.

1. When was the wheel invented? Why is it important to the history of transportation?

2. Select the one best answer. Transportation is:
   A. Airplanes, automobiles, and ships.
   B. Carrying people from one place to another.
   C. Moving people and goods from one place to another.
   D. Vehicles.

3. Match each statement with the right type of transportation.
   _____ Flying on American Airlines to Chicago.
   _____ Riding a bicycle to school.
   _____ Riding in the family car on a vacation.
   _____ Riding in a taxi to the train station.
   _____ Riding the city bus to work.
   A. Personal transportation.
   B. Public transportation.
   C. Commercial transportation.

4. The process of assigning a time to travel is called _____.

5. The process of placing cargo in a secure place before loading it on a vehicle is called _____.

6. The four modes of transportation are _____, _____, _____, and _____.

7. The two types of land transportation vehicles are _____ and _____.

8. Describe the two main components of a transportation system.

9. The five vehicular systems are _____, _____, _____, _____, and _____.

10. Roads and terminals are part of the _____ system in transportation.

Activity 4A

Agriculture and Bio-Related Technology

Agriculture and Bio-Related Technology, Textbook, page 628
This is a structured activity introducing students to food processing. They will see how an agricultural product is changed to an everyday food product. The students will make peanut butter from roasted peanuts, peanut oil, and salt. Note: If salted peanuts are used, omit the salt from the ingredients. At the end of the activity, the students may be treated to peanut butter and crackers.

Introduction

Many agricultural products are processed into foods people eat. One of these foods everyone knows about is peanut butter. Peanuts have been around for centuries. They have been found in mummy tombs in Peru. Peanut butter, however, is a product of recent history. A famous physician, Dr. John Kellogg, developed it in 1890. He also invented cornflakes.

Dr. Kellogg developed peanut butter as a protein substitute for patients with no teeth. Abrose Straub developed the first machine to make peanut butter in 1903. Later, the famous agricultural scientist, Dr. George Washington Carver, developed an improved version of peanut butter. In 1922, the commercial production of peanut butter was developed. This process kept the oil from separating in the peanut butter. Today, more than half the American peanut crop is processed into peanut butter.

Equipment and Ingredients

➤ Measuring cup
➤ Food blender
➤ 1 cup roasted, unsalted, shelled peanuts
➤ 1½–3 tablespoons peanut or safflower oil
➤ ½ teaspoon salt

Making Peanut Butter

1. Place the ingredients into a food blender.
2. Blend the ingredients in the food blender. Add small amounts of oil to make the peanut butter smooth.
3. Store in an airtight container in the refrigerator. Use within two weeks.
   Makes about 1½ cups of peanut butter.

Activity 4B

Construction Technology

Introduction

There are structures everywhere you look. Some of these structures are houses, schools, hospitals, stores, and factories. Around them are streets and roads, parking lots, and power lines. Underground, there are storm sewers, water mains, and gas lines. Elsewhere, there are airports, railroad lines, dams, and bridges. These are all part of our constructed world. They are the results of people using construction technology. In this activity, teams in your class are going to work together to construct a structure. They will use construction technology to build a model of a storage shed.

Safety Notes

Keep fingers 6´´ away from saw blades. Do not attempt to make miter cuts or crosscuts freehand; use the miter box. Extreme care is important when using the utility knife. Replace dull blades on utility knives.

Equipment and Supplies

➤ Scale lumber and building materials:
  - ➤ ¼´´ x ⅝´´ x 16´´ pine (scale 2 x 4 x 8)
  - ➤ ¼´´ x 1´´ x 16´´ pine (scale 2 x 6 x 8)
  - ➤ ⅛´´ x ⅝´´ x 16´´ pine (scale 1 x 4 x 8)
  - ➤ 8´´ x 16´´ 6-ply poster board (scale ¼´´ x 4´ x 8´ plywood)
  - ➤ 8´´ x 16´´ matboard (scale ½´´ x 4´ x 8´ plywood)
➤ Rules
➤ Squares
➤ Miter boxes
➤ Backsaws and coping saws
➤ Utility knives
➤ Hammers
➤ ⅝´´ x 18 brads
➤ Adhesives

Construction Technology, Textbook, pages 629–634
This is a structured activity introducing students to common residential construction practices. The activity is designed for group work. Each group follows a prescribed procedure to construct a major part of the structure. The parts are then assembled into a finished structure. At the start of the activity, divide the class into groups, as described in the activity. Show the class a completed model. Then, have each group carefully read the instructions for completing their part of the activity. The students should complete these steps, exercising care to follow the safety practices demonstrated during the class. The "input" material should be prepared before the activity is introduced. This includes scale 2 × 4s, 1 × 4s, and plywood sheets. The sizes of these materials are listed on page 629 of the textbook and are as follows:
- 8′ 2 × 4s = 1/4″ × 5/8″ × 16″
- 8′ 2 × 6s = 1/4″ × 1″ × 16″
- 8′ 1 × 4s = 1/8″ × 5/8″ × 16″
- 1/4″ × 4′ × 8′ plywood = 8″ × 16″ x 6 ply poster board.
- 1/2″ × 4′ × 8′ plywood = 8″ × 16″ mat board.
Student worksheets to support this activity can be found on pages 83–84 in the Student Activity Manual.

Procedure

Your teacher will divide the class into groups to construct the various parts of the shed. The main responsibilities of each group will be as follows:

➤ Group 1—Front wall.

➤ Group 2—Left sidewall.

➤ Group 3—Right sidewall.

➤ Group 4—Rear wall.

➤ Group 5—Floor.

➤ Group 6—Rafters.

Each group should complete the following steps:

1. Carefully study the plans provided:

➤ **Figure 4B-1**—Pictorial view.

➤ **Figure 4B-2**—Floor plan.

➤ **Figure 4B-3**—Floor joist plan.

➤ **Figure 4B-4**—Front elevation.

➤ **Figure 4B-5**—Left side elevation.

➤ **Figure 4B-6**—Right side elevation.

➤ **Figure 4B-7**—Rear elevation.

Figure 4B-1. This is a pictorial drawing of the storage shed.

Figure 4B-2. Here is the floor plan.

Figure 4B-5. This is the left side elevation.

Figure 4B-3. This is the floor joist plan.

Figure 4B-6. Here is the right side elevation.

Figure 4B-4. Here is the front elevation.

Figure 4B-7. This is the rear elevation.

2. Prepare a materials list for the section your group is to build. You may need to study several of the drawings to determine the sizes of all materials. List each part, the number needed, the size shown on the drawings, and the scaled size of material you will need. For example, your list might look as follows:

No.	Part	Actual size	Scaled size
8	studs	2 x 4 x 6´	¼´´ x ⅝´´ x 12´´

NOTE: The actual length is divided by six to determine its scaled length.

3. Get the materials your group needs to make the shed section.

4. Cut all materials to their correct sizes.

> **Safety**
> Be careful when using hand tools with cutting edges. Never touch the cutting edge with your hand. Cut away from any part of the body. Carry sharp edged and pointed tools turned downward and away from the body. Never carry sharp tools in your pockets. Store tools not in use. Always check with your instructor for safety instructions with any tool.

5. Have your teacher check the materials.

6. Assemble the assigned shed section according to the drawings. Be careful to observe the following precautions:

Group 1—Front wall. See **Figure 4B-8.**

a. The tie plate is ⅝´´ shorter than the top plate on both sides of the door. This lets the side section tie plate overlap and connect the sides to the front.

b. There are cripple (short) studs under the door header.

Group 2—Left sidewall. See **Figure 4B-9.**

a. There are cripple studs between the window header and sill.

b. The drawing appears to call for three studs on each side of the window. Actually, there are two studs separated by a space.

c. The tie plate extends ⅝´´ beyond the end of the top plate.

Group 3—Right sidewall. See **Figure 4B-10.**

a. This wall is different from the left sidewall.

b. The tie plate extends ⅝´´ beyond the end of the top plate.

Figure 4B-8. This is the front wall assembly drawing.

Figure 4B-9. Here is the left sidewall assembly drawing.

Group 4—Rear wall. See **Figure 4B-11.**

a. The tie plate is ⅝″ shorter than the top plate on both sides of the door. This allows the side section tie plate to overlap and connect the sides to the front.

b. There are cripple studs between the window header and sill.

c. The drawing appears to call for three studs on each side of the window. Actually, there are two studs separated by a space.

Figure 4B-10. This is the right sidewall assembly drawing.

Figure 4B-11. Here is the rear wall assembly drawing.

Group 5—Floor. See **Figure 4B-3.**

a. The floor joists are fabricated first.

b. The floor joist assembly is then covered with ½˝ plywood (matboard).

Group 6—Rafters. See **Figure 4B-12.**

a. Four sets of rafters have ¼˝ plywood (poster board) gussets on only one side. Be sure you make two left side and two right side rafters.

b. All other rafters have gussets on both sides of the joist.

Figure 4B-12. This is the rafter assembly drawing. See **Figure 4B-7.**

7. Have your teacher check your constructed sections or parts.

8. Assemble the shed.

a. Secure the rear wall to the floor.

b. Secure the right and left walls to the floor.

c. Secure the front wall to the floor.

d. Secure the rafters and ridge board in proper position.

9. Finish the shed by applying the following:

a. Siding.

b. Roof sheathing.

c. Shingles (abrasive paper).

10. Fabricate and install a door. (optional)

11. Install a window—a plastic square. (optional)

12. Paint the shed. (optional)

Activity 4C

Energy Conversion Technology

Standards for Technological Knowledge

12 13

Energy Conversion Technology, Textbook, pages 635–636, and Student Activity Manual, pages 87–88
This is a structured experiment presenting one of the many energy conversion processes. The students should be given the appropriate supplies and equipment. The procedure for the experiment should be demonstrated. Then, the class should be divided into groups. Each group should complete the experiment with one color of can. The groups will share their results with the class so a comparison between color and light-to-heat energy conversion can be made.

Supplies and Equipment
• Heat lamp
• Large (42 oz.) juice cans or No. 10 cans (from school cafeteria)
• Spray cans of various colors of paint—black, red, blue, green, yellow, brown, and white
• Thermometer reading up to 200°F
• Rubber grommets with a hole the diameter of the thermometer (Corks with a hole in the center can be substituted.)
• Data recording sheet (included)
• Laboratory summary sheet (included)

Introduction

Energy is defined as the ability to do work. Much of this work involves converting energy from one form to another. Water is stored behind a dam (potential energy). When the water is allowed to fall through pipes to a turbine (kinetic energy), the energy of falling water turns the turbine (mechanical energy). The turbine converts the motion into electrical energy. The electrical energy is transmitted over power lines, which have resistance. The resistance produces heat (thermal energy). At the destination, the electrical energy is converted to light, heat, or mechanical energy. In this activity, you will explore one type of energy conversion—light energy into heat energy.

Equipment and Supplies

➤ Heat lamp test stand

➤ Large (42 oz.) juice cans or #10 cans (from the school cafeteria) that have been painted various colors (black, red, blue, green, yellow, brown, and white). These cans should each have a hole in the center of the bottom. A rubber grommet with a hole the diameter of the thermometer should be in each hole.

➤ Thermometer reading up to 200°F

(continued)

Procedure

1. Obtain a heat lamp test stand.

2. Obtain a painted juice or vegetable can.

3. Record the distances and time that will be used in the experiment. Your teacher will establish these parameters.

4. Set up the experiment as demonstrated.

5. Take the reading at distance #1 for each time.

6. Allow the can to cool.

7. Repeat steps 5 and 6 for distances #2 and #3.

8. Report your results to the class.

9. Record the results from other groups on the Laboratory Sheet.

10. Complete a Laboratory Summary Sheet.

Energy Conversion Technology *(continued)*

Procedure

Preparing for the Activity
Before the activity is introduced, the following steps must be completed:
1. Obtain a heat lamp for each group.
2. Construct a lamp holder (socket, switch, and bracket) for each group. This lamp holder should hold the lamp parallel to the cans and be movable to the three positions specified for the experiment.
3. Obtain juice or vegetable cans.
4. Establish the distances and times to be used in the experiment. These will vary with the heat lamp used. You will want to use times and distances showing significant results. *Experiment* to establish these times and distances. (Start with 12″, 18″, and 24″, for 5, 10, and 15 minutes.)

Conducting the Activity
1. Divide the class into groups of two to three students.
2. Assign each group a color and task, such as the following:
"Determine the heat absorption of your color, in terms of the intensity of light (distance from the can to the light) and the time the light is exposed to the surface."
3. Have each group complete the following steps:
   A. Select an appropriate can.
   B. Locate the laboratory sheet in the Student Activity Manual.
   C. Set up the experiment as demonstrated. Note: Demonstrate the proper, safe way to conduct the experiment.
   D. Take the reading at distance #1 for each can.
   E. Repeat steps 3C and 3D for distances #2 and #3.
   F. Compare results of the experiment by reporting their results.
   G. Record the results from other groups on the laboratory sheet.
   H. Complete a laboratory summary sheet.
Student worksheets to support this activity can be found on pages 87–88 in the Student Activity Manual.

Activity 4D

Information and Communication Technology

Standards for Technological Knowledge

12 13

Information and Communication Technology, Textbook, pages 637–640, and Student Activity Manual, pages 89–93

Introduction

We communicate daily. Some of our communication is verbal (speech). People talk to other people. Some is visual. We write messages for others to read.

Often, we use a technological device to help us deliver our message. We may use printing presses or cameras to produce two-dimensional media, printed pages or photographs. Other times, we use electrical signals or electromagnetic waves. We use the radio, television, telephone, teletypewriter, or telegraph. These devices move a message from the source to the receiver. This activity will let you build a telegraph system. Then, you will use this technological system to convey a message.

This is a structured activity that can be used to study communication technology and custom manufacturing. It challenges the students to first construct simple telegraphs and then communicate messages using their communication devices. The activity contains a specific set of procedures the students are to follow. The students can, then, develop codes to communicate messages. Students may work individually or in groups of two or three students. Half of the individuals or groups should make sending units, and the other half should make receivers. To implement the activity, you will need the equipment and supplies listed on page 637 of the textbook. Each student or group of students should be given a set of supplies. The students are then challenged to complete the following steps:
1. Study the plans for the sender or receiver.
2. Follow the procedures and plans to construct the device.

Equipment and Supplies

- Two ¾´´ x 3½´´ x 5´´ pine boards
- Two ⅝´´ x 5½´´ x 28-gauge sheet steel
- Six ½´´ brads
- One 6d box nail
- Twelve ⅜´´ x No. 6 pan head sheet metal screws
- Two mini buzzers—Radio Shack No. 273-055 or equivalent
- Four quick wire disconnect—Radio Shack No. 274-0315 or equivalent
- One 9-volt battery and holder
- 25´ 3-conductor wire or 75´ single conductor wire.

Safety

- Be careful when using tin snips. Keep your free hand well away from the cutting edges.
- Edges of sheet metal may be sharp or jagged. Handle sheet metal with caution.

(continued)

➤ In this activity, you will use only a 9-volt power supply. There is no danger from such low voltage. Still, it is possible to get an uncomfortable shock under some conditions. Do not touch bare wires or terminals.

➤ Always work carefully with tools and materials. Do not clown around or attempt practical jokes. This can cause injury to yourself or others.

➤ If you are uncertain about what to do, ask your instructor. Don't guess!

➤ DO NOT attempt to connect the telegraph stations to any power source other than the 9-volt battery.

Procedure

Your teacher will divide you into teams of four students. Each team will be split up into two groups:

➤ Group A will build telegraph station No. 1.

➤ Group B will be responsible for building telegraph station No. 2.

How to Build

Each member of every group should complete the following steps:

1. Carefully study the plans for the telegraph set. (Group A—See **Figure 4D-1.** Group B—See **Figure 4D-2.**)

2. Select a piece of lumber for the base of the set.

3. Lay out the location of the key, buzzer, and wire disconnects.

4. Select a piece of 28-gauge sheet steel to make the telegraph key.

5. Lay out the pattern on the metal. See **Figure 4D-3.**

6. Cut the metal to size.

7. Punch the two screw holes.

8. Bend the metal to form the key. See **Figure 4D-4.**

9. Drive the 6d box nail in the correct position. Allow 1″ of it to remain above the surface of the wood.

10. Attach the telegraph key, buzzer, and wire disconnects.

11. Wire the telegraph station according to the original drawing.

Information and Communication Technology *(continued)*

3. Join another student or group of students who made the other type of device to develop a complete telegraph (hard-wired) communication system.

4. Develop a code to complete the message assignment on page 89 of the Student Activity Manual. Student worksheets to support this activity can be found on pages 89–93 in the Student Activity Manual.

Figure 4D-1. Group A will construct this station. Note that it is different from the one Group B is building.

Figure 4D-2. Group B will construct this station. Note that it is different from the one Group A is building.

Figure 4D-3. This is the pattern for the telegraph key.

Figure 4D-4. The telegraph key should be bent to this shape.

Sending the Message

1. Wire telegraph station No. 1 to telegraph station No. 2. See **Figure 4D-5.**

2. Complete an assignment using the telegraph system to communicate messages between Group A and Group B.

Telegraph Station 1

Telegraph Station 2

Setup for the Telegraph Stations

Figure 4D-5. Wire the two telegraph stations together as shown.

Activity 4E

Manufacturing Technology

Manufacturing is a major context in which technology is used. It changes the form of materials to make them worth more. This may be done using custom techniques. People can make a product all by themselves. They can make it for their own use. More often, however, a manufacturing system called *continuous manufacture* is used. The product is made on a production line. In this activity, you will use such a line. Working on the line, you and your classmates will change the form of materials. You will change strips of wood into a CD holder. See **Figure 4E-1.**

During the activity, notice the use of the resources. See if you can see tools, materials, energy, information, time, and people being used. Materials are specified in a bill of materials, **Figure 4E-2.** Also, note that most people will think the CD holder is worth more than the strips of wood and nails used to make it.

Figure 4E-1. This is an isometric drawing of the CD holder.

Equipment and Supplies

➤ Wood strips
   ➤ ¼″ x ¾″ x random length (side strip stock)
   ➤ ¼″ x 1⅜″ x random length (bottom strip stock)
   ➤ ¾″ x 5″ x random length (end stock)
➤ ⅝″ x 18 wire brads
➤ Backsaw or band saw
➤ Scroll saw
➤ Hammer
➤ Brace or drill press
➤ ¾ auger or speed bit

Qty	Description	Size
2	Ends	3/4" x 5" x 5"
4	Side strips	1/4" x 3/4" x 10"
2	Bottom strips	3/4" x 1 3/8" x 10"
24	Wire nails	3/4 x 18

Figure 4E-2. Here is the bill of materials for the CD holder.

Manufacturing Technology, Textbook, pages 641–644
This is a structured activity challenging the students to use drawings and operation sheets to plan for and produce a manufactured product. The activity is organized around a specific set of procedures the students are to follow. Each student can produce a product for himself or herself. This, however, will be very time-consuming. A better approach is to use a teacher-designed production line to produce the product. The line should have the same three branches shown on the operation process chart on page 642 of the textbook. Each of these branches should feed an assembly line. Another approach is to divide the class into groups. Each group should work like a job shop. Group 1 will produce the ends, Group 2 will produce the side strips, and Group 3 will produce the end strips. The parts from the "job shops" will be assembled into products at the end of the activity. In all cases, students should follow appropriate tool and machine procedures as demonstrated in the class. Before introducing the activity, it is necessary to prepare the "input" material for the line. This will require pine strips as follows: 3/4″ × 5″ × random length (end strips), 1/4″ × 3/4″ × random

(continued)

Figure 4E-3. This is the operation process chart. How many steps are there to produce the CD holder?

➤ Disc or oscillating sander
➤ Abrasive paper
➤ Sanding block
➤ Special tooling:
  ➤ End cutoff jig
  ➤ Side strip cutoff jig
  ➤ Bottom strip cutoff jig
  ➤ End hole drilling jig

Manufacturing Technology (continued)

length (side strips), and 1/4″ × 1 1/2″ × random length (bottom strips). Remember, the goal is to introduce the students to a manufacturing system. Specific skills are not important, and rotating students through jobs is not necessary. Also, tool and material nomenclature is an inappropriate focus for this section.

This activity, like the rubber band car activity in Section 2, is designed to be less structured than the game activity provided in Section 2 of the textbook. The students should be provided the drawings for the CD holder and challenged to complete the following steps:

1. Develop a bill of materials for the product.
2. Prepare a procedure sheet for building the product.

(continued)

Safety on the Band Saw

➤ Keep your hands away from the path of the blade.

➤ Keep your fingers at least 2´´ away from the blade at all times. Use a fixture to hold small workpieces.

➤ Upper guide assembly should always be ¼´´–½´´ above the stock. This reduces the amount of blade exposed.

➤ Push stock forward rather than to the side.

➤ Work only within the band saw's capacity. Thick stock should be fed more slowly than thin stock.

➤ Students in the lab should observe the safety zone around the band saw. A broken blade will occasionally "climb" out to the right of the operator.

➤ Should the blade break, step aside and disconnect power to the machine.

Procedure

1. Study the operation process chart. See **Figure 4E-3.** Determine the steps needed to make the CD holder.

2. Read the explanation of the major steps to determine how to complete each step.

   E-1 Set up a band saw or backsaw (miter box and backsaw) to cut a 5´´ piece from the end stock. A special cutoff jig can be built to guide the backsaw during the cutoff operation.

   E-2 Drill two ³/₄´´ holes in the end. The center of the holes are 1″ from the top edge of the end. Locate the holes' centers 1¹/₄″ from one another.

   E-3 Draw a line connecting the tops of the two holes and another connecting the bottoms of the holes.

   E-4 Insert a scroll saw blade in one of the ³/₄″ holes. Tighten the blade in the saw. Cut out the marked section between the two holes.

   E-5 Smooth and straighten the saw cuts with a file. **CAUTION: Do not damage the curved portion at the ends of the slots.**

   I-1 Inspect the part for size and quality of the finger slot.

Manufacturing Technology *(continued)*

3. List any restrictions they face, such as CD case size, number of CDs the holder will hold, and cost of the solution.
4. Build a prototype of the holder.
5. Discuss the procedures they used, ways to build the product more efficiently, and other similar topics.

E-6 Sand the edges of the end parts on a disc or oscillating sander or with abrasive paper and a sanding block. **CAUTION: Do not round the edges because the side strips must mount flat on these surfaces.**

E-7 Sand the ends of the end parts on a disc or oscillating sander or with abrasive paper and a sanding block. **CAUTION: Do not round the ends because the bottom strips must mount flat on these surfaces.**

E-8 Sand the faces with an oscillating sander to remove mill marks and smooth the surfaces.

I-2 Inspect the results of the sanding steps. Look for square, flat edges and ends.

S-1; B-1 Set up a band or miter saw to cut a 10″ piece from the stock for the end strip **(S-1)** or the bottom strip **(B-1)**. A special cutoff jig can be built to guide a backsaw during the cutoff operation.

S-2; B-2 Sand the faces with an oscillating sander to remove mill marks and smooth the surfaces. **CAUTION: Be sure not to round these surfaces because one face must fit flat against the ends of the holder.**

S-3; B-3 Sand the edges of the end parts on a disc or oscillating sander or with abrasive paper and a sanding block.

S-4; B-4 Sand the ends of the end parts on a disc or oscillating sander or with abrasive paper and a sanding block. **CAUTION: Do not round the ends because they must be flush with the ends after assembly.**

I-3; I-4 Inspect the results of the sanding steps. Look for square, flat edges and sides.

A-1 Locate the bottom strips. Nail both ends of each strip with two ³⁄₄″ x 18 wire nails. NOTE: A locating fixture may be used to speed this operation.

A-2 Locate the side strips. Nail both ends of each strip with two ³⁄₄″ x 18 wire nails. NOTE: A locating fixture may be used to speed this operation.

I-5 Inspect the assembled product. Route any defective products to a rework station or scrap.

A-3 Apply appropriate finishing material by brushing, wiping, or dipping.

I-6 Inspect the final product.

Activity 4F

Medical Technology

The Challenge

Develop a device that will allow people to pick objects up off the floor or ground without bending over.

Background

Some older people and many individuals who suffer from arthritis, back problems, and other conditions find it very difficult to bend over to pick objects up off the ground. Bioengineering and technology can address this problem. Devices can be developed to allow people to grasp objects at their feet without bending over.

Design Brief

Develop a device or system people can use to grasp and lift objects at floor level without bending over, using the following materials:

➤ Wood strips
➤ Small diameter metal rod or wire
➤ Masking or duct tape
➤ Garden hose
➤ Juice cans
➤ String
➤ Wood glue
➤ Wire nails or brads
➤ Bolts and nuts
➤ Poster board

Medical Technology, Textbook, pages 645–646
This is an open-ended activity based on a design brief. The students should be presented with the design problem and the overall procedures. Both of these elements are provided on pages 645–646 of the textbook. The students will be challenged to design a device allowing a person to pick objects up without bending over. This device will help people with arthritis, back injuries, and other conditions that make bending over difficult. The students should use the design process presented in Section 3 to complete this activity. They can work individually or in small groups to complete the activity.

Procedure

Address the design challenge by completing the following steps:

1. List the design limitations.
2. Develop several possible solutions to the problem.
3. Select a promising solution.
4. Refine the solution.
5. Build a prototype of the solution.
6. Test and evaluate the solution.
7. Indicate ways to improve the solution.

Activity 4G

Transportation Technology

Introduction

Every day, millions of people and countless tons of cargo are moved. They are transported from one place to another. Technological systems are used to move these goods and people. Vehicles use energy to haul cargo. People feel power applied to engines moving these vehicles over land, across water, and through the air.

These transportation systems include vehicles and pathways. The pathways allow the vehicles to move from one place to another. They may be highways, railroad tracks, rivers, or oceans. Air and space provide pathways for airplanes and spacecraft.

People design each of these systems. The systems allow us to extend our potential for movement. This activity will allow you to design and test a vehicle for a transportation system. You will use your creative ability and manual skill to design and test a boat hull.

Equipment and Supplies

➤ Waterway or long narrow channel that holds water

➤ 2″ x 4″ x 9″ foam block (hull)

➤ 2½″ x 3¾″ x 28-gauge sheet metal (keel)

➤ Small eye screw

➤ Sail material:
  ➤ 6″ x 6″ square of fabric
  ➤ Two ³⁄₁₆″ x 6″ dowels
  ➤ One ³⁄₈″ x 7¾″ dowel

➤ Rule and square

➤ Scratch awl

➤ Tin snips

➤ Mill file

➤ Needle nose pliers

➤ Tailor's chalk or fine tipped marker

➤ Fabric glue or needle and thread

Transportation Technology, Textbook, pages 647–650, and Student Activity Manual, pages 101–102
This is a structured activity introducing students to vehicle design. They will develop and test boat hulls for effectiveness. To implement this activity, divide the class into groups of two to four students. Provide each group with the supplies listed with the activity on pages 647–648 of the textbook. Have each group carefully follow the procedure listed in their textbooks to build boat hulls, sails, and keels. The groups should test their hulls in a waterway.

➤ Hot wire foam material cutter or band saw

➤ Coping saw

➤ Rasp or "surform"

➤ Coarse abrasive paper

➤ Compressed air (15 psi) or a fan and cardboard tube

➤ Stopwatch or other timepiece that reads in seconds

> **Safety**
> The tools and equipment needed for this activity can cause injuries, if not handled properly. Your instructor will provide safety rules. He or she will also demonstrate safe procedures. Do not use any tool or piece of equipment unless you know how to do so safely.

Procedure

Making Hulls

1. Your teacher will divide you into groups. Each group should have three students.

2. Each group will build three basic hulls. See **Figure 4G-1.**

   a. Student 1 will build a round bottom hull.

   b. Student 2 will build a V bottom hull.

   c. Student 3 will build a flat bottom hull.

3. Carefully watch your teacher demonstrate safe techniques for cutting and shaping foam material.

4. Each group should get three foam blocks.

5. Each member of the group will complete the following steps:

   a. Design a hull meeting his or her assigned shape.

   b. Cut the foam block to the basic shape, using a hot wire cutter or band saw.

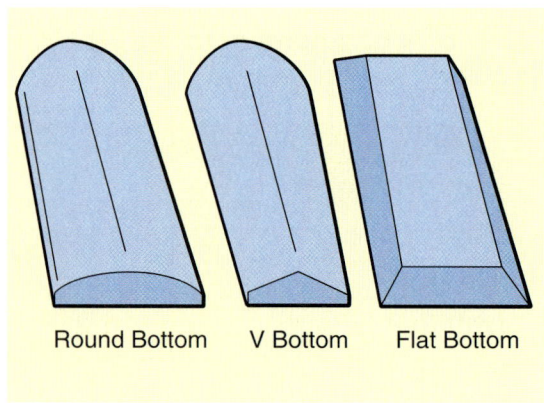

Round Bottom   V Bottom   Flat Bottom

Figure 4G-1. These are the boat hull designs. Each group will make one of each design.

c. Smooth the hull shape using a rasp, surform, or abrasive paper. Be sure both sides of the centerline are alike.

Making Sails

1. Select the materials listed under equipment and supplies for each sail.

2. Lay out and mark the fabric, using tailor's chalk or a fine tipped marker. See **Figure 4G-2.**

3. Fold the fabric at each end to form a sleeve.

4. Sew or glue the fabric in place.

5. Cut two pieces of ³⁄₁₆″ dowels, 6″ long, using a coping saw.

6. Insert the dowels in the top and bottom sleeves of the sail.

7. Cut a piece of ³⁄₈″ dowel, 7³⁄₄″ long, using a coping saw.

8. Shape the dowel to make the mast. See **Figure 4G-3.**

9. Insert the mast in the sail. See **Figure 4G-4.**

Making a Keel

1. Select a piece of 28-gauge sheet metal.

2. Lay out the keel. See **Figure 4G-5.**

3. Cut out the keel, using a pair of tin snips.

4. File all edges with a mill file to remove burrs.

Figure 4G-2. Make the sail from cloth and dowels.

Figure 4G-3. The mast should be tapered at one end for insertion into the foam hull.

Figure 4G-4. Assemble the sail as shown.

Assembling the Boat

1. Insert the keel into the center of the bottom of the boat.

2. Attach the sail to the top of the boat.

3. Slightly open the eye of the eye screw with a pair of pliers.

4. Install an eye screw in one side of the boat. This will attach to a string in the waterway. The string will keep the boat traveling in a straight line in the waterway.

Testing the Boats

1. Your teacher will provide the class with a waterway in which to test the boats. The waterway will have a string or fishing line stretched along its length.

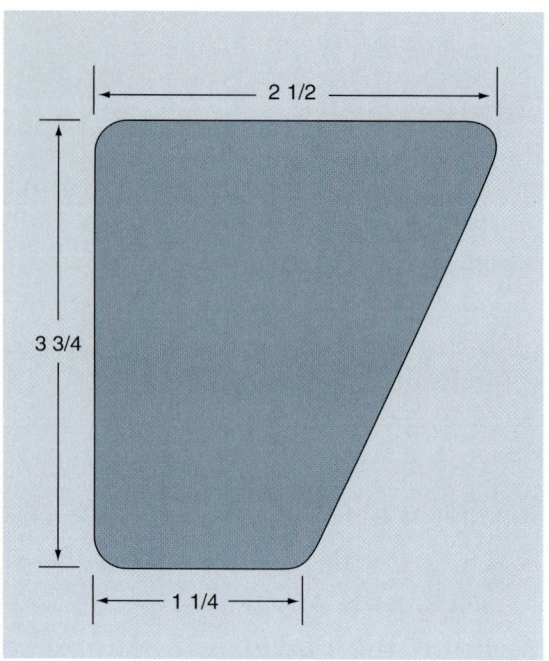

Figure 4G-5. Shape the keel out of sheet metal.

2. Place the boat at one end of the waterway.

3. Hook the screw eye over the guideline.

4. Give the sail a continuous blast of 15-psi compressed air or use a fan to blow air through a cardboard tube.

5. Clock the time it takes the boat to reach the end of the waterway.

6. Record the hull type of the boat.

7. Log the time it took the boat to reach the other end.

8. Determine the average travel time for each of the following groups:

    a. The flat bottom boats.

    b. The round bottom boats.

    c. The V bottom boats.

9. Compare the results and determine which shape is most efficient.

Transportation plays an important role in our lives. Without modern transportation, it would be difficult for us to travel and to use materials and goods from around the world.

Section 5
Technology and Society

Technology is very much a part of our lives. The more we control and change our environment, the more it affects how we live. Let's consider how technology affects our values. Suppose you had to live in a tent because your community had to move often. Would you value the same possessions you value now? Consider the things you have in your bedroom. How many of them would you want to keep if you were moving every month? Would you prefer a bed or a sleeping bag? Do you think you would want a chest of drawers or a duffel bag?

You may live all your life in one community. If you move, moving vans, trains, and airplanes are available so you can take all your belongings. You have read in the preceding chapters about how technology has become interwoven with our day-to-day living.

Not all the effects of technology have been good. Technology has had some negative impacts as well, and we have to look ahead and decide what is best for the future. We are constantly learning how to use technology to our advantage and prevent additional damage to our environment. In the following two chapters, you will have a chance to discuss the impacts of technology. You will learn how society is looking at the effects of technology on the environment.

26 Technological Impacts

27 Technology and the Future

Technology Headline

Speech Recognition Software

Have you been waiting for the day when you no longer have to sit at your computer for hours to type out your homework assignments? That day may be here sooner than you think! With new advancements in speech recognition software, it is already possible to have a computer that talks, listens, and understands. The latest products will allow your computer to follow your voice commands and reply to you in English, or even in Spanish.

The possibilities of this emerging technology are almost endless. Simply by asking a question out loud, you will be able to find out what time it is or what your schedule holds for the day. You will be able to play games without using your hands and open various documents with a simple command.

A major advantage of the latest speech recognition software is that you will not have to train your computer to understand you. You will be able to speak naturally. In addition, you will be able to add your own words to your computer's vocabulary.

Text will easily be converted to spoken words, so your computer will be able to read Internet sites or your own documents to you. Conversely, you will simply have to speak your term paper, and it will be typed out for you. This software will also help in translating from one language to another. You will be able to speak in one language and have the computer repeat your words back to you in another language.

Speech recognition technologies have other applications as well. They are being used as voice recognition security systems for secured transactions, such as with ATM machines. Research is also being done for use of these technologies in the military, health care, and personal vehicles. All of these applications, and certainly many others, would benefit from the ability to perform hands-free, voice-activated actions.

Other uses for speech recognition technologies will continue to emerge. It is expected that this software will become widely used in personal computers in the near future. It may not be long before you never have to type again!

Discussion Starter
Survey the class to determine if any of the students have ever experienced negative impacts of technology. Allow the students to report their personal experiences and the feelings they have concerning them. Help the students focus on the need to be responsible consumers of technology and gain the ability to evaluate the impacts technology has on society. If the students have not had many negative experiences with technology, have them interview their parents or other adults to learn of ways in which technology has had negative impacts. Have them describe their findings in class the next day.

Group Activity
Arrange the students in learning groups and ask them to brainstorm lists of all the ways society uses technology. The group with the most answers wins. Explain to the class that technology can have both positive and negative impacts on society and that it constantly changes the way we live.

Writing Assignment
Have the students write a few paragraphs about the ways in which they can act responsibly when using technology. Ask them to think about ways they can increase the benefits technology offers, while decreasing the harm it also can cause.

Chapter 26
Technological Impacts

Did You Know?

➤ Charles Kettering said, "We should all be concerned about the future because we will spend the rest of our lives there."

➤ Nearly 40 percent of the people in developing countries live in cities. This places large demands for jobs, housing, and public services.

Objectives

The information given in this chapter will help you do the following:

➤ Suggest some reasons for technological change.

➤ Explain how technology has made life better for humankind.

➤ Discuss some undesirable effects of technology relating to resources, the environment, and people.

Key Words

These words are used in this chapter. Do you know what they mean?

acid rain

air pollution

engineered material

exhaustible resource

global village

inexhaustible resource

noise pollution

nonrenewable resource

productive

renewable resource

scarcity

soil erosion

technological unemployment

water pollution

Chapter Outline
Did You Know?
Objectives
Key Words
The World of Change
Technology and Change
  Materials and Change
  Tools and Change
  Communication and Change
  Transportation and Change
  Construction and Change
  Production and Change
Negative Impacts
  Impacts on Resources
  Impacts on the Environment
  Impacts on People
Summary
Curricular Connections
Activities
Test Your Knowledge

Resource
Transparency 26-1: Technology Shapes and Is Shaped by the World

Research
Have the students find examples of and information about the vehicles used to travel to the West in the 1800s.

Brainstorm
Have the students select a technology they use daily and list its impacts.

Technology affects everyone. It impacts the way they live, think, and act. For example, many people feel they must be in constant contact with other people. This feeling is much different than the feelings people have had at other times in history. This new attitude is a result of television, cellular phones, and other rapid communication devices. People also think of time and distance differently than they once did. A trip across the country in the 1800s took weeks or even months. Travelers on the Oregon Trail took up to six months to travel from Missouri to Oregon. Now, we can drive the distance in a few days. You can fly from St. Louis, Missouri to Portland, Oregon in a matter of a few hours. What was once a trip for only the adventurous and brave is now routine for many people. Changes in transportation technology have changed this attitude about travel and distance. See **Figure 26-1.**

Let's consider a simple technology and its impact on people and the surroundings. Think about riding a bicycle to school or a friend's house. Using a technological device (the bicycle) can make the trip easier and quicker. Consider, however, the bicycle's effects on the surroundings. The bicycle operates best on a smooth path with a hard surface. The bike path uses land. This land now cannot be used for other purposes. The bicycle needs a storage place at home. It must also have a space at school or the friend's house. The storage may be a building or an outside bike rack. Both of these have some effect on the environment. They take up space and are built from natural resources.

Consider another option to make the trip to school or a friend's house. An automobile will get you there faster than the bicycle will. It will also protect you from the weather. The automobile can be used in hot or cold weather or when it is raining or snowing. Its impacts on the environment are much greater, however, than the bicycle's impacts. The path for the automobile must be wider than a bike path for the automobile to travel safely. The narrow bike path must be replaced with a street or highway. See

Traveling across the nation . . .

Weeks or Months   Days   Hours

Figure 26-1. Trips that once took weeks to complete now are done in a matter of hours.

TechnoFact: As our society and technology advance, the impacts of technology are increasingly more important considerations. A technology created solely for the good of a civilization still has the potential for serious negative impacts. The Nobel Prize was created because of Alfred Noble's fears about the negative impacts of his own invention. Alfred Noble created a controllable form of nitroglycerine. His invention, dynamite, was patented in 1867. Once he realized his invention could be used as a weapon, he used his riches to create the Noble Peace Prize.

Figure 26-2. The parking space for the vehicle takes up more space. To protect it, a garage is often built onto a home, or a parking garage is provided with an apartment building. These structures take up more space and use more materials than bicycle storage units. The exhaust from the automobile's engine can cause damage to the environment. The car makes noise as it moves from place to place.

These are just two examples of the impacts of technological devices on people and the environment. They show how using technology changes people and their surroundings. There are many more examples. Each application of technology has its own impacts.

The World of Change

All technological advancements cause change. They change the people who use them. People also cause technology to change. New technologies create new lifestyles and opinions about what is needed and good. For example, in earlier days, most men worked in stores and factories. The women stayed home and cared for the family. Housewives had time to prepare meals using fresh meats, fruits, and vegetables. As times changed, many women started working outside the home. This caused many technological changes in the workplace. Also, having less time to prepare meals created a need for processed foods and preprepared meals. Packaged and frozen dinners reduced the time needed to cook meals. Technology was used to reduce meal preparation time. It increased time available for other activities, such as work, recreation, and family events.

For another example, consider the physical condition of people. In earlier history, people did a lot of manual labor. The types of jobs and work around the home kept these people in good physical condition. Now, many people spend most of their time in a chair at a desk or in front of a computer. Children spend less time playing games requiring running and other physical exercise. They entertain themselves with video games and television. Many modern people are out of shape and overweight. This has given rise to the development and use of exercise equipment. This technology was almost unheard of a century ago.

Figure 26-2. Building roads uses land that now can no longer be used for other uses.

TechnoFact: Kirkpatrick Macmillan is credited with building the first pedal bicycle in 1839. He developed pedals linked to levers, attached to the back wheel. This rear wheel drive bicycle was the first of its kind.

TechnoFact: Clarence Birdseye invented the process for quick-freezing foods. He developed the idea by watching Canadians freeze their fresh fish by leaving them outside. His original machine, invented in 1924, froze fish between metal plates. Today, the concept is used to freeze many kinds of fruits and vegetables as well.

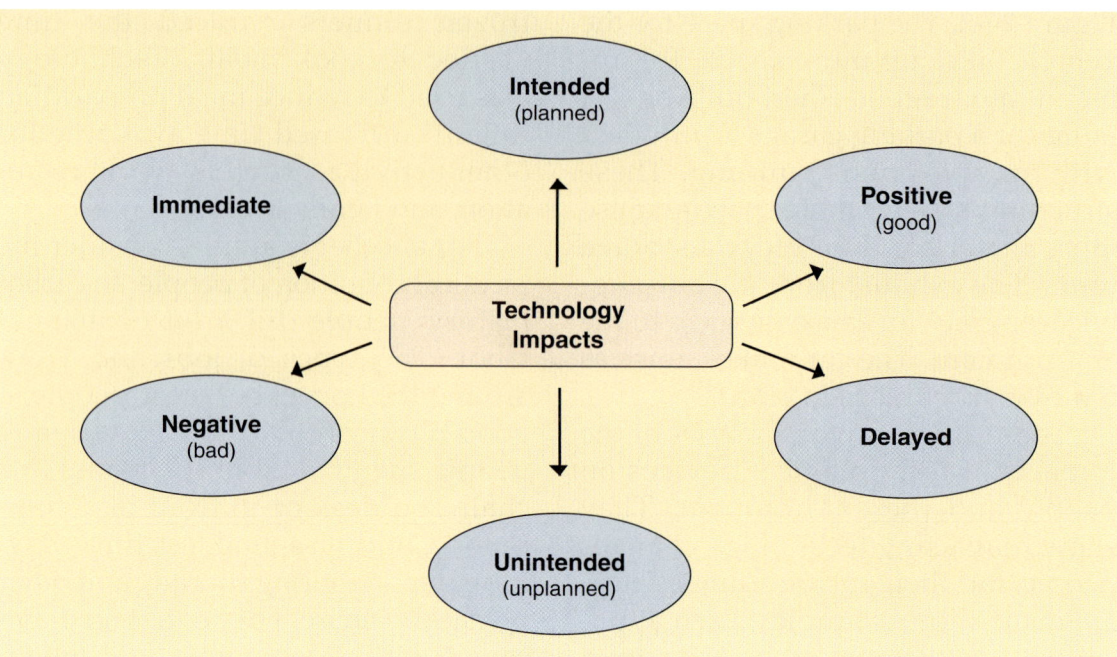

Figure 26-3. All technological changes have impacts. Some are good, and some are bad.

Some of the changes technology causes are good. Others are not. Some of the effects have been intended. They were planned. Other effects have been unhappy surprises. They were unforeseen and unplanned. Some impacts are recognized immediately, while others happen at a later date. See **Figure 26-3.**

It is important to realize that developing and using technology will produce change. Why do we change the way we do things? What is the reason we develop and use technological products and services? Usually, there are many reasons for change:

➤ We want to improve the world around us and make our countries, cities, and homes better places to live.

➤ We want to do work more easily and apply energy more efficiently.

Most of us want to be able to process information and material more quickly and with less effort.

➤ We want to travel more easily and receive information more quickly. People want to be connected to other people and places around the nation and world.

➤ We want more time to enjoy ourselves. This extra time is called *leisure.* Most of us want to have a variety of ways to spend our free time.

➤ We want to help others. People want to help other people live better lives with less poverty.

Technology has given us many good things. Because of technology we have better food, better houses, better ways to get information, and better ways to travel. See **Figure 26-4.**

TechnoFact: In 1996, the German Fraunhofer Institut was awarded a patent for its compression technique, known as *MP3.* The development of MP3 (MPEG-1 Layer 3) technology has had many positive and negative impacts. MP3s are a convenient way to transfer music files, but the MP3 compression also makes it easy to create high-quality pirated music.

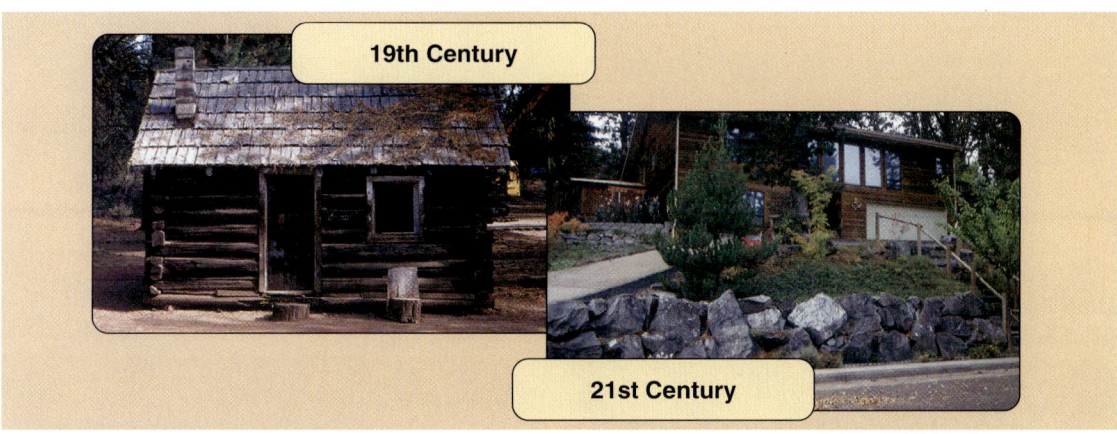

19th Century

21st Century

Figure 26-4. Think of the way people lived in these homes. How has technology changed our quality of life over time?

Standards for Technological Knowledge

6

Research
Have the students create a visual presentation showing how the style and construction of homes have changed throughout history.

Brainstorm
Ask the students how technologies in the home have changed since frontier days.

Because of technology, we have better ways to communicate with family and friends. Our health is better because we have better nutrition (diets). Better medicines help cure illnesses and treat physical conditions.

Technology and Change

All technology is interrelated. New materials allow us to build better products and buildings. Knowledge to work with these new materials requires new ways to communicate with workers. Also, new products require new ways to repair and service them. This requires new training programs. People have to be taught how to use these new products. Therefore, owners' manuals are developed, and users' videos are prepared. With each new advancement, new production and communication techniques are required.

Technological change is everywhere. It impacts us in every facet of our lives. Let's look at some of these changes.

Materials and Change

At one time, people used only the material they found around them. See **Figure 26-5.** These people used trees, stones, and clay. Plant fibers and animal parts were vital materials. Water provided a major source of power.

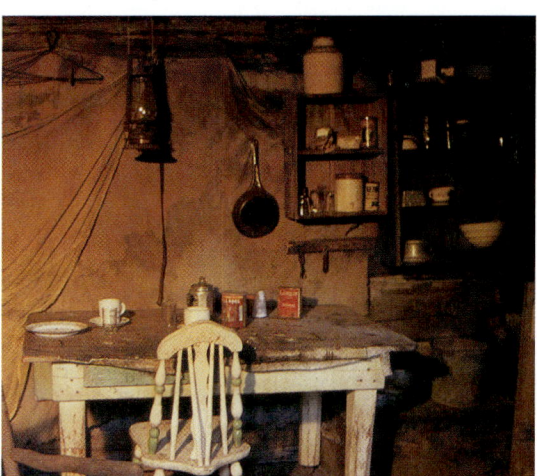

Figure 26-5. Look at this picture of a log cabin interior. What materials do you see being used? How is this different from a modern home?

TechnoFact: Technology has changed the way we treat illnesses and diseases. Many concepts that were developed in the past have been revisited because of modern technological advances. For example, chemotherapy was first developed in 1891 to kill germs in the body. Today, it is used to kill cancer cells.

Brainstorm
Have the students list products, currently made of plastics, that at one time were made of wood, metal, or glass.

Research
Have the students locate an image of a tool used in the past and an image of the same tool today.

Figure 26-6. This modern green bean picking machine has replaced labor-intensive handpicking.

Wood, coal, and animal oils were the fuels people used to produce heat, light, and power. Humans were tied to the natural world in which they lived.

Today, we have materials about which our ancestors never knew. Technology has allowed us to use natural resources in new ways. We can smelt (refine) metals, laminate wood particles together into sheets, and produce complex ceramic materials. Also, we have learned to make new materials. We have created a new class of materials called *engineered materials.* These materials are commonly called *plastics.* They are developed to meet a specific need. This is in contrast to designing a product around an existing material, such as wood or steel. Today, the need can be identified, and then a material (plastic or composite) can be developed to meet the special need.

Plastic was once considered a cheap, inferior material. Through technological advancements, however, today, many things are made of plastic. Wrinkle free, easy care clothing is made of fibers developed from plastic. Automobile and aircraft parts are made of composites (plastic-based materials). Sheet plastics are the shatterproof "glass" in storm doors. Human body parts can be replaced with tough, long lasting, plastic parts.

Tools and Change

Have you ever visited a museum showing old tools and machines? If so, you have seen how different they are from the tools we use today. Over time, people have developed new and improved tools and machines. These technological advancements have allowed people to do several things:

➤ Reduce the amount of labor needed to complete a job. See **Figure 26-6.**

➤ Produce higher quality, more consistent products.

➤ Reduce the scrap and waste created in the process.

These new machines let people build better products more cheaply.

TechnoFact: The textile and clothing industries manufacture many of their products using engineered, or synthetic, materials. In 1884, the first synthetic fiber was created from plant cellulose to take the place of silk. This artificial silk was named *rayon* and is still heavily used today.

Standards for Technological Knowledge

6

Research
Have the students find examples of the technologies in Figure 26-7.

Discussion
Ask the students how life would be different with communication changes.

Figure 26-7. This graphic shows how developing the printed message has changed over time.

They have made work on farms, in factories, and on construction sites easier. People can do more work with less effort using modern machines.

Communication and Change

In early times, people communicated by word-of-mouth. They told each other what they knew and felt. Later, the printed word became a way of communicating. People who could read had access to new information.

In recent years, communication technology has changed drastically. Today, it is possible to touch a series of numbers on a telephone and speak to people anywhere in the world. News from around the world enters our homes on radio or television waves or on cable. Signals bouncing off satellites orbiting in space beam sound and pictures into our homes.

Likewise, the way printed communication is produced has changed dramatically. At one time, type was set, pages were laid out, printing plates were produced, and pages were printed. This took large machines and many trained people. Today, the printed word is changed from ideas into pages with computer systems. What once took large offices with many people can now be done in a home office. See **Figure 26-7.** Anyone can now produce printed messages almost anywhere.

Communication media has become a major part of our lives. It is used in businesses and industry to keep people informed. See **Figure 26-8.** Radio and television entertain us. The

TechnoFact: Desktop publishing software first appeared in the mid- to late 1980s. Companies like Aldus (later acquired by Adobe) and Quark created the software to help those in the publishing industry prepare page layouts. Today, both professionals and home users frequently use desktop publishing software.

Standards for Technological Knowledge

Discussion
Ask the students what the term *global village* means.

Extend
Have the students make a chart of transportation advances.

Figure 26-8. Portable telephones can now be used in factories for more efficient communication. (Northern Telecom)

Internet allows people to gather information from sites around the world with ease. It allows people to "chat" with people they have never met.

Changes in communication technology have created a feeling of a *global village.* We are closely connected to everyone on the globe. This is similar to how our early ancestors were closely connected to everyone in their villages. This closeness makes floods, storms, and wars more personal and real.

Transportation and Change

In early times, people seldom traveled more than a few miles from their birthplace. Rivers and dirt trails were the avenues of travel. People seldom traveled for fun.

Before 1900, "getting away" probably meant hitching a horse to a buggy and taking a drive in the country. Now, it may mean a quick trip to a far-away place. It can be a getaway at some distant resort or park. The trip may start with a bus or automobile ride to an airport. Then, a plane can whisk you across the country in a matter of hours. People travel more often and with greater ease than ever before.

Transportation technology has shrunk the world. Technological advancements moved people from the steam engine to the jet engine in little more than 100 years. The speed of travel changed from less than 50 mph to more than 500 mph. What was once a long trip is now a short hop. Today, in the time it once took to travel 1000 miles, rockets have carried astronauts to the Moon and back to Earth. See **Figure 26-9.**

Fast trains make travel between cities quick and comfortable. For example, the Eurostar Italia moves people in Italy from Florence to Rome at speeds up to 185 mph. The 200-mile trip takes about 90 minutes.

Figure 26-9. Spacecraft allow people to explore space around Earth. (NASA)

TechnoFact: High-speed rail systems using magnetic levitation (maglev) have been developed all around the world. The maglev trains are suspended above the tracks by magnetic forces. This suspension causes no friction and allows the trains to travel at very high speeds, in excess of 200 miles per hour.

Figure 26-10. Notice the use of copper, wood, and ceramic materials on this roof.

Construction and Change

Technology has changed the types of structures we build. It has also changed the way we build them. At one time, construction projects were limited to the materials at hand. If people had lots of stones, they built stone structures. Places that had forests allowed people to build wood structures. Sod houses were built on plains without trees.

Also, construction techniques used lots of labor. Most of the work was done by hand. People cut, shaped, lifted, and assembled materials into buildings and other structures. Technology, however, has changed all this. We are building many more structures in all parts of the world. New materials allow us to build all kinds of structures in almost any location. See **Figure 26-10.** Also, new building techniques allow people to build very unique structures. For example, light-weight materials and new techniques are being used to build the International Space Station. See **Figure 26-11.**

Production and Change

At one time, people worked most of their waking hours. They hunted, gathered plants, cooked meals, and did other productive work. They had little time, however, to enjoy themselves. Over time, technology changed this. People could do their work in less time. This is called being more *productive.* It allows us to produce more goods or services in less time. Being more productive allows people to have better lives with less work. Also, it allows people to have leisure time to do whatever they want.

Productivity also contributes in other ways. In our society, people are paid for the work they complete. With

TechnoFact: The use of prefabricated panels may seem like a new procedure in construction, but it has been around longer than one might think. Prefabricated panels consist of wall sections manufactured in a factory setting and then combined at the construction site. This concept has been used for over 150 years. The Crystal Palace in London, built in 1851, was one of the first large prefabricated buildings.

Brainstorm
Have the students list the positive and negative impacts of several products.

Extend
Hold a debate on the use of new technologies.

Figure 26-11. Here is a view of the International Space Station. (NASA)

this income, we can purchase services and products others produce. Their work supports the efforts of other workers.

Production also allows people to feel they are of value. Their work is needed. They are doing something that helps themselves, their family, and other people.

Negative Impacts

In the previous discussion, the positive aspects of technology were discussed. Some of the ways technology has improved our lives were presented. There are thousands of other ways

technology has made life better. Not all impacts of technology, however, are positive. The use of technology can also have negative impacts on resources, the environment, and people. The progress and use of technology presents moral concerns. Individuals often question whether the use of some technologies is ethically acceptable.

This is not because technology is bad. Rather, it is because we have not always understood technology. At times, we have not used technology wisely. Technologically educated people should use facts collected to evaluate and understand trends in order to recognize the positive or negative effects of a technology. They

TechnoFact: The responsible disposal of the products we use is an important concern for people today. The trash and waste our society generates is substantial and can have very negative impacts on our environment. The people of New York City alone generate enough waste daily to fill the Empire State Building.

Figure 26-12. These wood beams are made from a renewable resource—trees.

should be able to fulfill their personal and public responsibilities to assess technology.

Impacts on Resources

One negative impact of technology is it can create *scarcity.* Materials are made from natural resources. Many of these resources have a limited supply. When they are used up, there will be no more. These are called *nonrenewable resources* or

exhaustible resources. For example, iron ore, coal, and petroleum are examples of nonrenewable resources. Other materials can be replaced. Farm crops, forests, and fish are examples of resources that can be replaced. They are called *renewable resources.* See **Figure 26-12.** Still other resources have an inexhaustible supply. Wind, solar energy, and ocean currents are examples of *inexhaustible resources.* To deal with the possible scarcity of resources, people should do the following:

➤ Reduce the use of nonrenewable resources in products and energy conversion activities.

➤ Develop and use renewable resources.

➤ Promote the use of inexhaustible resources, particularly in energy conversion technologies. See **Figure 26-13.**

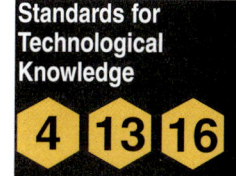

Standards for Technological Knowledge

4 13 16

Resource
Transparency 26-2: Energy Resources

Research
Have the students find a new technology that reduces the natural resource demands of automobiles.

Extend
Create a display showing the students ways to reduce the use of heating fuels in homes.

Figure 26-13. The solar cell (left) and the wind turbine (right) are examples of using inexhaustible energy sources.

TechnoFact: To combat some of the negative effects of automobiles, many car manufacturers are designing concept and prototype vehicles that have a lesser impact on the environment. Many of these vehicles use fuel cells. A fuel cell generates electricity from the chemical reaction between hydrogen and oxygen. One advantage of a fuel cell is that water is the only emission. However, hydrogen fuel is very hard to store and transport.

➤ Recycle used materials to produce new materials.

For example, consider our use of petroleum. The United States imports about half of its petroleum from foreign countries. About one-third of all oil is used in transportation activities. Much of this fuel is burned in automobiles.

People need to be aware of the impacts of their use of transportation technologies. They can address their consumption of gasoline by demanding automobiles that have more efficient engines and lighter bodies. People can use mass transportation to reduce their use of personal automobiles. Electric and hybrid (combination fossil fuel and electric power) vehicles will lessen the impacts of travel on the environment. See **Figure 26-14.**

Likewise, people can reduce the use of natural gas and petroleum in their homes. Taking advantage of the

Figure 26-15. The plastic wrap on this house will seal air leaks that allow heat to escape from the building during the winter.

heating effect of winter sunlight with windows facing south can reduce fuel use. Likewise, putting more insulation in walls and ceilings can reduce heat loss. Also, installing windows with two and three panes of glazing (glass) in them reduces energy use. Sealing up cracks in walls and around doors and windows reduces heating and cooling requirements. See **Figure 26-15.**

Impacts on the Environment

Some technologies can be used for the benefit of our environment. They can be used to repair damage caused by natural disasters, as well as to break down waste from the use of many products and systems. Modern construction technologies and landscaping practices can be used to lessen the effects of earthquakes and major storms. In addition, new ways of reducing waste production can help in repairing the environment.

Figure 26-14. Using this battery-powered bus can help reduce the air pollution in a city.

TechnoFact: The Environmental Protection Agency (EPA) has established national standards for six major air pollutants: nitrogen dioxide (NO_2), ozone (O_3), sulfur dioxide (SO_2), particulate matter (PM), carbon monoxide (CO), and lead (Pb). The EPA measures and records the amounts of each of these air pollutants at centers around the United States. From 1999–2001, the level of each of the six pollutants was decreasing.

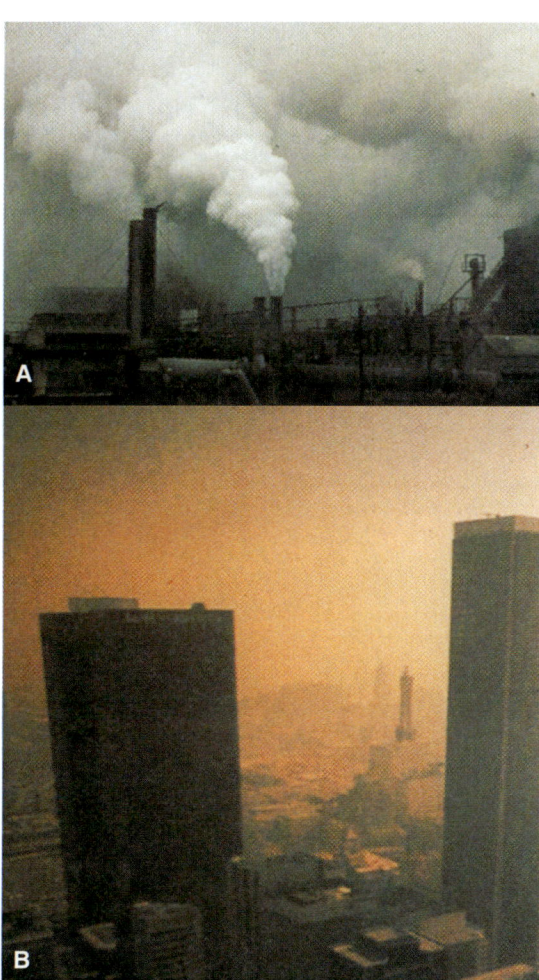

Figure 26-16. A—These are factory smokestacks from years ago. Scrubbers and other technologies take many of the pollutants out of burning processes. B—Auto exhaust gases cause smog problems, like these in Los Angeles. (American Petroleum Institute)

Not all the changes technology causes, however, are good for the environment. Decisions to develop and use technologies frequently put environmental and financial concerns in direct opposition with one another. Some technological activities that have improved how we live have also have damaged our surroundings. This includes the following types of damage to the environment:

➤ Air pollution.
➤ Water pollution.
➤ Noise pollution.
➤ Soil erosion.

Air Pollution

Some technological activities put dust, fumes, smoke, gases, and other materials into the air. This damage is called *air pollution.* It can cause forests to lose their leaves, metals to corrode, and people to suffer illnesses.

Nature causes some pollution. Winds stir up plant pollen and dust. Plants and animals release gases into the air. The pollution human technological activities cause, however, can be more serious. This pollution is released in large quantities.

Burning fossil fuels in power plants and vehicles are some of the greatest sources of air pollution. See **Figure 26-16.** Burning these fuels releases many chemicals into the air.

Chemicals can be harmful when they are introduced into the air. Industrial accidents and spraying crops with insecticides can cause air pollution. Some pollutants in the air are captured by moisture and return to Earth in what is known as *acid rain.* This rain can cause major damage to forests and farm crops.

Water Pollution

Disposal of wastes from various technological activities has caused *water pollution.* The most common

Standards for Technological Knowledge

4 **5** **13**

Resource
Reproducible Master 26-1: Global Warming

Extend
Have the students create a model of a watershed and show the collection of runoff water.

TechnoFact: Sounds and noise are commonly measured in decibels (dB). The decibel scale ranges from 0, the weakest sound heard by the human ear, to 194, the loudest sound possible. According to the National Institute for Occupational Safety and Health (NIOSH), exposures to sounds over 85 dB may cause hearing loss. NIOSH has generated a list of estimated work-related noises, including normal conversation (60dB), a ringing telephone (80dB), a chain saw (110dB), and a rocket launch (180dB).

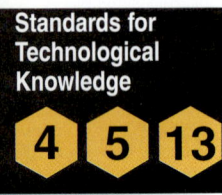

Standards for Technological Knowledge

4 5 13

Research
Have the students find out how techniques like no-till farming, terraces, and buffer zones help control soil erosion.

Research
Have the students research the building techniques used to control sound in recording studios.

Discussion
Discuss the machines and processes that were first invented during the Industrial Revolution.

Figure 26-17. This waste lagoon is used to treat industrial waste. (American Petroleum Institute)

contaminants are wastes from factories, individuals, and farms. See **Figure 26-17.** Factories use water to cool and clean materials. People use sewer systems. Runoff from farmland contains fertilizers, weed killers, and insecticides. A common household pollutant is laundry detergent, which enters water supplies through sewers. People applying too much fertilizer to lawns and landscape plants also can cause water pollution. Polluted waters, besides being unfit to drink, are harmful to both humans and wildlife. Pollution particularly affects fish and waterfowl.

Noise Pollution

Another unwanted output of technological activities is *noise pollution.* This is actually "silence pollution."

Unwanted noise pollutes the silence. This noise can be from factories, transportation vehicles, musical concerts, and other loud technologies.

To deal with noise pollution, the government has passed laws and regulations controlling the levels of noise allowed. Also, some workers wear ear protection to protect their hearing. Motor vehicles are required to use mufflers to reduce noise.

Soil Erosion

Many human activities can cause *soil erosion.* Inefficient farming operations can allow rain and irrigation water to wash away valuable soil. Poorly planned earthmoving activities on construction sites and in mining can cause erosion. Human carelessness can cause forest and range fires,

TechnoFact: The development of new subdivisions often includes the use of retention ponds to help control the water runoff. Many new suburban subdivisions are built on land that was previously used for the purposes of farming. The water runoff was controlled by underground tiles and a series of ditches. However, the homes and lawns built on the farmland disrupt the previous flow of water. The designers create the subdivision in a way that allows water to flow into the retention pond, where it can be held or sent into the series of tiles or ditches.

Standards for Technological Knowledge

4 **5** **13**

Discussion
Discuss how technological unemployment affects communities.

Research
Have the students investigate how products are shipped from manufacturing centers around the world to the places they are sold.

Figure 26-18. Carefully planned mining activities can reduce soil erosion and water pollution.

which can leave land exposed to erosion. Tracks made by recreation or off-road vehicles have damaged desert lands and croplands.

Proper planning in many outdoor technological activities can reduce soil erosion. No-till farming, terraces, and grass buffer zones can help preserve farm soil. Grading mine sites can reduce water runoff damaging adjacent land. See **Figure 26-18.** Allowing natural fires to maintain forestland and rangeland health can reduce the erosion damage major fires cause. See **Figure 26-19.**

Impacts on People

Technology affects everyone. It can make life better, but it can also have negative impacts. These impacts can range from impacts on a person's outlook on life to job opportunities.

Impacts on Jobs

Development of new technologies has not always benefited all people equally. Many technologies have replaced people in the workforce. Before the Industrial Revolution, skilled craft workers designed and made most products. They worked in small shops that, in many cases, they owned. The Industrial Revolution changed all this. Designers developed products for other people to make. Mass production machines replaced the high skill of the workers. The skills moved from the workers to the machines. Machine operators with less skill did the same task over and over all day long. Work was long and hard. Often, the jobs were boring and had few personal rewards.

Now, computers and robots are replacing many machine operators. This shift has led to a demand for new

TechnoFact: In 1980, China designated five areas as Special Economic Zones (SEZ). The roles of the SEZs are to attract foreign companies into establishing manufacturing plants in China. The SEZs are given more independence and tax incentives than other areas in the country. The SEZs include Shenzhen, Shantou, Zhuhai, Hainan Province, and Xiamen.

Standards for Technological Knowledge

4 **13**

Extend
Have the students interview their grandparents or great-grandparents about the technological changes they have seen in their lifetimes and the impacts of those changes.

Figure 26-19. This land a forest fire has swept is susceptible to erosion.

types of workers: technologists and technicians. These workers need to be better educated and better trained. The low skill, low education jobs are rapidly disappearing because of technology. This has led to *technological unemployment,* which means people with little education have lost their jobs to computer-driven machines and robots. Also, there is another impact called *technological unemployability.* This means many people cannot become employed because they lack the proper education. No matter how many jobs are available, they cannot find work. These people do not have the skills needed for the available jobs. They are unemployable. These people are the high school dropouts who have few skills and lack the knowledge needed for modern jobs.

Transportation and communication technologies have also added to the negative impacts on people's jobs. With modern transportation systems, low skill work can be moved to less developed countries. See **Figure 26-20.** These countries have many people who will work for low wages. Communication technology allows for designs to be developed in one country. The products can be made in another country. Transportation technology allows the output to be shipped anywhere on the globe.

Jobs in the highly developed world require well educated people. These people must be willing to continue their education through training programs and in schools. Modern workers must have personal goals and drive. They must be interested in

TechnoFact: Books and the Internet are great resources for understanding the technology used in our lives. One printed resource that may be of value is *The New Way Things Work* by David Macaulay. An Internet site that examines the workings of technology is www.howstuffworks.com.

Standards for Technological Knowledge

4 **13**

Discussion
Ask the students how people can overcome the fear of technology.

Discussion
Discuss the ways technology influences economic, political, and cultural issues.

Figure 26-20. Modern transportation systems allow jobs and products to move freely around the world.

growing with their job and advancing to new jobs. The keys for job success are education and flexibility.

Impacts on People's Lives

Technology can also affect how a person sees the world. Many people are somewhat uncertain or fearful of technology. Technology means change, and some people do not like change. Today, some people say, "If people were supposed to fly, they would have wings." This and similar statements express a fear of flying. These people grew up with the automobile. They feel safe traveling in a car, but unsafe in an airplane. Some people ignore the accident records telling us air travel is many times safer than automobile travel.

Another example is the fear of nuclear-powered electric plants. Most of the arguments against them use fear. Few people understand how these plants work or the level of danger these plants pose. The discussion would be better if knowledge, rather than fear, was used. We may end up at the same decision about the use of these electric plants. It would, however, be a decision based on information. The road to reducing the fear of technology is to understand its operation. Knowledge of technology can erase many fears people have about its use.

The development and use of technology influence economic, political, and cultural issues. Some technology systems have been used to both inform and influence society.

TechnoFact: The Federal Emergency Management Agency (FEMA) is an independent agency that has many tasks. Those tasks include coordinating the federal response to disasters and training people in emergency preparedness. FEMA was created by an executive order from President Carter in 1979. The order merged a number of agencies, including Federal Disaster Assistance Administration, the National Fire Prevention and Control Administration, and the Federal Insurance Administration into one central agency.

Technology Explained

earthquake protection: Actions people take to make buildings safer during natural disasters.

The structures people build are not immune to natural forces. For example, earthquakes can strike suddenly, violently, and without warning, **Figure A.** In early days, builders relied on the weight of the structure to protect it. Often, the structures were not bolted to the foundations. The walls were only nailed to the floor systems. History, however, has shown that this protection is inadequate.

To reduce the risk of damage and injury, people can do a number of things in their homes. These actions can

Figure A. Here are two views of the damage an earthquake can cause. (Federal Emergency Management Agency)

reduce the dangers of serious injury or loss of life. Some of these actions include fastening shelves and bookcases securely to walls, **Figure B.** Wood strips or metal bars should be installed to keep

objects from vibrating off shelves. Large or heavy objects should be placed on lower shelves. Items such as bottled foods, glass, and china should be stored in closed cabinets that have doors with latches.

Standards for Technological Knowledge

4 **13**

Brainstorm
Have the students list ways they could protect their houses and belongings from earthquake damage.

Discussion
Discuss the school's and the community's plans for different emergencies.

Technology also affects the way people of different cultures live, the kinds of work they do, and the decisions they have to make.

Technology affects different cultures in various ways. Although technology and information is shared globally today more than ever, all technology is not available to all cultures. The cultural groups in third-world countries may not have access to electricity, so technologies such as television, computers, and the Internet are

not available. Cultures also choose to use, or not use, technology. For example, the Amish culture chooses to not use automobiles and electricity.

Summary

Technology has contributed to our way of life. It has given us many useful products and services. Because of technology, we live better. We have more comfortable homes and better

Heavy items, such as pictures and mirrors, should be hung away from beds and areas where people sit. Pictures and hanging plants should be hung off securely installed screw eyes. The eyes should be closed to keep the objects from falling.

In addition, water heaters should be secured by strapping them to the wall, **Figure C.** They should also be bolted to the floor. All overhead lights should be braced so they will not swing in large arcs. Gas lines to heating units and other appliances should have flexible connections.

Homes can be altered for earthquake protection in a number of ways. The structure should be securely bolted to the foundation. The super-structures should be secured to the foundation with straps. Freestanding chimneys have metal liners installed. They should be strapped to the side of the house. Any deep cracks in ceilings or foundations should be filled.

Secure the bookcase to the wall with angle brackets.

Add wood strips or steel rods to keep books from falling forward.

Figure B. This is how to make a bookcase earthquake resistant.

Strap the unit to the wall studs.

Add flexible gas line connection.

Figure C. Here is how to make a hot water heater earthquake resistant.

health care, and we can travel and communicate with more ease.

Technological advancement is not, however, without a price. There are negative impacts. Poor use of technology can cause air and water pollution. Poor farming, forestry, and mining practices can damage soil. Jobs change or disappear as a result of technological advancements.

We can never go back to a life with a low level of technology. Most people would not want to do that. We need to use our technology wisely, however, to eliminate or reduce its undesirable effects. We need to keep ourselves prepared to produce and use new technologies.

Extend
Have the students create a program that would help your school deal with the impacts of waste and garbage.

Curricular Connections

Social Studies

Collect news articles about the impacts of technology on people and society. Arrange them in terms of positive and negative impacts. Develop a poster communicating your findings.

Science

Collect samples of water from areas in the community. Test them for purity and chart your results.

Mathematics

Research the length of the workday for ten-year periods (decades), starting in 1800. Graph the results. Complete the same activity for average income (wages) from factory workers, starting in 1900.

Standards for Technological Knowledge

Activities

1. On your way to school or home, list examples of pollution you see or smell. Describe what caused them and how they can be reduced.

2. Prepare a scrapbook or poster showing ways technology is benefiting humankind. Use the Internet, newspapers, and magazines as sources of information.

3. Prepare a scrapbook or poster showing types and sources of pollution. Use the Internet, newspapers, and magazines as sources of information.

4. Visit a farm or forest and discuss soil erosion problems with the owner.

5. Choose a culture different from your own and research ways technology has influenced its development. Write a short report on your findings.

TechnoFact: Many technological decisions are not easy choices to make. The use of nuclear energy, human cloning, and stem cell research are all tough technological choices. Government organizations and the U.S. Congress are often the ones that make such choices. When tough decisions are being made, write to your senators and representatives to make your feelings known.

Test Your Knowledge

Do not write in this book. Place your answers to this test on a separate sheet of paper.

1. What are five reasons to develop and use new technologies?
2. Describe the importance of engineered materials.
3. Give two examples of benefits people get from new and improved tools.
4. Paraphrase the meaning of a global village.
5. Summarize the main advancements in transportation over the past century.
6. How has technology changed the types of buildings we can build?
7. Producing more in less time means you are more _____.
8. Identify three things people can do to avoid scarcity of resources.
9. List four types of negative impacts of technology.
10. Losing you job because of technology is called _____.

Answers to *Test Your Knowledge* Questions

1. Answers will vary. (examples: to improve the world around us, do work more easily, travel more easily, have more time to enjoy ourselves, help others)

2. These materials are developed to meet a specific need.

3. Answers will vary. (examples: a reduction in the amount of labor needed to complete a job; the production of higher quality, more consistent products)

4. The term *global village* refers to the feeling we are closely connected to everyone on the globe.

5. We can now travel much faster and much farther, in more comfort.

6. It has given us new materials and building techniques.

7. productive

8. Answers will vary. (examples: reduce the use of nonrenewable resources, develop and use renewable resources, promote the use of inexhaustible resources)

9. Answers will vary. (examples: air pollution, water pollution, noise pollution, soil erosion)

10. technological unemployment

Modular Activity

This activity develops the skills used in TSA's Environmental Challenge event.

Environmental Challenge

Activity Overview

In this activity, you will identify an environmental problem in your community, develop a solution to the problem, and implement the solution. You will then analyze the effectiveness of your solution and create a visual display and report outlining the project.

Materials

➤ Supplies for a visual display
➤ Three-ring binder
➤ Computer with word processing software

Background Information

Identifying a problem. Begin by identifying at least three environmental problems. Use research techniques to identify these problems. Some methods you can use to learn about environmental problems include:

➤ Read articles dealing with environmental issues in your local newspaper.
➤ Contact local environmental organizations.
➤ Interview people at your school.

You may want to consider problems related to the following topics:

➤ Air pollution issues, such as automobile exhaust, smokestack emissions, and cigarette smoke.
➤ Water pollution issues, such as a polluted creek or stream.
➤ Land issues, such as littering in public areas and construction development in natural areas.
➤ Animal protection issues, such as a hazardous situation for native animals.
➤ Waste issues, such as lack of recycling at homes, businesses, or schools.

For each problem you identify, use brainstorming techniques to list possible solutions. Select a problem-solution combination you will be able to research and implement effectively.

Research. Research the problem using books, periodicals, and the Internet. Clearly define the problem. What is causing the problem? What hazards are created by the problem? What types of solutions have others used for similar problems?

Implementing your solution. If your solution depends on involving others, you will need to find volunteers or create a method of informing people of your solution. You may need to contact individuals or post printed flyers. Before implementing your solution, be sure to consider how you will analyze it.

Analyzing your solution. Analyze your solution by comparing the situation before the solution was implemented to the situation after the solution was implemented. If possible, select several evaluation criteria. These may include a survey of opinions or measurable values. Photographs may also be helpful in analyzing your solution.

Visual display. Your visual display may include poster boards, models, and other items. Consider the principles of design, such as balance, proportion, contrast, and rhythm, in the design of your display. Begin the display by brainstorming and preparing a rough sketch. Include elements to attract the viewer's attention. Select font types and sizes for easy readability.

Guidelines

➤ Your visual display must include (1) your method of identifying the problem, (2) a statement of your solution, (3) your research, (4) your steps in producing a solution, (5) your implementation of the solution, and (6) your analysis of the solution.

➤ Your report must be created using a word processor, must be presented in a three-ring binder, and cannot exceed ten pages. Your report must include a cover page, table of contents, method of identifying the problem, statement of the solution, research, implementation of the solution, and analysis of implementation.

Evaluation Criteria

Your project will be evaluated using the following criteria:

➤ Problem identification

➤ Solution functionality, practicality, and impact

➤ Visual display

➤ Report

Chapter 27
Technology and the Future

Did You Know?

➤ Science fiction writers often create imaginary future worlds in their stories. Some writers have actually been successful at predicting the future. Gene Roddenberry is one such writer. His *Star Trek* television series and movies have shown devices that, years later, have been invented in some form. The communicators used in *Star Trek* are very similar to the cell phones used today. The computer used in the ship has functions similar to the capabilities of our current computers. Other science fiction books and movies feature technologies like homing beacons and sensors, which are now realities. Today, some filmmakers even hire futurists so their films are as accurate as possible.

Objectives

The information given in this chapter will help you do the following:

➤ Explain the role of a futurist.

➤ Define *forecasting*.

➤ Describe the four methods of forecasting.

➤ Recognize several potential future technologies that may exist in each of the seven applications of technology.

Key Words

These words are used in this chapter. Do you know what they mean?

aquaculture
biomass fuel
Delphi Method
expert survey
forecast
futurist
hydroponics
hypersonic flight
manufactured panel
model
scenario
trend

Chapter Outline
Did You Know?
Objectives
Key Words
Futurists
Forecasting
  Trending
  Surveying
  Creating Scenarios
  Modeling
The Future of Technology
  Agriculture and the Future
  Construction and the Future
  Energy and the Future
  Communication and the Future
  Medicine and the Future
  Manufacturing and the Future
  Transportation and the Future
Summary
Curricular Connections
Activities
Test Your Knowledge

Technology is constantly changing the way we live our lives. See **Figure 27-1**. The new, high-tech devices will soon be commonplace or even obsolete. Look around and imagine what your life would be like without the technology you have. You would be affected in ways you probably cannot imagine.

It is often helpful, when looking into the future, to look back on the past to understand how far we have come. Imagine that instead of being born when you were, you were born about 200 years earlier, in 1800. Life would be very different in many obvious ways. There would be no Internet or television. Automobiles and microwave ovens did not exist. There are many things you take for granted that would not be around. It would be another six years until your parents could drink coffee from a coffeepot that strained the grounds. You would be 44 years old before you were able to communicate with people in other

Figure 27-2. The horse and surrey is a mode of travel from many years ago. You can still find it used today in Amish communities.

towns through the telegraph. More than likely, you would not even know anyone from another town. Even if the town was only ten miles away, it would take four hours to get there by wagon. See **Figure 27-2**. You would not need to go shopping for designer clothing because the sewing machine was not invented until 1833. You would be 93 years old before your clothes had zippers, and you would not have lived to see airplanes, rockets, frozen food, or nylon fabric.

If that was life 200 years ago, imagine life 200 years from now. Would you even know how to use the technology of the time, if you were transported 200 years into the future? What types of technology would people have? See **Figure 27-3**. Would people still be using computers? Do you think cars would still have wheels and use gasoline? Would there even be cars? We know life in the future will be different. The world is always changing. It has even been said that the only thing constant is change. There is no way to know the exact answers to all our questions about the future. There

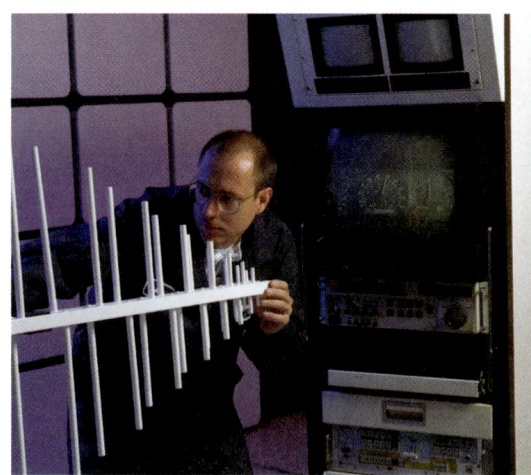

Figure 27-1. Technologies, like electronic devices, are becoming more advanced. (Visteon)

TechnoFact: The first battery was invented in 1800 by Alessandro Volta. He created the battery by stacking layers of silver and zinc disks, separating the dissimilar metals with brine-moistened cardboard. The year 1900 marked the invention of both the escalator and the rigid airship (made infamous by the *Hindenburg*).

Figure 27-3. This concept drawing of a space elevator is one possible future technology. (NASA's Marshall Space Flight Center and Science@NASA)

are, however, methods of predicting what will come in the future.

Futurists

Have you ever tried to predict the future? Maybe you predicted your score on a test, or you may have speculated which teams would make the play-offs. If you have done anything like this, you may be a beginning futurist. *Futurists* are people who spend their time studying the future. They are not fortune-tellers and do not have psychic powers. These people are much more scientific. They use data and information about what is happening now to create descriptions of what may be coming in the future.

Futurists are found in many careers and fields. Economists are often futurists. They use the current events in the financial markets to predict what will

happen to the economy. Science fiction writers are futurists. They write about the future and technologies people will be using in many years. A futurist does not, however, have to look years into the future. Some futurists only look into the future several hours or days. One of the most common kinds of futurist is a meteorologist. These futurists predict the weather hours into the future. A long-range prediction for a meteorologist is one week.

The job of a futurist is very important. There are very practical uses for the reasonable guesses of futurists. The predictions of futurists help change some courses of action. If a futurist sees that the economy will go into a recession, lawmakers and consumers can make changes to help the economy. Another use of a futurist is to determine the jobs that will be needed in the future. A futurist may help a college or university change the classes they offer, so the schools can prepare students for the jobs that will be needed when the students graduate.

Forecasting

The job of the futurist is to make predictions. The predictions are known as forecasts. *Forecasts* are estimates of the possible future. A weather forecast is one example. It is not 100 percent accurate. This forecast is the forecaster's most scientific guess. It can change and be incorrect at times.

Weather forecasts, like all forecasts, are more accurate for the near

Research
Have the students find organizations and societies that create future and forecasting publications.

Discussion
Discuss careers that involve forecasting the future.

TechnoFact: The World Futures Society is a nonprofit educational and scientific organization that examines developments of the future. The society has over thirty thousand members worldwide, in more than eighty countries. Membership in the World Futures Society includes a subscription to the *Futurist* magazine, which is a collection of ideas about current trends and possibilities of the future.

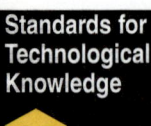
Figure Discussion
Discuss the uses for long-range, mid-range, and short-range forecasts. Refer to Figure 27-4.

Discussion
Ask the students how they have heard the term *trend* used.

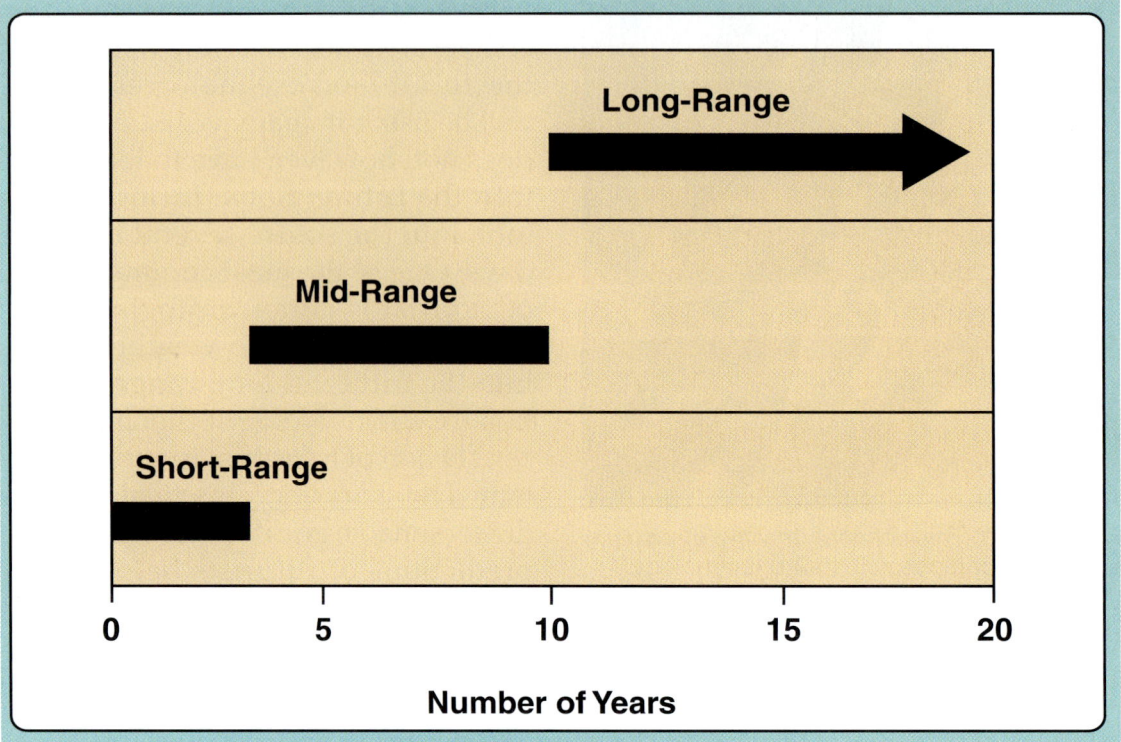

Figure 27-4. This graph shows the range of forecasts.

future. A forecast for the use of electric automobiles in the next two weeks is fairly easy to develop. A forecast for the use of these automobiles in 50 years is much more difficult. There are basically three types of forecasts: short-range, mid-range, and long-range. In most types of forecasting, short-range forecasts are projections for the next 2 years. See **Figure 27-4.** Mid-range forecasts may look into the future about 10 years. Long-range forecasts may extend through the next 50 years. It is hard to even imagine what life will be like in 50 years, let alone to try to predict the technology that will be used.

Because it would be hard to simply guess what the future holds, futurists used different methods of forecasting.

These methods help the futurists more accurately predict what the futures holds. The most popular methods include the following techniques:

➤ Trending.

➤ Surveying.

➤ Creating scenarios.

➤ Modeling.

Trending

Trending is a very popular method of forecasting. It is used in many types of forecasts and is based on trends. A *trend* is an activity that occurred in the past, is occurring at the present, and has an outcome expected to occur in the future. Trends are patterns of technological activities showing a tendency

TechnoFact: Many businesses use methods of forecasting in the ordering of products and supplies. When a retail store orders its goods, it must predict what people will purchase in the next few days or weeks. It uses trends and upcoming events to help forecast the consumers' needs.

Figure 27-5. Futurists use past and present events to forecast the trends of the future.

or taking a general direction. They are used to offer guidance in determining if a product or system should be used. See **Figure 27-5.** Futurists examine trends and determine in which direction they are moving. They also estimate how long the trends will last and monitor possible consequences of technological progress. For example, in the past, cellular phones were much larger in size than they are today. This trend suggests cellular phones will be even smaller in the future. There is a limit, however, to how small these phones can be and still work. A futurist will estimate the year the cellular phones will reach this size limit.

Surveying

There are several ways futurists conduct surveys. The first is called an *expert survey*. In this method, several experts on certain topics are asked for their opinions. They may be asked a series of questions or asked to write an essay on their thoughts about the future. This method results in a number of different outlooks on the future.

Some surveys, however, are used to develop a single view on the future. One type of survey used for this is the *Delphi Method.* The RAND Corporation developed the Delphi Method to forecast the future. The Delphi Method begins by sending a number of experts a survey about the future. These experts respond to the questions and send the surveys back. All the responses are gathered and sent back to the experts, along with a new survey. The experts again complete and return the surveys. This process continues until they all reach agreement on the final conclusion.

Creating Scenarios

A *scenario* is an outline of a series of events. See **Figure 27-6.** The events can be either real or imagined. Futurists create scenarios for planned and proposed events. A scenario is divided into three sections. The first is the explanation of the event. A possible future event may be that there is not enough farmland to grow crops the growing population needs. The next section of the scenario is a possible solution to the problem. A solution to the problem of less farmland might be to develop aquaculture. *Aquaculture* is the growing of plants in the ocean. The last section of a scenario is the expected outcome of the solution. In this example, the outcome may be that the use of aquaculture has replaced the need for crops that previously required farmland.

This scenario contains only one possible solution and outcome. Most

TechnoFact: The RAND Corporation is an independent, nonprofit "think-tank" aiming at improving public policy in a number of different areas. Its key research areas are child policy, civil and criminal justice, community and regional issues, drug policy, education, health, international policy, infrastructure, methodology, national security, population and aging, science and technology, and terrorism. Project RAND was created in 1945 as an outgrowth of World War II.

futurists create several scenarios for each event. By creating many scenarios, they are able to choose the best outcome. They can then suggest the best solution to the problem.

Modeling

Models are helpful in controlling events in the future. They are used to change the outcome of a system or process. Computer, mathematical, and graphic models are all used in forecasting the future. Weather forecasting often uses computer models. These models are used to show the movements of warm and cold fronts. Computers are also used to calculate large

Figure 27-7. Weather stations use computers to calculate mathematical models. (NASA)

mathematical models. See **Figure 27-7.** Economists use mathematical models to show how the economy will change over time. The data from mathematical models is often put into graphs.

Graphs are helpful because they are usually easy to understand. It is also easy to compare two different graphs. Futurists will often create two different graphs by changing a single variable. By making different graphs, the researcher can ask, "What if?" For example, futurists may be studying the likelihood of developing cars that steer automatically. They may develop one graph showing the likelihood if cars are invented that ride on rails. Another graph may be created to show what would occur if cars steered with a Global Positioning Satellite (GPS) system.

Graphic models can also help organize the future. Charts are used to plan future events. Building contractors use charts to plan the construction of a

> By the year 2040, city planners found that our major cities needed to solve serious problems. Moving about in the central city was a problem. Streets were clogged with traffic. Expressways had not helped the problem to any degree. Even worse, air pollution from the traffic was threatening the health of citizens.
>
> Planners decided on a bold plan to reduce traffic. First, all personal vehicle traffic was banned from the cities. Mass transit systems replaced the automobile. Only trucks delivering goods, foodstuffs, and services were allowed within the city limits. Bicycle traffic was allowed anywhere.
>
> Neighborhood shopping areas were located where they were within easy walking distance for shoppers. Streets were no longer congested with traffic. Air pollution was reduced. Many commuters were attracted back to the central city by the quality of life they found there.

Figure 27-6. A scenario can be an imagined event. It looks at related happenings and attempts to predict what the solution might be. Usually, several scenarios will be written for the same set of events.

building. See **Figure 27-8.** These contractors list all the different processes that must be completed in the form of a timeline. They use the chart to organize and control the process.

The Future of Technology

When we think of the future, it is common to think of the technology that will be used. You may wonder what types of houses you will live in or what foods you will eat. You may think about the vehicles you will drive or the medications that will be available. Every application of technology will see changes in the future.

Agriculture and the Future

Agriculture is the application of technology dealing mostly with growing foods we eat. Throughout history, agriculture has become much more advanced. It will continue to become more advanced in the future.

There are several areas of advancement we can forecast for the future of agriculture:

➤ The farming and harvesting of crops will be done in new places. Currently, crops are mainly grown in soil, on land. In the future, they will be grown without soil, using *hydroponics.* They will also be grown in the ocean, a method known as aquaculture. Crops will probably be grown in space, if people begin to live there for long durations of time.

➤ Crops will be genetically altered. One reason for this is so they can grow without the use of pesticides. They will be altered to withstand bugs and other pests. They will also be altered to provide additional vitamins and minerals.

➤ The process of farming will become more automated and computer-based. See **Figure 27-9.** Farmers will be able to farm more land in less time. Computers will analyze and process data to help select the

Standards for Technological Knowledge

5

Research
Have the students research the way hydroponics are used today.

Extend
Hold a debate about the use of genetically altered foods.

June	July	Aug	Sep	Oct	Nov	Dec	Jan
Documents					Drywall		
	Survey/Grade					Flooring	
		Site work/Foundation					Interior trim
				Rough carpentry			Final inspection
			Plumbing				
				HVAC			
				Electrical			
				Roofing			

Figure 27-8. This chart is an example model used to organize the construction of a home.

TechnoFact: The first documented hybrid plant was created over three hundred years ago. Thomas Fairchild cross-pollinated a sweet william with a carnation and created a plant known as "Fairchild's mule."

Figure 27-9. Today, many types of farm equipment are equipped with computers and GPS systems. In the future, the computers will be more advanced and control more functions. (NASA's Marshall Space Flight Center and Science@NASA)

correct crops, apply fertilizers, and harvest outputs.

➤ The technology will exist to genetically engineer animals. The purpose of the engineering will most likely be to create animals producing high yields, with less feed.

Construction and the Future

In the past, constructed works were built mainly from wood, stone, and steel. In the future, this might be much different. Future buildings may be built with new lightweight materials, such as plastics. While building materials are a large change, other changes will include the following:

➤ Buildings will be built to take advantage of computer technology. Homes will be networked to allow computers to be placed in any room. They will also be built so computers can control the rooms. A central computer will control the lights, heating, cooling, music, and many other features.

➤ Because cities are getting larger and there is less land, buildings will be constructed in the sea and underground.

➤ Homes will be built with *manufactured panels,* which will include the interior and exterior surfaces of the walls. See **Figure 27-10.** The electrical and plumbing will be pre-built inside the panel. Robots will be able to assemble the pieces to complete the construction of the home.

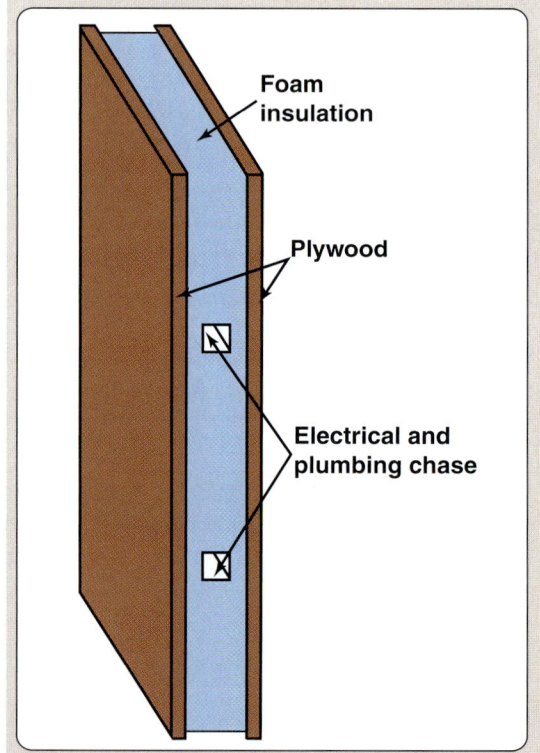

Foam insulation

Plywood

Electrical and plumbing chase

Figure 27-10. Building panels can be used to replace typical frame construction.

TechnoFact: A central computer controls automated, or "smart," homes. The use of a computer raises several issues for potential "smart" home owners. One issue is that of viruses. If the computer is linked to the outside world, it is susceptible to a virus, which could cause the system, or house, to shut down. The second issue is security. The systems would use either wired or wireless communications, both of which could be open to hackers. The hackers could literally take control of the house. These issues will have to be resolved before smart homes become commonplace.

Figure 27-11. Wind energy is one form of energy that will be used more in the future. (AMEC Wind)

Energy and the Future

The energy that will be used in the future will be much different than what is used today. The current trend shows the world will run out of fossil fuels. In order to keep this from happening, several events will take place in the future:

➤ Energy conservation will be a major design factor. Products will be designed to use as little energy as possible. Homes will be better insulated and will use energy more efficiently.

➤ New types of fuels will be used as alternatives to fossil fuels. *Biomass fuels* are an example of a new type of fuel. They are mainly plants that absorb energy from sunlight when they grow. The plants (for example, trees, corn, and sugar cane) release the energy when they are burned. Biomass fuels, which can also include garbage, are cleaner and cheaper than fossil fuels.

➤ Other existing sources of energy will be used more in the future. Solar, wind, and nuclear energy will be used more widely. See **Figure 27-11.** Tidal energy will be collected from the waves of the ocean.

Communication and the Future

The years since the mid-1980s have seen great advancements in the areas of information and communication, with the evolution of the personal computer (PC). In the years to come, those advancements will be made obsolete. Computers will become more powerful and smaller in the future. The spaces for the storage of information will also become smaller. Entire libraries of information will be held on devices the size of an eraser. The future world of communication will include technologies like the following:

➤ The widespread use of satellites for many forms of communication. Satellites will be used for televisions, radios, and telephones.

➤ Videophones, interactive video screens, and virtual reality. These technologies will be common in homes. See **Figure 27-12.**

➤ Printers and fax machines able to print in three dimensions.

TechnoFact: Nanotechnology is a rapidly emerging technology that may have a profound impact on the future. Nanotechnology is the building of devices on the scale of a nanometer, which is 0.000000001 meters. A human hair has a width of 50,000 nanometers. In the future, computer CPUs may be as small as 100 nanometers and as powerful as millions of today's computers.

➤ Voice recognition. It will replace the keyboards on computers.

Medicine and the Future

Medical technology is an area of great concern. Due to advances in medical technology and medicine, people are living longer than they ever have in the past. The longer people live, the more important medical technologies are. As people age, they need devices like eyeglasses and hearing aids. Because of age or illness, some people even need new organs. In the future, doctors will be able to create new organs for those in need. They will also be able to do the following procedures:

➤ Take three-dimensional, color X rays of patients. **See Figure 27-13.**

➤ Replace limbs with prosthetics that use electricity to function like human limbs.

Figure 27-12. Virtual reality will be used for common activities. (NASA)

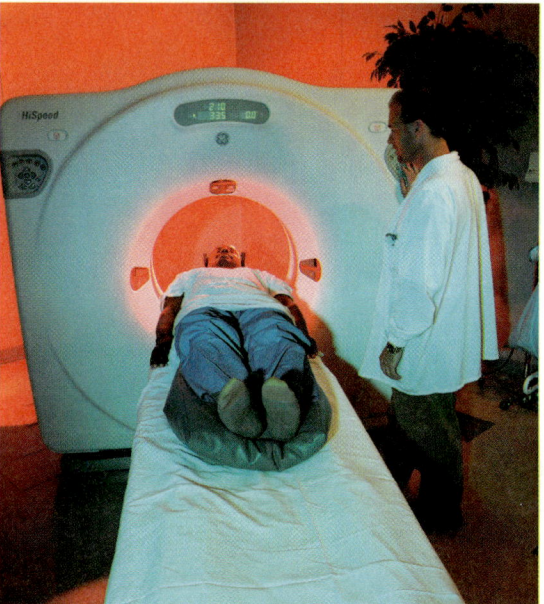

Figure 27-13. Digital imaging will advance, and doctors will be able to review more advanced images of the body. (NASA)

➤ Implant chips containing medical records into patients. The patients will be scanned to find all their information. This will be very helpful in emergency situations, when doctors do not have time to locate the patients' records.

Manufacturing and the Future

The manufacturing of goods will be different in the future. It will be much more automated than it is now. Robots and machines will replace low skilled workers. They will be used to complete repetitive or dangerous tasks. The use of these robots and machines will also create jobs for people who can design and repair robots. Other advances in manufacturing will include the following changes:

TechnoFact: Telemedicine, which is clinical medicine performed when the doctor and patient are in two separate places, connected with telecommunications, will be utilized in the future. Telesurgery, a division of telemedicine, will allow the world's best surgeons to operate on patients around the world without leaving their offices.

➤ Vehicles designed to routinely travel between Earth and outer space. See **Figure 27-14.** Outer space will become a vacation spot for many travelers.

➤ Airplanes that can fly at hypersonic speeds. *Hypersonic flight* occurs at five times the speed of sound.

➤ Cars guided by rails. This will decrease the amount of traffic accidents. It will also increase the speed at which cars can travel.

Figure 27-14. Reusable space vehicles will make space travel more economical. (NASA)

➤ New materials will change the way products are manufactured. They will be lighter and stronger than current materials.

➤ Products will become smaller and "smarter." More products will contain computers and computer chips. This will require products to be manufactured on a smaller scale.

Transportation and the Future

There are several problems facing transportation in the future. One is the growing congestion on roadways. Designers will design solutions to this problem. High-speed commuter trains and vehicles better designed for city traffic will be developed. Future transportation vehicles will also include the following:

Summary

The present is much different from the past. We know technology will continue to change, and the future will be very different than today. It is interesting to try to predict the future. Predictions of the future are known as forecasts. Making these predictions is the main job of futurists. Futurists use several different methods to forecast the future. They base their forecasts on the events of the present and how the events will impact the future.

Forecasting helps us plan for the future. We are able to control and shape the future by making changes today. If a forecast shows we will have a problem in the future, we can take steps to make sure the problem never arises. Changes will occur in all areas of technology, and we must prepare for them. We cannot stop the changes, but we can control them.

TechnoFact: The careers in forecasting the future are many. Most companies and industries have people who spend part, if not all, of their work time investigating the future. Company recruiters must predict the future needs of the company to hire the right people. Educators must look into the future in order to present ideas and concepts that are up-to-date and relevant. Analysts forecast the future in order to predict which companies will survive and what changes need to be made.

Curricular Connections

Language Arts

Write a creative essay about the technology that will be used in the future.

Social Studies

Write a short paper on the way society will change in the future. Explain where and how people will live and work. Also, discuss how people will communicate with each other.

Standards for Technological Knowledge

3

Activities

1. Imagine your life 50 years from now. Choose an event you are forecasting will happen in your life. Create a scenario of this event.

2. Select a career in forecasting. Write an essay on the requirements, duties, and education required for the job.

3. Choose a future technology from one of the contexts of technology. Create a display showing how the technology will function and be used.

4. Prepare a report on an emerging and innovative technology. You may choose one of the emerging technologies described in this chapter or research elsewhere to find a topic. Use a variety of sources in preparing your report—websites, magazines, and television programs are good sources of information on emerging technologies.

TechnoFact: Several companies have been designing commercial space stations to be placed in orbit when space tourism becomes an everyday reality.

Test Your Knowledge

Do not write in this book. Place your answers to this test on a separate sheet of paper.

1. Identify five ways your life would be different if you were born 200 years ago.
2. A _____ is someone who forecasts certain technological developments.
3. Give three examples of careers in which futurists may be found.
4. A _____ forecast is a projection about two to ten years into the future.
5. Name four ways to forecast the future.
6. A trend is an activity that:
   A. Occurred in the past.
   B. Is occurring in the present.
   C. Has an outcome expected to occur in the future.
   D. All of the above.
7. The Delphi Method is one example of a model. True or false.
8. Summarize the process of creating scenarios.
9. List three examples of potential future technologies in the area of agriculture.
10. Buildings will be built exactly the same in the future as they are today. True or false.

**Technological
Impacts, Textbook,
pages 692–693**
This is an activity
allowing students to
explore how the tech-
nology they use
affects society. It is a
one to two day activ-
ity that will work best
if the students work
individually or in
pairs. They will select
everyday objects
from around the
classroom and
explore what hap-
pens to the objects
when they are dis-
carded. The students
will list the ways the
objects impact
society. To conclude
the activity, they will
select one of their
objects and create a
drawing of changes
that can be made to
decrease the nega-
tive impacts.

Activity 5A

Technological Impacts

Introduction

The technology we use impacts our daily lives. The clothes we wear, the types of vehicles we ride in, and the food we eat all require technological decisions. These technological decisions must be made with great care because they affect more than just our own lives. The technology we use affects society. When we use and dispose of technological products, they affect other people, resources, and the environment. We must consider the impacts of the use and disposal of each product of technology we find in our daily lives. This activity will give you an opportunity to consider the impacts of several products you use on a daily basis.

Materials and Supplies

➤ Pencil and paper
➤ Assorted classroom materials

Procedure

You will work by yourself or with one partner. During the activity, you will complete the following steps:

1. Prepare a form like the one on the next page. Your form should fill a piece of paper.

2. Select five objects from around the classroom, school, or your home. Try to pick objects that are different from each other (for example, a school bus, a stapler, and a peanut butter and jelly sandwich).

3. Complete the form by doing the following:

   a. List each item in the first column.

   b. Record the positive and negative impacts on people, resources, and the environment that exist while the object is being used. Place a (P) after the positive impacts and an (N) after the negative ones.

Standards for
Technological
Knowledge

13

    c. List the positive and negative impacts on people, resources, and the environment that exist when the object is disposed of at the end of its life cycle. Place a (P) after the positive impacts and an (N) after the negative ones.

4. Select one of the items and think of ways to decrease the negative impacts.

5. Create a report including the following items:

    a. A description of the item.

    b. A drawing of the improved product.

    c. An explanation of the changes made.

Technological Impact Survey						
	Use			**Disposal**		
Item	**People**	**Resources**	**Environment**	**People**	**Resources**	**Environment**
1						
2						
3						
4						
5						

Technical Terms

A

acetate: a thin plastic used in product and architectural models.

acid rain: pollutants in the air captured by moisture and returned to Earth.

acoustical property: a quality of a material governing how it reacts to sound waves. Materials absorbing the waves are said to be *insulators,* while those carrying sound are called *transmitters.*

acrylic: a plastic material that comes in sheets and can be purchased at either a hardware or hobby store.

active solar system: a system using moving parts to capture and use solar energy. It can be used to provide hot water and heating for homes.

adaptation: when inventions are used for something other than the purposes for which they were intended.

advertising: the act of making a product or business known to the public. Advertisements can be placed in newspapers and magazines, aired on radio or television, or displayed on billboards and signs.

aesthetic: the appearance or design of a product. Texture and color are aesthetic features.

agricultural knowledge: the knowledge about using machines and systems to raise and process foods.

agricultural technology: developing and using devices and systems to plant, grow, and harvest crops. It also includes raising livestock for food and other useful products.

agriculture: using materials, information, and machines to produce the food and natural fibers needed to maintain life.

air pollution: dust, fumes, smoke, gases, and other materials that damage the air.

air transportation: a mode of transportation using aircraft to move people and cargo to their destinations.

alloy: a mixture of two or more elements.

alphabet of lines: technical lines used in multiview drawings.

animal husbandry: the breeding, feeding, and training of animals.

animal science: the science of crossbreeding livestock.

anode: the negative side of a battery.

apartment: units owned by a company or individual and rented to the people living in them.

appearance: a characteristic people see and have feelings toward.

aquaculture: the growing and harvesting of fish, shellfish, and aquatic plants in controlled conditions. It uses ponds, instead of soil, to grow its crop.

architectural drawing: a drawing showing buildings and structures that will be constructed.

artificial ecosystem: a human-made complex reproducing some facets of the natural environment.

artistic design: a type of design used to solve problems chosen by the artist. The solution may be a piece of pottery, drawing, symphony, or pop song. This type of design focuses on two aspects of design: content and form.

aspect of design: the design criteria. Some of the common aspects designers consider are appearance, function, human factors, production, and finances.

assembling: the act of putting parts together.

assembly drawing: a drawing made to show how to assemble a product.

assess: to decide or determine.

assessment: the final stage in the testing of a solution.

audience assessment: gathering information about a group of people to be reached by a message.

audio recording: the recording of sound, such as talking, singing, or music, on tapes, records, or CDs.

B

balance: the weight of the elements on each side of the design. There are two types of balance: symmetrical and asymmetrical.

baler: a machine that picks up a windrow (band of hay) and conveys it into a baling chamber, where the hay is compressed into a cube.

bar code reader: a system used to read numerical codes on packages and tags.

bill of materials: a list of all the parts needed to make one product. It gives the part name, size, and quantity of material to be used.

bioenergy: energy from organic matter. Biochemicals, biofuels, and biopower are three ways bioenergy is used.

biogas: a gas produced by processing animal and plant waste.

biomass: the sum of all organic matter in an area.

biomass fuel: a new type of fuel that will be used as an alternative to fossil fuels.

biotechnology: a part of technology dealing with using biological agents in an industrial process to produce goods or services.

boundary: a property line established by a survey.

box: a shape that can be either a cube or rectangle. A cube is a three-dimensional square. All sides of the cube are the same length. A rectangle is very similar to a cube. The difference is that a rectangle has at least one side that is a different length.

brainstorming: a group process for solving a problem. The group is asked to think of as many solutions as possible in a short time. The group members are also encouraged to expand on each others' ideas. They are discouraged from criticizing or evaluating any of the ideas at this stage.

brand name: the title of a line of products.

bristol board: similar to common poster board, this product can easily be curved, but is not very rigid. Usually found in sheets $1/16''$ thick, it is smooth and shiny on both sides.

broadcast message: communication carried by radio and television stations.

Bronze Age: the stage in human history that took place after the Stone Age. During this time, farming developed, villages and towns started to appear, and metal tools replaced stone ones.

building code: standards or regulations regarding the safe construction of a structure.

bullet point: a single piece of information, such as one item from a list.

C

canal lock: a device allowing ships to change elevation as they travel through a canal.

career ladder: the process one goes through to work his or her way up to better jobs through experience, additional training or education, and perseverance.

cargo: materials or products carried on transportation vehicles.

carrier: a channel for carrying electronically generated messages. It can be radio waves of the electromagnetic spectrum, a wire conductor, or optical fibers.

casting: a process in which an industrial material (usually metal or ceramic) is first made into a liquid. The liquid is poured into a mold, and the material is allowed, or caused, to solidify. The material is finally removed from the mold and finished.

cathode: the positive side of a battery.

central processing unit (CPU): the part of the computer processing and interpreting all information.

ceramic: inorganic (never living) matter made up of crystals. Types include clay-based, refractories, and glass.

challenge: an obstacle or goal needing to be met.

chart: a graphic aid used to model how a set of information is arranged or highlight a series of events. Charts can also show sets of data.

chemical converter: a technology that converts energy in the molecular structure of substances to another energy form.

chemical energy: a reaction between two substances when mixed. For example, when petroleum and oxygen are mixed, they will burn rapidly, if ignited.

chemical processing: using chemicals to change the form of materials. The chemicals change the structure of the material's molecules.

chemical property: a quality controlling how a material will react to chemicals. For example, some materials will corrode or form rust. Others will resist corrosion.

chipboard: often used for backers in calendars, notepads, and packages of paper. It is gray in color and is more rigid than bristol board. The thickness ranges from $1/32"$ to $1/8"$.

civil structure: a structure designed for public use, including roads, bridges, monuments, and other public projects, such as sewers and pipelines.

clay: a natural material used to create models.

client: a customer who buys designers' work.

clinical test: a test conducted in three phases to decide if a drug is safe and effective.

closed-loop system: a control system that uses feedback and is a built-in part of the device. Components within the device monitor conditions and make adjustments to produce the desired output.

coal: a solid form of fossil fuel comprised mainly of carbon.

color: a property of light. The light reflected off an object determines the color.

combine: a machine used to harvest grains.

commercial building: a building used for professional offices, shopping centers, supermarkets, lodging establishments, or repair facilities.

communication: the act of exchanging ideas, information, and opinions.

communication and information technology: developing and using devices and systems to gather, process, and share information and ideas.

communication satellite: broadcasting stations in outer space that can receive messages from Earth and then relay those messages back to Earth.

communication technology: the use of equipment and systems to send and receive information.

compact disc read-only memory (CD-ROM): a compact disc that stores information and can be accessed, but not altered, by the user.

company: economic organizations that change resources into products and services.

compass: a drawing tool used to create circles and arcs.

competition: when two or more companies sell the same types of products and strive to gain customers.

competitor: companies selling the same types of products.

composite: a solid material combining two or more materials, yet each material retains its own properties. Concrete is an example. It combines cement, sand, and gravel, but none of the ingredients change.

comprehensive: a complete sketch describing the arrangement of the elements, including type, illustration, and white space, when preparing to produce a published message.

computer: a programmable electronic device that can store, retrieve, and process data.

computer-aided design (CAD): a type of design using computers to create and store technical drawings for products.

computerized tomography (CT) scanner: a scanner that rotates around the patient's body, as a computer processes the data and creates a cross-sectional image of the body part being scanned.

computer model: a three-dimensional view generated by a computer screen. Some computers can actually "test" the product before it is built.

computer test: a form of testing that can generate very accurate data. It requires a computer model and testing or simulation software.

conclusion: the findings of the research after all data has been gathered and analyzed.

conditioning: an action altering and improving the internal structure of materials. This action will change the properties of the material.

condominium: a living unit joined together with another, but owned by separate families or individuals.

conductor: a material or object permitting an electric current to flow easily.

cone: a shape that is round at one end and comes to a point at the other.

conservation: making better use of the available supplies of any material.

constraint: a limitation or boundary for a design. It tells designers how far they can go with a design.

construction: the process of using manufactured goods and industrial materials to build structures on a site.

construction knowledge: the knowledge about using machines and systems to erect buildings and other structures.

construction technology: using systems and processes to erect structures on the sites where they will be used.

consumer: a user of a product.

consumer safety: a solution tested by government agencies to ensure they are safe for public use.

contact paper: a vinyl material that is sticky on one side. To give a model a certain look, the non-sticky side is printed with a color, pattern, or texture and applied onto another material.

container shipping: using sealed containers to group and contain items for bulk shipping.

content: the topic, information, or emotion an artist is trying to communicate. The content is often the problem being solved.

continuous manufacture: a system in which the parts move down a line. Workers at different stations complete specific tasks. The product takes shape, as it moves down the line. Completed parts are put together to form the finished product.

contractor: one who hires workers and directs building processes.

contrast: a juxtaposition adding variety to a design. Value and color are often used to draw attention to a certain part of the design.

control: a subsystem enabling a vehicle to change speed and direction.

conveyor: a stationary structure that moves material. The structure supports rollers or belts over which the material is transported from one place to another.

cornflakes: a breakfast cereal processed from corn.

corrugated cardboard: a very sturdy material ranging in sizes from $1/8''$ to $1/2''$. It has outer sheets of heavy paper, and the inner core is made of heavy paper bent into ridges.

cost: the money spent to design the product and the money that will be spent manufacturing the solution.

creative thinking: the human ability to think in creative ways.

criterion: an element of the problem needing to be solved. Criteria are some of the requirements, or parameters, placed on the development of the product or system.

crop: grain, vegetables, or fruit that has been agriculturally cultivated.

cultivator: a machine used on farmland to remove weeds and open the soil for water.

custom manufacturing: a process in which products are designed and built to individual specifications.

cylinder: a round shaft with one circle on each end.

D

data: a collection of facts, numbers, and ideas.

decoding: putting a meaning to a message. Decoding is understanding the message so proper action can be taken.

Delphi Method: a survey developed by the RAND Corporation to forecast the future.

descriptive knowledge: the knowledge of relationships, values, shapes, and forms used to describe objects and events.

descriptive research: information gathered by measuring and describing products and events. It describes something as it is.

design: a common technique used to create products or systems. It is a plan for a device or product. The process of design begins with defining a problem. It also involves gathering information, developing and testing, finding solutions, retesting, and selecting the best solution.

design brief: an outline created by the designer to guide the entire design process. It includes the requirements, or the criteria and constraints, of the design.

design process: a series of steps to design a new product or system. These steps lead a designer from the problem to the solution.

design sheet: created by designers, it acts as a summary of possible solutions to a design problem.

design style: a period of design in which many of the same appearance features are evident.

desired output: something produced that had been hoped for.

detail drawing: generally, an orthographic (two-dimensional) drawing giving the size and shape of an individual part. Manufacturing workers will use this drawing to make the part.

detailed sketch: a sketch containing detailed information on size and shape of the product. It will show length, width, squareness, and roundness.

diagnose: to identify medical problems.

die: a set of metal blocks used to cut out, form, and stamp material.

digital videodisc (DVD): a compact disc used for storing information, such as high-resolution video material.

digitize: to put into the form of a series of numbers, so it can be processed by the computer.

dimension: a measurement showing three different types of information: size, location, and shape.

dimensioning: a process using two types of lines in a drawing: extension and dimension lines.

discovery: the act of first noticing something that occurs naturally.

discrimination: An expression of a biased attitude, act, or behavior toward another person.

disease: any change interfering with the normal functioning of the body.

distance multiplier: a simple machine, such as a lever, that can change the amount of movement applied to it. A small motion applied at the end of its arm will create a greater movement at the end applied to the load.

double sided tape: tape that is sticky on both sides and used to stick two materials on top of each other.

downlink: the signal relayed back from Earth stations.

drawing board: a smooth surface on which to draw. It includes one straight edge for the T square to move along.

drilling machine: a machine tool with a rotating drill feeding into the workpiece and producing a hole.

drip irrigation: a type of irrigation that uses main lines to bring water near the plants. Individual tubes or emitters bring water from the main lines to each plant.

drug: any substance used to prevent, diagnose, or treat diseases.

ductility: the ability of a material to be pulled, stretched, or hammered without breaking.

durability: the amount of time a solution will work.

E

earthquake protection: actions people take to make buildings safer during natural disasters.

electrical energy: the energy of moving electrons. The movement can be caused by lightning, batteries, or generators. Moving electrons create an electric current.

electrical property: the quality controlling a material's reaction to electrical current.

electricity: the movement of electrons through materials called *conductors.*

electrocardiograph (EKG): a device used to produce a visual record of the heart's electrical activity.

electromagnetic induction: a physical principle causing a flow of electrons (current) when a wire moves through a magnetic field.

electromagnetic radiation: energy moving through space in waves. Heat and light from the Sun are two examples.

elegant solution: a product meeting a human need in the simplest, most direct way.

elements of design: sizes and shapes developed using rough sketches.

encoding: all the tasks that are part of designing a message to be sent through a communication system. The message may be a symbol, sound, or motion.

endoscope: a narrow, flexible tube, containing a number of fiber-optic fibers, allowing a physician to look inside the body.

energy: the ability to do work. Energy comes from inexhaustible, renewable, and exhaustible sources. It allows technological systems to operate.

energy and power technology: developing and using systems and processes to convert, transmit, and use energy.

energy converter: a device that changes one type of energy into a different energy form.

energy knowledge: the knowledge about using machines and systems to convert, transmit, and apply energy.

engineer: a person who designs processes, products, and structures.

engineered material: an new class of materials, commonly called *plastics,* developed to meet a specific need.

engineering design: a plan to create solutions for problems. Architects, fashion designers, industrial engineers, and chemists are examples of engineering designers. Their jobs involve two aspects of design: form and function.

engineering drawing: a sketch used to illustrate products that will be manufactured.

engineering material: solid matter which has a set, rigid structure. Solids maintain this structure without support from a container. Metals, ceramics, polymers, and composites are the four major types of engineering materials.

equality: A state in which all people are alike in status.

ergonomics: human factors engineering. It is used to make sure people can use the design solutions.

evaluation grid: a chart used to record data. The possible solutions are listed across the top of the grid, and the aspects of the design are listed along the side. This forms a grid and is used to make the choice objective.

exhaustible resource: a material that, once used, can never be replaced. Examples are fossil fuels, such as coal and petroleum.

experimental research: 1. the kind of research scientists conduct, which structures activities so changes or improvements can be measured. **2.** using tests to learn new information. Designers use this research to learn about materials and procedures.

expert survey: a survey in which several experts on certain topics are asked for their opinions.

external constraint: a limit people other than the designer and client set. These constraints can be technical, financial, time, or production constraints.

external selection: a selection someone outside of the design process makes.

F

feedback: the process of giving back data on how well a system is operating. It provides information so adjustments can be made to improve the system's outputs. In the design process, feedback is the use of information from a later step to improve an earlier step. The purpose of feedback is to improve the operation of any system.

feedback system: a control system. Control mechanisms compare information about what is happening to what is desired, and then they adjust the systems to make the desired outcomes more probable.

ferry: transportation by boat used to move people and vehicles across bodies of water.

field test: a test conducted in the environments in which the solutions will be used.

film messages: photographs and transparencies, movies, slides, and filmstrips used for entertainment or for presenting information.

finance: money needed to pay for other inputs (resources) of a system.

financial invention: inventions created to make money. These inventions often make things quicker or easier to do.

finishing: protecting and beautifying the surface of a material.

firing: using heat to condition ceramic materials.

First World country: a region that has resources (land, knowledge, people, materials, and machines) and a desire (values and acceptance) for technology.

fission: a nuclear reaction that causes the atom's nucleus to split apart. In turn, this causes other atoms to split.

flame cutting: using burning gases to melt away unwanted materials.

flexible manufacturing: a system that uses complex machines computers control. It can produce small lots like *intermittent manufacturing*, but it uses *continuous manufacturing* actions.

flexography: a type of printing done from a rubber-like plate with raised letters on it.

flight controls: a series of movable control surfaces that can be adjusted to cause an airplane to climb, dive, or turn.

flight simulator: a computer-controlled device allowing pilots to practice flying an airplane without actually using a plane.

flooring: wood, carpeting, linoleum, or ceramic tile used to cover the subfloor.

fluid converter: a device that converts moving fluids, such as air and water, to another form of energy.

foam: a type of sculpting material that is sturdy, but easy to shape.

foam core board: a sturdy material that has an inner layer of foam, covered on each side with paper. It is used for final mock-ups and can be used for some prototypes.

force multiplier: the quality of a simple machine that makes it capable of exerting more force to a load than is applied to the simple machine.

forecast: to make a prediction.

forestry: the growing of trees for commercial use, such as lumber and timber products, paper and pulp, and chips and fibers.

forging: forming hot material by heating it and shaping it with a hammer.

form: 1. a temporary structure built to contain concrete until the concrete hardens. 2. the space an object takes up.

format: the size and shape of a message carrier. The carrier can be a flyer, billboard, tabloid, regular newspaper, or magazine.

forming: squeezing or stretching materials into the desired shape. It also includes bending, shaping, stamping, and crushing.

formula: the most common type of mathematical model. Formulas help us to understand complex math, science, and technology concepts.

fossil fuels: remains of once living matter. Dead plants or animals partially decayed become a material high in carbon. The carbon in fossil fuels makes them burn easily.

foundation: the part of a structure tying it to the ground. It also supports the weight of the structure.

four-cycle engine: a heat engine having a power stroke every fourth revolution of the crankshaft.

freighter: an oceangoing vessel designed to carry products and materials.

fuel cell: an energy converter that converts chemicals directly into electrical energy.

function: a characteristic describing how the solution works.

fusion: a nuclear reaction causing parts of hydrogen atoms to fuse (join).

futurist: a person who uses different methods of forecasting to predict the future.

G

gas turbine: a type of engine that creates power from high velocity gases leaving the engine.

gene splicing: the process of using enzymes to cut the DNA chain at any point and splice them back together with more desirable parts.

genetic engineering: selective breeding and pollinating, which allows people to develop plants and animals with desirable traits.

geothermal energy: energy originating deep inside the Earth in the form of heat.

global village: the world viewed as a community in which distance has been drastically reduced by electronic media.

goal: a reason or purpose for a system.

grain drill: a machine developed in the early 1700s to plant seeds. It is also called a *seed drill.*

graph: 1. an illustration showing how two or more items compare to each other. The most common types are line, bar, and pie graphs. **2.** a visual representation of data.

graphic communication: a message, such as a picture, graph, photograph, or word, placed on a flat surface.

graphic model: a visual representation on paper. These models are used to organize and communicate information.

graphic organizer: a diagram that helps to organize thoughts and develop ideas.

grinding and sanding machine: a machine tool that uses abrasives to cut materials for the workpiece.

guidance: a subsystem of a transportation vehicle that receives information needed to operate the vehicle.

H

hardboard: made by compressing and rolling wood fibers together, it is used in a number of ways in models. It is dark brown in color, very hard, and stiff.

hardware: equipment and components making up an entire computer system.

harrow: a frame with teeth, dragged over the ground to give the soil tilth.

harassment: An offensive and unwelcome action against another person.

hazardous material: Any substance that exposes people to a health risk.

hazardous waste: An unwanted by-product of technology that exposes people to a health risk.

heat energy: a form of energy. It is present in the increased activity of molecules in a heated substance.

heat engine: an energy converter that converts energy, such as gasoline, into heat, and then converts the energy from the heat into mechanical energy.

high-rise building: a multi-story residential or commercial building that has a skeleton frame.

historical research: information gathered from already existing information.

hot-melt glue: a type of glue used in modeling. It dries fast and clear. A glue gun is needed to heat and apply the glue. It can be used on most materials, except certain foams.

human engineering: the management of humans that examines the way people interact with the solution.

human factors engineering: ergonomics. It is used to make sure people can use the solutions.

humanities knowledge: knowledge about the society around us. It includes religious beliefs, governmental organizations, and the history of families, communities, and the world.

hydroelectric: making electricity using waterpower.

hydroponics: growing plants in nutrient solutions without soil.

hypersonic flight: the ability to travel in the air at five times the speed of sound.

Hypertext Markup Language (HTML): a special type of computer database system used to create documents on the World Wide Web. The system links objects (text, pictures, music, and programs) to each other.

I

ideation: creating a number of new ideas to solve a problem.

illustration board: a type of paperboard material used by designers that comes in a variety of colors.

immunization: the process of systematically vaccinating people through a series of shots to prevent disease.

inclined plane: a surface placed at an angle to a horizontal surface. It makes the moving of a heavy object from one level to another possible with less force than lifting straight up.

Industrial Age: a period in which most Western countries changed from rural to urban, as new machines and sources of power were developed to support the industry of the time.

industrial building: structures that house companies making products and providing important services.

Industrial Revolution: the time period from 1750 to 1850. During this time, many machines and devices were invented, including the steam engine and telegraph.

industry: a number of companies producing competing products.

inexhaustible resource: a resource that has an endless supply.

information: data that has been sorted and arranged. It is facts and opinions people receive during daily life from a variety of sources.

Information Age: the period of time in which technology changed rapidly. The computer and microchip were developed.

information and communication knowledge: the knowledge about using machines and systems to collect, process, and exchange information and ideas.

infrastructure: basic facilities (water, sewage, and power systems) needed for a building.

in-house design team: a group of people working for a company. The designers specialize in designing products for only that company.

innovation: the process of altering an existing product or system to improve it.

inoculate: to vaccinate people through a series of shots in order to promote natural resistance to specific diseases.

input: a resource used by a system to meet the identified goal.

input device: any piece of hardware, such as a scanner, that can be used to enter information into a computer.

inspection: the act of checking parts and products for quality.

instrument landing system (ILS): a navigational aid used to help pilots guide aircraft onto runways for safe landings.

insulation: a material that resists heat passage.

intended output: the reason for which the system was designed. These outputs were produced in response to human needs and wants.

intensive care: the area of a medical facility where seriously ill people receive constant care and monitoring.

intermittent manufacturing: a system that produces parts of a finished product in batches. Different parts of a product are built at different stations. If more than one process is required on a part, the whole batch will move to the next station.

intermodal transportation: a system of travel using more than one transportation system. For example, an air traveler may use land transportation to get to an airport and an escalator or people mover (moving sidewalk) to get to the plane.

internal combustion engine: a heat engine that burns fuel inside its cylinders.

internal selection: a selection of the best solution by the design team, after reviewing and evaluating designs using a set of criteria.

internal source: an improvement coming from inside the design company.

invention: a new and unique product created by an inventor.

invention process: the use of imagination and knowledge to turn ideas into devices, products, and systems.

Iron Age: the period from 1000 to 500 B.C. People learned to smelt iron, and metal became widely available in the Western world. During this time, new metals led to better agriculture, new building techniques, and improved ships.

irrigation: artificial watering to maintain plant growth in areas too dry for successful farming.

isometric sketch: a sketch showing the front, top, and sides of an object, just as the eye sees them.

J

jet engine: an engine that obtains oxygen for thrust.

joist: a parallel beam extending from one wall to the other to support the weight of the floor, furniture, and people that will be on them.

K

kinetic energy: the energy an object has because of its motion.

knowledge: specific information known about various subjects.

L

laboratory: an area where temperature, moisture, and amount of light can be controlled.

landscape plant: a plant grown in a controlled environment.

land transportation: a transportation system operating on or beneath the earth's surface.

language: the signs and symbols people use to communicate with one another.

laser: a device that emits a beam of coherent, monochromatic light.

layout: the process of arranging printed material and graphic illustrations and photographs on a page to prepare the material for print.

leisure invention: an invention created for the pleasure of inventing. This type of invention is often very creative and patented.

lever: a mechanism or simple machine that multiplies the force applied to it. It consists of a long arm, to which the force can be applied, and a fulcrum (pivot point), on which the arm rotates.

library research: research used to get background information on a problem. Books, catalogs, magazines, and the Internet may all be used for this type of research.

line: the shortest distance between two points. In design, it is described as a stretched dot.

load: the physical placement of cargo and people on a vehicle.

M

machine: a device or mechanism that changes the amount, speed, or direction of a force. A machine is also a framework to which mechanisms or tools can be attached to make work more efficient.

machineable wax: a very dense material cut and shaped in milling machines and lathes. It is used for small models or parts of prototypes.

machine tool: a machine that makes other machines. Machine tools change raw materials into parts; these parts later become machines and other products.

machining: using a tool to cut chips of material from the workpiece.

magnetic property: the quality describing a material's reaction to magnetic forces.

magnetic resonance imaging (MRI): a device that uses magnetic waves, rather than X rays, to create an image.

manufactured panel: a panel that will include the interior and exterior surfaces of the wall.

manufacturing: changing raw materials into useful products for public use.

manufacturing knowledge: the knowledge about using machines and systems to convert natural materials into products.

manufacturing technology: developing and using systems and processes to convert materials into products in a factory.

market: a group of people buying and selling a product.

market research: research performed in person, over the phone, or through the mail that helps designers determine what types of people (markets) like their products.

matboard: museum board. Matboard comes in many colors and a variety of textures.

material: a substance from which useful products or items are made. Technological systems use three major types of materials: substances that provide energy, such as petroleum, coal, wind, or falling water; liquids, gases, and non-rigid solids that support life, such as air, water, and fertilizers; and industrial engineering materials, which are solids with a rigid structure and are the basis for all products having a set form.

material property: a characteristic that affects how a material reacts to outside conditions. There are seven groups of properties: physical, mechanical, chemical, thermal, electrical and magnetic, optical, and acoustical.

mathematical model: the use of symbols, along with numbers and letters, to help us understand complex math, science, and technology concepts.

mechanical converter: a device that converts kinetic energy to another form of energy.

mechanical drawing: a very accurate and precise drawing used by a designer. There are two main classes of drawings: orthographic and pictorial. Orthographic drawings are also known as *multi-view drawings*. Pictorial drawings are created to look three-dimensional.

mechanical energy: the energy present in moving bodies. It is sometimes called *kinetic energy*.

mechanical processing: changing material by cutting, crushing, pounding, or grinding into a new form.

mechanical property: the quality of a material that affects how it reacts to mechanical force and loads. This property affects how the material will react to twisting, pulling, and squeezing forces.

mechanical system: a system that provides convenience and comfort inside structures that provide shelter.

mechanism: a basic device that will control or add power to a tool. Mechanisms are designed to multiply the force applied or the distance traveled. The six mechanisms are the lever, wheel and axle, pulley, inclined plane, wedge, and screw.

medical knowledge: the knowledge about using machines and systems to treat diseases and maintain the health of living beings.

medical technology: developing and using devices and systems promoting health and curing illnesses.

medium: any material that will carry a message from a sender to a receiver.

memory: the place where the information is stored on the computer.

metal: inorganic (never living) material which is usually in a solid form. Other marks of a metal are its opacity, ductility, and conductivity.

middle manager: a person who has a management level position below the officers of the company or corporation. Some middle managers are supervisors and heads of departments.

milling and sawing machine: a machine that uses a motor to make a straight or circular saw blade move.

mock-up: an appearance model of a product. It shows how the object will look in real life, but it does not operate. A mock-up is made of easily worked materials, such as clay, wood, or cardboard.

model: a three-dimensional copy of a new product.

modeling: predicting future events by charting what is likely to happen. The person decides what the major event is likely to be. Then, he or she sets down, or "charts," the minor events leading up to the major event.

molding: shaping a part or product with the use of a mold.

monitor: the primary device for displaying information from the computer.

mounting tape: a thin piece of foam that is sticky on both sides.

multiview drawing: a drawing that uses several views to describe the object. It is also called an *orthographic drawing*.

N

natural gas: mostly made up of a gas called *methane*. This gas is a simple chemical compound made up of carbon and hydrogen atoms. Pumped from underground, it is generally used for heating buildings and producing electric power.

natural resource: a material that appears in nature.

natural system: everything that appears in nature without human interference.

need: a requirement to live, such as shelter, clothing, and food.

noise pollution: unwanted noise from factories, transportation vehicles, musical concerts, and other loud technologies.

nonrenewable: a material that, once used, can never be replaced.

nonrenewable resource: a material made from natural resources, many with a limited supply, that, once used up, cannot be replaced.

nuclear energy: heat released when atoms are split.

O

oblique drawing: a sketch used to show objects in which one view is the most important.

oblique sketch: a drawing similar to an isometric sketch, except only the side view is at an angle. The angle is usually 45°. It can, however, be any angle. These sketches show the front view in its true shape.

ocean liner: a large ship designed to carry passengers on sea voyages.

open-loop system: a feedback system that has external controls and requires human intervention.

operating system: a group of devices that manage the computer system by controlling the hardware and software. It manages such things as the processor, memory, and disc space. It provides a consistent way for application software to deal with the hardware.

operative management: managers who help plan the day-to-day operations of their businesses. They also set up work schedules and supervise production workers, salespeople, bookkeepers, and other employees.

opportunity: the use of a new or existing product or system in a new way.

optical property: the quality governing a material's reaction to light. Some materials absorb light, some reflect light, and others allow the light to pass through.

orthographic drawing: a drawing that has several views used to describe the object. It is also called a *multiview drawing*.

output: the result of any system.

output device: a piece of hardware allowing the user to get information from a computer.

outsourcing: sending designs outside of the company to be built.

P

paraffin wax: a mineral wax that can be carved for small models. It can also be heated and cast into a mold.

passive solar system: a system that does not use moving parts to capture and use solar energy.

patch: a small piece of software that fixes a known problem.

pathologist: a medical professional that examines tissues to determine if it is normal or diseased.

peanut butter: nuts processed into a spread.

peat: decayed matter compressed into solid material. It is used to heat homes in some parts of the world.

performance: how well a solution works. It also can be the quality of a picture or sound.

personal problem: a problem affecting an individual. Emotional problems, low self-esteem, and academic problems are all examples of personal problems.

perspective drawing: a drawing most like what the eye sees. These drawings use vanishing points and vertical lines to convey the image.

pest control: a spray used to control insects and weeds that may damage farm crops.

petroleum: the liquid form of fossil fuel comprised mainly of carbon and hydrogen.

photographic system: a method of storing visual material by capturing it on film and paper that has been made sensitive to light. A camera captures the picture and exposes the film. After developing, the film is used in an enlarger to project the pictures onto light-sensitive paper.

physical model: an actual three-dimensional replica of the design. These models are built so the designers and clients are able to see what the design will look like.

physical property: a basic feature of material, such as density, moisture content, and smoothness.

pictorial drawing: a drawing of an object that appears as the eye sees the object. The three major types of pictorial drawings are isometric, oblique, and perspective drawings.

pilot run: the use of a manufacturing system to produce test products.

pipeline: a large-diameter pipe usually laid underground. It is designed to move liquids and loose, solid material from one place to another.

pivot sprinkler: a sprinkler that uses one long line attached at one end to a water source. The line is constantly moving very slowly in a circle. Sprinkler or mist heads apply the water as the line pivots.

planing and shaping machine: a machine that cuts metal using a single-point tool. Both machines produce a flat cut on the surface of the work.

plant: to place seeds in the ground to grow.

plant science: the science of cross-pollinating crops.

plastic: synthetic (human-made) organic materials designed and produced to meet specific needs.

plastic cement: a very strong adhesive used on plastic pieces. It dries clear and fast.

plotter: a printer able to print on large paper.

plow: a blade-shaped plowshare that cuts, lifts, and turns over the soil.

plywood: a material made by gluing thin sheets of wood together. It is often used as a base for models and normally used in places that will not be seen.

pollution: an output harming the environment. Types of pollution are air, water, soil, noise, and visual.

polymer: organic (once living) matter usually made from natural gas or petroleum. Wood can, however, also be changed into a polymer.

polystyrene: a foam used in construction as sheet insulation. It can be cut and shaped with everything from saws to sandpaper, but hot glue and other types of model glue will melt it.

polyurethane: commonly known as *urethane foam*. It is often used for larger or thicker models because it breaks easier if it is thinly shaped.

potential energy: energy at rest, but able to do work.

power: 1. the amount of work done in a set period of time, or the rate at which work is done. **2.** the rate at which energy is changed from one form to another or moved from one place to another.

prevent: to keep something from happening.

primary processing: converting raw materials into industrial materials.

printing system: the methods and machines used to produce words and pictures on untreated paper and other materials.

problem: anything that can be made better through change or improvement.

problem solving: the process of beginning with a problem and ending with a solution. Once the cause of the problem has been recognized, the next step is to fix it and check if the solution works.

problem statement: a clearly defined, open-ended statement stating a problem, not a solution.

process: the actions taken to put resources to use. A process includes the steps taken to produce products and services.

processing: changing material to make it more useful.

produce: to make or grow.

product design: the creation of products meeting human needs and wants.

production: the building of a solution.

productive: 1. producing something useful for the betterment and enjoyment of life. **2.** accomplishing more work in less time.

profit: the amount of money left over after the company's bills are paid.

propeller: a rotating, multibladed device that drives a vehicle by moving air or water.

property: a characteristic of a material. There are seven properties: physical, mechanical, chemical, thermal, electrical and magnetic, optical, and acoustical.

proportion: the size and shape of an object.

propulsion: a vehicle subsystem that provides the means to move a vehicle.

prototype: a working model of a product. It is made to test the design. Usually, a prototype is made from the material that will be used in the manufactured product.

public building: a building designed to meet public needs. The building is usually paid for with tax money.

published message: a message produced by printing. Published messages are produced in newspapers, magazines, books, and greeting cards.

pulley: a small wheel with a grooved rim. A rope is run through the wheel to lift weights. Sometimes several of them are used together to multiply applied force. A pulley can also be a wheel (attached to a shaft) that transfers motion by way of a belt.

pyramid: a shape that comes to a point at one end and a square at the other end.

Q

quality: in manufacturing, the degree or grade of excellence in a product.

questioning: a process in which the designer asks the question why things are done the way they are in the existing product.

R

radiant energy: the energy (movement) of atoms present in sunlight, fire, and any matter.

radiology: the use of electromagnetic waves and high frequency sounds to diagnose diseases and injuries.

rafter: a sloping member of a roof frame that runs between the ridge board and the outer walls of a building.

recall: a request that those who purchased a product bring it back because a safety problem was found.

receive: to convert incoming radio waves into perceptible signals.

receiver: an electrical or electronic device that gathers and processes messages generated by a transmitter.

refined sketch: a freehand drawing that adds detail and develops design ideas shown in rough sketches. The purpose is to narrow down and improve the promising solutions presented in the rough sketches.

regulation: a rules or law made by a government organization.

reliability: The extent to which a product, service, or system yields the same results every time it is used; dependability.

rendering: a colored or shaded sketch. It shows the final appearance of a product.

renewable resource: a material that can be replaced.

research: to scientifically seek and discover facts. The three major types of research are library, survey, and experimental research.

research process: a set of steps used to find all the facts about a problem. It is also called the *scientific method.*

residential building: a structure used for living.

rhythm: the effect of motion, created with repetition of elements.

roadway: a permanent structure used for vehicle transportation.

rocket engine: a reaction engine based on a principle of physics. Solid fuel and liquid fuel are the two major types of rocket engines.

rough: an initial sketch showing basic ideas.

rough sketch: a pictorial (picture-like) free-hand drawing showing only basic ideas of the size and shape of a product. These sketches are done quickly and without detail to capture ideas that come to the designer.

route: an invisible path determining the way of travel from one place to another.

S

safety: the act of making a product safe for consumers to use, for workers to produce, and for the environment.

salary: money earned from work and paid on a weekly or monthly basis.

scale: a specialized type of ruler used to make measurements.

scarcity: the state of being in short supply. It is a condition in which there is not enough of an item to meet the demand.

scenario: a method of "futuring," in which the person sets up a series of related events that might happen in the future. Then, the consequences (results) of these events are given.

schedule: a time set for travel.

schematic: the use of pictures and symbols, rather than words, to show a process.

schematic drawing: a drawing used to show the relation of parts to one another and how the product flows.

scientific discovery: the act of first noticing something that occurs naturally, usually by a scientist or researcher.

scientific knowledge: knowledge about the world. It explains the laws and principles governing the universe.

scientific method: 1. a way of discovering new scientific information or testing old scientific theories. **2.** a set of steps used to find all the facts about a problem.

screw: a simple machine in which an inclined plane is wrapped around a shaft. A screw is a force multiplier used to fasten parts.

secondary processing: changing industrial materials into usable products.

separation: a type of process that removes unwanted portions of a workpiece.

shading: the use of marking made within outlines to suggest degrees of light and dark in a drawing. It relies on a light source and helps to show how the object will look in either sunlight or interior lighting. Shading also allows the designer to see the shape and form of the object. It adds depth and dimension to a drawing to make it appear more realistic.

shadowing: The use of shadows placed on the opposite side of the light source and following the rough shape of the object being drawn to add depth and dimension to a flat sketch.

shape: the space made by enclosing a line. It can be flat or three-dimensional and have a variety of lengths and widths.

shearing: a separation action that removes excess material by breaking it into two parts. Scissors, tin snips, and shears are shearing tools.

shearing machine: a device that slices materials into parts by using opposed edges to cut the workpiece. Common shearing machines are sheet metal shears, punch presses, and paper cutters.

shipping lane: a sea lane. The regular route ships take over the water.

simple machine: a basic device controlling or adding power to a tool.

sketch: a tool that helps designers communicate their ideas on paper.

social invention: an invention created to help the inventor or other people. These inventions make our lives better and easier and impact society.

social problem: a problem dealing with society. Illiteracy, violence, and crime are examples of social problems. Some social problems are the results of natural disasters, such as hurricanes, floods, and tornados.

soil erosion: the wearing away of the earth's surface, primarily due to water and wind.

solar cell: a device that converts the energy from sunlight into electric energy.

solar converter: a device that converts energy from the Sun to another energy form.

solar energy: energy of the Sun given off as heat and light.

solid model: the most complete three-dimensional computer model. These models are used to represent the insides and outsides of designs.

space transportation: a new mode of transportation that uses unmanned and manned flights to explore the universe.

specifications: written directions for the builder that controls the quality of work and materials on a construction project.

specification sheet: a list describing items in detail that cannot be shown in drawings.

sphere: a perfectly round object.

spin-off: inventions that have been adapted to solve other problems.

spray adhesive: a type of glue applied by spraying it out of an aerosol can.

sprinkler: a device used to control and distribute water.

standardization: a process used to make standard parts that can be used for more than one product.

step-down transformer: a transformer reducing the voltage of electric current.

step-up transformer: a transformer increasing the voltage of electric current.

Stone Age: the period of time up to 3500 B.C., when the use of stones was the main material for making tools.

storage: a space in which items are placed for safe keeping.

storyboard: the "layout" for a television commercial. It consists of sketches of each shot, accompanied by the script for the commercial.

structure: a vehicle subsystem that provides a rigid framework to protect the vehicle's contents and support other systems.

stud: a vertical member of a wall frame. Usually made of wood 2" x 4" or 2" x 6" thick and spaced 16" or 24" apart.

superstructure: the part of a structure placed on and above the foundation.

surface model: a three-dimensional computer model showing how a certain design will look. It can be shown in different lights and angles, but because the inside is hollow, it cannot be used to show interior details.

surgery: a medical procedure used to remove diseased organs, repair broken bones, and stop bleeding.

surveying: measuring to determine boundaries of lots and acreage.

survey research: information gathered to find people's reactions to designs and events.

suspension: the subsystem that maintains the vehicle on a pathway.

swather: a machine that cuts and windrows the hay in one pass over the field.

system: a set of related parts. Together, they form a whole, designed to accomplish some purpose.

system design: the organization of different parts to solve a problem.

T

table: a graphic model made of rows and columns, often used to compare sets of data.

tablet: a device that has a pad and puck and is used with the CAD system.

tanker: large ocean-going vessel carrying liquid materials, such as petroleum and chemicals.

technical drawing: a drawing used by engineers and architects to show how products are to be made or buildings are to be built. These drawings can also show how communities are to be developed or parks are to be landscaped.

technician: a specialist who works with an engineer or a scientist. In engineering occupations, technicians work between the engineer and skilled artisan.

technological knowledge: knowledge of the human-built world used to design, produce, and use tools and materials.

technological problem: a problem that can affect individuals and groups of people. These problems can be solved with devices or systems.

technological system: an organized grouping of inputs, processes, outputs, feedback, and goals. The purpose of such a system is to produce goods and services.

technological unemployment: a situation in which people with little education lose their jobs to computer-driven machines and robots.

technologist: a specialist in a manufacturing enterprise or some other enterprise. He or she works under an engineer or a scientist.

technology: the use of knowledge, tools, and systems to make life easier and better.

telecommunication system: a system of exchanging information over a distance.

template: a piece of plastic with shapes and symbols cut into it that designers use to trace around the edge.

terminal: a support system of the transportation system used to make travel more convenient. Services at a terminal may include food areas, rest rooms, and shops.

test: an experiment or examination used to generate results.

texture: the surface of an object. It can be both the appearance and feel of the product.

thermal conductivity: the measurement of heat moving through a material.

thermal converter: a device that converts heat energy to another form.

thermal energy: energy from the Sun's heat.

thermal expansion: a state in which materials are heated and become longer and wider.

thermal processing: using heat to change material into a more useful form.

thermal property: a material's reaction to heat. Heat may expand some materials. Certain materials allow heat to travel through them easily. Others resist movement of heat.

thermoforming: a forming process that produces plastic parts of all shapes.

Third World country: a region where technological advancement is slow due to poor resources and an absence of human will.

30°-60° triangle: a triangle that has one 90° corner, one 30° corner, and one 60° corner.

Thrombe wall: an 8"–16" thick masonry wall with a layer of glass mounted about 1" in front of the wall.

thumbnail sketch: a small, rough sketch used to generate ideas.

tidal energy: energy that uses the mechanical energy of moving water. The gravitational pull of the Moon drives the tides.

tillage: the operation of breaking and pulverizing soil.

time: a key resource in developing and operating technological systems, measured in machine time, work hours, product life, or service intervals.

tool: a manufactured device designed to perform specific tasks. Generally, it means an instrument worked by hand.

tooling: a special device that holds a part while it is being manufactured. It is designed to make manufacturing more efficient.

tower crane: a self-raising crane used to lift materials on high-rise building projects.

tractor: a machine that provides power to pull all types of farm equipment.

trade-off: a preference or an exchange for one feature or idea in favor of another.

transformer: a device that can increase or decrease the voltage (force) of electric current.

transmission: the act of moving energy to where it performs work.

transmit: to send a message from one person or place to another.

transmitter: an electrical or electronic device that develops and sends messages over air waves.

transmitting: any method of moving a message or signal between a sender and a receiver.

transportation: the movement of people and materials from one place to another.

transportation knowledge: the knowledge about using machines and systems to move people and cargo.

transportation technology: developing and using devices and systems to move people and cargo from an origin point to a destination.

treat: to cure diseases, heal injuries, and ease symptoms.

trend: the general course of events. In technology, a trend is the direction taken by inventions and the producing of new products and machines.

T square: a tool made up of two pieces, the head and blade. It is used to draw all horizontal lines in a drawing.

tugboat: a strongly built, powerful boat used for towing and pushing loads in the water.

turbine: a device that transforms energy from a flowing stream of fluid into rotating mechanical energy.

turning machine: a machine tool that uses a stationary tool to cut a material rotating around an axis.

U

ultrasonic: having a high frequency sound.

undesirable output: scrap, waste, and pollution. These outputs are produced by most technological systems and can be harmful to the environment.

unit train: a train that carries only one product to a single destination. For example, a train may carry only coal from a mine to an electrical power station.

unity: a condition of artistic harmony. It is the opposite of contrast. Unity helps the design blend together.

unload: the removal of cargo from a vehicle, or passengers disembarking at their destination.

uplink: the upward signal sent from Earth stations.

upper management: managers who set company goals and plot courses of action. They direct the operations of their companies' departments. Upper management can include the president, vice presidents, Chief Executive Officer (CEO), and Chief Operating Officer (COO).

use: to select an appropriate product and determine which technological service to put into action.

utility: a service, such as electricity, water, or fuel, brought into a building.

V

vaccination: a series of shots to prevent disease.

vaccine: a special drug created to prevent diseases.

value: the degree of light and dark in a design.

variable: any object that can be changed.

veneer: a thin sheet of wood that comes in many different widths. It is used to give the appearance of a high quality wood.

videocassette recorder (VCR): a magnetic tape recorder used to record and play back video images and sound.

video recording: capturing motion pictures to be replayed through a television set. One method is to place the material on magnetic tape.

vocabulary: words developed to describe new technological devices, tools, and actions.

W

wage: money earned for each hour of work completed.

want: something people desire. Wants are not needed to survive, but they help make life better and easier. A want can be small or large and can make life more enjoyable.

water pollution: waste from factories, individuals, and farms that contaminates the water.

water transportation: transportation through, and supported by, water. This includes ships, sailboats, rafts, barges, tugboats, and submarines.

water turbine: a device that uses the energy of moving water to make electricity.

wave communication: electronic communication that uses radio waves of the electromagnetic spectrum to carry signals. Sounds or pictures are changed into pulses of electrical energy. These pulses are carried through the air by the electromagnetic waves. They are picked up at a distant point and changed back to sound or pictures.

wave energy: energy that uses the forces present in the coming and going of waves near the shore. Both of these sources can power electricity-generating equipment.

wax: a material that can come from several different sources (animals, vegetables, or minerals). It is used for sculpting and molding material.

wedge: a simple machine that combines two inclined planes. It is the basis for a number of hand tools, including the knife, chisel, and ax.

wellness: a state of personal well-being.

wheel and axle: one of the six simple machines. It is a type of lever. The axle, or shaft, is the fulcrum (pivot point), and the wheel is the lever.

white glue: a safe, water-based adhesive that is very easy to use. It can be used on paper products, foam, and even some wood.

wind energy: an indirect form of solar energy.

wireframe model: a three-dimensional model that uses lines to represent the edges of objects.

wood glue: an adhesive used to glue pieces of wood. It dries quickly (in 20–30 minutes) and is easy to use.

work: the application of force applied to move an object.

worker: a person who produces products and services that have been designed.

working drawing: the most complete drawing produced by a designer and used by engineers and architects to display all the information needed to build the solution. These drawings also show how the products are assembled.

World Wide Web: a computer system of Internet servers that uses specially formatted documents. The special format allows users to easily jump from one document to another.

X

X-ray: a type of machine that examines, treats, or photographs with short, electromagnetic waves that can pass through solid materials.

Index

Abilities, 177
Abrasives, 93
Academic connections. *Refer to chart on page 17*
Acetate, 348
Acid rain, 667
Acoustical properties, 120, 123
Acrylic, 348
Actions
    assessing, 30
    designing, 29–30
    producing, 30
    technology and, 29–30
    using, 30
Active solar systems, 504–506
Actuating, 576
Adaptations, 225
Adhesives, 352–353, 485
    spray, 352
Adobe, 479
Advertising, 576
Aesthetic features, 278
Aesthetics, 320
Agricultural crops, 437–440
Agricultural knowledge, 28
Agricultural technology, 69–70, 436–462
Agriculture, 432, 436, 452–455
    biotechnology and, 452–455
    the future and, 685–686
    quality control in, 462
    technology in, 440–450
    types of, 437
        special, 452
Air, 610
    pollution, 667
    transportation, 614
Airlines, 614
Alloys, 111
Alphabet of lines, 394
Alternate energy sources, 143–146
Alternating current, 497
Alternators, 496
Amorphous materials, 112

Amplifying, 102, 527–528
    media, 103
Anaerobic generation, 136
Analyzing data, 278–280
Animal husbandry, 437
Animal science, 436
Annealing, 565
Anodes, 402
Antibiotics, 452
Antibodies, 452
Apartments, 470
Appearance, 314–315, 368
Application software, 548
Aquaculture, 452, 683
Aqueducts, 372
Aquifers, 445
Architectural drawings, 387–388
Arms, 381
Artificial ecosystems, 459–462
Artistic design, 235
Aseptic surgery, 584
Aspects of design, 314–317
Aspiration, 68
Assembling, 565–567
Assembly drawings, 389–391
Assessing, 29–30, 362–363
    solutions, 375–376
Assessments, 246, 362
    audience, 538
Audible sound, 519
Audience assessment, 538
Audio recordings, 527
    systems, 523, 527–528
Aviation craft, 614
Axes, isometric, 295
Axles and wheels, 96
Backfilling, 486
Backs, 381
Balance, 323
Balers, 449
Ball peen hammers, 93
Bar code readers, 326–327

Barges, 613
Baseboard trim, 484
Batch manufacturing, 569
Batteries, 501–502
Batts, 483
Beacons, boundary, 527
Bead board, 350
Behavior, 165
Berries, 439
Berry farms, 439
Betas, 341
Bikes, stationary, 587
Bills of materials, 384–385
Binders, 112
Biochemicals, 146
Bioenergy, 146
Biofuels, 146
Biogas, 136–137
Biomass, 138, 145
    energy, 145–146
    fuels, 687
Biometrics, 433
Biopower, 146
Biopsies, 595
Biotechnology, 452–455
    quality control in, 462
Blanching, 69
Blast freezing, 458
Blending, 68
Blocking out shapes, 295–299
Blue board, 350
Blue-collar workers, 170–171
Blueprints, 382
Boards, ridge, 481
Bodies of water, navigable, 613
Body scanners, 592–593
Border lines, 394
Boundaries, 476
Boundary beacons, 527
Boxes, 295–296
Brainstorming, 288–289
Braking systems, 621
Brand names, 415
Brass, 108
Bristol board, 347
British thermal units, 133
Broad thinking, 286
Broadcast messages, 543–545
Bronze Age, 58–59
Browsers, 529
Budgets, 262

Building
    codes, 420
    frameworks, 479–481
    preparation for, 474–477
Buildings, 469–472
    commercial, 469, 471
    high-rise, 226–227
    industrial, 469, 471
    multistoried, 470
    public, 469, 472
    religious, 469, 472
    residential, 469–470
Bullet points, 280
Burrs, 93
Cables, glass fiber, 524
CAD. See Computer-aided design
Canal locks, 372–373
Canning, 457–458
Capillaries, 584
Carbon, 138
Carboniferous period, 138
Cardboard, 347
Careers
    advancing in, 178
    information, 177–178
    ladders, 178
Cargo, 607
Carriers, 522
    image, 540
Casting, 561
    mold, 562
    molding and, 560–562
CAT. See Computerized axial tomography
Catalysts, 453
Cathedrals, 472
Catheters, 598
Cathodes, 502
Cattle ranches, 450
CD-ROMs. See Compact discs read-only
    memory
Ceilings, 484
Cells
    fuel, 502–503
    photovoltaic, 503
    solar, 503
Centerlines, 395
Centimeters, 87
Central processing units (CPUs), 546
Ceramics, 111–112
Challenges, 220

Change
  communication and, 661–662
  construction and, 663
  materials and, 659–660
  production and, 663–664
  technology and, 659–664
  tools and, 660–661
  transportation and, 662
  the world of, 657–659
Characteristics, 364–369
Charts, 335
  flow process, 572
Chemical converters, 495, 501–503
Chemical deposits, 116
Chemical energy, 135
Chemical formulas, 337
Chemical preservation, 457, 459
Chemical processes, 120
Chemical processing, 118, 560
Chemical properties, 120, 122
Chemotherapy, 598
Chickens, 450
Chipboard, 347
Chucks, 101
Churches, 472
CIM. *See* Computer-integrated manufacturing
Circuits, 497
Circumferences, 96
Cities, 466
Citrus fruits, 439
Civil structures, 472–473
Classes of levers, 95
Claw hammers, 92
Clay, 350
Clay-based materials, 112
Cleaning up, 484
Clearing sites, 485
Clients, 260, 312
Climate controls, 616
Climbers, stair, 587
Climbing frames, 474
Clinical tests, 601
Closed-loop systems, 48
Coal, 139–140
Codes, building, 420
Coherent light, 102
Coke, 108
Cole plants, 439
Color casts, 252

Colors, 320, 322
  intermediate, 322
  primary, 322
  secondary, 322
Combinations of shapes, 298–299
Combines, 447
Combustion, 38
  external, engines, 501
  internal, engines, 498–499
Commercial buildings, 469, 471
Commercial transportation, 608
Communicating, 517
  designs, 246–248
  solutions, 384–404
Communication, 332–333, 432, 516–517
  change and, 661–662
  documents, 384–387
  electronic, 519. *See also* Wave
    communication
  the future and, 687–688
  graphic, 518
  graphic vs. electronic, 520–521
  information and, 516–517
  knowledge, 28
  machine-to-machine, 521, 533–534
  machine-to-people, 521, 532–533
  messages, 537–545
  people-to-machine, 521, 532
  people-to-people, 521–532
  process, 534–537
  satellites, 525
  systems, 521–534
    quality control in, 548
  technology, 70, 516–548
  wave, 518–519
Compact discs read-only memory (CD-ROMs), 528
Companies, 173
Compasses, 400
Competition, 413–415
Competitors, 419
Complex language, 82
Composites, 111, 113, 660
Compost, 146
Comprehensives, 539
Compression strokes, 498
Computer-aided design (CAD), 401
  tools, 399, 401–402
Computer-integrated manufacturing (CIM), 534

Computerized axial tomography (CAT), 593
Computerized tomography (CT), 593
Computers, 529
    applications, 163–164
    hardware, 545–547
    models, 337–339
    personal, 545–546
    software, 547–548
    systems, 523, 528–529
    testing, 370
Concentrated power systems, 506
Concentrated solar power technologies, 506
Conceptual models, 339–342
Conclusions, 280
    making, 280–281
Concrete, 113
Conditioning, 565
Condominiums, 470
Conductivity, thermal, 122
Conductors, 122, 509
Cones, 295, 298
Connections to other subjects. *Refer to chart on page 17*
Conservation, 510
Conserving energy, 509–511
Consolidation, 50
Constraints, 262
    establishing, 262–263
    external, 263
    financial, 262
    production, 262
    technical, 262
    time, 262
Constructed works, 469
Constructing structures, 473–486
Construction, 432, 469
    change and, 663
    the future and, 686
    knowledge, 28
    lines, 394
    process, 473
    technologists, 175
    technology, 70, 468–486
Consumers, 417–418
    safety, 368
Contact paper, 348
Container shipping, 50–51
Content, 235
Contexts
    functional, 56, 68–71
    historical, 56–63
    societal, 56, 63–67
    technological, 56–71, 432
        agricultural and related technology, 436–462
        communication technology, 516–548
        construction technology, 468–486
        energy conversion technology, 494–511
        information technology, 516–548
        manufacturing technology, 558–577
        medical technology, 584–601
        transportation technology, 606–625
Continuous manufacturing, 569–570
Contractors, 476
Contrast, 323
Control, 616, 620–621
    pest, 445
    quality
        in agriculture, 462
        in biotechnology, 462
        in communication systems, 548
        in energy systems, 509
        in power systems, 509
        in transportation systems, 624–625
    systems, 620
Controlling, 45, 576
    quality, 573–574
Controls
    climate, 616
    flight, 182–183
Conversion
    converters and, 494–507
    energy, 432
        technology, 494–511
Converters
    chemical, 495, 501–503
    conversion and, 494–507
    energy, 495
    fluid, 495, 506–507
    mechanical, 495–497
    solar, 495, 503–506
    thermal, 495, 497–501
Conveyors, 611
Cooperation, 165
Copy, writing, 538
Corn, 438
Cornflakes, 418–419
Corrugated cardboard, 347
Costs, 367

CPUs. *See* Central processing units
Cranes, 474–475
Creating
    scenarios, 682–684
    solutions, 240–242, 286–307
    technology, 210
        communicating solutions, 384–404
        creating solutions, 286–307
        the design process, 234–248
        identifying problems, 256–265
        improving solutions, 412–422
        innovations, 214–227
        inventions, 214–227
        modeling solutions, 332–357
        refining solutions, 312–327
        researching problems, 270–281
        selecting solutions, 312–327
        testing solutions, 362–376
Creative thinking, 220
Criteria, 261–262, 363–364
Crops, 436
    agricultural, 437–440
    forage, 439–440
Crystals, 111
CT. *See* Computerized tomography
Cultivators, 444
Curing, 459
Curled edges, 93
Current, 503
    alternating, 497
Curricular connections. *Refer to chart on page 17*
Curtain walls, 226
Custom manufacturing, 569
Cutaway shots, 408
Cutting
    flame, 564
    tools, 89–90
Cylinders, 295, 297
Dairies, 450
Data, 154–155, 275, 516
    analyzing, 278–280, 375
    gathering, 275–278, 375
    numerical, 516
    processing, 154, 517
    sheets, 149
Deadlines, 181
Debugging, 31
Decay, 138

Decking, 481
Decoding, 536–537
Decorating, 484
Dehydration, 457–458
Delivering published messages, 541
Delphi Method, 683
Dental hygienists, 591
Dentists, 591
Dependability, 165
Deposits, chemical, 116
Descriptive knowledge, 27, 152
Descriptive research, 158
Design, 30. *See also* Problems, Solutions
    artistic, 235
    aspects of, 314–317
    briefs, 239–240, 264–265
    budgets, 262
    communicating, 246–248
    definition of, 234–235
    elements of, 292, 320–323
    engineering, 235–237
    environments of, 278
    features, 300
    functions of, 277–278
    graphic, 551–552
    principles of, 323–324
    process, 234–248
        steps of, 237–248
    product, 236–237
    sheets, 317–318
    sketching. *See* Rough sketching
    style, 314
    system, 236
    teams, 312–313
    types of, 235–237
    web page, 553–554
Designing, 29–30
    film messages, 541–542
    published messages, 538–539
    tooling, 573
Desired outputs, 38
Destinations, 607
Destructive tests, 364
Detailed drawings, 389
Detailed sketches, 242, 324–325
Diagnosing, 591
Diagnosis, 592–595
Diagnostic equipment, 592
Diagnostic technologies, 595

Dies, 563
    making, 171
Diesel-electric locomotives, 618
Differentials, 96
Digging sticks, 108
Digital photography, 251–252
Digital videodiscs (DVDs), 520
Digitizing, 528
Dimensioning, 395
Dimensions, 325, 395–396
Dipping, 567
Directing, 45
Disc plows, 442
Discoveries, 217. *See also* Inventions
Discrimination, 165
Discs, 442
    compact, read-only memory, 528
Diseases, 453, 591
Dishes, 525
Distance multipliers, 95
Distances, measuring, 86–87
Distributing film messages, 543
DNA, recombinant, 454
Documents, communication, 384–387
Door trim, 484
Double sided tape, 353
Downlinks, 526
Drafting systems, 523, 532
Drag, 619
Drawing
    boards, 399
    classifications, 392–399
    design features, 300
    tools, 399–402
Drawings, 384–388. *See also* Sketches
    architectural, 387–388
    assembly, 389–391
    detailed, 389
    engineering, 387–388
    isometric, 396–397
    mechanical, 247, 385
        making, 402–403
    multiview, 392
    oblique, 396–398
    one–view, 393
    orthographic, 294, 392–396
    perspective, 396, 398–399
    pictorial, 295, 392, 396–399
    schematic, 389, 391–392
    technical, 31

three-view, 393
    two-view, 393
    working, 389–392
Dressing. *See* Planing
Drilling
    machines, 100–101
    tools, 90–91
Drills, 444
Drip irrigation, 447
Drives, creating, 485–486
Drugs, 599
    treatment with, 599–601
Ductility, 122
Durability, 365
DVDs. *See* Digital videodiscs
Earthquake protection, 672–673
Eaves, 481
Economics, 367
Ecosystems, artificial, 459–462
E-ink, 21
EKG machines. *See* Electrocardiograph
    machines
Elastomers, 113
Electric generators, 496
Electric power systems, 497
Electrical energy, 135
Electrical fixtures, 484
Electrical properties, 120, 122–123
Electrical transmission, 509
Electricity, 315, 509
Electrocardiograph (EKG) machines, 594
Electrochemical processes, 120
Electromagnetic induction, 497
Electromagnetic radiation, 518–519
Electronic communication, 519. *See also*
    Wave communication
    vs. graphic communication, 520–521
Electronic ink, 21
Electronic paper, 21
Electroplating, 567
Electrostatic reproduction, 530
Elegant solutions, 306–307
Elements of design, 292, 320–323
Elevators, 183
Emergency centers, 596
Emergency rooms, 596
Employer expectations, 165
Enclosing
    ceilings, 484
    roofs, 482

structures, 481–483
wall materials, 481–482
walls, 484
Encoding, 534–535
Endoscopes, 594–595
Energy, 40, 42, 130, 132–133
biomass, 145–146
chemical, 135
conserving, 509–511
conversion, 432
technology, 494–511
converters, 495
electrical, 135
forms of, 134–135
the future and, 687
geothermal, 124–125, 144–145
heat, 134–135
kinetic, 132
knowledge, 28
mechanical, 135, 143, 495
nature of, 134–137
nuclear, 135, 138, 141–142
potential, 132
power and, 133
radiant, 134
solar, 143
sources, 137–138, 146
alternate, 143–146
Sun as, 135–137
supplies, 138–142
systems, 509
technology, 70
technology and, 42, 130–146
thermal, 143
tidal, 143–144
transmission, 507–509
using, 509–511
wave, 143–144
wind, 143
work and, 131–133
Engineered materials, 660
Engineering
design, 235–237
drawings, 387–388
genetic, 454
human, 366–369
human factors, 315–316
materials, 110–113
Engineers, 175

Engines, 441
external combustion, 501
four cycle, 498
heat, 497
internal combustion, 498–499
jet, 499
rocket, 499–501
liquid fuel, 500
Environment, impacts on the, 666–669
Environmental challenge, 676–677
Environments, testing, 370–371
Enzymes, 453
E-paper, 21
Equality, 165
Equipment, 589
arranging, 573
collecting, 374
diagnostic, 592
harvesting, 441, 447–449
irrigation, 441, 445–447
medical, 596–599
pest control, 441, 444–445
planting, 441, 443–444
power, 440–442
selecting, 572–573
storage, 441, 450
tillage, 440, 442–443
Ergonomic environment, 381
Ergonomics, 316, 381
Erosion, 667–669
Etching, 530
Ethics, work, 165
Evaluating results, 375
Evaluation grids, 318–319
Exercise, 586–588
Exhaust strokes, 498
Exhaustible fuels, 141
Exhaustible material resources, 114
Exhaustible resources, 114, 116–118, 665
Expansion, 122
Expectations, employer, 165
Expendable mold casting, 562
Experimental models, 339–342
Experimental research, 159, 271, 274–275
Experimentation, 333–334
Experimenting, 221
Expert surveys, 683
External combustion engines, 501
External constraints, 263

External selection, 312–313
Extras, 542–543
Eyes, 381
Faceprints, 433
Facial recognition systems, 433
Farming, 114. *See also* Agriculture
Farms
    berry, 439
    horse, 450
    poultry, 450
    swine, 450
    truck, 439
Features, drawing design, 300
Feedback, 38
    systems, 48–49
Ferries, 613
Fiber optics, 524
Fiberglass, 113
Field tests, 370
Filberts, 439
Film messages, 541
    designing, 541–542
    distributing, 543
    preparing to produce, 542–543
    producing, 541–543
Final sketching, 291
Finances, 182, 314, 317
Financial constraints, 262
Financial inventions, 222–223
Financial resources, 40
    technology and, 42
Finishing, 567–569
    insides, 484–485
    interiors, 483–485
    sites, 485–486
Firebrick, 112
Firing, 565
First World countries, 67
First-class levers, 95
First-line supervision, 175
Fission, 142
Fixed path systems, 610
Fixed path transportation, 610–612
Fixtures, 484
Flaking, 58
Flame cutting, 564
Flexible manufacturing, 569–571
Flexography, 530
Flight
    controls, 182–183

hypersonic, 689
    simulators, 622–623
Floating cities, 79
Flooring, 484
Floors, 479
    framing, 480
Flow process charts, 572
Fluids, 508
    converters, 495, 506–507
    transmission, 508–509
Foam, 350
Foam core board, 347
Food
    preservation, 456–459
    processing, 456
    production, 455–459
    technologists, 175
Forage crops, 439–440
Force multipliers, 95
Forecasting, 681–685
Forecasts, 681
Forestry, 114, 452
Forging, 563
Formats, 538
Forming, 562–563
    roll, 563
Forms, 235–236, 320–321
Formulas, 336
    chemical, 337
45° triangles, 400
Fossil fuels, 138
Fossilization, 141
Found objects, 349
Foundations, 477
    setting, 477–479
Four cycle engines, 498
Fractions, 86
Framed superstructures, 479
Frames, climbing, 474
Frameworks, 479–481
Framing
    floor, 480
    roof, 481
    wall, 481
Freedom Ship, 79
Freezing, 457–458
    blast, 458
Freight, 611
Freighters, 613
Fruits, 439

Fuel cells, 502–503
Fuels. *See* Energy sources
    biomass, 687
    fossil, 138
Function
    aspect of design, 236, 314–315
    characteristic, 364–366, 413
    reason for improvement, 415–416
Functional contexts, 56, 68–71
Furrows, 446
Fusion, 142
Future
    agriculture and the, 685–686
    communication and the, 687–688
    construction and the, 686
    energy and the, 687
    manufacturing and the, 688–689
    medicine and the, 688
    technology and the, 680–689
    transportation and the, 689
Futurists, 681
Gables, 572
Galvanizing, 568
Gardens, 437
Gas
    natural, 141
    turbines, 499
Gene splicing, 454
General aviation craft, 614
Generators, 496
Genetic engineering, 454
Geosynchronous orbits, 526
Geothermal energy, 124–125, 144–145
Geysers, 145
Girders, 480
Glass, 112
Glaze, 568
Glazing, 666. *See also* Windows
Glide slopes, 526
Global society, 65
Global village, 521, 662
Glue, 352. *See also* Adhesives
Goals, 37–38, 176–178
    personal, 176–177
    technology and, 49
Government organizations, 419–421
Grading ground, 485–486
Grain drills, 444

Grains, 436, 438
Graphic communication, 518
    vs. electronic communication, 520–521
Graphic design, 551–552
Graphic models, 335–336
Graphic organizers, 289
Graphic systems, 518
Graphs, 279, 335
Gravity tables, 69
Gravure printing, 541
Grids, evaluation, 318–319
Grinding, 69
    machines, 100–101
Gripping tools, 91–92
Grits, 418
Ground, 485–486
Ground-effect machines, 619
Guidance, 616, 620
    systems, 620
Gyms, home, 588
Gypsum, 484
Hacksaws, 89
Hamlets, 466
Hammers, 92–93
    ball peen, 93
    claw, 92
    shingling, 92
    sledge, 92
Harassment, 165
Hardboard, 345
Hardware, 172, 545. *See also* Machines,
    Materials, Tools
    computer, 545–547
Harrows, 442
Harvesting, 447
    equipment, 441, 447–449
Hazardous materials and wastes, 149
Hazelnuts, 439
Headers, 447–448, 481
Health, 584–601
Heat
    energy, 134–135
    engines, 497
Heating systems, 503–504
Herbicides, 445, 454
Hidden lines, 394–395
High-rise buildings, 226–227
Historical context, 56–63

Historical research, 158
History
 of inventions, 214–217
 technology and, 25–26
Hoists, 475
Holding tools, 91
Home gyms, 588
Homes, 469–470
Horse farms, 450
Hot links, 529
Hot springs, 145
Hot-melt glue, 352
Hovercrafts, 619
HTML. *See* Hypertext markup language
Human engineering, 366–369
Human factors, 314–317
 engineering, 315–316
Humanities knowledge, 27, 152
Husbandry, animal, 437
Hybrid vehicles, 666
Hydroelectric plants, 141
Hydroponics, 452, 685
Hygienists, dental, 591
Hypersonic flight, 689
Hypertext markup language (HTML), 529
Hypotheses, 274
Ideas, 286–287
Ideation, 287–291
Identifying problems, 219, 238–240, 256–265
Illnesses, 591
Illustration board, 347
Image carriers, 540
Imaging, magnetic resonance, 593–594
Immunization, 590
 prevention through, 590–591
Impacts
 on the environment, 666–669
 on jobs, 669–671
 negative, 664–672
 on people, 669–672
 on people's lives, 671–672
 on resources, 665–666
 technological, 656–673
Improvements
 making, 421–422
 reasons for, 413–416
 sources of, 416–421
Improving solutions, 248, 412–422
Inclined planes, 97–98
Individual telecommunication, 523

Individuals, technology and, 65–66
Inducing, 497
Induction, electromagnetic, 497
Industrial activities, 471
Industrial Age, 60–62
Industrial buildings, 469, 471
Industrial materials, 110
Industrial Revolution, 60, 215
Industrial stock, 120
Industrial structures. *See* Industrial
 buildings
Industrial technologists, 175
Industrialized countries, 67
Industries, 174
Inexhaustible resources, 665
Information, 154–155, 278, 516–517. *See also*
 Knowledge
 Age, 62–63, 157, 545
  jobs in the, 172–173
 career, 177–178
 collecting, 220
 communication and, 516–517
 gaining, 158–160
 knowledge, 28
 location, 325
 processing, 517
 Revolution, 62
 shape, 325
 size, 325
 technology, 70, 152–160, 516–548
Infrastructures, 476
In-house design teams, 312–313
Injuries, 591
Innovations, 224–225
 inventions and, 214–227, 230–231
Inoculations, 585
Input devices, 546
Inputs, 37–43
 finishing, 484–485
Inspections, 572–573
Installing
 fixtures, 484
 flooring, 484
 trim, 484
 utilities, 482–483
Instincts, 82
Instrument landing systems (ILSs), 526–527
Insulating, 483
Insulation, 483
Insulators, 122

Intaglio, 530
    printing, 541
Intake strokes, 498
Intended outputs, 45
Intensive care, 596
Interactive television, 211
Interests, 176–178
Interiors, 483–485
Intermediate colors, 322
Intermittent manufacturing, 569–570
Intermodal transportation, 614–615
Internal combustion engines, 498–499
Internal selection, 313–314
Internal sources of improvement, 421
Internet, 523, 528–529. *See also* Computers
Intervention radiology, 598
Invasive diagnostic technologies, 595
Inventing, 222–223
Inventions, 157, 214, 217–218
    financial, 222–223
    history of, 214–217
    innovations and, 214–227, 230–231
    leisure, 222–223
    process, 219–221
    social, 222–223
    teams, 223–224
Inventors, 221–222
Iron Age, 59–60
Irradiation, 457, 459
Irrigation, 445
    drip, 447
    equipment, 441, 445–447
Isometric axes, 295
Isometric drawings, 302, 396–397
Isometric sketches, 302
Jet engines, 499
Jobs, 474
    impacts on, 669–671
    in the Information Age, 172–173
    people and, 176–178
    technology and, 170–176
Joists, 480
Joules, 133
Judgments, 82
Jump cuts, 409
Kilometers, 87
Kinetic energy, 132
Knowledge, 40, 42, 154–155
    agricultural, 28
    body of, 154

communication, 28
construction, 28
descriptive, 27, 152
energy, 28
gaining, 158–160
humanities, 27, 152
information, 28
manufacturing, 28
medical, 28
scientific, 27, 152, 155–157
technological, 28, 152, 157
technology and, 26–28, 42
transportation, 28
Laboratories, 370
Land, 610
    transportation, 610–612
Landing systems, 526–527
Landscape plants, 460–461
Landscaping, 485–486
Lanes, shipping, 613
Language, 27
    complex, 82
    hypertext markup, 529
    technology and, 30–31
Language arts connections. *Refer to chart on page 17*
Lasers, 102–103
Layouts, 540
Leaders, 390–391
Legs, 381
Leisure, 658
    inventions, 222–223
Letterpress printing, 530, 540
Leveling ground, 485–486
Levers, 94–95
    classes of, 95
Library research, 271–273
Life expectancy, 365
Lift, 182
Light, 528
Light energy, 134
Linear motion, 101
Lines, 320
    alphabet of, 394
    border, 394
    construction, 394
    hidden, 394–395
    object, 394
    types of, 394–395
Links, 529

Liquid fuel rocket engines, 500
Lithography, 157
    offset, 530, 540
Lives, impacts on people's, 671–672
Livestock, 450–452
Loadbearing walls, 226
Loading, 609
Location
    dimensions, 395
    information, 325
Locks, canal, 372–373
Locomotives, 618
Machineable wax, 351
Machines, 40–41, 84, 98–102, 217–218
    drilling, 100–101
    electrocardiograph, 594
    grinding, 100–101
    ground-effect, 619
    milling, 100–101
    motion. *See* Mechanical energy
    planing, 100–101
    rowing, 587
    sanding, 100–101
    sawing, 100–101
    shaping, 100–101
    shearing, 100–102
    simple, 94
    technology and, 41–42
    tools, 99
    turning, 100
    X-ray, 593
Machine-to-machine communication, 521,
    533–534
Machine-to-people communication, 521,
    532–533
Machining, 563
    ultrasonic, 565
Maglev. *See* Magnetic levitation
Magnetic levitation (maglev), 619
Magnetic properties, 120, 122–123
Magnetic resonance imaging (MRI), 593–594
Maize, 438
Mallets, 92
    plastic, 93
    rubber, 93
Management, 45
    operative, 175
    processes, 43, 45
    upper, 174

Managers, 45, 174, 576
    middle, 175
Managing time, 181
Manufactured panels, 686
Manufacturing, 120, 432, 558–569
    batch, 569
    challenge, 580–581
    computer-integrated, 534
    continuous, 569–570
    custom, 569
    flexible, 569–571
    the future and, 688–689
    intermittent, 569–570
    knowledge, 28
    systems, 569–571
        developing, 571–575
    technologists, 175
    technology, 70–71, 558–577
Market research, 273
Marketing products, 576–577
Marketplace, 182
Markets, 273–274, 367
Mass telecommunication, 524
Matboard, 347
Material properties, 366
Material resources, 113–118
Material safety data sheets, 149
Material sheets, 386
Materials, 41, 110
    amorphous, 112
    bills of, 384–385
    change and, 659–660
    clay-based, 112
    collecting, 374
    engineered, 660
    engineering, 110–113
    hazardous, 149
    industrial, 110
    modeling, 342–353
    molding, 349–351
    natural, 109
    opaque, 123
    properties of, 120–124
    raw, 118–120
    sculpting, 349–351
    sheet, 344–349, 386
        metal, 348–349
        paper, 345–347
        plastic, 348
        wood, 344–345

storage, 123
technology and, 108–125
translucent, 123
transparent, 123
types of, 110–111
Maternity wards, 596
Mathematical models, 336–337
Mathematics connections. *Refer to chart on page 17*
Measuring
distances, 86–87
relationships, 88
systems, 87
tools, 85–88
Mechanical converters, 495–497
Mechanical drawings, 247, 385
making, 402–403
tools, 399–402
Mechanical energy, 135, 143, 495
Mechanical processing, 118–119, 560
Mechanical properties, 120–122
Mechanical systems, 482
Mechanical transmission, 508
Mechanisms, 84, 94–98
Media, 518
Medical equipment, 596–599
Medical floors, 596
Medical knowledge, 28
Medical technologists, 175, 591
Medical technology, 71, 584–601
Medicine, 432
the future and, 688
goals of, 584–601
technology in, 591–601
Memory, 547
Messages
broadcast, 543
producing, 543–545
communication, 537–545
film, 541
designing, 541–542
distributing, 543
preparing to produce, 542–543
producing, 541–543
published, 537
delivering, 541
designing, 538–539
preparing to produce, 539–540
producing, 537–541

Metal sheet materials, 348–349
Metals, 111
Methane, 141
Metric system. *See* Measuring systems
Metropolises, 466
Microorganisms, 453
Microwaves, 524
Middle managers, 175
Millimeters, 87
Milling machines, 100–101
Mineral ores, 116
Mines
open pit, 117
underground, 118
Minimum tillage, 442
Mock-ups, 243, 339–341
Modeling, 682, 684–685
clay, 350
materials, 342–353
solutions, 242–244, 332–357
tools, 353–354
Models, 242, 332, 684
computer, 337–339
definition of, 332
graphic, 335–336
making, 354–357
mathematical, 336–337
physical, 339–342
reasons for using, 332–335
solid, 339
surface, 338
types of, 335–342
wireframe, 338
Modes of travel, 610
Mold casting, 562
Molding, 560–562
materials, 349–351
Molds, 351, 560
making, 171
Money, 182. *See also* Finances, Financial resources
technology and, 168–182
Monitors, 547
Monochromatic light, 102
Moral judgments, 82
Mosaic, 529
Mosques, 472
Motion, 508. *See also* Mechanical energy
Motors, 498. *See also* Engines

Mounting tape, 353
Movable type, 530
Moving, 609
MRI, 594. *See also* Magnetic resonance
    imaging
Multipliers, 95
Multiview drawings, 392
Museum board. *See* Matboard
Natural gas, 141
Natural materials, 109. *See also* Natural resources
Natural resources, 39, 109
        growing, 114–116
        harvesting, 114–116
        locating, 114
        technology and, 41
Natural systems, 36
Nature, 114
Navigable bodies of water, 613
Necks, 381
Needs, 256
        societal, 64
Negative impacts, 664–672
Negatives, 530
Nodal points, 433
Noise pollution, 667–668
Non-destructive tests, 364
Nonfood, 440
Noninvasive diagnostic equipment, 592
Nonrenewable fuels, 141
Nonrenewable resources, 665
Nuclear energy, 135, 138, 141–142
Numerical data, 516
Nurses, 591
Nutrition, 659
Nuts, 439. *See also* Peanut butter
Object lines, 394
Oblique drawings, 396–398
Oblique sketches, 302–303
Ocean liners, 613
Odometers, 86
Offset lithography, 530, 540
One-shot mold casting, 562
One-view drawings, 393
Opaque materials, 123
Open pit mines, 117
Open-loop systems, 48
Operating
        rooms, 596
        systems, 548
Operational models, 339–342

Operations, 572
        selecting, 571
        sequencing, 571–572
Operative management, 175
Opportunities, 257
        problems and, 256–258
Optical properties, 120, 123
Orbits, geosynchronous, 526
Orchards, 439
Ores, 116
Organization, 334–335
Organizations, government, 419–421
Organizers, graphic, 289
Organizing, 45, 576
Origins, 607
Orthographic drawings, 294, 392–396
Orthographic projection, 392
Orthographic sketches, 294
Outlines, adding, 299–300
Outpatient treatment centers, 596
Output devices, 547
Outputs, 38, 45
        desired, 38
        intended, 45
        system, 45–48
        undesirable, 46
Outsourcing, 317
Packets, 529
Painting, 484
Panels, manufactured, 686
Paper
        contact, 348
        sheet materials, 345–347
Paraffin wax, 351
Passive solar systems, 503–505
Patches, 415
Pathologists, 595
Pathology units, 596
PCs. *See* Personal computers
Peanut butter, 68
        making, 68–69
Peat, 138
Pediatric floors, 596
Penstocks, 507
People, 39
        impacts on, 669–672
        jobs and, 176–178
        lives of, 671–672
        movers, 612
        technology and, 40–41, 168–182

People-to-machine communication, 521, 532
People-to-people communication, 521–532
Performance, 365
Permanent mold casting, 562
Personal computers (PCs), 545–546
Personal goals, 176–177
Personal problems, 258
Personal protection, 589
Personal transportation, 608
Perspective drawings, 396, 398–399
Perspective sketches, 303–306
Perspectives
    three-point, 399
    two-point, 398
Pest control, 445
    equipment, 441, 444–445
Pesticides, 445
Petroleum, 140
Pharmacies, 596
Pharmacists, 591
Photographic systems, 523, 531–532
Photography, 251–252
Photosynthesis, 143
Photovoltaic cells, 503
Physical models, 339–342
Physical properties, 120–121
Physical therapy rooms, 596
Physical traits, 177
Physicians, 591
Pickling, 459
Pictorial drawings, 295, 392, 396–399
Pigs, 450
Pilings, 478
Pilot runs, 575
Pipelines, 611
Pivot sprinklers, 447
Placebos, 601
Planes, inclined, 97–98
Planing, 485
    machines, 100–101
Planning, 45, 576
Plant science, 436
Planting, 436
    equipment, 441, 443–444
Plants
    cole, 439
    landscape, 460–461
Plaster of paris, 562
Plastic, 62, 106, 113, 482, 660
    cement, 353
    sheet materials, 348

Plates, 481
    sill, 480
Plot models, 339–342
Plotters, 402
Plowing, 441
Plumbing fixtures, 484
Plywood, 345
Point of sale terminals, 326
Polishing tools, 93–94
Pollution, 46–47
    air, 667
    noise, 667–668
    water, 667–668
Polymers, 111–113
    synthetic, 113
Polystyrene, 350
Polyurethane, 351
Potential energy, 132
Poultry, 450
Pounding tools, 92–93
Power, 133
    concentrated systems, 506
    electric systems, 497
    energy and, 133
    equipment, 440–442
    strokes, 498
    systems, 509
    technology, 70
    transmission, 507–509
Precision, 87
Preliminary sketching. *See* Rough sketching
Presentation models, 339–342
Presenting solutions, 403–404
Preservation
    chemical, 457, 459
    food, 456–459
Presidents, 174
Preventing, 585
Prevention
    programs, 585–591
    through immunization, 590–591
Primary colors, 322
Primary processes, 559–560
Primary processing, 560
Primary tools, 85
Printing, 537
    intaglio, 541
    letterpress, 530, 540
    relief, 530
    screen, 530, 541
    systems, 523, 529–531

Problems, 219, 257. *See also* Design, Solutions
    identifying, 219–220, 238–240, 256–265
    listing, 275. *See also* Identifying problems
    opportunities and, 256–258
    personal, 258
    recognizing, 258–261
    researching, 240, 270–281
    social, 258
    solving, 218–219
    statements, 261
    technological, 258
    types of, 258
Process charts, 572
Processes, 38, 218
    chemical, 120
    communication, 534–537
    construction, 473
    design, 234–248
        steps of, 237–248
    electrochemical, 120
    invention, 219–221
    management, 43, 45
    primary, 559–560
    production, 43
    research, 275–281
    secondary, 560–569
    sketching, 294–300
    system, 43–45
    transportation, 608–610
Processing, 118, 456
    chemical, 118, 560
    data, 517
    food, 456
    information, 517
    mechanical, 118–119, 560
    primary, 560
    products of, 120
    secondary, 559–560
    thermal, 118–120, 560
Processing units, central, 546
Producing, 29–30
    broadcast messages, 543–545
    communication messages, 537–545
    film messages, 541–543
    products, 182, 575–576
    published messages, 537–541
    services, 182

Production, 314, 317
    change and, 663–664
    constraints, 262
    food, 455–459
    processes, 43–45
    video, 407–409
Productive activities, 471
Productivity, 663
Products, 46, 218
    design, 236–237
    marketing, 576–577
    producing, 182, 575–576
Profit, 413–414
Programs, 546, 548
    prevention, 585–591
    wellness, 585–589
Propellers, 244–245
Properties, 120
    acoustical, 120, 123
    chemical, 120, 122
    electrical, 120, 122–123
    magnetic, 120, 122–123
    material, 366
    of materials, 120–124
    mechanical, 120–122
    optical, 120, 123
    physical, 120–121
    thermal, 120, 122
Proportion, 323–324
Propulsion, 616–618
    systems, 617
Protection
    earthquake, 672–673
    personal, 589
Prototypes, 243–244, 341–342
    models, 339–342
Public buildings, 469, 472
Public transportation, 608, 623
Published messages, 537
    delivering, 541
    designing, 538–539
    preparing to produce, 539–540
    producing, 537–541
Pulleys, 96–97
Pulling equipment. *See* Power equipment
Purposes, determining, 372
Pyramids, 295, 298
Quality, 573
    controlling, 573–574
        in agriculture, 462

in biotechnology, 462
in communication systems, 548
in energy systems, 509
in power systems, 509
in transportation systems, 624–625
of work sheets, 386
Questioning, 289–291
Radiant energy, 134
Radiating, 525
Radiation, 459
electromagnetic, 518–519
Radio waves, 518
Radiology, 592
intervention, 598
therapeutic, 598
treatment, 598
units, 596
Rafters, 481
Railroads, 610
RAM. *See* Random-access memory
Ranches, 450
Random-access memory (RAM), 547
Rational thought, 82
Raw materials, 118–120
Read-only memory (ROM), 547
compact discs, 528
Recalls, 416
Receivers, 522
Receiving, 536
Reciprocating motion, 508
Recognition software, 655
Recognition systems, 433
Recognizing problems, 258–261
Recombinant DNA, 454
Recorders, videocassette, 528
Recording
systems, 527–528
audio, 527
video, 527
Refined sketches, 293, 539
Refined sketching, 291
Refining
solutions, 242, 319–324
selecting and, 312–326
techniques, 300–302
Refractories, 112
Refrigeration, 457–458
Regulating technology, 253
Regulations, 420
Rehearsing, 543

Relationships, measuring, 88
Relief printing, 530, 540
Religious buildings, 469, 472
Renderings, 242, 324
Renewable material resources, 114
Renewable resources, 114–116, 665
Reproduction, electrostatic, 530
Research, 270
descriptive, 158
experimental, 159, 271, 274–275
historical, 158
library, 271–273
market, 273
process, 274–281
survey, 271, 273–274
types of, 270–275
Researching problems, 240, 270–281
Residential buildings, 469–470
Resonance imaging, magnetic, 593–594
Resources, 64
exhaustible, 114, 116–118, 665
material, 114
financial. *See* Financial resources
impacts on, 665–666
inexhaustible, 665
natural, 109
growing, 114–116
harvesting, 114–116
locating, 114
nonrenewable, 665
obtaining, 113–118
purchasing, 182
renewable, 114–116, 665
material, 114
ripe, 115–116
technology and, 78
energy, 130–146
information, 152–160
materials, 108–125
money, 168–182
people, 168–182
time, 168–182
tools, 82–103
Respectful behavior, 165
Results, evaluating, 375
Retrieval, 537
Retrieving, 537
Reviewing, 221
Rhythm, 323
Ridge boards, 481

Ridges, 481
Ripe resources, 115–116
Roadways, 618
Rocket engines, 499–501
    liquid fuel, 500
Roll forming, 563
ROM. *See* Read-only memory
Roofs, 479, 481
    enclosing, 482
    framing, 481
Rough sketches, 291–292
Rough sketching, 241, 291
Roughing in, 482
Roughs, 539
Routine diagnostic equipment, 592
Routing, 609
Rowing machines, 587
Rulers, 86
Safety, 368, 413, 416
    consumer, 368
    data sheets, 149
Salaries, 170
Sanding machines, 100–101
Satellites, 525
Sawing
    machines, 100–101
    tools, 89–90
Scale models, 339–342
Scales, 400
Scanners
    body, 592–593
    tomography, 593
Scarcity, 665
Scenarios, 683
    creating, 682–684
Scheduling, 609
Schematic drawings, 389, 391–392
Schematics, 336
Science, 436
    connections. *Refer to chart on page 17*
Scientific discoveries, 217
Scientific knowledge, 27, 152, 155–157
Scientific method, 275
Scrap, 46
Screen printing, 530, 541
Screws, 98
Scripts, 542
Scriptwriting, 542
Scroll saws, 101
Sculpting materials, 349–351
Sea-lanes, 613

Secondary colors, 322
Secondary processes, 560–569
Secondary processing, 559–560
Second-class levers, 95
Seed drills, 444
Selecting solutions, 242, 312–314
    refining and, 312–326
Selection
    external, 312–313
    internal, 313–314
Semiconductors, 503
Semiskilled workers, 171
Separating, 563–565
Separation, 563
Services, 182
Shades, 322
Shading, 300–301
Shadowing, 301–302
Shape
    dimensions, 395
    information, 325
Shapes, 320–321
    blocking out, 295–299
    combinations of, 298–299
Shaping machines, 100–101
Shearing, 564
    machines, 100–102
    tools, 90
Sheathing, 481
Sheet materials, 344–349
    metal, 348–349
    paper, 345–347
    plastic, 348
    wood, 344–345
Shingling hammers, 92
Shipping
    container, 50–51
    lanes, 613
Shutter lag, 252
SI Metric system, 87
Sickles, 436
Sill plates, 480
Simple machines, 94
Simulating, 623
Simulators, 622
    flight, 622–623
Sites
    clearing, 485
    finishing, 485–486
    preparing, 477
60 Hz alternating current, 497

Size
 dimensions, 395
 information, 325
Sketches, 291. *See also* Drawings
 detailed, 242, 324–325
 isometric, 302
 oblique, 302–303
 orthographic, 294
 perspective, 303–306
 refined, 293, 539
 rough, 291–292
 thumbnail, 292
 types of, 302–306
Sketching, 291–306
 process, 294–300
 refined, 291
 rough, 291
Skilled workers, 170–171
Sledge hammers, 92
Slewing gear, 474
Slicing tools, 90
Slopes, glide, 526
Smelting, 660
Social inventions, 222–223
Social problems, 258
Social studies connections. *Refer to chart on page 17*
Societal context, 56, 63–67
Societal needs, 64
Society
 changing, 168–169
 global, 65
 scope of, 65
 technology and, 66–67, 652, 656–673, 680–689
Software, 548
 computer, 547–548
 speech recognition, 655
Soil, 460
 erosion, 667–669
Solar cells, 503
Solar converters, 495, 503–506
Solar energy, 143
Solar heating systems, 503–504
Solar power technologies, 506
Solar rooms, 505
Solar systems
 active, 504–506
 passive, 503–505
Solariums, 505

Solid models, 339
Solid set sprinklers, 447
Solutions. *See also* Design, Problems
 assessing, 375–376
 communicating, 384–404
 creating, 240–242, 286–307
 elegant, 306–307
 improving, 248, 412–422
 modeling, 242–244, 332–357
 performance of, 277
 presenting, 403–404
 refining, 242, 319–324
  selecting and, 312–326
 selecting, 242, 312–314
  refining and, 312–326
 testing, 244–246, 362–376
 users of, 276–277
Sound, 528
 audible, 519
Space, 610
 transportation, 614
Spec sheets. *See* Specification sheets
Specification sheets, 384, 386–387
Specifications, 475
Speech recognition software, 655
Spheres, 295, 297–298
Spin-offs, 225
Splicing genes, 454
Sports, 588–589
Spray adhesives, 352
Sprinklers, 446
 pivot, 447
 solid set, 447
Squares, 88
Stair climbers, 587
Standard stock, 120, 560
Standardization, 421
Stationary bikes, 587
Statistics, 516
Step-down transformers, 509
Step-up transformers, 509
Stock
 industrial, 120
 standard, 120, 560
Stone Age, 57–58
Storage, 572, 610
 equipment, 441, 450
 material, 123
 retrieval and, 537
 tool, 102

Storing, 537
Storyboards, 407, 544
Strokes, 498
Structural challenge, 489–490
Structure, 615–616
Structures. *See also* Buildings
    civil, 472–473
        constructing, 473–486
        enclosing, 481–483
Studs, 481
Study models, 339–342
Style, 314
Styli, 527
Subroutines, 548
Substructures, 477
Sun, 135–137
Superstructures, 477
    framed, 479
Supervision, first-line, 175
Support systems, 621–624
Surface models, 338
Surgery, 598
    aseptic, 584
Surgical floors, 596
Surveying, 477, 682–683
Surveys
    expert, 683
    research, 271, 273–274
Suspension, 616, 618–620
    systems, 618
Swathers, 449
Swine, 450
Synagogues, 472
Synthetic drugs, 599
Synthetic polymers, 113
Systems, 34, 36–49
    active solar, 504–506
    braking, 621
    closed-loop, 48
    communication, 521–534
        quality control in, 548
    computer, 523, 528–529
    concentrated power, 506
    control, 620
        technology, 379–380
    definition of, 36
    design, 236
    drafting, 523, 532
    electric power, 497

energy, 509
facial recognition, 433
feedback, 48–49
fixed path, 610
graphic, 518
guidance, 620
heating, 503–504
inputs, 39
instrument landing, 526–527
Internet, 523, 528–529
manufacturing, 569–571
    developing, 571–575
mechanical, 482
natural, 36
open-loop, 48
operating, 548
outputs, 45–48
parts of, 37–49
passive solar, 503–505
photographic, 523, 531–532
power, 509
printing, 523, 529–531
processes, 43–45
propulsion, 617
recording, 523, 527–528
support, 621–624
suspension, 618
technological, 37
telecommunication, 523–526
testing, 574–575
transportation, 615–624
    quality control in, 624–625
variable path, 610
vehicle, 615–624
wave, 518
T squares, 399
Tables, 336
Tablets, 402
Talents, 176
Tankers, 613
Tape, 353
Teams
    design, 312–313
    invention, 223–224
Technical constraints, 262
Technical drawings, 31
Technicians, 175
Technological actions, 29–30
Technological Age, 63

Technological contexts, 432
    agricultural and related technology,
      436–462
    communication technology, 516–548
    construction technology, 468–486
    energy conversion technology, 494–511
    information technology, 516–548
    manufacturing technology, 558–577
    medical technology, 584–601
    transportation technology, 606–625
Technological impacts, 656–673
Technological knowledge, 28, 152, 157
Technological problems, 258
Technological systems, 37
Technological unemployability, 670
Technological unemployment, 670
Technologists, 175
    medical, 591
Technology
    action and, 29–30
    agricultural, 69–70, 436–462
    in agriculture, 440–450
    change and, 659–664
    communication, 70, 516–548
    construction, 70, 468–486
    contexts of, 56–71
    creating, 210
        communicating solutions, 384–404
        creating solutions, 286–307
        the design process, 234–248
        identifying problems, 256–265
        improving solutions, 412–422
        innovations, 214–227
        inventions, 214–227
        modeling solutions, 332–357
        refining solutions, 312–327
        researching problems, 270–281
        selecting solutions, 312–327
        testing solutions, 362–376
    definition of, 22, 24–31, 56
    diagnosis and, 592–595
    diagnostic, 595
    energy, 70
        conversion, 494–511
    energy and, 42, 130–146
    exercise and, 586–588
    financial resources and, 42
    the future and, 680–689
    the future of, 685–689
    goals and, 49

    history and, 25–26
    the individual and, 65–66
    information, 70, 152–160, 516–548
    jobs and, 170–176
    knowledge and, 26–28, 42
    language and, 30–31
    machines and, 41–42
    manufacturing, 70–71, 558–577
    materials and, 108–125
    medical, 71, 584–601
    in medicine, 591–601
    money and, 168–182
    natural resources and, 41
    people and, 40–41, 168–182
    power, 70
    regulating, 253
    resources and, 78
        energy, 130–146
        information, 152–160
        materials, 108–125
        money, 168–182
        people, 168–182
        time, 168–182
        tools, 82–103
    scope of, 20, 24–31, 36–49, 56–71
    society and, 66–67, 652, 656–673,
      680–689
    solar power, 506
    sports and, 588–589
    as a system, 36–49
    system control, 379–380
    time and, 42–43, 168–182
    tools and, 82–103
    transportation, 71, 606–625
    treatment and, 596–601
    wellness and, 586–589
Telecommunication
    individual, 523
    mass, 524
    systems, 523–526
Telegraphs, 523
Telephones, 524
Television, interactive, 211
Tempering, 565
Templates, 400
Terminals, 622
Testing, 362–363
    computer, 370
    environments, 370–371
    methods, 375

Testing *(continued)*
    procedure, 371–375
    solutions, 244–246, 362–376
    systems, 574–575
Tests, 362
    choosing, 372–374
    clinical, 601
    conducting, 375
    destructive, 364
    field, 370
    non-destructive, 364
Texture, 320, 322–323
Therapeutic radiology, 598
Therapy rooms, physical, 596
Thermal conditioning, 565
Thermal conductivity, 122
Thermal converters, 495, 497–501
Thermal energy, 143
Thermal expansion, 122
Thermal processing, 118–120, 560
Thermal properties, 120, 122
Thermoforming, 563
Thermoplastics, 113
Thermosets, 113
Thermostats, 38
Thinking
    broad, 286
    creatively, 220
Third World countries, 67
Third-class levers, 95
30°-60° triangles, 400
Thought, rational, 82
Thoughtware, 172, 286–287. *See also*
    Knowledge
Three-point perspectives, 399
Three-view drawings, 393
Threshers, 447–448
Thrombe walls, 504
Throttles, 621
Thumbnail sketches, 292
Tidal energy, 143–144
Tillage, 442
    equipment, 440, 442–443
Time, 40, 42, 178–181
    constraints, 262
    managing, 181
    technology and, 42–43, 168–182
Tints, 322
Tomography scanners, 593

Tooling, 573
Toolmakers, 83–84
Tools, 84–94
    change and, 660–661
    computer-aided design, 399, 401–402
    cutting, 89–90
    drawing, 399–402
    drilling, 90–91
    gripping, 91–92
    holding, 91
    machine, 99
    measuring, 85–88
    modeling, 353–354
    polishing, 93–94
    pounding, 92–93
    primary, 85
    sawing, 89–90
    shearing, 90
    slicing, 90
    storage, 102
    technology and, 82–103
    traditional, 399–401
    turning, 91–92
Towboats, 613
Tower cranes, 474–475
Towns, 466
Toxicity, 601
Traction devices, 441
Tractors, 441
Trade-offs, 375
Traditional tools, 399–401
Trains, 611
Traits, 176–178
    physical, 177
Transformers, 509
Translucent materials, 123
Transmission, 507
    electrical, 509
    energy, 507–509
    fluid, 508–509
    mechanical, 508
    power, 507–509
Transmitters, 521
Transmitting, 535–536
Transparent materials, 123
Transportation, 432, 572, 607
    air, 614
    change and, 662
    commercial, 608

definition of, 607–608
fixed path, 610–612
the future and, 689
intermodal, 614–615
knowledge, 28
land, 610–612
modes of, 610–615
personal, 608
processes, 608–610
public, 608, 623
space, 614
systems, 615–624
    quality control in, 624–625
technology, 71, 606–625
types of, 608
variable path, 612
water, 612–613
Travel, modes of, 610
Treadmills, 587
Treating, 591
Treatment
  centers, 596
  with drugs, 599–601
  with medical equipment, 596–599
  radiology, 598
  technology and, 596–601
  with vaccines, 599–601
Trenches, 478
Trending, 682–683
Trends, 682
Trim, 484
Truck farms, 439
Tubing, 509
Tugboats, 613
Turbines, 496, 510–511
  gas, 499
  water, 507
  wind, 506–507
Turkeys, 450
Turning
  machines, 100
  tools, 91–92
Two-point perspectives, 398
Two-view drawings, 393
Type, movable, 530
U.S. Conventional system, 87
Ultrasonic machining, 565
Ultrasonics, 592
Ultrasound, 594

Underground mines, 118
Undesirable outputs, 46
Unemployability, technological, 670
Unemployment, technological, 670
Unit trains, 611
Unity, 323
Unloading, 609
Unskilled workers, 171
Uplinks, 526
Upper management, 174
Useufacturing, 558
Using, 29–30
Utilities, 482
  installing, 482–483
Vaccinations, 590
Vaccines, 590, 599
  treatment with, 599–601
Value, 320–321
Valuing, 64
Vanishing points, 304
Variable path systems, 610
Variable path transportation, 612
Variable path vehicles, 612
Variables, 370
VCRs. *See* Videocassette recorders
Vegetables, 438–439
Vehicles
  hybrid, 666
  systems, 615–624
  variable path, 612
Veneer, 113, 345
Venues, 588
Vice presidents, 174
Video production, 407–409
Video recordings, 527
  systems, 523, 527–528
Videocassette recorders (VCRs), 528
Videodiscs, digital, 520
Village, global, 521
Villages, 466
Vinyl. *See* Plastic
Visualizing objects, 294–295
Vocabulary, 31
Voltage, 509
Wages, 170
Walks, creating, 485–486
Walls, 479
  materials, 481–482
  curtain, 226

Walls *(continued)*
    framing, 481
    interior, 484
    loadbearing, 226
    Thrombe, 504
Wants, 256
Waste, 46
    hazardous, 149
Water, 610
    bodies of navigable, 613
    pollution, 667–668
    transportation, 612–613
    turbines, 507
Waterpower, 138, 141
Waterwheels, 141
Wave communication, 518–519
Wave energy, 143–144
Wave systems, 518
Waves, 518
Wax, 351
Web page design, 553–554
Wedges, 98
Weeds, 445
Welding, 566
Wellness, 585
    exercise and, 586–588
    programs, 585–589
    sports and, 588–589
    technology and, 586–589

Wheels and axles, 96
White balance, 408
White glue, 352
White-collar workers, 170–171
Wind, 136
    energy, 143
    turbines, 506–507
Windows, 505, 510
    trim, 484
Windrows, 449
Wireframe models, 338
Wires. *See* Conductors
Wood
    glue, 352
    sheet materials, 344–345
Work, 132
    changing society and, 168–169
    energy and, 131–133
    ethic, 165
Workers, 170–171, 175
Working
    drawings, 389–392
    sketching, 291
World Wide Web (WWW), 529
Wrists, 381
Writing copy, 538
WWW. *See* World Wide Web
Xerography, 541
X-ray machines, 593